Introduction to Nanophotonics

Nanophotonics is where photonics merges with nanoscience and nanotechnology, and where spatial confinement considerably modifies light propagation and light–matter interaction. Describing the basic phenomena, principles, experimental advances and potential impact of nanophotonics, this graduate-level textbook is ideal for students in physics, optical and electronic engineering and materials science.

The textbook highlights practical issues, material properties and device feasibility, and includes the basic optical properties of metals, semiconductors and dielectrics. Mathematics is kept to a minimum and theoretical issues are reduced to a conceptual level. Each chapter ends in problems so readers can monitor their understanding of the material presented.

The introductory quantum theory of solids and size effects in semiconductors is considered to give a parallel discussion of wave optics and wave mechanics of nanostructures. The physical and historical interplay of wave optics and quantum mechanics is traced. Nanoplasmonics, an essential part of modern photonics, is also included.

Sergey V. Gaponenko is Head of the Laboratory for Nano-optics at the Stepanov Institute of Physics, National Academy of Sciences of Belarus. He is also Chairman of the Association of Lasers and Optics and Vice-president of the Laser Association.

Introduction to Nanophotonics

Sergey V. Gaponenko
National Academy of Sciences, Belarus

CAMBRIDGE
UNIVERSITY PRESS

University Printing House, Cambridge CB2 8BS, United Kingdom

One Liberty Plaza, 20th Floor, New York, NY 10006, USA

477 Williamstown Road, Port Melbourne, VIC 3207, Australia

314-321, 3rd Floor, Plot 3, Splendor Forum, Jasola District Centre, New Delhi - 110025, India

79 Anson Road, #06-04/06, Singapore 079906

Cambridge University Press is part of the University of Cambridge.

It furthers the University's mission by disseminating knowledge in the pursuit of education, learning and research at the highest international levels of excellence.

www.cambridge.org
Information on this title: www.cambridge.org/9780521763752

First published 2010

A catalogue record for this publication is available from the British Library

ISBN 978-0-521-76375-2 Hardback

To Olga

Contents

Preface

It is an extraordinary paradox of Nature that, being seemingly the only creatures capable of understanding its harmony, we naively attempt to chase its very essence through our daily experience based on mass-point mechanics and ray optics, while its elusive structure is mainly contained in wave phenomena. It may be nanophotonics where many pathways happily merge that promises not only mental satisfaction in our scientific quest but also an extra bonus in the form of new technologies and devices.

In this book I have tried to give a consistent description of the basic physical phenomena, principles, experimental advances and potential impact of light propagation, emission, absorption, and scattering in complex nanostructures. Introductory quantum theory of solids and quantum confinement effects are considered to give a parallel discussion of wave optics and wave mechanics of complex structures as well as to outline the beneficial result of combined electron wave and light wave confinements in a single device. Properties of metal nanostructures with unprecedented capability to concentrate light and enhance its emission and scattering are discussed in detail.

Keeping mathematics to a reasonable minimum and reducing theoretical issues to a conceptual level, the book is aimed at assisting diploma and senior students in physics, optical and electronic engineering and material science. The contents include a vast diversity of phenomena from guiding and localization of light in complex dielectrics to single molecule detection by surface enhanced spectroscopy. The physical and historical interplay of wave optics and quantum mechanics is traced whenever possible to highlight the internal concordance inherent in physics and nature. Nanophotonics is presented as an open field of science and technology which has been conceived as an organic junction of quantum mechanics, quantum electrodynamics, optical physics, material science and engineering to offer an impressive impact on information and communication technology.

The book is principally based upon scientific experience the author gained while working at the Institute of Molecular and Atomic Physics in Minsk, Belarus, in the decade from 1997 to 2007. I am indebted to many colleagues from this institute for the creative atmosphere and high research grade. I gratefully acknowledge the fruitful cooperation and ongoing discussions with many colleagues in Belarus, Russia and other countries with special thanks to the European network of excellence "PHOREMOST" (Nanophotonics to realize molecular scale technologies) which has been organized and successfully driven for several years by Clivia Sotomayor Torres within the 6th Framework Programme of the European Union. Many of my PhD students have made their theses in nanophotonics and their results have been included in this book. I would specially acknowledge that Chapter 3 has been seriously influenced by cooperation with Sergey Zhukovsky and Chapter 16 has been written based on continuous discussions with Dmitry Guzatov. I am grateful to these colleagues as well as to Dmitry Mogilevtsev, Maxim Ermolenko, Andrey Lutich, Maxim Gaponenko, and Andrey Nemilentsau for reading selected chapters and critical comments on their style and content. My colleague and friend Andrey Lavrinenko made a strong influence on my understanding of wave phenomena in complex structures and kindly provided the cover image for this book based on his calculations of light propagation in a photonic crystal with guiding defects. Great efforts by Tamara Chystaya for arranging the compuscript of the book are deeply appreciated.

This book would never have been accomplished without fruitful cooperation with Cambridge University Press, mainly with John Fowler, Lindsay Barnes and Caroline Brown. I am also indebted to the referees for encouraging comments and helpful advice in the early stages of this book project.

S. V. Gaponenko
Minsk, 2009

Notations and acronyms

A	amplitude
a	length, radius, width
a^0	$= 5.292\ldots \cdot 10^{-11}$ m, atomic length unit
a_{B}	Bohr radius of a hydrogen atom, $a_{\mathrm{B}} \approx a^0$ holds
a_{B}^*	Bohr radius of an exciton
$\mathbf{a}$	acceleration
$\mathbf{a_i}$	elementary translation vectors
a_{L}	crystal lattice constant
$\mathbf{b_i}$	elementary translation vectors in reciprocal space
b_i	reciprocal lattice constants
$\mathbf{B}$	magnetic induction vector
C	cross-section
C_{abs}	absorption cross-section
C_{ext}	extinction cross-section
C_{scat}	scattering cross-section
c	$= 299\,792\,458\ \mathrm{ms}^{-1}$, speed of light in vacuum
D	density of modes, density of states
D	diffusion coefficient
$\mathbf{D}$	electric displacement vector
d	thickness
d	dimensionality
e	$= 1.6021892\ldots \cdot 10^{-19}$ C, elementary electric charge
E	a particle energy
E_{c}	energy at the bottom of the conduction band
E_{F}	Fermi energy
E_{g}	band gap energy
E_{v}	energy at the top of the valence band
$\mathbf{E}$, E	electric field vector, amplitude
F	distribution function
$\mathbf{F}$	force
f	volume fraction
G	generator of a fractal structure
$\mathbf{H}$	magnetic field vector, Hamiltonian operator
h	$= 6.626069 \cdot 10^{-34}$ J · s, Planck constant
$\hbar$	$= h/2\pi$
I	intensity

$\mathbf{J}$	electric current density
k	wave number
$\mathbf{k}$, $\mathbf{K}$	wave vector
k_B	$= 1.380662\ldots \cdot 10^{-23}$ J/K, Boltzman constant
ℓ	mean free path
$\mathbf{L}$	angular momentum
L	length
m	mass
m_0	$= 9.109534\ldots \cdot 10^{-31}$ kg, an electron's rest mass
m^*	effective mass
M	exciton mass
$\mathbf{M}$	magnetic polarization vector
n	the principal quantum number
n	refractive index
n_1	real part of refractive index
n_2	imaginary part of refractive index
$p, \mathbf{p}$	momentum
$\mathbf{P}$	electric polarization vector
Q	efficiency factor
R	reflection coefficient for intensity
R, r	radius
r	reflection coefficient for amplitude
$\mathbf{r}$	radius vector
Ry	≈ 13.60 eV, Rydberg constant, Rydberg energy
Ry*	exciton Rydberg energy
$\mathbf{S}$	Poynting vector
t	transmission coefficient for amplitude
t	time
T	period of oscillations
T	temperature
T	transmission coefficient for intensity
$\mathbf{T}$	translation vector
U, u	potential energy
v, $\mathbf{v}$	velocity
v_g, $\mathbf{v_g}$	group velocity
V	volume
W	light energy
W_{abs}	light energy absorption rate
W_{ext}	light energy extinction rate
W_{scat}	light energy scattering rate
$W(\mathbf{r})$	spontaneous emission rate at point $\mathbf{r}$
W_0	spontaneous emission rate in vacuum
Y_{lm}	spherical Bessel functions
x, y, z	Cartesian coordinates

α	polarizability
α_{abs}	absorption coefficient
Γ	scattering rate
ε	relative dielectric permittivity
ε_0	$= 8.8541878 \cdot 10^{-12}$ F/m, dielectric constant (the dielectric permittivity of a vacuum)
κ	evanescence parameter
λ	wavelength
μ	relative magnetic permeability
$\mathrm{\mu}$	dipole moment, chemical potential
$\mathrm{\mu}_0$	$= 1.256637 \cdot 10^{-6}$ H/m, magnetic permeability of vacuum
μ_{eh}	electron–hole reduced effective mass
ν	frequency
ρ	electric charge density
ρ	material resistivity per unit area and unit length
σ	conductivity
τ	decay constant, scattering time, phase time
Φ	potential
φ	phase
χ_{nl}	roots of Bessel functions
χ	susceptibility
Ψ	time-dependent wave function
ψ	time-independent wave function
ω	circular frequency
ω_p	plasma circular frequency

AAAS	American Association for the Advancement of Science
AIP	American Institute of Physics
bcc	body-centered cubic (lattice)
CCD	charge coupled device
CD	compact disk
CIE	Comission Internationale de l'Eclairage (International Commission for Illumination)
CMOS	complementary metal-oxide-semiconductor (notation for modern microelectronics technology platform)
CNDO/S	complete neglect of differential orbital, spectroscopic version (a quantum chemical technique)
cw	continuous wave
DOS	density of states
EM	electromagnetic
fcc	face-centered cubic (lattice)
FTIR	frustrated total internal reflection
IR	infrared
LED	light emitting diode

LDOS	local density of states
MBE	molecular beam epitaxy
MOCVD	metal-organic chemical vapor deposition
MOVPE	metal-organic vapor phase epitaxy
NA	numerical aperture
RBG	red-blue-green
SEF	surface enhanced fluorescence
SEM	scanning electron microscopy
SERS	surface enhanced Raman scattering
SNOM	scanning near field optical microscope
SOI	silicon on insulator
SPP	surface plasmon polariton
TE	transverse electric (mode)
TEM	transmission electron microscopy
TIR	total internal reflection
TM	transverse magnetic (mode)
UV	ultraviolet

1 Introduction

1.1 Light and matter on a nanometer scale

The notion of "photonics" implies the science and technology related to generation, absorption, emission, harvesting, processing of light and their applications in various devices. Light is electromagnetic radiation available for direct human perception, in the wavelength range from approximately 400 to approximately 700 nanometers. Typically, adjacent far ultraviolet and near infrared ranges are also involved to give the approximate range of electromagnetic radiation from 100 nanometers to 1–2 micrometers as the subject of photonics. If the space has certain inhomogeneities on a similar scale to the wavelength of the light, then multiple scattering and interference phenomena arise modifying the propagation of light waves. Light scattering is the necessary prerequisite for vision. Shining colors in soap bubbles and thin films of gasoline on a wet road after rain are primary experiences of light-wave interference everybody gains in early childhood. To modifiy the conditions for light propagation, inhomogeneities in space which are not negligible as compared to the wavelength of the light, i.e. starting from the size range 10–100 nm to a few micrometers, become important. Space inhomogeneity for light waves implies inhomogeneity in dielectric permittivity.

Matter is formed from atoms which in turn can be subdivided into nuclei and electrons. An elementary atom of hydrogen has a radius for the first electron orbital of 0.053 nm. Atoms may form molecules and solids. Many typical organic molecules have sizes of the order of 1 nm. Typical crystalline solids feature a lattice period of approximately 0.5 nm. Interaction of light with matter actually reduces to the processes involved in the electron subsystem of molecules and solids. Therefore to understand light–matter interactions, electron properties must be examined in detail. Electrons are viewed as objects possessing wave properties in terms of wavelength, and corpuscular properties in terms of mass and charge. If an electron has gained kinetic energy as a result of acceleration in an electric field between a couple of plates with voltage 1 V (e.g. generated in a silicon photocell), then its kinetic energy of 1 eV results in an electron de Broglie wavelength close to 1 nm. For kinetic energies corresponding to a characteristic value of $k_B T = 27$ meV at room temperature, the electron de Broglie wavelength in solids is of the order of 10 nm. Here k_B is the Boltzman constant and T is temperature. When space inhomogeneities present which are not negligible as compared to the electron wavelength then scattering and interference of electrons develop modifying in many instances the interaction of light with matter. Space inhomogeneities for electrons means inhomogeneity in charge or mass displacement, electric or magnetic field variations.

1.2 What is nanophotonics?

Nanophotonics is the recently emerged, but already well defined, field of science and technology aimed at establishing and using the peculiar properties of light and light–matter interaction in various nanostructures. Since it is the spatial confinement of light waves in complex media and electron waves in various nanostructured solids that determine multiple physical phenomena in nanophotonics, it is possible to characterize *nanophotonics as the science and technology of confined light waves and electron waves*. It can be tentatively divided into four sections.

The **first** section of nanophotonics is *electron confinement effects on the optical properties of matter*; mainly semiconductor and dielectric materials. These phenomena are typically referred to as *quantum confinement* effects since manifestations of *wavy properties of electrons are typically labeled as quantum phenomena*. The net optical manifestations of these effects are size-dependent optical absorption spectra, emission spectra and transition probabilities for solid matter purposefully structured on the scale of a few nanometers. Their potential applications are the variety of optical components with size-controlled, tuned and adjustable parameters including emitters, filters, lasers and components, optical switches, electro-optical modulators etc. This sub-field of nanophotonics has become the subject of systematic research since the 1970s. Different issues related to electron confinement phenomena and their optical manifestations have already been the subject of several books [1–3].

The **second** section of nanophotonics constitutes *light wave confinement phenomena* in structured dielectrics, including the fine concept of photonic solids in which light wave propagation is controlled in a similar manner to electron waves in solids. This subfield of nanophotonics is principally *classical* in its essence, i.e. it is based entirely on *wave optics* and does not imply any notion beyond classical Maxwell equations. It actually dates back to early identification of light interference phenomena by Isaac Newton in the eighteenth century and to genuine prediction by Lord Rayleigh of the remarkable reflective properties of periodic media in the 1880s. The main practical outcome for this field is ingenious photonic circuitry development, from ultrasmall but high-quality cavities to ultracompact waveguides. Different issues of light confinement phenomena and near-field optics have already been included in several books [4–8].

The **third** section of nanophotonics is essentially the *quantum optics of nanostructures*. It deals with modified light–matter interaction in nanostructures with confined light waves. Spontaneous emission and scattering of light essentially modifies and becomes controllable since spontaneous photon emission and spontaneous photon scattering can be promoted or inhibited by engineering photon density of states often referred to as electromagnetic mode density. The ultimate case of this modified light–matter interaction is the development of confined light–matter states in microcavities and photonic crystals. This section of nanophotonics is relatively new. It has its root in the seminal paper by E. Purcell in 1946 predicting the modified spontaneous decay rate of a quantum sytem (e.g. an atom) in a cavity. In the 1970s V. P. Bykov suggested freezing spontaneous decay of excited atoms in a periodic structure where no means is available to carry off the emitted radiation. Eventually these ideas evolved into the concept of *photonic crystals* and *photonic solids* through seminal papers in the 1980s by E. Yablonovitch who suggested thresholdless lasing in a device with inhibited spontaneous emission of radiation, and by S. John who indicated possible light wave localization in disordered structures. Modified light–matter interaction in microcavities and photonics crystals is the subject of a few books [9–11].

The **fourth** section of nanophotonics is optics and optical engineering based on *metal-dielectric nanostructures*. Typically metals are not considered an important subject in optical research and engineering. Our experience mainly reduces to everyday observation of ourselves in aluminium mirrors. However metal-dielectric nanostructures feature a number of amazing properties resulting from the development of electron excitations at metal–dielectric interfaces called *surface plasmons*. Structuring of metal-dielectric composites on the nanoscale (10–100 nm) makes surface effects dominant. Actually, the optical properties of metal nanoparticles have been used for many centuries in stained glass but nowadays the study of metal-dielectric composites in optics has evolved into the well-defined field of *nanoplasmonics*. High concentrations of electromagnetic radiation and modification of the rates of quantum transitions in the near vicinity of metallic singularities result in novel light emitting devices and ultrasensitive spectral analysis with ultimate detection of a single molecule by means of Raman scattering. A few issues of nanoplasmonics have been considered in books [12,13].

1.3 Where are the photons in nanophotonics and in this book?

The author takes the approach whereby temptation to use the term "photon" is purposefully avoided unless the concept of light quanta is essential in order to understand the phenomenon in question. It is anticipated by many scientists and has been clearly outlined by W. Lamb in his seminal paper entitled "Anti-photon" [14]. In nanophotonics photons become necessary when trying to understand the emission of light by an excited quantum system and the scattering of light when light frequencies change (Raman scattering). Then, eventually, *quantum electrodynamics* comes to the stage. Not all phenomena of light propagation need the involvement of photons and the vast majority of light absorption phenomena can be treated in a semiclassical way when the matter is described in terms of quantum mechanics (more accurately speaking, *wave mechanics*), but light is understood as classical electromagnetic waves.

The rest of the book is organized as follows. Fifteen chapters, from Chapter 2 to Chapter 16, are organized in the form of two large parts.

Part I is entitled "*Electrons and electromagnetic waves in nanostructures*" and contains Chapters 2 to 12. It considers electrons in complex media and nanostructures in terms of *wave mechanics*, and electromagnetic radiation in complex media and nanostructures in terms of *wave optics*. Parallel consideration of wave phenomena in the theory of matter and in the theory of light in complex structures is pursued purposefully to highlight the conformity of wave phenomena in nature, both for electrons in matter and for classical waves in complex media. This harmony did result in amazing interactions between wave optics and wave mechanics in the past century. At first, in the 1920s, wave optics stimulated basic ideas of wave mechanics to conform mechanics with optics, not only in the classical parts (geometrical ray optics versus classical mass-point mechanics), but also in the other part, i.e. wave mechanics versus wave optics. Furthermore, dedicated advances in wave mechanics in explaining electron properties of solids stimulated, in the 1980s, a systematic transfer of its results to wave optics to provide a fine concept of photonic solids. Part I includes basic properties of quantum particles and light waves (Chapter 2), systematic analysis of textbook problems from quantum mechanics and wave optics revealing conformity of the main laws and formulas in these two seemingly unlinked fields of physics (Chapter 3), introduction to electron theory and optics of solids and quantum confinement phenomena in nanostructures (Chapter 4), consideration of semiconductor (Chapter 5) and metal (Chapter 6) nanoparticles, properties of light in periodic (Chapter 7) and non-periodic (Chapter 8) dielectric media, brief description of optical nano-circuitry (Chapter 9), tunneling of light (Chapter 10) and principal properties of metal-dielectric nanostructures (Chapter 11). Chapter 12 summarizes parallelism in electronic and optical phenomena based on wave properties of electrons and light.

Part II is entitled "*Light–matter interaction in nanostructures*" and contains four Chapters, from Chapter 13 to 16. It gives a brief introduction to quantum electrodynamics (Chapter 13), discussion of modification of spontaneous decay rates and spontaneous scattering rates resulting from modified density of photon states in nanostructures (Chapter 14), as well as brief consideration of light–matter states beyond the perturbational approach in microcavities and photonic crystals (Chapter 15). Finally, in Chapter 16, plasmonic enhancement of luminescence and Raman scattering is considered as the bright example of light–matter interaction engineered on a nanoscale.

The author hopes the reader will be successfully introduced into the amazing world of nanophotonics by going through this book. The book is written in a style appropriate to senior students at university, the more advanced reader being referred to original and review articles cited therein. The reader is expected to enjoy the beauty of the photonic nanoworld and to be capable of contributing properly to the "bright future" outlined in the Strategic Research Agenda [15] with the nanophotonic roadmap of how to get there, highlighted recently by the European network of excellence "Nanophotonics to realize molecular scale technologies" (PhoREMOST) [16].

References

[1] J.H. Davies. *The Physics of Low-Dimensional Semiconductors* (Cambridge: Cambridge University Press, 1998).
[2] C. Klingshirn. *Semiconductor Optics* (Berlin: Springer, 1995).
[3] S.V. Gaponenko. *Optical Properties of Semiconductor Nanocrystals* (Cambridge: Cambridge University Press, 1998).
[4] M. Born and E. Wolf. *Principles of Optics* (New York: MacMillan, 1964).

[5] A. Yariv and P. Yeh. *Optical Waves in Crystals* (New York: John Wiley and Sons, 1984).

[6] J. D. Joannopoulos, R. D. Meade and J. N. Winn. *Photonic Crystals: Molding the Flow of Light* (Princeton: University Press, 1995).

[7] L. Novotny and B. Hecht. *Principles of Nano-Optics*. (Cambridge: Cambridge University Press, 2007).

[8] A. M. Zheltikov. *Optics of Microstructured Fibers* (Moscow: Nauka, 2004) – in Russian.

[9] Y. Yamamoto, F. Tassone and H. Cao. *Semiconductor Cavity Quantum Electrodynamics* (Berlin: Springer-Verlag, 2000).

[10] V. P. Bykov. *Radiation of Atoms in a Resonant Environment* (Singapore: World Scientific, 1993).

[11] D. S. Mogilevtsev and S. Ya. Kilin. *Quantum Optics Methods of Structured Reservoirs*. (Minsk: Belorusskaya Nauka, 2007) – *in Russian.*

[12] U. Kreibig and M. Vollmer. *Optical Properties of Metal Clusters* (Berlin: Springer 1995).

[13] S. A. Maier. *Plasmonics: Fundamentals and Applications* (Berlin: Springer Verlag, 2007).

[14] W. E. Lamb. Anti-photon. *Appl. Phys. B*, **60** (1995), 77–80.

[15] *Towards a Bright Future for Europe. Strategic Research Agenda in Photonics* (Düsseldorf: VDI Technologiezentrum, 2006).

[16] *Emerging Nanophotonics. PhoREMOST Network of Excellence* (Cork: Tyndall National Institute, 2008), www.phoremost.org.

PART I

ELECTRONS AND ELECTROMAGNETIC WAVES IN NANOSTRUCTURES

2 Basic properties of electromagnetic waves and quantum particles

It seemed as if Nature had realized one and the same law twice by entirely different means.

Erwin Schrödinger (Nobel lecture, 1933)

2.1 Wavelengths and dispersion laws

A classical wave is described in terms of its frequency ν, amplitude A and wave vector $\mathbf{k}$. Along with ν, circular frequency $\omega = 2\pi\nu$ is often used. Frequencies ν and ω are related to the period T of oscillations

$$T = \frac{1}{\nu} = \frac{2\pi}{\omega}. \tag{2.1}$$

The wavelength λ can be defined as a distance over which the wave travels during a single period T or as a distance between two points obeying the same phase of oscillations. It is related to the wave vector $\mathbf{k}$ as,

$$k = |\mathbf{k}| = \frac{2\pi}{\lambda}, \tag{2.2}$$

where k is referred to as the wave number. The wave vector direction coincides with the direction of the phase motion. The propagation speed of the wave v can be considered in terms of phase velocity and group velocity. Phase velocity decribes the speed of motion of a plane with constant phase:

$$v = \frac{\lambda}{T} = \lambda\nu = \frac{\omega}{k}. \tag{2.3}$$

Group velocity v_g describes the speed and direction of energy transfer in the course of a wave process. It reads as a vector in a 2- and 3-dimensional case,

$$\mathbf{v}_g = \frac{d\omega}{d\mathbf{k}} \tag{2.4}$$

and reduces to a scalar value in a one-dimensional problem, i.e.

$$v_g = \frac{d\omega}{dk}. \tag{2.5}$$

In the particular case of a linear $\omega(k)$ relation, phase and group velocities coincide. In isotropic media the group velocity has direction, coinciding with the wave vector.

For electromagnetic waves in a vacuum the group and phase velocities are equal to $c = 299\,792\,458\ \mathrm{ms}^{-1}$.

Within the scale of electromagnetic waves the visible part occupies a rather narrow band of wavelengths from 400 nm (violet light) to 700 nm (red light). Typically adjacent near-ultraviolet (UV) and near-infrared (IR) bands are also considered as the subject of optical science to form the optical range of electromagnetic waves from 100-200 nm to several micrometers. For example, the most efficient conversion of electrical energy into light occurs in a gas of mercury atoms at 253 nm, which is the basic light source in luminescent lamps (finally converted into the visible by means of luminophores, for example Na vapor in street lamps). Similar or even shorter light wavelengths are used in modern microelectronics technology to get the image of desirable microchip circuitry onto a silicon wafer with submicron resolution. From the IR side, 1550 nm is probably the most representative electromagnetic wavelength since it is the principal wavelength of optical communication, falling within the transparency of commercial silica waveguides and the erbium amplifiers band to provide low-loss propagation over long distances and amplification options to compensate for losses. It is due to this favorable combination of silica and erbium properties that everyone can enjoy worldwide PC-networking.

Quantum particles, e.g. electrons, are believed to exhibit wave properties with wave vector and wavelength being related to their momentum **p** in accordance with the relations first proposed by de Broglie in 1923 [1]

$$\mathbf{p} = \hbar \mathbf{k}, \quad \lambda = \frac{h}{p} \tag{2.6}$$

Here $h = 6.626069 \times 10^{-34}$ Js is the Planck constant and $\hbar \equiv h/2\pi$. Kinetic energy E is related to wave number as

$$E = \frac{p^2}{2m} = \frac{\hbar^2 k^2}{2m}. \tag{2.7}$$

In terms of kinetic energy E, the particle de Broglie wavelength, using Eq. (2.6), is

$$\lambda = \frac{h}{\sqrt{2mE}}. \tag{2.8}$$

The last relation is instructive in obtaining an immediate intuitive idea about the typical wavelength scale inherent in the electron world. For example, if an electron with rest mass $m_0 = 9.109534 \times 10^{-31}$ kg has gained kinetic energy $E = 10$ eV while being accelerated in a vacuum between a couple of electrodes with potential difference 10 V, its de Broglie wavelength equals $3.88 \times 10^{-10}\ \mathrm{m} = 0.388$ nm.

We arrive at the following important conclusion. When considering spatial structuring of matter on the nanometer scale, one has electron confinement phenomena at the very short end of the scale (of the order of 1 nm) and confinement of light waves at the very long end of the scale (of the order of 1000 nm). Accordingly, the whole variety of optical manifestations of electron and light wave spatial confinement in nanostructured materials, composites, devices constitute the field of *nanophotonics*.

Relations $\omega(k)$ between frequency ω and wave number k for classical waves and $E(p)$ between kinetic energy E and momentum p for a particle are called the *dispersion laws*.

When considering electromagnetic waves in a vacuum, in accordance with Eq. (2.3) one has the linear dispersion law,

$$\omega = ck. \tag{2.9}$$

By adding the $\hbar$ factor in both parts of Eq. (2.9) the $\omega(k)$ function readily transforms into $E(p)$ recalling Eq. (2.6) along with the other cornerstone relation proposed by Max Planck in 1900 [2],

$$E = \hbar\omega \tag{2.10}$$

where E now stands for the energy quantum of electromagnetic radiation, the *photon* energy. Then we can replace the dispersion law, Eq. (2.9), by

$$E = pc \tag{2.11}$$

referring to this as "the photon dispersion law".

Using electronvolts as convenient energy units, one has a typical range of photon energies from approximately 1 eV to 10 eV to be considered in optics, with 1.770 eV and 3.097 eV corresponding to the visible limits of 700 and 400 nm, respectively. The photon wavelength – photon energy relation in a vacuum reads,

$$E = \frac{hc}{\lambda}, \quad E[\text{electronvolts}] = \frac{1239.85\ldots}{\lambda[\text{nanometers}]} \tag{2.12}$$

which is a consequence of Eqs. (2.11) and (2.6). To compare electronvolts to joules, $1\ \text{eV} = 1.602\,189 \times 10^{-19}$ J where the coefficient equals the electron charge value in coulombs.

In a medium other than a vacuum c should be replaced by $c/n(\omega)$ with $n(\omega)$ being *the index of refraction* and Eqs. (2.10) and (2.12) read,

$$\omega = ck/n(\omega), \quad E = pc/n(\omega). \tag{2.13}$$

Figure 2.1 represents the dispersion laws for electromagnetic waves (photons) and for electrons. A straight line inherent in electromagnetic waves in a vacuum changes slope in a given homogeneous medium in accordance with the refractive index n. A parabolic $E(p)$ function for a particle with mass m, in accordance with Eq. (2.7), exhibits different steepness depending on the mass of the particle under consideration.

In a complex medium, refractive index n becomes wavelength dependent (typically reducing with increasing wavelength) and the dispersion law becomes non-linear. Then absolute values of phase and group velocities diverge. In anisotropic media, not only absolute v and v_g values but also directions of phase and energy transfer do not coincide. Notably, the refractive index for electromagnetic waves in a large variety of continuous dielectric media is a quite stable material parameter within the optical range. For all known dielectric and semiconductor materials, refractive index ranges from $n = 1$ for a vacuum to $n = 4$ for germanium monocrystals (in the near IR). For a given material, relative n variation typically measures about 10% within the optical range. Under conditions of high powered optical excitation, when a considerable portion of the electrons experience

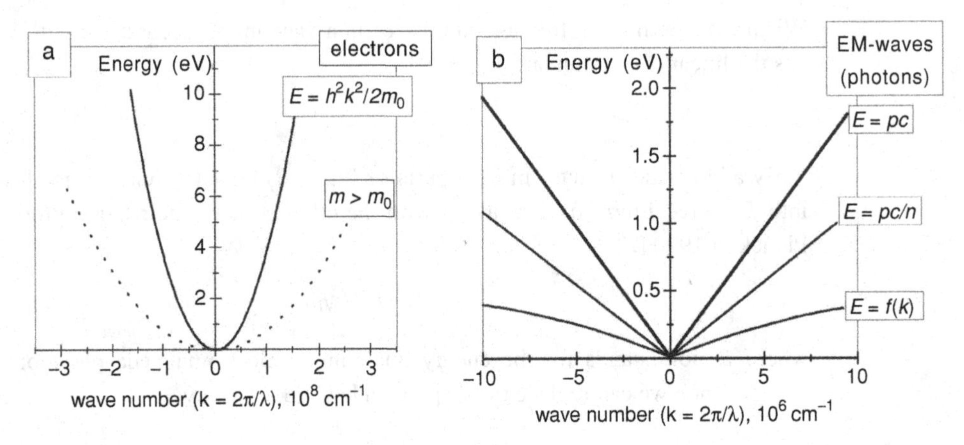

Fig. 2.1 **Dispersion curves (energy versus wave number) for (a) electrons and (b) electromagnetic waves (photons)**

upward transitions to higher states, refractive index can hardly be modified by a few percent as compared to its original value. Even $\Delta n/n = 0.01$ is considered to be distinguished nonlinear-optical behavior. Under the same conditions, the absorption coefficient may exhibit a ten-fold decrease (optical absorption saturation). Under the condition of an external electric field of the order of 10^5 V/cm, 0.1% variation in n is considered as pronounced electro-optical performance.

In complex media, quantum particle motion does not necessarily reduce to a simple parabolic $E(p)$ function, i.e. a particle sometimes cannot be ascribed a constant mass. This results in a generalized definition of particle mass,

$$m^{-1} = \frac{\mathrm{d}^2 E}{\mathrm{d}p^2}, \tag{2.14}$$

which for parabolic law coincides with the usual constant mass. An electron in a periodic lattice features multiple extrema in $E(p)$ dependence. Since every continuous function can be expanded in a series like,

$$E(p) = E(p_0) + (p - p_0)\left.\frac{\mathrm{d}E}{\mathrm{d}p}\right|_{p=p_0} + \frac{1}{2}(p - p_0)^2 \left.\frac{\mathrm{d}^2 E}{\mathrm{d}p^2}\right|_{p=p_0} + \cdots, \tag{2.15}$$

one can count energy and momentum from a given extremum point to get the parabolic $E(p)$ relation provided that the first derivative at the extremum point equals zero, and cutting off the terms with derivatives higher than two. In a periodic anisotropic medium the tensor of *effective mass* is used with components determined as,

$$\frac{1}{m_{ij}*} = \frac{\partial^2 E}{\partial p_i \partial p_j}. \tag{2.16}$$

The tensor feature of effective mass means the acceleration the particle experiences under the action of an external force may not coincide with the direction of the force.

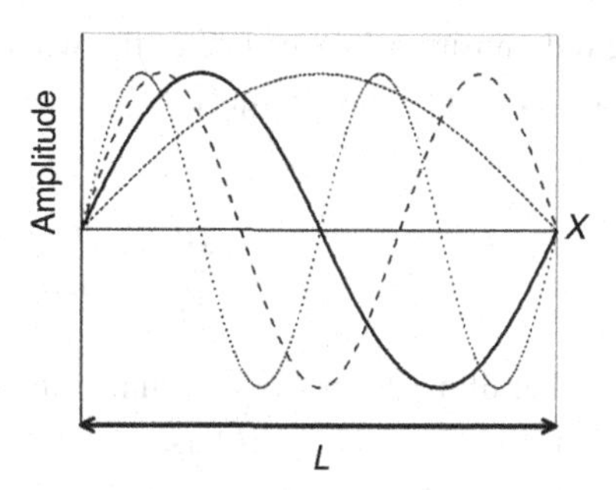

Fig. 2.2 Normal modes inside a cube. The first four modes are shown with wave numbers $k = \pi/L$, $2\pi/L$, $3\pi/L$, $4\pi/L$.

2.2 Density of states

Every type of wave features an important property that can be considered in terms of certain conservation laws. Every type of wave obeys a finite number of *modes* within a finite volume and a finite range of either of the values of frequency, wave number, wavelength. Accordingly, since quantum particles are considered as waves, within a given finite volume a particle can exist in a finite number of *states,* characterized by the defined values of parameters such as energy, momentum, wavelength, wave number.

To prove the above statements and to derive specific expressions for density of modes and density of states we have to recall the consideration first advanced by Rayleigh in 1900. Consider plane waves (*normal modes*) within a cube with side length L. We shall count how many modes lie within an interval k, $k + \mathrm{d}k$. A portion of these modes with longer wavelengths is shown in Fig. 2.2. These are modes having nodes on the cube borders. Wavelengths obey a raw $L/2, 2L/2, \ldots, nL/2$ where $n = 1, 2, 3, \ldots$. Accordingly, wave numbers will constitute a series,

$$k_x = n_x\frac{\pi}{L}, \quad k_y = n_y\frac{\pi}{L}, \quad k_z = n_z\frac{\pi}{L}. \tag{2.17}$$

One can see the modes under consideration form a discrete set in k-space, every couple of neighboring modes having spacing,

$$\Delta k_x = \Delta k_y = \Delta k_z = \frac{\pi}{L}. \tag{2.18}$$

This means every mode occupies in k-space the volume,

$$V_k = \left(\frac{\pi}{L}\right)^3. \tag{2.19}$$

Let us count the number of modes for all directions within the interval $[k, k + \mathrm{d}k]$, i.e. the number of modes contained in a spherical shell between a sphere with radius k, and a sphere with radius $k + \mathrm{d}k$. Taking the volume of such a layer,

$$\mathrm{d}V_k = 4\pi k^2 \mathrm{d}k \tag{2.20}$$

and keeping only positive ks (i.e. $1/8$ of the whole layer volume) by dividing the volume into the volume of a single mode Eq. (2.19) we arrive at the number of modes in the interval $[k, k + \mathrm{d}k]$,

$$\frac{\mathrm{d}V_k}{V_k} = \frac{L^3}{2\pi^2} k^2 \mathrm{d}k. \tag{2.21}$$

To get the number of modes per unit volume, one has to divide Eq. (2.21) by L^3. Then we can introduce density of modes $D(k)$ as,

$$D(k)\mathrm{d}k \equiv \frac{\mathrm{d}V_k}{V_k L^3} = \frac{k^2}{2\pi^2}\mathrm{d}k. \tag{2.22}$$

Applying a similar consideration to 2- and 1-dimensional problems we can finally write:

$$D_3(k) = \frac{k^2}{2\pi^2}, \quad D_2(k) = \frac{k}{2\pi}, \quad D_1(k) = \frac{1}{\pi}. \tag{2.23}$$

Note our consideration does not imply any specific type of wave and therefore is general for all types of waves. Using the $D(k)$ function one can readily derive density of modes (states) with respect to other wave parameters, e.g. energy, momentum, wavelength, frequency, using the following relationships:

$$D(\omega) = D(k)\frac{\mathrm{d}k}{\mathrm{d}\omega}, \quad D(E) = D(k)\frac{\mathrm{d}k}{\mathrm{d}E}, \quad D(p) = D(k)\frac{\mathrm{d}k}{\mathrm{d}p}, \quad D(\lambda) = D(k)\frac{\mathrm{d}k}{\mathrm{d}\lambda}. \tag{2.24}$$

One can see that unlike $D(k)$, the functions $D(\omega)$, $D(E)$, and $D(p)$ are different for different waves because of the specific dispersion law in each case. For electromagnetic waves the $D(\omega)$ relation is often considered in various optical problems. According to Eqs. (2.23) and (2.24) with coefficient 2 to account for the transverse nature of electromagnetic waves allowing two possible polarizations, it reads for 3-dimensional space,

$$D_3^{\gamma}(\omega) = \frac{\omega^2}{\pi^2 c^3}. \tag{2.25}$$

Density of *modes* for electromagnetic waves is often referred to as the *photon density of states*.

For quantum particles with a finite mass m Eqs. (2.23) and (2.24) with Eqs. (2.6) and (2.7) give:

$$D_3^{e}(E) = \frac{8\pi m^{3/2} E^{1/2}}{2^{1/2} h^3}, \quad D_3(p) = \frac{4\pi p^2}{h^3}. \tag{2.26}$$

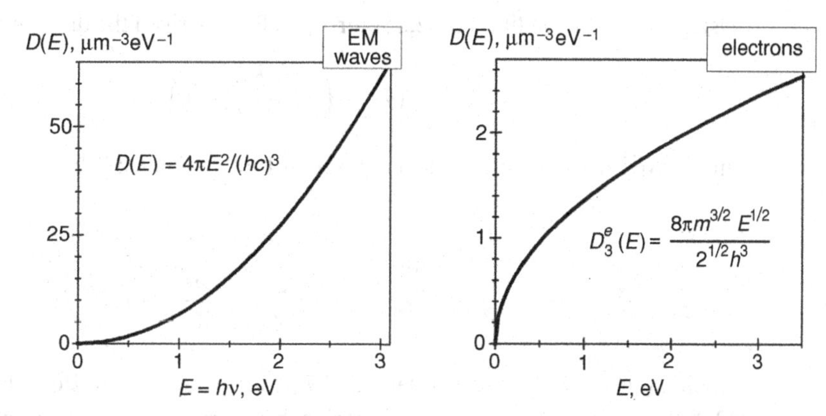

Fig. 2.3 **The density of modes (states) for electromagnetic waves and for electrons in a three-dimensional space. The energy scale for electromagnetic waves has been used in accordance with the relation $E = h\nu$ to enable direct comparison with the electron density of states.**

These expressions are valid for all particles with mass m. For electrons two possible spin orientations additionally increase the number of states by a factor of two. Plots of density of modes for electromagnetic waves and density of states for electrons are presented in Fig. 2.3.

The very concept of density of states belongs to a few basic concepts of quantum physics and plays an important role in nanophotonics. Therefore it is reasonable to recall a few major steps in the formulation and application of this concept in quantum physics.

At the very dawn of quantum physics in 1900, when considering the problem of black body radiation, J. W. Strutt (Lord Rayleigh) proposed counting normal modes in a box to arrive at a reasonable wavelength dependence of the radiation spectrum for longer wavelengths [3]. This formula has later been referred to as the Rayleigh–Jeans law. Rayleigh's approach had not gained relevant attention at that time. In the same period, Max Planck proposed the complete formula for the full spectral range with no counting modes involved [2]. He considered electromagnetic energy radiated by a cavity to obtain the term $\omega^2/\pi^2 c^3$, formally coinciding with the density of modes (see Eq. (2.21)). Simultaneously Planck introduced $E = \hbar\omega$ for an electromagnetic energy portion thus announcing the beginning of the quantum era in physics. Untill 1924 counting modes has not been systematically used in theoretical physics by other authors. It was S. N. Bose who understood that Rayleigh's idea should be considered among the basic concepts of the emerging theory [4]. Bose considered the equilibrium electromagnetic radiation as an equilibrium gas of electromagnetic quanta (later on called *photons* [5]). Then energy density contained in the interval $d\omega$ can be calculated as,

$$U(\omega)\mathrm{d}\omega = E(\omega)D(\omega)F(\omega)\mathrm{d}\omega, \tag{2.27}$$

where $E = \hbar\omega$ is the energy carried by a single quantum, $D(\omega)$ is the density of available *states*, and $F(\omega)$ is the distribution function which describes how these states are populated.

Using an expression like Eq. (2.25) for $D(\omega)$ Bose derived the distribution function,

$$F(\omega) = \left(\exp\frac{\hbar\omega}{k_{\mathrm{B}}T} - 1\right)^{-1}, \tag{2.28}$$

and arrived at the basic formula proposed in 1900 by Planck [2]:

$$U(\omega) = \hbar\omega\frac{\omega^2}{\pi^2c^3}\frac{1}{\exp\dfrac{\hbar\omega}{k_{\mathrm{B}}T} - 1}, \tag{2.29}$$

where k_{B} is the Boltzman constant, and T is temperature. The Bose distribution function (2.28) was derived based on the assumption that every state can be populated by an unlimited number of quanta. Bose's approach immediately ascribed to Rayleigh's modes the meaning of *quantum states* thus announcing the beginning of quantum statistical physics.

Remarkably, the young and unknown (at that time) Bose did send his manuscript entitled "Planck's Gesetz und Lichtquantenhypothese" [4], not to a scientific magazine, but to A. Einstein, as a highly reputed expert. Einstein recognized the solid value of the approach proposed. Not only did he recommend the paper to Zeitschrift für Physik, but he provided comments on its distinguished scientific value. Very soon Einstein applied this approach to gaseous atoms [6], i.e. assigning the concept of quantum states and distribution function to particles of matter.

Note that density of modes expressions Eq. (2.23) does not depend on the specific box shape used to count the standing waves. This statement has been rigorously proved mathematically by H. Weyl [7]. For acoustic waves it means nobody can hear the shape of a drum [7]. The number of modes within a given frequency range is defined by the volume and the dimensionality of space only.

2.3 Maxwell and Helmholtz equations

An electromagnetic field is described by a set of four vectors: electric field vector **E**, magnetic field vector **H**, electric displacement vector **D**, and magnetic induction vector **B**. Vectors **E** and **H** are independent whereas the other two, vectors, **D** and **B**, are related to the first two vectors via material equations (in the MKS system)

$$\mathbf{D} = \varepsilon\varepsilon_0\mathbf{E} = \varepsilon_0\mathbf{E} + \mathbf{P}, \tag{2.30}$$

$$\mathbf{B} = \mu\mu_0\mathbf{H} = \mu_0\mathbf{H} + \mathbf{M}. \tag{2.31}$$

Here ε is the dimensionless relative dielectric permittivity of the medium under consideration, μ is the dimensionless relative magnetic permeability of the medium, ε_0 and μ_0 are the permittivity and the permeability of the vacuum, **P** and **M** are the electric and the magnetic polarizations, respectively. In the general case of an anisotropic medium, permittivity and permeability are tensors. For isotropic media these tensors reduce to scalars. Vectors **E**, **H**,

D and **B** satisfy Maxwell's equations:

$$\nabla \times \mathbf{E} = -\frac{\partial \mathbf{B}}{\partial t}, \tag{2.32}$$

$$\nabla \times \mathbf{H} = \frac{\partial \mathbf{D}}{\partial t} + \mathbf{J}, \tag{2.33}$$

$$\nabla \cdot \mathbf{D} = \rho, \tag{2.34}$$

$$\nabla \cdot \mathbf{B} = 0, \tag{2.35}$$

where **J** is the electric current density (in amperes per square meter), ρ is the electric charge density (in coulombs per cubic meter). In a medium without charges and currents ($\mathbf{J} = 0$, $\rho = 0$) the set of Maxwell's equations leads to a pair of *wave equations* for **E** and **H** (see, e.g. Ref. [8])

$$\nabla^2 \mathbf{E} - \mu\varepsilon\mu_0\varepsilon_0 \frac{\partial^2 \mathbf{E}}{\partial t^2} = 0, \quad \nabla^2 \mathbf{H} - \mu\varepsilon\mu_0\varepsilon_0 \frac{\partial^2 \mathbf{H}}{\partial t^2} = 0, \tag{2.36}$$

with the known solutions in the form of plane waves,

$$\psi = \mathrm{e}^{i(\omega t - \mathbf{k}\cdot\mathbf{r})}, \tag{2.37}$$

with the wave number expressed as,

$$k = |\mathbf{k}| = \omega\sqrt{\varepsilon\varepsilon_0\mu\mu_0}. \tag{2.38}$$

The phase velocity of a wave (2.3) then reads,

$$v = \frac{\omega}{k} = \frac{1}{\sqrt{\varepsilon\varepsilon_0\mu\mu_0}}. \tag{2.39}$$

For a vacuum $\varepsilon = 1$, $\mu = 1$ hold and the phase velocity is,

$$v_{\text{vacuum}} = c = \frac{1}{\sqrt{\varepsilon_0\mu_0}}. \tag{2.40}$$

For media other than a vacuum we can write,

$$v = c/n, \tag{2.41}$$

with index of refraction n,

$$n = \sqrt{\varepsilon\mu}. \tag{2.42}$$

Finally, for non-magnetic media with $\mu = 1$ one simply has $n = \sqrt{\varepsilon}$.

Using c and n notations, the wave equations (2.36) obey a compact form,

$$\nabla^2 \mathbf{E} - \frac{n^2}{c^2}\frac{\partial^2 \mathbf{E}}{\partial t^2} = 0, \quad \nabla^2 \mathbf{H} - \frac{n^2}{c^2}\frac{\partial^2 \mathbf{H}}{\partial t^2} = 0. \tag{2.43}$$

For further consideration it is convenient to introduce the scalar $E(\mathbf{r})$ and $H(\mathbf{r})$ functions using relations,

$$\mathbf{E}(\mathbf{r}, t) = \mathbf{p}_1 E(\mathbf{r}, t), \quad \mathbf{H}(\mathbf{r}, t) = \mathbf{p}_2 H(\mathbf{r}, t), \tag{2.44}$$

with $\mathbf{p}_1$ and $\mathbf{p}_2$ being the unit vectors. Consider the case of a plane monochromatic wave with frequency ω. The time-dependent part is well defined as (tracing electric field only for convenience),

$$E(\mathbf{r}, t) = E(\mathbf{r})e^{i\omega t}. \tag{2.45}$$

Therefore Eq. (2.43) can be simplified to get a scalar equation for the time-independent function,

$$\nabla^2 E(\mathbf{r}) + \frac{n^2(\mathbf{r})}{c^2}\omega^2 E(\mathbf{r}) = 0, \tag{2.46}$$

or

$$\nabla^2 E(\mathbf{r}) + k^2 E(\mathbf{r}) = 0, \quad k = \frac{n(\mathbf{r})}{c}\omega. \tag{2.47}$$

This equation is called the scalar *Helmholtz equation*. It describes the space distribution of an electric field when an electromagnetic wave with frequency ω propagates in a medium with index of refraction $n(\mathbf{r})$. It is clear that a similar consideration holds for the $\mathbf{H}(\mathbf{r}, t)$ counterpart.

2.4 Phase space, density of states and uncertainty relation

We have to ascribe to the phase space the certain physical structure that is completely extraneous to classical dynamics.

Max Planck, 1916 [9]

In fact, counting modes as proposed by Rayleigh for radiation did not gain serious acknowledgement among his contemporaries. Nevertheless, the idea of a finite number of states for a quantum particle within a finite phase space (coordinate–momentum) volume $\Delta x \Delta y \Delta z \Delta p_x \Delta p_y \Delta p_z$ was proposed by Planck in 1906 [10]. It did, however, play an important role in formulation of the foundations of quantum mechanics and became readily adopted by many physicists in the beginning of the twentieth century. Unlike classical mechanics where a particle state in phase space is unambiguously represented by a point with coordinates (x, p), in wave mechanics relevant to the atomic world every state of a particle gains in the phase space a cell whose volume equals h^s where s is the geometrical dimensionality of space. This very profound idea was advanced by Plank based on intuition prior to the clear formulation of the wave properties of quantum particles. In this section, we shall show how Planck's idea of cellular phase space structure in the atomic world merges in conformity with Rayleigh's consideration of modes in a cavity.

Have a look at the denominator of Eq. (2.26). The h^3 factor there indicates the measure of an elementary h^3 cell in the phase space for 3-dimensional space. To count the number of states for a quantum particle within a region of phase space $\mathrm{d}p\mathrm{d}V$ we need to count the number of h^3cells in a spherical layer with radius p and thickness $\mathrm{d}p$ with volume $\mathrm{d}V$,

$$\mathrm{d}N = \frac{4\pi p^2}{h^3}\mathrm{d}p\mathrm{d}V \equiv D_3(p)\mathrm{d}p\mathrm{d}V. \tag{2.48}$$

Thus we immediately arrive at an expression for $D_3(p)$ coinciding with Eq. (2.26). Two further aspects should be mentioned in the context of our consideration. First, when counting modes in a cavity we considered standing waves only with positive coordinates and wave numbers, whereas in Eq. (2.48) we used momenta within the full solid angle to get $4\pi p^2$ for the square of the sphere with radius p. This peculiar circumstance is typically omitted in handbooks on quantum mechanics and statistical physics. Second, we note once again, that Eq. (2.48) has been derived based on a postulate of an elementary h^3 cell, whereas Eq. (2.26) was obtained by accurate counting modes combined with a postulate of $p = \hbar k$ for a quantum particle.

Once we adopt the concept of an elementary cell in phase space, *the Heisenberg uncertainty relation* becomes obvious. No knowledge can be gained about particle parameters, momentum and coordinates, with any accuracy better than a single cell. Heisenberg himself recognized that the uncertainty relation merges reasonably with the idea of the elementary phase space cell. In 1927, in his paper "*Über den anschaulichen Inhalt der quantentheoretischen Kinematik und Mekhanik*"[1] he wrote the uncertainty relation "accurately expresses the facts which earlier were attempted to be described by means of decomposition of the phase space into cells with size h^3" [11]. Later, in 1932, he considered in detail the density of states in terms of the number of elementary phase space cells [12].

We now show how Rayleigh's counting modes evolves to the Heisenberg uncertainty relation. Consider the cavity size L to be a measure of the coordinate uncertainty, i.e. $\Delta x = L$. Consider the spacing between neighboring modes in k-space as a measure of the wave number uncertainty, i.e. $\Delta k = \pi/L = \Delta k = \pi/\Delta x$. We arrive at the relation,

$$\Delta x \Delta k = \pi. \tag{2.49}$$

Now using $p = \hbar k$ (Eq. 2.6) and multiplying both parts by the factor $\hbar$ we get,

$$\Delta p \Delta x = \pi \hbar, \tag{2.50}$$

which coincides with the Heisenberg uncertainty relation,

$$\Delta p \Delta x \geq h \tag{2.51}$$

with an accuracy of factor 2.

To summarise, we have seen in this section that Rayleigh's approach of counting modes in a finite cavity combined with the idea of wave properties of material particles, $p = \hbar k$, leads to the Heisenberg uncertainty relation and to Plank's idea of a discrete cellular structure of the phase space on a microscale.

[1] "On the descriptive nature of quantum-theoretical kinematics and mechanics".

2.5 Wave function and the Schrödinger equation

In wave mechanics, a single-particle or a many-particle state is described by a wave function Ψ, which depends on the number of degrees of freedom available for a particle or a system of particles. For a single particle Ψ depends on time and three coordinates (x, y, z), whereas for a couple of particles, Ψ depends on time and six coordinates $(x_1, y_1, z_1, x_2, y_2, z_2)$. A wave function is believed to give the probability of a system to be found in certain places in space. The value

$$|\Psi(\xi)|^2 \mathrm{d}\xi = \Psi^*(\xi)\Psi(\xi)\mathrm{d}\xi \tag{2.52}$$

is proportional to the probability of finding, in measurements, coordinates of the particle or the system of particles in the range $[\xi, \xi + \mathrm{d}\xi]$, ξ being the set of all coordinates of the particles in the system, i.e.

$$\mathrm{d}\xi = \mathrm{d}x_1\mathrm{d}y_1\mathrm{d}z_1\mathrm{d}x_2\mathrm{d}y_2, \mathrm{d}z_2 \ldots \mathrm{d}x_n\mathrm{d}y_n\mathrm{d}z_n, \tag{2.53}$$

where n is the number of particles in the system. The probabilistic interpretation of a wave function dates back to 1926 when it was proposed by Max Born [13]. It constitutes the major postulate of wave mechanics.

The wave function can be normalized in such a way as to have,

$$\int |\Psi(\xi)|^2 \, \mathrm{d}\xi = 1. \tag{2.54}$$

Then $|\Psi|^2\mathrm{d}\xi$ equals the probability $\mathrm{d}W(\xi)$ that ξ belongs to the interval $[\xi, \xi + \mathrm{d}\xi]$. A normalized wave function is always determined with the accuracy of the factor $\mathrm{e}^{i\alpha}$, where α is an arbitrary real number. This ambiguity has no effect on physical results since all physical values are determined by a product $\Psi\Psi^*$.

The second basic postulate of wave mechanics is the *superposition principle*. This states that if a quantum system can be in states described by the functions Ψ_1 and Ψ_2, then it can be in any state described by a linear combination of Ψ_1 and Ψ_2, i.e.

$$\Psi = a_1\Psi_1 + a_2\Psi_2, \tag{2.55}$$

where a_1 and a_2 are arbitrary complex numbers.

To know the wave function of a particle or a system one needs to solve the *Schrödinger equation*. This equation constitutes the third postulate of wave mechanics. It states,

$$\mathbf{H}\Psi = i\hbar\frac{\partial\Psi}{\partial t}, \tag{2.56}$$

where $\mathbf{H}$ is the system *Hamiltonian*. The time-independent Hamiltonian coincides with the energy operator. For a single particle it becomes,

$$\mathbf{H} = -\frac{\hbar^2}{2m}\nabla^2 + U(\mathbf{r}), \tag{2.57}$$

where $-\frac{\hbar^2}{2m}\nabla^2$ is the kinetic energy operator and $U(\mathbf{r})$ is the potential energy of a particle. For an n-particle system the Hamiltonian is given by,

$$\mathbf{H} = -\sum_{i=1}^{n} \frac{\hbar^2}{2m_i}\nabla_i^2 + U(\mathbf{r}_1, \mathbf{r}_2, \ldots, \mathbf{r}_n). \tag{2.58}$$

If the Hamiltonian does not depend on time, the time and space variables can be separated, i.e.

$$\Psi(\xi, t) = \psi(\xi)\varphi(t). \tag{2.59}$$

In this case the time-dependent equation (2.56) reduces to the *steady-state equation*,

$$\mathbf{H}\psi(\xi) = E\psi(\xi), \tag{2.60}$$

where E is a constant value. The steady-state Schrödinger equation is a problem for eigenfunctions and eigenvalues of the Hamiltonian operator $\mathbf{H}$. Values of E are the energy values of a system in state $\psi(\xi)$. States with a defined energy $E$are called steady states. Equation (2.56) was proposed in 1926 by Erwin Schrödinger [14].

Consider a particle in a one-dimensional space with a constant potential energy U_0. Equation (2.60) takes the form,

$$-\frac{\hbar^2}{2m}\frac{\mathrm{d}^2\psi(x)}{\mathrm{d}x^2} + U_0\psi(x) = E\psi(x), \tag{2.61}$$

or, in a more convenient form,

$$\frac{\mathrm{d}^2\psi(x)}{\mathrm{d}x^2} + \frac{2m}{\hbar^2}(E - U_0)\psi(x) = 0. \tag{2.62}$$

Introducing the notation of a wave number,

$$k^2 = \frac{2m(E - U_0)}{\hbar^2}, \tag{2.63}$$

we arrive at a compact form,

$$\frac{\mathrm{d}^2\psi(x)}{\mathrm{d}x^2} + k^2\psi(x) = 0. \tag{2.64}$$

The latter equation resembles that known for a pendulum, for harmonic motion and for an *LC*-circuit. It has the general solution,

$$\psi(x) = A\exp(ikx) + B\exp(-ikx), \tag{2.65}$$

where A, B are constants to be derived based on the problem conditions. Equation (2.65) can be written in the other form inherent in a plane harmonic wave,

$$\psi = A'\sin kx + B'\cos kx. \tag{2.66}$$

Equation (2.63) gives,

$$E - U_0 = \frac{\hbar^2 k^2}{2m} = \frac{p^2}{2m}. \tag{2.67}$$

The difference $E - U_0$ is the particle kinetic energy, whereas E is the particle full energy. The de Broglie wavelength of a particle is,

$$\lambda = \frac{2\pi}{k} = \frac{2\pi\hbar}{\sqrt{2m(E-U_0)}}. \tag{2.68}$$

2.6 Quantum particle in complex potentials

In this section we consider how spatial confinements modify the motion and the wave function of a quantum particle. The reader is referred to Refs. [15–18] for more detail.

A rectangular well with infinite walls

The steady-state Shrödinger equation for a particle in a rectangular potential well with infinite walls (Fig. 2.4a) has the form,

$$-\frac{\hbar^2}{2m}\frac{\mathrm{d}^2\psi(x)}{\mathrm{d}x^2} + U(x)\psi(x) = E\psi(x), \tag{2.69}$$

where $U(x)$ is a potential box with width a, i.e.

$$U(x) = \begin{cases} 0 & \text{for } |x| < a/2 \\ \infty & \text{for } |x| > a/2 \end{cases}. \tag{2.70}$$

Based on the symmetry of the problem one can foresee odd and even solutions. The symmetry of the potential,

$$U(x) = U(-x)$$

results in a symmetry of the probability density,

$$|\psi(x)|^2 = |\psi(-x)|^2,$$

whence,

$$\psi(x) = \pm\psi(-x),$$

and so we arrive at two independent solutions with different parity. The odd and even types of solutions are,

$$\psi^- = \frac{\sqrt{2}}{a}\cos\frac{\pi n}{a}x \quad (n = 1, 3, 5, \ldots), \tag{2.71}$$

$$\psi^+ = \frac{\sqrt{2}}{a}\sin\frac{\pi n}{a}x \quad (n = 2, 4, 6, \ldots). \tag{2.72}$$

The energy spectrum consists of a set of discrete levels,

$$E_n = \frac{\pi^2\hbar^2}{2ma^2}n^2, \quad (n = 1, 2, 3, \ldots). \tag{2.73}$$

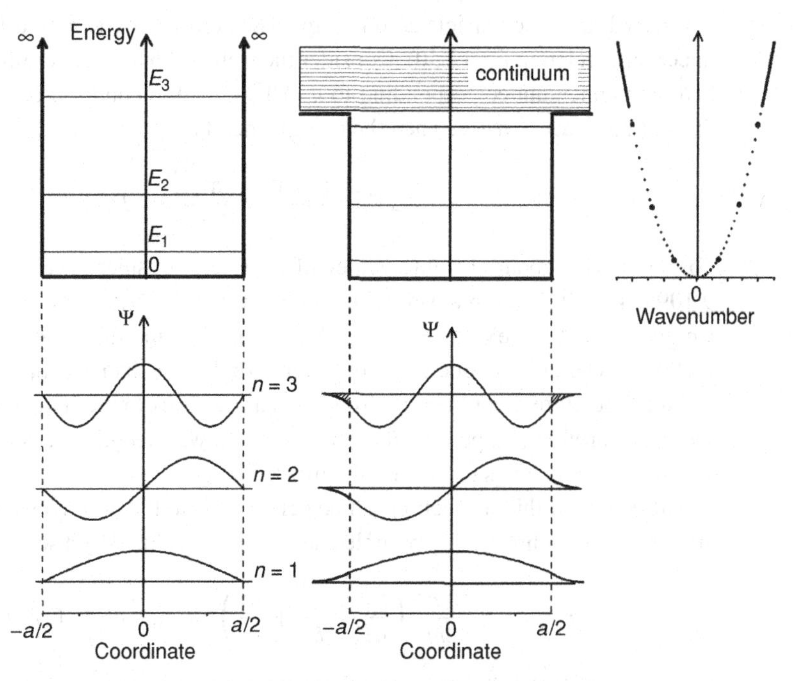

Fig. 2.4 **One-dimensional potential well with infinite (left) and finite (middle) walls, the first three states, corresponding to $n = 1, 2$ and 3, and the dispersion law in the case of the finite well (right). In the case of infinite walls, the energy states obey a series $E_n \sim n^2$ and the wave functions vanish at the walls. The total number of states is infinite. The probability of finding a particle inside the well is exactly equal to unity. In the case of finite walls, the states with energy higher than U_0 correspond to infinite motion and form a continuum. At least one state always exists within the well.**

We can ignore the case of $n = 0$ since this means there is no particle inside the well. The first three levels and corresponding wave functions are shown in Fig. 2.4 (left). Spacing between neighboring levels,

$$\Delta E_n = E_{n+1} - E_n = \frac{\pi^2 \hbar^2}{2ma^2}(2n + 1), \tag{2.74}$$

grows monotonically with n. For every state the wave function equals zero at the walls. The total probability of finding a particle inside the well equals 1.

Note, Eq. (2.73) gives the values of *kinetic* energy. Using Eqs. (2.6) and (2.7) we can write expressions for particle momentum, wave number and wavelength:

$$p_n = \frac{\pi\hbar}{a}n, \quad k_n = \frac{\pi}{a}n, \quad \lambda_n = \frac{2a}{n}. \tag{2.75}$$

Note that wavelengths correspond to integer numbers of $\lambda/2$ inside the well.

The minimal particle energy,

$$E_1 = \frac{\pi^2 \hbar^2}{2ma^2} \tag{2.76}$$

is referred to as the particle *zero energy*. The zero energy can be estimated based on the uncertainty relation Eq. (2.50). Consider the well width a to be coordinate uncertainty Δx, and the momentum corresponding to $n = 1$ to be momentum uncertainty Δp. According to Eq. (2.50) $\Delta p = \pi\hbar/a$. Then the energy uncertainty,

$$\Delta E = \frac{(\Delta p)^2}{2m} = \frac{\pi^2\hbar^2}{2ma^2} \equiv E_1. \tag{2.77}$$

To get an idea about absolute values of energies, consider an electron with mass $m_0 = 9.109534 \times 10^{-31}$ kg in a well with width 2 nm and infinite walls. Then for energy levels we get $E_1 = 0.094$ eV, $E_2 = 0.376$ eV, ..., and the minimal energy spacing $E_2 - E_1 = 0.282$ eV. The latter is one order of magnitude higher than the $k_B T = 0.027$ eV value at room temperature. If a transition from the state with E_1 to the state with E_2 is promoted by means absorption of a photon, the corresponding wavelength of electromagnetic radiation, $\lambda = 4394$ nm, belongs to the middle infrared range.

The results of this subsection can be generalized to the 3-dimensional case. If a well has sizes a, b, c in 3 directions a particle energy will be defined by a set of 3 quantum numbers:

$$E_{n_1 n_2 n_3} = \frac{\pi^2\hbar^2}{2m}\left(\frac{n_1^2}{a^2} + \frac{n_2^2}{b^2} + \frac{n_3^2}{c^2}\right), n_1, n_2, n_3 = 1, 2, 3, \ldots . \tag{2.78}$$

One can see that different sets of quantum numbers (and accordingly, different wave functions) may give the same value of energy. Such states are called *degenerate states*.

A rectangular well with finite barriers

In the case of finite walls (Fig. 2.4 (middle)), the states with energy higher than U_0 correspond to infinite motion and form a continuum. At least one state always exists within the well. The total number of discrete states is determined by the well width and height. The parameters in the figure correspond to the three states inside the well. Unlike the case of infinite walls, the wave functions extend to the classically forbidden regions $|x| > a/2$. Wave functions are no longer equal to zero at the walls but extend outside the walls exponentially. This means a probability arises of finding a particle outside the well. Extension of wave functions outside the well grows with n value. The probability of finding a particle inside the well is always less than unity and decreases with increasing E_n. A relation between E and k (dispersion law) in the case of a free particle has the form $E = \hbar^2 k^2/2m$ (dashed curve in Fig. 2.4 (right)). In the case of the finite potential well, a part of the dispersion curve relevant to confined states is replaced by discrete points (solid, line and points in the figure).

Particle wavelength, in the case of finite walls, gets longer as compared to the well with the same width but with infinite walls. Accordingly, the energy levels are lower than in the case of infinite walls. The total number of states inside the well is controlled by the relation,

$$a\sqrt{2mU_0} > \pi\hbar(n-1). \tag{2.79}$$

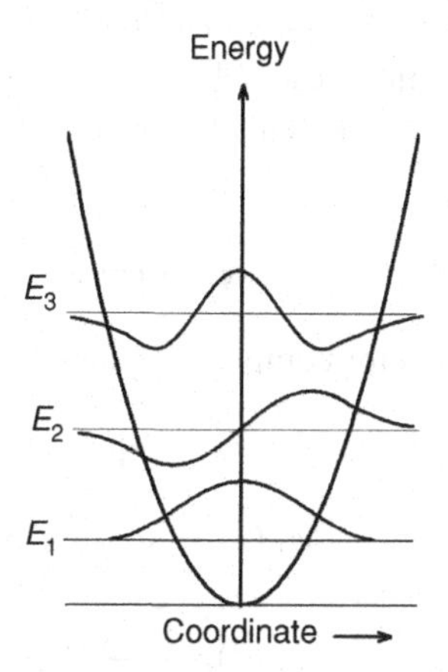

Fig. 2.5 **Quantum harmonic oscillator: the shape of the potential, the first three energy levels** $E_0 = {}^1/_2\hbar\omega$, $E_1 = {}^3/_2\hbar\omega$, $E_2 = {}^5/_2\hbar\omega$ **and the corresponding wave functions.**

This holds for $n = 1$ for any combination of a, m, U_0, i.e. at least one state always exists inside the well. The maximal number of levels is equal to maximal n for which Eq. (2.79) still holds. For deeper levels (smaller n) Eq. (2.73) can be used as a reasonable approximation.

In the case of an asymmetric potential well with different walls (U_0 at the left side and U_1 at the right side) the number of states inside the well is controlled by the relation,

$$a\sqrt{2mU_0} > \pi\hbar(n - \tfrac{1}{2}) - \hbar \arcsin\sqrt{U_0/U_1}. \tag{2.80}$$

This relation does not necessarily hold for $n = 1$. For a combination of a, m, U_0 that give a small $a\sqrt{2mU_0}$ value there might be no state inside the well.

The latter conclusion is valid in a more general formulation. In the case of a one-dimensional problem, a symmetric potential of any shape gives rise to at least a single localized state, provided that $U(\infty) = U(-\infty)$ holds and there is a minimum between $U(\infty)$, $U(-\infty)$. In the case that $U(\infty) \neq U(-\infty)$, a localized state may not occur.

In the case of 2- and 3-dimensional problems, a localized state may not occur even for symmetric potential wells. In other words, a particle is not necessarily "captured" by a 2- and 3-dimensional well. This is the case for narrow (smaller a) and shallow (lower U_0) wells.

Quantum harmonic oscillator

A harmonic oscillator is a quantum particle moving in a field with potential being a square function of coordinate. In a one-dimensional problem this means (Fig. 2.5),

$$U(x) = {}^1/_2 m\omega^2 x^2. \tag{2.81}$$

The steady-state Schrödinger equation has the form,

$$\nabla^2\psi(x) + \left(k^2 - \lambda^2 x^2\right)\psi(x) = 0, \tag{2.82}$$

где $k^2 = 2mE/\hbar^2$, $\lambda = m\omega/\hbar$. As in a rectangular well with infinite walls, the energy spectrum of a particle resembles an infinite number of discrete states with energy values E_n. The symmetry of potential gives rise to odd and even solutions. The general solution reads,

$$\psi_n(x) = u_n(x)\exp\left(-\lambda x^2/2\right), \tag{2.83}$$

where $u_n(x)$ stands for complex polynomials. The first three solutions have the form:

$$\begin{aligned}\psi_0(x) &= \exp\left(-\lambda x^2/2\right),\\ \psi_1(x) &= \sqrt{2\lambda}\cdot x\exp\left(-\frac{1}{2}\lambda x^2\right),\\ \psi_2(x) &= \frac{1}{\sqrt{2}}\left(1-2\lambda x^2\right)\exp\left(-\frac{1}{2}\lambda x^2\right).\end{aligned} \tag{2.84}$$

These are shown in Figure 2.5. The number of nodes in the wave function equals n. The energy values are,

$$E_n = \hbar\omega(n + 1/2), \quad \text{where } n = 0, 1, 2, \ldots. \tag{2.85}$$

The zero energy of a quantum harmonic oscillator corresponds to $n = 0$ and is,

$$E_0 = \hbar\omega/2. \tag{2.86}$$

The principal feature of a quantum harmonic oscillator is the same spacing, $\Delta E = \hbar\omega$, between all pairs of neighboring energy levels.

The model of a quantum harmonic oscillator is very instructive for many problems in quantum physics. It provides an approximate energy spectrum for many complex potentials near local minima. Since every continuous function $U(x)$ in the vicinity of an extremum point x_0 can be expanded in a series,

$$U(x) = U(x_0) + \frac{1}{2}\frac{d^2U}{dx^2}(x - x_0)^2 + \cdots, \tag{2.87}$$

every physical system features harmonic oscillator properties near the potential minimum. At moderate temperatures, the harmonic approximation works for oscillations of nuclei and atomic cores in molecules and solids. Therefore the multitude of crystal lattice oscillations is treated as a set of harmonic oscillators with the quanta of these oscillations called *phonons*. In this case the finite zero-energy phenomenon has a very profound physical meaning. It means that lattice oscillations never stop, even at absolute zero temperature. These zero oscillations have been detected experimentally and provide explanations as to why light atoms e.g. He, never form a solid phase.

In Chapter 12 we shall see that quantum electrodynamics treats an electromagnetic field as a set of harmonic oscillators to justify the concept of light quanta – *photons*. Zero oscillations in this case mean the energy of an electromagnetic field never goes to zero.

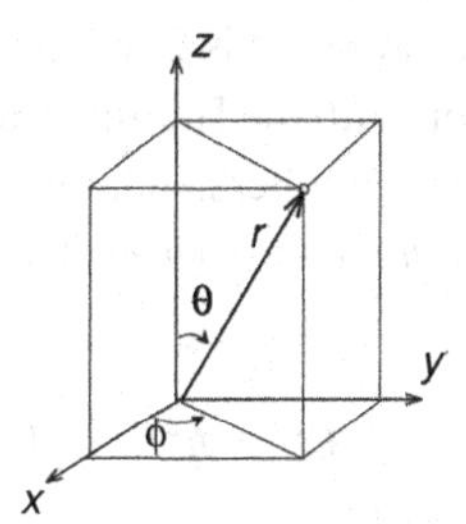

Fig. 2.6 Spherical coordinates.

Particle in a spherically symmetric potential

Spherically symmetric potentials arise in many problems of quantum physics where interactions occur that depend on interparticle distance. A non-exhaustive list includes electrons in atoms, impurity centers in solids, excitons in crystals. The model of a spherical potential box in spite of its seeming abstractiveness is crucially important for electrons in quantum dots.

In the case of a spherically symmetric potential $U(r)$ we deal with a Hamiltonian,

$$\mathbf{H} = -\frac{\hbar^2}{2m}\nabla^2 + U(r), \tag{2.88}$$

where $r = \sqrt{x^2 + y^2 + z^2}$. Taking into account the symmetry of the problem, it is reasonable to consider it in spherical coordinates, r, ϑ, and φ (Fig. 2.6):

$$x = r\sin\vartheta\cos\varphi, \quad y = r\sin\vartheta\sin\varphi, \quad z = r\cos\vartheta. \tag{2.89}$$

In spherical coordinates Hamiltonian (2.88) reads,

$$\mathbf{H} = -\frac{\hbar^2}{2mr^2}\frac{\partial}{\partial r}\left(r^2\frac{\partial}{\partial r}\right) - \frac{\hbar^2\Lambda}{2mr^2} + U(r), \tag{2.90}$$

where the Λ operator is,

$$\Lambda = \frac{1}{\sin\vartheta}\left[\frac{\partial}{\partial\vartheta}\left(\sin\vartheta\frac{\partial}{\partial\vartheta}\right) + \frac{1}{\sin\vartheta}\frac{\partial^2}{\partial\varphi^2}\right]. \tag{2.91}$$

We shall skip mathematical details and highlight only the principal results that arise from the spherical symmetry of the potential. In this case, the wave function can be separated into functions of r, ϑ, and φ:

$$\psi = R(r)\,\Theta(\vartheta)\,\Phi(\varphi), \tag{2.92}$$

and can be written in the form,

$$\Psi_{n,l,m}(r,\vartheta,\varphi) = \frac{u_{n,l}(r)}{r}Y_{lm}(\vartheta,\varphi), \tag{2.93}$$

where Y_{lm} are the *spherical Bessel functions*, and $U(r)$ satisfies an equation

$$-\frac{\hbar^2}{2m}\frac{\mathrm{d}^2u}{\mathrm{d}r^2} + \left[U(r) + \frac{\hbar^2}{2mr^2}l(l+1)\right]u = Eu. \tag{2.94}$$

To obtain the energy values, it is now possible to consider the one-dimensional Eq. (2.94) instead of the equation with the Hamiltonian (2.90). The state of the system is characterized by the three quantum numbers, namely, the *principal quantum number* n, *the orbital number* l, and *the magnetic number* m. The *orbital quantum number* determines the angular momentum value $\mathbf{L}$:

$$\mathbf{L}^2 = \hbar^2 l(l+1), \quad l = 0, 1, 2, 3, \ldots \tag{2.95}$$

The *magnetic quantum number* determines the L component parallel to the z-axis,

$$L_z = \hbar m, \quad m = 0, \pm 1, \pm 2, \ldots \pm l. \tag{2.96}$$

Every state with a certain l value is $(2l+1)$ - degenerate according to $2l+1$ values of m. The states corresponding to different l values are usually denoted as s-, p-, d-, f- and g-states, and further in alphabetical order. States with zero angular momentum ($l = 0$) are referred to as s-states, states with $l = 1$ are denoted as p-states and so on. The parity of states corresponds to the parity of the l value, because the radial function is not sensitive to inversion (r remains the same after inversion) and the spherical function after inversion transforms as follows:

$$Y_{lm}(\vartheta, \varphi) \rightarrow (-1)^l Y_{lm}(\vartheta, \varphi).$$

The specific values of energy are determined by a $U(r)$ function. Consider a simplest case, corresponding to a spherically symmetric potential well with an infinite barrier, i.e.,

$$U(r) = \begin{cases} 0 & \text{for } r \leq a \\ \infty & \text{for } r > a \end{cases}. \tag{2.97}$$

In this case energy values are expressed as follows:

$$E_{nl} = \frac{\hbar^2 \chi_{nl}^2}{2ma^2}, \tag{2.98}$$

where χ_{nl} are roots of the spherical Bessel functions with n being the number of the root and l being the order of the function. Table 1.1 lists x_{nl} values for several n, l values. Note that for $l = 0$ these values are equal to πn ($n = 1, 2, 3, \ldots$) and Eq. (2.98) converges with the relevant expression in the case of a one-dimensional box (Eq. (2.73)). This results from the fact that for $l = 0$, Eq. (2.94) for the radial function $u(r)$ reduces to Eq. (2.69) with the potential (2.70). To summarize, a particle in a spherical well possesses the set of energy levels $1s, 2s, 3s, \ldots,$ coinciding with energies of a particle in a rectangular one-dimensional well, and additional levels $1p, 1d, 1f, \ldots, 2p, 2d, 2f, \ldots,$ that arise due to spherical symmetry of the well (Fig. 2.7).

In the case of the spherical well with finite potential, U_0, Eq. (2.98) can be considered as a good approximation only if U_0 is large enough, namely for $U_0 \gg \hbar^2/8ma^2$. The right side of this inequality is a consequence of the uncertainly relation. In the

Table 2.1. Roots of the Bessel functions χ_{nl}

l	$n = 1$	$n = 2$	$n = 3$
0	3.142 (π)	6.283 (2π)	9.425 (3π)
1	4.493	7.725	10.904
2	5.764	9.095	12.323
3	6.988	10.417	
4	8.183	11.705	
5	9.356		
6	10.513		
7	11.657		

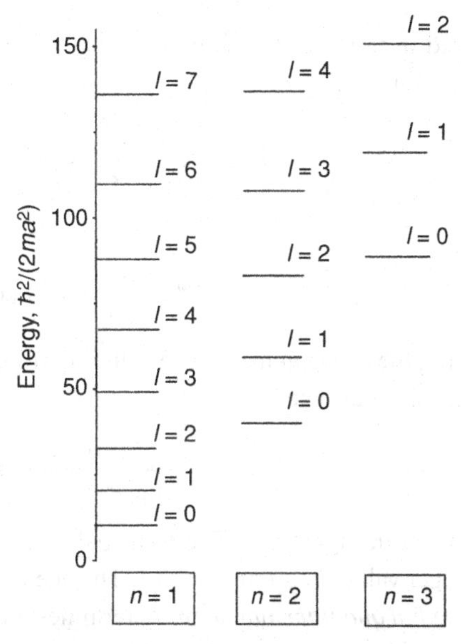

Fig. 2.7 Energy levels of a particle in a spherical well with infinite barriers. Energy is scaled in the dimensionless units of $\chi_{nl}^2 = E_{nl}(\hbar^2/2ma^2)^{-1}$, where χ_{nl} values are the roots of the Bessel functions listed in Table 2.1. The states are classified by the principal quantum number, n, and by the orbital quantum number, l. Every state is $(2l+1)$ - degenerate. For $l = 0$ (so-called s-states) $x_{n0} = \pi n$ holds, and corresponding energies obey a series derived for a particle in a rectangular well.

case when,

$$U_0 = U_{0\,\mathrm{min}} = \frac{\pi^2 \hbar^2}{8ma^2},$$

exactly one state exists within the well. For $U_0 < U_{0\,\mathrm{min}}$ no state exists in the well at all. This is an important difference of the three-dimensional case as compared to the one-dimensional problem.

Electron in Coulomb potential

For an electron with charge e interacting with another particle with the same charge,[2]

$$U(r) = -\frac{e^2}{r}, \tag{2.99}$$

the equation for the radial part of the wave function can be written as

$$\left[\frac{d^2}{d\rho} + \varepsilon + \frac{2}{\rho} - \frac{l(l+1)}{\rho^2}\right] u(\rho) = 0. \tag{2.100}$$

The dimensionless argument and energy,

$$\rho = \frac{r}{a^0}, \qquad \varepsilon = \frac{E}{E^0}$$

are expressed in terms of the so-called *atomic length unit* a^0 and *atomic energy unit* E^0 given by (in SI units),

$$a^0 = 4\pi\varepsilon_0 \frac{\hbar^2}{m_0 e^2} \approx 5.292 \cdot 10^{-2}\ \text{nm} \tag{2.101}$$

and

$$E^0 = \frac{e^2}{2a^0} \approx 13.60\,\text{eV}, \tag{2.102}$$

with m_0 being the electron mass. The solution of Eq. (2.100) leads to the following result.

Energy levels obey a series,

$$\varepsilon = -\frac{1}{(n_r + l + 1)^2} \equiv -\frac{1}{n^2}, \tag{2.103}$$

which is shown in Figure 2.8. The principal quantum number is $n = n_r + l + 1$. It takes positive integer values beginning with 1. The energy is unambiguously defined by a given n value. The *radial quantum number* n_r determines the quantity of nodes of the corresponding wave function. For every n value, exactly n states exist differing in l which runs from 0 to $(n-1)$. Additionally, for every given l value, $(2l+1)$ - degeneracy occurs with respect to $m = 0, \pm 1, \pm 2, \ldots$ Therefore the total degeneracy is,

$$\sum_{l=0}^{n-1} (2l+1) = n^2.$$

For $n = 1$, $l = 0$ (1s-state), the wave function obeys a spherical symmetry with a^0 corresponding to the most probable distance where an electron can be found. Therefore, the relevant value in real atom-like structures is called the "*Bohr radius*".

So far, idealized elementary problems have been examined. Now we are in a position to deal with the simplest real quantum mechanical object, i.e., the *hydrogen atom*, consisting

[2] It is assumed that the particle interacting with the electron is fixed in space forming the origin of coordinates, or that it is so heavy compared with an electron that an electron can not disturb this particle by its charge.

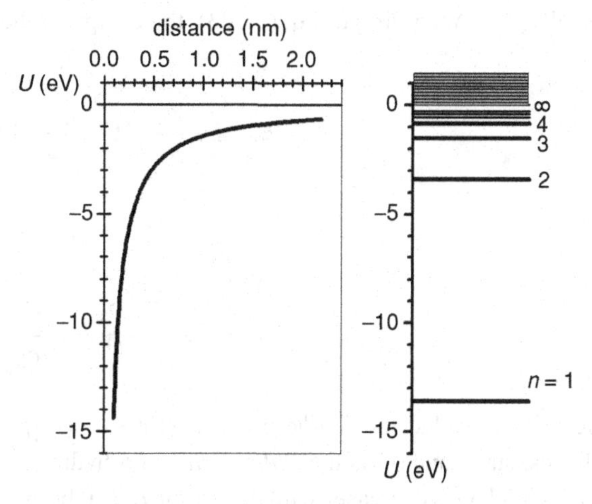

Fig. 2.8 **The Coulomb potential $-e^2/r$ (left) and energy levels of an electron. For $E < 0$ the energy spectrum consists of discrete levels $E_n = -E^0/n^2$, $E^0 = 13.60$ eV, each level being n^2-degenerate. For $E > 0$ a particle exhibits an infinite motion with a continuous energy spectrum.**

of a proton with mass M_0 and an electron. The relevant Schrödinger equation is the two-particle equation with the Hamiltonian,

$$\mathbf{H} = -\frac{\hbar^2}{2M_0}\nabla_\mathrm{p}^2 - \frac{\hbar^2}{2m_0}\nabla_\mathrm{e}^2 - \frac{e^2}{|\mathbf{r}_\mathrm{p} - \mathbf{r}_\mathrm{e}|}, \tag{2.104}$$

where $\mathbf{r}_\mathrm{p}$ and $\mathbf{r}_\mathrm{e}$ are the radius-vectors of the proton and electron, and p and e indices in the ∇^2 operator denote differentiation with respect to the proton and electron coordinates, respectively. We introduce a relative radius-vector $\mathbf{r}$ and a radius-vector of the center of mass as follows:

$$\mathbf{r} = \mathbf{r}_\mathrm{p} - \mathbf{r}_\mathrm{e}, \quad \mathbf{R} = \frac{m_0\mathbf{r}_\mathrm{e} + M_0\mathbf{r}_\mathrm{p}}{m_0 + M_0}, \tag{2.105}$$

and use the full mass and the reduced mass of the system, M and μ,

$$M = m_0 + M_0, \quad \mu = \frac{m_0 M_0}{m_0 + M_0}. \tag{2.106}$$

The Hamiltonian (2.104) then reads,

$$\mathbf{H} = -\frac{\hbar^2}{2M}\nabla_R^2 - \frac{\hbar^2}{2\mu}\nabla_r^2 - \frac{e^2}{r}. \tag{2.107}$$

One can see Eq. (2.107) diverges into the Hamiltonian of a free particle with mass M and the Hamiltonian of a particle with mass μ in the potential $-e^2/r$. The former describes an infinite center-of-mass motion of the two-particle atom, whereas the latter gives rise to internal states. According to Eq. (2.103), the energy of these states can be written as,

$$E_n = -\frac{\mathrm{Ry}}{n^2} \quad \text{for } E < 0 \tag{2.108}$$

with,

$$\mathrm{Ry} = \frac{e^2}{2a_\mathrm{B}}, \quad a_\mathrm{B} = \frac{\hbar^2}{\mu e^2}. \tag{2.109}$$

Here Ry is called the *Rydberg constant* and corresponds to the *ionization energy* of the lowest state, and a_B is the *Bohr radius* of a hydrogen atom. The distance between the neighboring levels decreases with n, and for $E > 0$ both electron and proton experience an infinite motion.

One can see that the energy spectrum and Bohr radius for a hydrogen atom differ from the relevant values of a single-particle problem by the μ/m_e coefficient. In the case under consideration this coefficient is 0.9995. For this reason expressions (2.101) and (2.102) are widely used instead of the exact values (2.109). This is reasonable in the case of a proton and an electron but should be used with care for other hydrogen-like systems. For example, in a positronium atom, consisting of an electron and a positron with equal masses, the explicit values (2.109) should be used.

The problems of a particle in a spherical potential well and of the hydrogen atom are very important for further consideration. The former is used to model an electron and a hole in a nanocrystal, and the latter is essential for excitons in a bulk crystal and in nanocrystals as well. Furthermore, the example of a two-particle problem is a precursor to the general approach used for many-body systems. It contains a transition from the many-particle problem (proton and electron) to the one-particle problem by means of renormalization of mass (reduced mass μ instead of M_0 and m_0) and a differentiation between the collective behavior (center-of-mass translational motion) and the single-particle motion in some effective field. This approach has far-reaching consequences resulting in the concepts of effective mass and of quasiparticles, to be presented in Chapter 4.

Problems

1. Find the wavelength where the energy of an electron and a photon coincide, i.e. the crossover point for corresponding dispersion laws. Consider the electromagnetic wave range to which this crossover point corresponds (optical, radio, microwave etc.). Discuss which physical processes occur at this point.

2. Derive 1- and 2-dimensional density of states for a particle with mass m. Evaluate the correlation between the space dimensionality and density of modes dependence on particle energy and momentum.

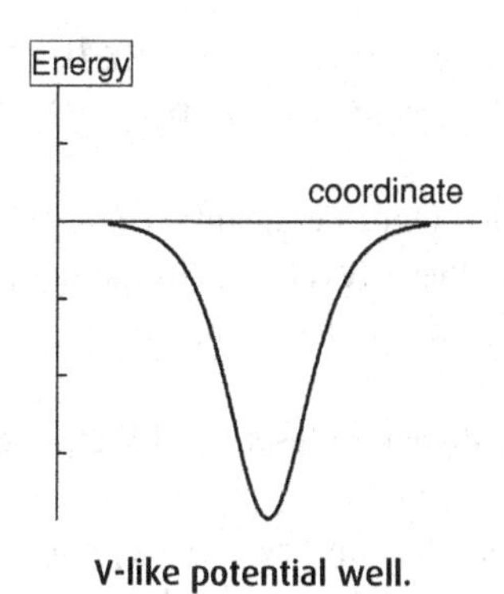

Fig. 2.9 V-like potential well.

3. The Bose–Einstein distribution function for molecules and atoms is,

$$F(E) = \left(\exp\frac{E-\mu}{k_B T} - 1\right)^{-1} \tag{2.110}$$

where E is the kinetic energy and μ is the *chemical potential.* Consider and explain the difference between this distribution and Eq. (2.28).

4. Compare the Helmholtz (Eq. (2.47)) and the steady-state Schrödinger (Eq. (2.62)) equations. Examine the similarities and differences.

5. Find the wavelength dependence $U(\lambda)$ for the energy density contained in equilibrium electromagnetic radiation. Use Eq. (2.29) and the relation,

$$U(\lambda) = U(\omega)\frac{d\omega}{d\lambda}. \tag{2.111}$$

Observe the difference in $U(\omega)$ and $U(\lambda)$ functions and explain why the relation (2.111) should be used in this consideration.

6. Compare the wavelengths a particle has in free space and in a potential $U(x) = \text{const.}$ Compare particle wavelengths for $U = 0$ and $U \neq 0$ with an electromagnetic wavelength for $n = 1$ and $n > 1$.

7. Compare Figure 2.2 and Figure 2.4 (left), find and discuss the similarities in these figures as well as in Eqs. (2.17) and (2.75).

8. Plot the dispersion law $E(p)$ or $E(k)$ for a particle in a rectangular well with infinite barriers.

9. Compare spacing between energy levels with growing quantum numbers for a particle in a rectangular box, harmonic oscillator and Coulomb potential, and outline the difference.

10. Using results for a finite rectangular well, a harmonic oscillator and Coulomb potential try to guess the energy spectrum of a V-like potential well shown in Figure 2.9.

11. Calculate the electromagnetic wavelength corresponding to the ionization energy of a hydrogen atom.

References

[1] L. de Broglie. Ondes et quanta. *Compt. Rend.*, **177** (1923), 507–509.
[2] M. Planck. Uber irreversible Strahlungsvorgange. *Ann. Phys.,* **1** (1900), 69–122.
[3] Lord Rayleigh. Remarks upon the Law of Complete Radiation. *Philos. Mag.*, **49** (1900), 539–540.
[4] S. N. Bose. Planck's Gesetz und Lichtquantenhypothese. *Zs. Physik*, **26** (1924), 178–181.
[5] G. N. Lewis. The conservation of photons. *Nature,* **118** (1926), 874–875.
[6] A. Einstein. Quantentheorie des einatomigen idealen Gases, Sitzungsber. *Preuss. Akad. Wiss.*, **22** (1924), 261–267.
[7] M. Schroeder. *Fractals, Chaos, Power Laws* (New York: Freeman and Co, 1990).
[8] A Yariv and P. Yeh. *Optical Waves in Crystals* (New York: John Wiley and Sons, 1984).
[9] M. Planck. Die physikalische Struktur des Phasenraumes. *Ann. Phys.,* **50** (1916), 385–395.
[10] M. Planck. *Vorlesungen über die Theorie der Wärmestrahlung* (Barth, Leipzig, 1906, 1913, 1919, 1921, 1923).
[11] W. Heisenberg. Uber den anschaulichen Inhalt der quantentheoretischen Kinematik und Mechanik. *Z. Phys.,* **43** (1927), 172–198.
[12] W. Heisenberg. *The Physical Principles of Quantum Theory. Supplement to the Russian Edition* (Leningrad: Gostekhizdat, 1932).
[13] M. Born. Quantenmechanik der Stossvorgange. *Zs. Physik,* **38** (1926), 803–827.
[14] E. Schrödinger. Quantisierung als Eigenwertproblem. *Ann. Physik*, **81** (1926), 109–139.
[15] L. I. Schiff. *Quantum Mechanics* (New York: McGraw-Hill, 1968).
[16] S. Flugge. *Practical Quantum Mechanics*, part I. (Berlin: Springer, 1971, Moscow: Mir, 1974).
[17] L. D. Landau and I. M. Lifshitz. *Quantum Mechanics* (Moscow: Nauka, 1989).
[18] A. Yariv. *An Introduction to Theory and Applications of Quantum Mechanics* (New York: Wiley & Sons, 1982).

3 Wave optics versus wave mechanics I

Let us return from optics to mechanics and explore the analogy to its fullest extent. In optics the old system of mechanics corresponds to intellectually operating with isolated mutually independent light rays. The new undulatory mechanics corresponds to the wave theory of light.

Erwin Schrödinger, Nobel lecture, 1933

In this chapter we shall see that electromagnetic waves and electrons feature a number of common properties under conditions of spatial confinement. Simple and familiar problems from introductory quantum mechanics and textbook wave optics are recalled in this chapter to emphasize the basic features of waves in spatially inhomogeneous media. Herewith we make a first step towards understanding the properties of electrons and electromagnetic waves in nanostructures and notice that these properties in many instances are counterparts. Different formulas and statements of this chapter can be found in handbooks on quantum mechanics [1–9] and wave optics [10–13]. A few textbooks on quantum mechanics do consider analogies of propagation and reflection phenomena in wave optics with those in wave mechanics [6–9].

3.1 Isomorphism of the Schrödinger and Helmholtz equations

In Chapter 2 we discussed that an electron in quantum mechanics is described by the wave function, the square of its absolute value giving the probability of finding an electron at a specific point in space. This function satisfies the *Schrödinger* equation (2.56) which is the second-order differential equation with respect to space and the first-order differential equation with respect to time. Electromagnetic waves are described by space- and time-dependent electric and magnetic fields which are, unlike electron wave function, directly observable and measurable. Electric and magnetic fields satisfy a couple of identical wave equations (both given in Eq. 2.36), each of which is the second-order differential equation with respect to space and time. Although time dependence in these equations differs, in the steady state when time dependence is eliminated, both equations reduce to a mathematically identical form, namely to the second-order differential equation with respect to coordinates. The similarity of quantum particles and electromagnetic waves was the subject of Problem 4 in Chapter 2. Recalling Eqs. (2.46), (2.56) and (2.57), restricting ourselves to the electric field only when examining electromagnetic waves, for simplicity[1] and using notation A for the amplitude of the electric field to discriminate it from particle energy E,we arrive at the pair of equations:

$$\nabla^2 A(\mathbf{r}) + \frac{n^2(\mathbf{r})}{c^2}\omega^2 A(\mathbf{r}) = 0 \tag{3.1}$$

and

$$\frac{\hbar^2}{2m}\nabla^2\psi(\mathbf{r}) + [E - U(\mathbf{r})]\psi(\mathbf{r}) = 0 \tag{3.2}$$

which are mathematically identical. This statement is usually referred to as isomorphism of the *Helmholtz* (3.1) and *Schrödinger* (3.2) equations. From a mathematical point of view, a more strict statement is "the steady-state *Schrödinger* equation is the differential equation known as the *Helmholtz* equation". However, typically the notation "*Helmholtz equation*" is applied to classical waves (not only to electromagnetic waves but, e.g. to *acoustic* waves as well), whereas the notation "*Schrödinger equation*" always belongs to the wealth of quantum physics. Indeed, physically Eqs. (3.1) and (3.2) describe different entities and contain different parameters of the matter and the field.

The purpose of further consideration is to trace the physical counterparts resulting from the mathematical isomorphism of the Schrödinger and Helmholtz equations. In the forthcoming sections, we shall consider a number of one-dimensional problems. In the one-dimensional case Eqs. (3.1) and (3.2) take the form,

$$\frac{\mathrm{d}^2 A(x)}{\mathrm{d}x^2} + \frac{n^2(x)}{c^2}\omega^2 A(x) = 0 \tag{3.3}$$

and

$$\frac{\mathrm{d}^2\psi(x)}{\mathrm{d}x^2} + \frac{2m}{\hbar^2}[E - U(x)]\psi(x) = 0. \tag{3.4}$$

Now the formal substitution,

$$\frac{n^2(x)}{c^2}\omega^2 \leftrightarrow \frac{2m}{\hbar^2}[E - U(x)] \tag{3.5}$$

merges Eqs. (3.3) and (3.4). In the case of space-independent refractive index, $n(x) = n_0 \equiv \text{const}$ and constant potential, $U(x) = U_0 \equiv \text{const}$, one can completely merge the Schrödinger and the Helmholtz equations to become,

$$\frac{\mathrm{d}^2 A(x)}{\mathrm{d}x^2} + k_{\mathrm{EM}}^2 A(x) = 0 \tag{3.6}$$

and

$$\frac{\mathrm{d}^2\psi(x)}{\mathrm{d}x^2} + k_{\mathrm{QM}}^2\psi(x) = 0, \tag{3.7}$$

[1] Keeping the electric field rather than the magnetic field in equations has another reason beyond simplicity. Most inhomogeneous media and structures to be considered hereafter in this book possess dielectric rather than magnetic inhomogeneity, therefore spatial confinement of the electric field is the primary physical issue under consideration.

where k_{EM} and k_{QM} stand for the electromagnetic and quantum mechanical wave numbers, respectively, and are expressed as,

$$k_{EM} = \frac{n_0 \omega}{c} \quad \text{and} \quad k_{QM} = \frac{1}{\hbar}\sqrt{2m(E - U_0)}, \tag{3.8}$$

for the classical and the quantum counterparts. Note, expressions (3.5) and (3.8) provide an instructive hint that space variation in refractive index may result in a similar effect to space variation in potential, more strictly in $E - U(x)$. Note that both values determine the speed of waves. In the case of classical waves we have $v = c/n$. In the case of a quantum particle the speed can be derived from,

$$\text{Kinetic energy} \equiv E - U_0 \equiv \frac{p^2}{2m}, \tag{3.9}$$

whence,

$$p \equiv mv = \sqrt{2m(E - U_0)} \Rightarrow v = \sqrt{\frac{2(E - U_0)}{m}}. \tag{3.10}$$

We shall see in the following sections a variety of counterparting of optical and quantum phenomena which are the consequences of (i) the wave nature of light and electrons, (ii) the existence of evanescent waves in so-called "classically forbidden" areas, and (iii) the interference of waves scattered over barriers/wells. Unlike traditional considerations in many textbooks on quantum mechanics our purpose is to emphasize the similarities between light and electrons, rather than to outline the differences between an electron compared to a solid ball in classical mechanics.

3.2 Propagation over wells and barriers

Potential and refraction steps

Consider a particle with energy E performing a motion over a potential semi-infinite step at the point $x = 0$ of height U_0 (Fig. 3.1, left panel). We assume the particle has met a barrier when moving from left to right along the x-direction. We expect that particle motion will be disturbed by the barrier so that at $x < 0$ we have to consider forward, and backward propagation, whereas for $x > 0$ only forward propagation remains, since there is no reason to expect backward propagation behind the barrier. We can then write the wave function as,

$$\psi_I(x) = A\exp(ik_1x) + B\exp(-ik_1x) \quad \text{for} \quad x < 0 \tag{3.11}$$

$$\psi_{II}(x) = C\exp(ik_2x) \quad \text{for} \quad x > 0, \tag{3.12}$$

where

$$k_1 = \frac{\sqrt{2mE}}{\hbar}, \quad k_2 = \frac{\sqrt{2m(E - U_0)}}{\hbar}, \tag{3.13}$$

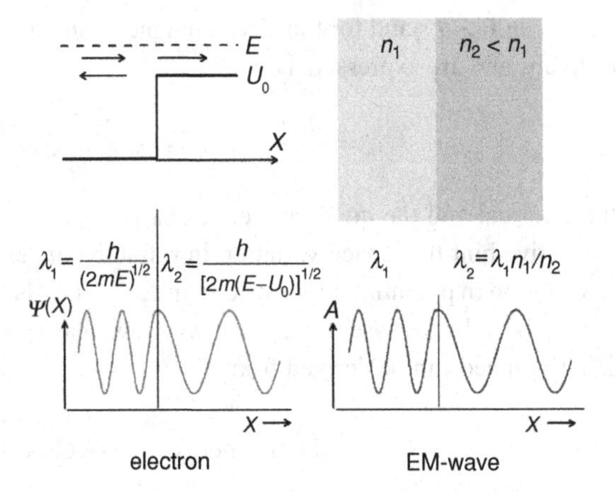

Fig. 3.1 **Propagation of a quantum particle (left) and an electromagnetic wave (right) over a potential step and refraction index step, respectively. In both cases partial reflection occurs with simultaneous change in wavelength at the step. An upward potential step in quantum mechanics corresponds to a downward refraction step in optics. Graphs for wave function and electric field amplitude do not account for the partial decrease in amplitude of transmitted waves.**

k_1 and k_2 are particle wave numbers in front of and behind the barrier, respectively. One notices immediately that Eqs. (3.13) mean that a particle de Broglie wavelength has been changed at the potential step from λ_0 to λ_1 as follows:

$$\lambda_1 = \frac{2\pi}{k_1} = \frac{h}{\sqrt{2mE}}, \quad \lambda_2 = \frac{2\pi}{k_2} = \frac{h}{\sqrt{2m(E-U_0)}}. \tag{3.14}$$

Let us now look for reflection, r and transmission, t coefficients for the wave function amplitudes, i.e.

$$r = B/A, \quad t = C/A, \tag{3.15}$$

which determine the ratio of backward and transmitted wave amplitudes. We start from Eqs. (3.11), (3.12) and apply the continuity condition to the wave functions and to their first derivatives at point $x = 0$,

$$\psi_I(0) = \psi_{II}(0) \Rightarrow A + B = C,$$
$$\psi_I'(0) = \psi_{II}'(0) \Rightarrow A - B = \frac{k_2}{k_1}C.$$

Then after some simple arithmetic we arrive at,

$$r = \frac{k_1 - k_2}{k_1 + k_2}, \quad t = \frac{2k_1}{k_1 + k_2}. \tag{3.16}$$

Using Eqs. (3.13) we can express r, t in terms of the two parameters of the problem, namely the particle full energy E, and the potential step U_0,

$$r = \frac{\sqrt{E} - \sqrt{E - U_0}}{\sqrt{E} + \sqrt{E - U_0}}, \quad t = \frac{2\sqrt{E}}{\sqrt{E} + \sqrt{E - U_0}}. \tag{3.17}$$

To obtain the full probabilities that can provide the relative flux portion of particles reflected from and transmitted over the potential step, one should recall that we need to know $|\psi(x)|^2 = \psi(x)\psi^*(x)$. Then we arrive at,

$$R = r^2 = \frac{(\sqrt{E} - \sqrt{E - U_0})^2}{(\sqrt{E} + \sqrt{E - U_0})^2}, \quad T = 1 - R = \frac{4\sqrt{E(E - U_0)}}{(\sqrt{E} + \sqrt{E - U_0})^2}. \tag{3.18}$$

To summarize, taking into account the probabilistic interpretation of wave function, a flux of particles moving with the same energy (wave number, wavelength) over a potential step experiences partial reflection and partial transmission with simultaneous modification in wavelength (wave number). The particle full energy E remains the same but the particle kinetic energy becomes smaller by the amount of the step. This consideration remains valid in the case where the particle moves from the right to the left. The reflection and transmission remain the same and the wavelength, wave number and energy will be modified in the reverse way.

Let us in the same manner consider a classical, say electromagnetic, wave with frequency ω, wave number k_1, wavelength λ_1, meeting a refraction index step from n_1 to $n_2 < n_1$ at the border of the two media.[2] Because of the mathematical identity of Eqs. (3.3) and (3.4), similar to Eqs. (3.11)–(3.16) we arrive at change in wave number, change in wavelength, partial transmission and reflection with the frequency remaining unperturbed at the border. Namely,

$$\begin{aligned} k_1 &= \frac{\omega}{c} n_1 \quad \text{transforms to} \quad k_2 = \frac{\omega}{c} n_2, \\ \lambda_1 &= \frac{2\pi}{k_1} \quad \text{transforms to} \quad \lambda_2 = \frac{2\pi}{k_2} = \lambda_1 \frac{n_1}{n_2}, \end{aligned} \tag{3.19}$$

and the amplitude transmission and reflection coefficients are expressed as in Eq. (3.16),

$$r = \frac{k_1 - k_2}{k_1 + k_2}, \quad t = \frac{2k_1}{k_1 + k_2}. \tag{3.20}$$

In optics, Eqs. (3.20) are usually used in terms of refraction indexes rather than wave numbers,

$$r = \frac{n_1 - n_2}{n_1 + n_2}, \quad t = \frac{2n_1}{n_1 + n_2}. \tag{3.21}$$

Equations (3.21) give the amplitude reflection and transmission coefficients. To obtain the intensity reflection coefficient, R and intensity transmission coefficient, T we need,

$$R = r^2 = \frac{(n_1 - n_2)^2}{(n_1 + n_2)^2}, \quad T = 1 - R = \frac{4n_1 n_2}{(n_1 + n_2)^2}. \tag{3.22}$$

These expressions are symmetrical with respect to changes $n_1 \leftrightarrow n_2$. One can see that although Eqs. (3.16) directly replicate their optical counterparts Eq. (3.20), Eqs. (3.18) yet differ from Eqs. (3.21) in a sense that Eq. (3.18) contain energies whereas Eqs. (3.22)

[2] For acoustic waves the description remains the same, with relative material density which determines the speed of sound instead of index of refraction determining the speed of light in optics.

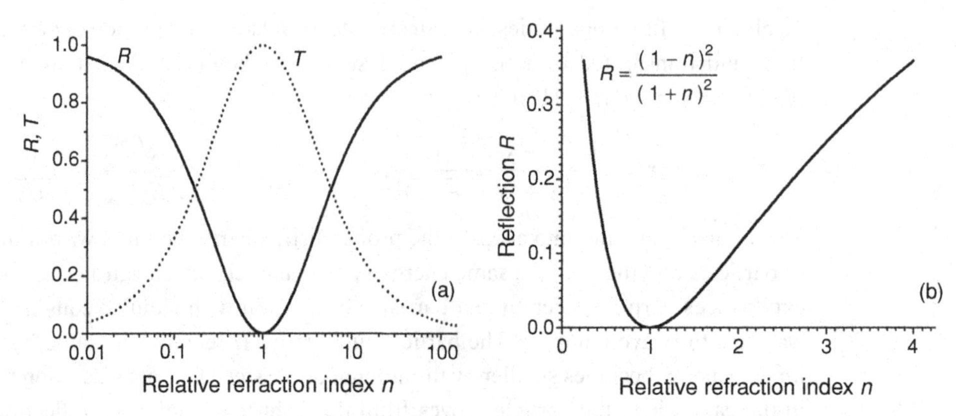

Fig. 3.2 **(a) Reflection and transmission coefficients at the potential step according to Eqs. (3.25), (3.26) and (b) reflection coefficient for $0.25 < n < 4$, which is relevant to electromagnetic waves within the optical range. Note the symmetry of the R and T functions for $n > 1$ and $n < 1$.**

contain only dimensionless indexes. This difference has no physical meaning and can be cancelled by introducing the relative refraction index indicating how many times the speed changes at the step. Recalling,

$$n_{\mathrm{EM}} = \frac{v_2}{v_1} = \frac{n_1}{n_2} = \frac{k_1}{k_2} \tag{3.23}$$

used in optics, we introduce its quantum mechanical counterpart,

$$n_{\mathrm{QM}} = \frac{v_2}{v_1} = \frac{k_2}{k_1} = \sqrt{\frac{E - U_0}{E}}, \tag{3.24}$$

which represents the dimensionless ratio of particle speed values in the two regions of space (in front of and over the potential step) in complete concordance with the relative refraction index notation in optics. Note that the upward potential step in quantum mechanics corresponds to the downward refraction step in optics. This asymmetry arises from the linear speed versus wave number relation in quantum mechnics and reciprocal speed versus wave number relation in optics. Notably, this asymmetry has no effect on formulas for R and T because these contain functions of n which are invariant with respect to substitution $n \to 1/n$.

With these notations the quantum mechanical reflection coefficient and the transmission coefficient become,

$$R_{\mathrm{QM}} = \frac{(1 - n_{\mathrm{QM}})^2}{(1 + n_{\mathrm{QM}})^2}, \quad T_{\mathrm{QM}} = \frac{4n_{\mathrm{QM}}}{(1 + n_{\mathrm{QM}})^2}, \tag{3.25}$$

merging completely with the optical counterparts,

$$R_{\mathrm{EM}} = \frac{(1 - n_{\mathrm{EM}})^2}{(1 + n_{\mathrm{EM}})^2}, \quad T_{\mathrm{EM}} = \frac{4n_{\mathrm{EM}}}{(1 + n_{\mathrm{EM}})^2}. \tag{3.26}$$

For very high $n \gg 1$ as well as for very low $n \ll 1$ the reflection coefficient tends to unity (Fig. 3.2(a)) whereas the transmission coefficient tends to zero. On a logarithmic scale with

Table 3.1. Index of refraction for selected solids. Wavelength is 0.632 μm if not specified. Note that materials with specified wavelengths absorb light in the visible. Materials are ordered according to growing n. Source: Yariv and Yeh [11]

Material	n	Material	n
Na_3AlF_6[14]	1.34	CuI	2.32
MgF_2	1.37	ZnS	2.35
LiF	1.39	Diamond	2.41
CaF_2	1.43	CdS	2.47
Fused silica (SiO_2)	1.46	ZnSe	2.50
BaF_2	1.47	CdSe 1 μm	2.55
KCl	1.49	As-S glass	2.61
NaCl	1.54	SiC	2.64
KBr	1.56	CdTe 10.6 μm	2.69
Mica	1.58	TiO_2 rutile	2.80
Polystyrene	1.59	ZnTe	2.98
MgO	1.74	GaP	3.31
Al_2O_3	1.77	GaAs 1.15 μm	3.37
Flint glass	1.90	InAs 10.6 μm	3.42
CuCl	1.96	Si 10.6 μm	3.42
AgCl	2.05	GaSb 10.6 μm	3.84
CuBr	2.10	InSb 10.6 μm	3.95
$LiNbO_3$	2.20	Ge 10.6 μm	4.00
TiO_2 polycrystalline	2.22	Te 10.6 μm	4.08

respect to n the functions $R(n)$ and $T(n)$ are *symmetrical*. This means the transmission and reflection coefficients are the same for $n_{12} = n_1/n_2$ and $n_{21} = n_2/n_1$, i.e. results for a barrier and a well of the same height/depth coincide.

For a quantum particle carrying electric charge, e.g. an electron, potential barriers and, accordingly, relative refraction indexes, may arise from other charged particles, e.g. ions in a vacuum, charged impurities in dielectrics. Then the potential barrier can actually rise to large values. This is not the case for electromagnetic waves. In the microwave range, artificial materials with n reaching 100 (sometimes in a rather narrow spectral window) are feasible. In the optical range, superior n values do not exceed 4 (Table 3.1). Furthermore, the highest refraction index values are inherent in materials which are not transparent in the visible. Accordingly, the reflection coefficient obeys the range from 0.04 (air/glass) to 0.35 (air/germanium).

Rectangular barriers and wells

In this subsection we consider the motion of a quantum particle over a finite rectangular barrier and well (Fig. 3.3) and compare the results with propagation of electromagnetic

Table 3.2. Index of refraction for selected liquids in the visible [12,15]

Material	n	Material	n
Methanol	1.328	CCl_4	1.461
Water	1.330	Toluene	1.499
Ethanol	1.361	Benzene	1.501
Cyclohexane	1.426	H_2S	1.629

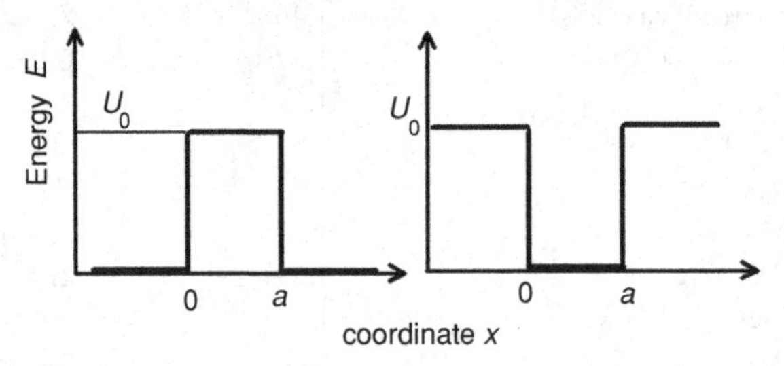

Fig. 3.3 **A rectangular potential barrier (left) and well (right).**

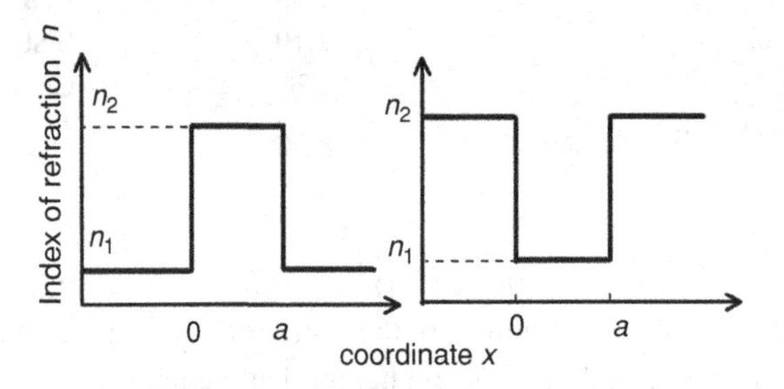

Fig. 3.4 **Refraction index profiles for a medium with a dielectric slab.**

waves through a dielectric slab which has a different refraction index n_2 as compared to the ambient medium n_1 (Fig. 3.4.).

We start from the Schrödinger equation (3.4) with the potential barrier $U(x)$,

$$U(x) = U_0 \quad \text{for} \quad 0 < x < a,$$
$$U(x) = 0 \quad \text{for} \quad x > a, x < a.$$

Similar to Eqs. (3.11) and (3.12) we write the wave function ψ_I in front of the barrier, ψ_{II} over the barrier, and ψ_{III} behind the barrier in the form,

$$\begin{aligned} \psi_I(x) &= A\exp(ik_1x) + B\exp(-ik_1x) \quad \text{for } x < 0, \\ \psi_{II}(x) &= C\exp(ik_2x) + D\exp(-ik_2x) \quad \text{for } 0 < x < a, \\ \psi_{III}(x) &= G\exp(ik_1x) \quad \text{for } x > a. \end{aligned} \tag{3.27}$$

Unlike a potential step, in the case of a barrier we consider that there are two waves in two different directions over the barrier between points 0 and a. As in the previous subsection, we apply the conditions of continuity for the wave function and its first derivative at the points of the potential steps $x = 0, x = a$, i.e.

$$\begin{aligned} \psi_I(0) &= \psi_{II}(0), \quad \psi_I'(0) = \psi_{II}'(0), \\ \psi_{II}(a) &= \psi_{III}(a), \quad \psi_{II}'(a) = \psi_{III}'(a), \end{aligned} \tag{3.28}$$

and arrive at four equations for coefficients A,B,C,D,G. We may assume $A = 1$ and then evaluate four variables from four equations, with special attention to $r = B$, which determines the amplitude of the reflected wave and $t = G$ which determines the amplitude of the transmitted wave. Alternatively we may look for $r = B/A$ and $t = G/A$ with no assumption on the A value. Then we go to $R = r^2$, and $T = 1 - R$ to get the probability density for the reflected and transmitted portions of a flux of quantum particles. The final results are,

$$R = \frac{\left(k_1^2 - k_2^2\right)\sin^2(ak_2)}{\left(k_1^2 - k_2^2\right)^2 \sin^2(ak_2) + 4k_1^2k_2^2}, \quad T = \frac{4k_1^2k_2^2}{\left(k_1^2 - k_2^2\right)^2 \sin^2(ak_2) + 4k_1^2k_2^2}. \tag{3.29}$$

Recalling the notations in Eq. (3.13) we can rewrite these expressions in terms of the barrier parameters U_0 and a:

$$\begin{aligned} R &= \frac{U_0^2 \sin^2\left(\frac{a}{\hbar}\sqrt{m(E - U_0)}\right)}{U_0^2 \sin^2\left(\frac{a}{\hbar}\sqrt{m(E - U_0)}\right) + 4E(E - U_0)}, \\ T &= \frac{4E(E - U_0)}{U_0^2 \sin^2\left(\frac{a}{\hbar}\sqrt{m(E - U_0)}\right) + 4E(E - U_0)}. \end{aligned} \tag{3.30}$$

Finally, using the relative refraction index introduced by Eq. (3.24) and Eqs. (3.30) take the form,

$$\begin{aligned} R &= \frac{\left(1 - n_{\mathrm{QM}}^2\right)^2 \sin^2\left(\frac{a}{\hbar}\sqrt{mE}n_{\mathrm{QM}}\right)}{\left(1 - n_{\mathrm{QM}}^2\right)^2 \sin^2\left(\frac{a}{\hbar}\sqrt{mE}n_{\mathrm{QM}}\right) + 4n_{\mathrm{QM}}^2}, \\ T &= \frac{4n_{\mathrm{QM}}^2}{\left(1 - n_{\mathrm{QM}}^2\right)^2 \sin^2\left(\frac{a}{\hbar}\sqrt{mE}n_{\mathrm{QM}}\right) + 4n_{\mathrm{QM}}^2}. \end{aligned} \tag{3.31}$$

The transmission coefficient $T(E)$ is plotted in Fig. 3.5 superimposed with the barrier. The dimensionless argument of the sine function in Eqs. (3.29)–(3.31),

$$ak_2 = \frac{a}{\hbar}\sqrt{m(E - U_0)} = \frac{a}{\hbar}\sqrt{mE}n_{\mathrm{QM}}, \tag{3.32}$$

has rather transparent physical meaning if we express it in terms of a particle de Broglie wavelength. It is 2π times the ratio of the barrier width and the de Broglie wavelength

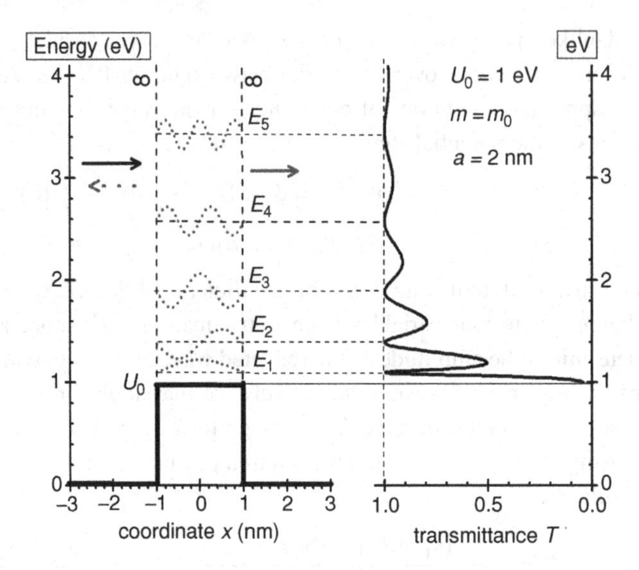

Fig. 3.5 **Electron motion over a potential barrier with height $U_0 = 1$ eV and width $a = 2$ nm. A particle with energy E higher than U_0 possesses finite probabilities of passing over, T as well as to be reflected back, $R = 1 - T$. Transmittance is the oscillating function of E (Eq. (3.30)). A particle does not "see" the barrier if its de Broglie wavelength over the barrier satisfies the condition $\lambda_N = 2a/N, N = 1, 2, 3, \ldots$ The first five energies of the reflectionless propagation are presented along with the wave functions of a particle over the barrier. Note that transmittance tends to unity for $E \to \infty$ and drops to a small but finite value at $E \to U_0$. Note there are different wavelengths and different amplitudes within and outside the barrier (not plotted in the figure).**

of a particle over the barrier, i.e. $2\pi a/\lambda_2 = 2\pi a n_{\mathrm{QM}}/\lambda_1$. We arrive at the effect which is normally discussed as non-trivial in many quantum mechanical textbooks. When the sine argument Eq. (3.32) equals an integer number of π, then the transmission coefficient T equals 1 and, accordingly, the reflection coefficient R equals zero (Eqs. 3.29, 3.30). For a particle wavelength over the barrier,

$$\frac{\lambda_2^{(N)}}{2} N = a, \quad N = 1, 2, 3, \ldots \tag{3.33}$$

i.e., for the particle wavelength outside the barrier,

$$\frac{\lambda_1^{(N)}}{2} N = a n_{\mathrm{QM}}, \quad N = 1, 2, 3, \ldots, \tag{3.33$'$}$$

particles travel over the barrier without reflection! Equation (3.33) means there should be an integer number of particle half-wavelength within the barrier width. The E_N set satisfies the condition (based on Eq. 3.33),

$$E_N - U_0 = \frac{\pi^2 \hbar^2}{2ma^2} N^2, \quad N = 1, 2, 3, \ldots. \tag{3.34}$$

One can recognize the right-hand part of Eq. (3.34) represents a set of energy levels for a particle in as an infinite well with width a. This is because this set also satisfies Eq. (3.33), forming standing waves in the well. The first five energy levels and the corresponding wave functions relevant to resonance propagation over a barrier are plotted in Figure 3.5. Note, the wavelength modified over the barrier is referred to rather than the original wavelength in front of the barrier. The reader familiar with wave optics will possibly recall this simple and evident condition of resonant Fabry–Perot modes for light passing through a dielectric film or thin plate. This means reflectionless propagation has resonant origin and arises from constructive interference of waves reflected at the potential steps. The interference increases the amplitude so that the output amplitude behind the barrier equals the input amplitude in front of it. "Local" reflection r and transmission t coefficients at the output step are given by Eq. (3.17) and therefore increase within the barrier $1/t$ times. The wave function for certain E_n values *accumulates* over the barrier as a light wave does inside a cavity. When speaking about a single particle, we can make the following statement: a particle travelling over a potential barrier can be found with higher probability within the barrier than outside if its energy belongs to the set E_n. The probability rises with the relative barrier height $(E - U_0)/E$, which is n_{QM}^2in our notation (3.24).

It is reasonable to evaluate which wavelength λ_N^0 a particle should originally possess to travel over the barrier without reflection. Simple arithmetic using Eqs. (3.33) and (2.8) gives,

$$\lambda_N^0 = \frac{a\lambda(U_0)}{\sqrt{\lambda^2(U_0)N^2/4 + a^2}}. \tag{3.35}$$

In the limit $E \to U_0$ the transmission coefficient tends to the constant value [1],

$$T_0 = \left(1 + \frac{mU_0a^2}{2\hbar^2}\right)^{-1}, \tag{3.36}$$

rather than to zero. In the forthcoming sections we shall see it tends to zero in the limit $E \to 0$. In the range $0 < E < U_0$ the transmittance is small, but finite, by means of tunneling underneath the barrier.

Based on the above consideration for a potential barrier we can immediately write transmission and reflection coefficients for a potential well. It is clear that Eqs. (3.27)–(3.29) hold equally for a barrier and a well. Equations (3.30) modify by formal substitutions,

$$E \to E - U_0, \quad E - U_0 \to E$$

and read,

$$\begin{aligned} R &= \frac{U_0^2 \sin^2\left(\frac{a}{\hbar}\sqrt{mE}\right)}{U_0^2 \sin^2\left(\frac{a}{\hbar}\sqrt{mE}\right) + 4E(E - U_0)}, \\ T &= \frac{4E(E - U_0)}{U_0^2 \sin^2\left(\frac{a}{\hbar}\sqrt{mE}\right) + 4E(E - U_0)}. \end{aligned} \tag{3.37}$$

A set of resonant particle energies providing reflectionless travel over a well strictly reduces to the energy spectrum of a particle in an infinite well with the same width,

$$E_N = \frac{\pi^2\hbar^2}{2ma^2}N^2, \quad N = 1, 2, 3, \ldots \tag{3.38}$$

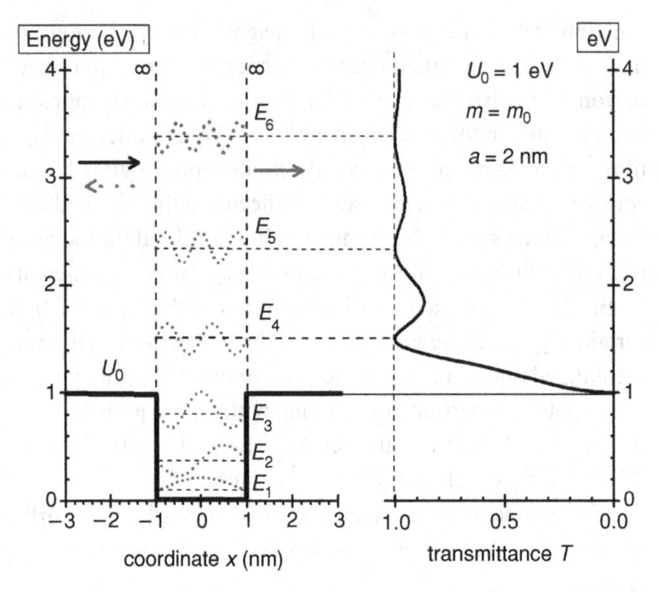

Fig. 3.6 **Electron motion over a potential well with finite depth $U_0 = 1$ eV and width $a = 2$ nm. A particle with energy higher than 1 eV possesses the finite probability of passing over the well, T (Eq. 3.35) and the finite probability of being reflected by the well, $R = 1 - T$. A particle does not "see" the barrier if its de Broglie wavelength over the well satisfies the condition $\lambda_N = 2a/N, N = 1, 2, 3, \ldots$ The first six energy levels E_N satisfying this condition are shown along with superimposed wavefunctions. The first 3 of them lie below the potential threshold U_0 and only for $N > 3$ does resonant propagation occur. Note different wavelengths and different amplitudes within and outside the barrier.**

completely coinciding with Eq. (2.73) from Chapter 2. However, unlike a potential barrier, for a potential well condition (3.38) is necessary but not sufficient. Only those energies given by Eq. (3.38) correspond to resonant propagation for which $E_N > U_0$ holds, otherwise, the states possess energies inside the well. For the parameters used in Figure 3.6 resonant propagation occurs for $N > 3$. Because of this energetic cutoff, the spectrum of resonant propagation over a well will always be more rare than the spectrum for the barrier with the same width a and potential drop U_0.

For a complex shape of barrier or well, expressions (3.30) and (3.37) modify. Instead of the parameters $a\sqrt{m(E - U_0)}/\hbar$ for the barrier and $a\sqrt{mE}/\hbar$ for the well, now the functions,

$$\frac{1}{\hbar}\int_0^a \sqrt{m(E - U(x)}\mathrm{d}x \tag{3.39}$$

should be used as the sine arguments. The latter expression can be presented in the form $a\sqrt{m(E - U(x_0))}/\hbar$ where x_0 lies in the interval $0 < x_0 < a$. Though the basic physical results remain the same, the straightforward discussion of interference of waves reflected at $x = 0$ and $x = a$ is not valid in this case.

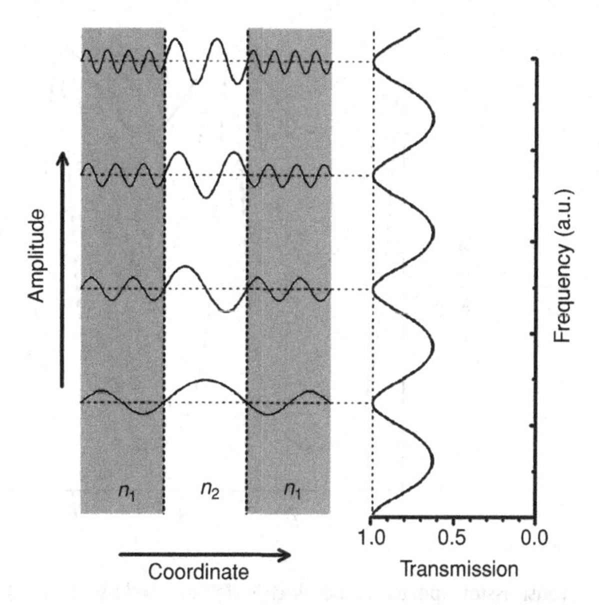

Fig. 3.7 **Propagation of electromagnetic waves in a dielectric slab with refractive index n_2 and thickness a embedded in ambient medium with refractive index n_1. The right panel represents the transmission spectrum according to Eq. (3.41). The left panel shows the slab and spatial distribution of wave amplitude $A(x)$ at frequencies corresponding to the first four maxima in transmission. Note, transmission maxima equal unity and occur for wavelengths λ_N satisfying the condition $\lambda_N/2N = n_2 a$. Notable is the higher amplitude for those wavelengths inside the slab compared with waves in the ambient medium. This demonstrates energy accumulation to provide reflectionless transmission. The refraction index ratio n_2/n_1 is requested for evaluation in Problem 4.**

In accordance with the title of this chapter, the optical counterparts of potential barriers and wells are to be identified.

We consider a plane-parallel dielectric plate (Fig. 3.7, left) with thickness a and refraction index $n_2 \neq n_1$ differing from that of the ambient medium. Every step in the refraction index gives rise to partial reflection of the incident electromagnetic wave; reflection r and transmission t coefficients for electric field amplitude are given by Eq. (3.21). For an incoming wave with amplitude A_0, the amplitude A of the outgoing wave can be written as,

$$A = A_0 t^2 e^{i\delta/2}(1 + r^2 e^{i\delta} + r^4 e^{2i\delta} + \cdots + r^{2N} e^{iN\delta} + \cdots), \tag{3.40}$$

where $\delta = 2k_2 a$ is the phase shift gained after a single roundtrip of the wave inside the plate. The infinite row in brackets accounts for multiple roundtrips of the wave inside the plate, each roundtrip resulting in the $q = r^2 e^{i\delta}$ factor for the amplitude. Note the series in brackets forms a decreasing geometrical sequence whose sum equals $1/(1-q)$. Taking into account also that $t^2 = 1 - r^2$ (see Eq. (3.21)) Eq. (3.40) reduces to,

$$A = A_0 \frac{(1 - r^2)}{1 - r^2 e^{i\delta}} e^{i\delta/2}. \tag{3.41}$$

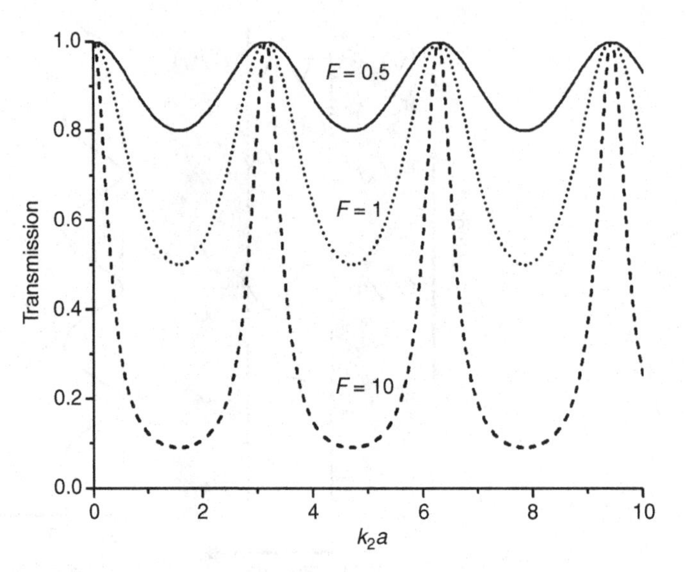

Fig. 3.8 **Transmission spectra of a dielectric slab defined by Eq. (3.40) for different finesse values.**

Now we can get the intensity transmission coefficient T,

$$T = \left|\frac{A}{A_0}\right|^2 = \frac{(1-r^2)^2}{|1-r^2 e^{i\delta}|^2} = \frac{(1-r^2)^2}{(1-r^2)^2 + 4r^2 \sin^2\left(\dfrac{\delta}{2}\right)} = \left[1 + \frac{4r^2}{(1-r^2)^2}\sin^2(k_2 a)\right]^{-1}. \tag{3.42}$$

In Eq. (3.40) the relationships $\exp(i\delta) = \cos\delta + i\sin\delta$ and $\sin^2(\delta/2) = (1-\cos\delta)/2$ have been applied.

The maximal transmission value is 1. This means transmission losses because of reflection are cancelled in resonances where $\sin(k_2 a) = 1$, i.e. $k_2 a = \pi N$ and,

$$\frac{\lambda_N}{2} N = n_2 a, \quad N = 1, 2, 3, \ldots$$

The latter condition coincides with Eq. (3.33′). The product of refractive index and geometrical length in optics is commonly referred to as the "*optical thickness*". Therefore the optical condition of reflectionless propagation sounds as though "the integer number of the original half-wavelength should fit the optical thickness of the slab". The minimal transmission value reads,

$$T_{\min} = \left[1 + \frac{4r^2}{(1-r^2)^2}\right]^{-1} = \frac{(1-r^2)^2}{(1+r^2)^2} = \left(\frac{1-R}{1+R}\right)^2.$$

The $\sin^2$ prefactor in Eq. (3.42),

$$F = \frac{4r^2}{(1-r^2)^2}$$

is referred to as *finesse*. It defines the sharpness of the transmission spectrum (Fig. 3.8). Finesse infinitely increases when $R = r^2 \to 1$.

Using an expression for r in terms of relative refraction index $n = n_2/n_1$ (see Eqs. (3.21) and (3.23)),

$$r = \frac{n-1}{n+1},$$

Eq. (3.42) reads,

$$T_{\mathrm{EM}} = \left[1 + \left(\frac{n^2-1}{2n}\right)^2 \sin^2(k_2 a)\right]^{-1}, \quad k_2 = \frac{\omega}{c} n_2. \tag{3.43}$$

The subscript "EM" is attached to T to remind us we are considering electromagnetic waves. Compare this result with the quantum mechanical counterpart,

$$T_{\mathrm{QM}} = \left[1 + \frac{U_0^2}{4E(E-U_0)} \sin^2(k_2 a)\right]^{-1}, \quad k_2 = \frac{1}{\hbar}\sqrt{m(E-U_0)}, \tag{3.44}$$

which is taken from Eq. (3.30) but transformed to emphasize structural similarity to (3.43). Indeed one can see comparing (3.44) with (3.43) that both transmissions are given by the expression $[1 + (\ldots)\sin^2(k_2 a)]^{-1}$and therefore in both cases we have unit transmission (and accordingly, reflectionless propagation) when an integer number of half-wavelengths inside the plate (barrier, well) fits its thickness (width) as expressed by Eq. (3.33). The sine prefactor in the case of the optical problem at first glance looks like a constant term. However, in fact $n = n(\omega)$ should be considered to account for the spectral dependence of refraction index for the materials under consideration. In the case of a quantum particle, the quadratic dependence of energy versus wave number does manifest itself via the specific form of the sine prefactor. Remarkably, that substitution of the "quantum mechanical relative refraction index" n_{QM} according to Eq. (3.24) transforms Eq. (3.44) into the expression,

$$T_{\mathrm{QM}} = \left[1 + \frac{\left(1-n_{\mathrm{QM}}^2\right)^2}{4n_{\mathrm{QM}}^2} \sin^2(k_2 a)\right]^{-1}, \quad \text{where } n_{\mathrm{QM}}^2 = (E-U_0)/E, \tag{3.45}$$

which completely resembles the optical counterpart (3.43). The same isomorphism can be demonstrated for reflection coefficients (see Problem 1).

Note the different approach we have used to calculate transmission in the optical problem compared with the quantum mechanical case. In the quantum mechanical problem the conditions of continuity of wave function and its first derivative have been used, whereas in the optical problem summation of amplitudes for every transmitted wave has been performed. In every case we have followed the traditional routes proposed in the relevant textbooks. The reader is asked to prove for themselves that in both cases both approaches lead to the same results (Problem 2).

The resonant conditions of the reflectionless propagation of classical waves and quantum particles reveal an important physical issue that is worth emphasizing. Total transmittance of a barrier or well occurs in spite of, and even as a result of, finite reflection at each of the two potential/refraction steps. And if, in spite of the finite reflection at the rear step, we find the initial incident amplitude of the wave and wave function recovered behind the rear step, this means the wave and wave function "accumulates" within a barrier or well. This phenomenon is clearly seen in Figures 3.5 to 3.7. Barriers and wells exhibit the property of a cavity in storing energy due to the non-zero *Q-factor*. The latter is a common property of every oscillating system in mechanics, electricity and optics. It is inherent for *pendulums*, *LC-circuits*, and *resonators*. It can be defined as the ratio of the energy stored in a system to that portion of energy the system loses in every period of oscillation. This notation will be recalled many times in the rest of the book. Here we just note the important time-related issue. Storing energy needs a certain period of time (namely, Q-times periods of oscillation) in terms of roundtrips of, say, a light wave between a planar cavity formed by a pair of mirrors (the refraction step is also a mirror). Therefore the "pictorial gallery" of wave functions and electromagnetic waves plotted in this section corresponds to the situation established after a certain period of time. Note that only steady-state equations and solutions have been examined. In this context recalling different time derivatives in the Schrödinger and Maxwell equations (first time derivative in quantum mechanics and second time derivative in electrodynamics) we may expect certain differences in non-steady-state optics versus non-steady-state quantum mechanics. Examining these differences may form a defined field of research in the near future, driven by progress in nanophotonics.

In this section, the refraction index in quantum mechanics Eq. (3.24) was introduced based on kinetic energy and particle velocity drop/rise at the upward/downward potential step, respectively. Because of the different dispersion laws for a particle and for a classic wave, a potential barrier appears to resemble a refraction drop rather than a step, and vice versa, a potential drop resembles a refraction upward step. Notably, the transmission and reflection formulas appear to be invariant with respect to substitution $n \to 1/n$. The question arises: is it more reasonable to redefine n_{QM} as inverse with respect to that defined by Eq. (3.24)? The potential upward step will correspond then to the upward refraction step and the potential drop will correspond to the refraction downward step. Wavelength modification at step/drop of potential/refraction will be completely alike, the transmission/reflection expression will remain unchanged since those are seen to conserve at $n \leftrightarrow 1/n$ substitution. However this seemingly more reasonable consideration in fact appears to be rather misleading as we move from one-dimensional geometry to at least two-dimensional space. Then Snell's refraction law must be recalled, which says it is the relative refraction index which controls the refraction angle, which in turn, is determined exclusively by the velocity change rather than wavelength change. The refraction angle β versus incident angle α in optics is,

$$\frac{\sin\alpha}{\sin\beta} = n_{\text{EM}}, \tag{3.46}$$

and remains similar in quantum mechanics,

$$\frac{\sin\alpha}{\sin\beta} = n_{\text{QM}}, \tag{3.47}$$

provided that the n_{QM} notation defined by Eq. (3.24) is used. This issue has been outlined by A. Sommerfeld [8]. The notation of Eq. (3.24) has also been used by D. I. Blokhintsev [9].

Potential and refraction barriers/steps compared

Strictly tracing the mathematical identity of Equations (3.3) and (3.4) in the very beginning of this chapter, one can see that it is the value $E - U(x)$ for a quantum particle that should be superimposed with $n^2(x) = \varepsilon(x)$ for electromagnetic waves. Plotting these values instead of the more conventional presentation given in Figures 3.3 and 3.4 results in a complete merging of electromagnetic and quantum mechanical representations of potential and refraction barriers and wells. Notably, in this case precise coincidence occurs for an upward step in refraction and a potential barrier, on one hand, and for a downward step in refraction and a downward step in potential, on the other hand.

In summary, propagation over potential steps, wells and barriers in quantum mechanics replicates the features and the laws of propagation of light in dielectric media with step-like profiles of refractive index. Specific features of particles and light waves are accounted for through the refractive index which includes a certain dependence of frequency versus wave number $\omega(k)$ for light waves and quadratic dependence $E(k)$ for particles. The upward potential step in quantum mechanics corresponds to a downward refraction step in optics, and a potential barrier (well) corresponds to a refraction well (barrier). This asymmetry arises from the linear speed versus wave number relation in quantum mechanics and the reciprocal speed versus wave number relation in optics. This asymmetry has no effect on the formulas for R and T because these contain functions of n which are invariant with respect to substitution $n \to 1/n$. The principal analogies and relevant formulas of electromagnetic versus quantum phenomena are listed in Table. 3.3.

3.3 Dielectric function of free electron gas and optical properties of metals

In this section, we digress to overview the basic electromagnetic response of a gas of charged particles to an oscillating electric field. The outcome of this will be to help in understanding where the optical analog to propagation beneath a potential barrier should be searched for.

An electric field **E** gives rise to polarization described by the **P** vector, and **D** vector in a medium. This can be written as (see Eq. 2.30),

$$\mathbf{D} = \varepsilon_0 \mathbf{E} + \mathbf{P} = \varepsilon_0 \varepsilon \mathbf{E}, \tag{3.48}$$

in terms of the relative dielectric permittivity of the medium ε. Our purpose is to evaluate the dielectric permittivity for a free gas of charged particles, say, electrons, with respect to as oscillating electric field,

$$E(t) = E_0 \exp(i\omega t).$$

This field causes the electron to experience an external force $\mathbf{F} = -e\mathbf{E}$, which in turn results in acceleration (hereafter one-dimensional consideration is used for simplicity),

$$\frac{\mathrm{d}^2 x(t)}{\mathrm{d}t^2} = -\frac{e}{m} E(t). \tag{3.49}$$

Deviation of a single charge in space gives rise to polarization $p = -ex$ which for particle density N forms the total polarization of the medium (per unit volume),

$$P = -exN. \tag{3.50}$$

Table 3.3. Wave optics versus quantum mechanics: propagation in free space, over potential/refraction step, well and barrier

Property	Quantum mechanics	Wave optics
Function	Wave function $\psi(x)$	Electric field $A(x)$
Continuous medium	Potential U_0	Refraction index n
Infinite motion	Wave number $k_{QM} = \sqrt{2m(E - U_0)}/\hbar$, Speed $v = \hbar k/m = n_{QM}\sqrt{2E/m}$	Wave number $k_{EM} = \omega n/c$, Speed $v = \omega/k = c/n$
Dispersion law in free space	$E = \hbar^2 k^2/2m$	$\omega = ck$
Basic equation	$\psi''(x) + k_{QM}^2 \psi(x) = 0$	$A''(x) + k_{EM}^2 A(x) = 0$
Relative refraction index	$n_{QM} = v_2/v_1 = k_2/k_1 = \sqrt{(E - U_0)/E}$	$n_{EM} = v_2/v_1 = n_1/n_2 = k_1/k_2$
Dispersion law in a continuous medium	$E = \hbar^2 k^2 n_{QM}^2 (2m)^{-1}$	$\omega = ck/n$
Potential/refraction step*	$R_{QM} = \frac{(1 - n_{QM})^2}{(1 + n_{QM})^2}, \quad T_{QM} = \frac{4n_{QM}}{(1 + n_{QM})^2}$	$R_{EM} = \frac{(1 - n_{EM})^2}{(1 + n_{EM})^2}, \quad T_{EM} = \frac{4n_{EM}}{(1 + n_{EM})^2}$
Potential/refraction barrier/well*	$T_{QM} = \left[1 + \frac{(1 - n_{QM}^2)^2}{4n_{QM}^2} \sin^2(k_2 a)\right]^{-1}, \quad k_2 = \frac{1}{\hbar}\sqrt{2m(E - U_0)}$	$T_{EM} = \left[1 + \frac{(n_{EM}^2 - 1)^2}{4n_{EM}^2} \sin^2(k_2 a)\right]^{-1}, \quad k_2 = \frac{\omega}{c} n_2$

* Upward potential step in quantum mechanics (QM) corresponds to downward refraction step in optics, potential barrier (well) correspond to refraction well (barrier). This asymmetry arises from the linear speed versus wave number relation in QM and reciprocal speed versus wave number relation in optics. This asymmetry has no effect on formulas for R and T, because these contain functions of n which are invariant with respect to substitution $n \to 1/n$.

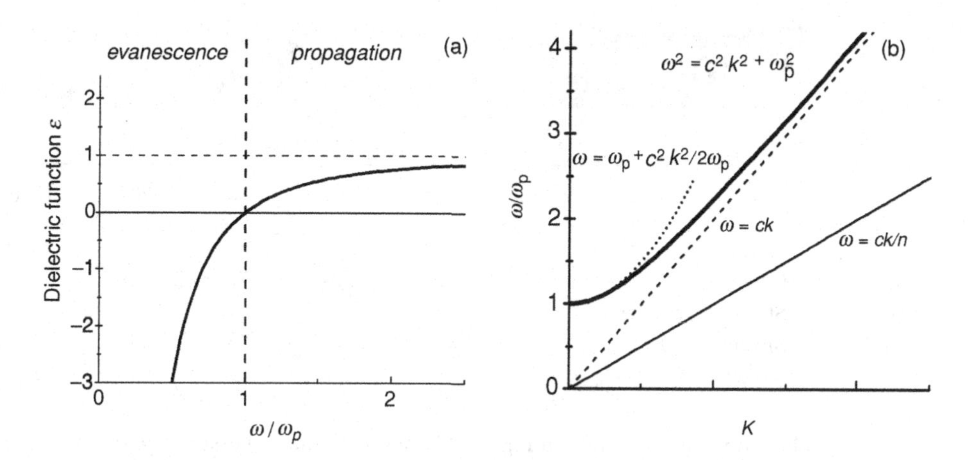

Fig. 3.9 **Properties of a gas of charged particles. (a) Dielectric function. (b) Dispersion law.**

Then the dielectric permittivity ε can be found from Eqs. (3.48) and (3.49) as,

$$\varepsilon = \frac{D}{\varepsilon_0 E} = 1 + \frac{P}{\varepsilon_0 E} = 1 - \frac{exN}{\varepsilon_0 E}. \tag{3.51}$$

Therefore we need to solve Eq. (3.49) and then substitute the x value into Eq. (3.51). The solution of Eq. (3.49) has the form of a function with the same time dependence as $E(t)$:

$$x(t) = x_0 \exp(i\omega t). \tag{3.52}$$

Taking its second derivative and substituting into Eq. (3.50) the solution reads,

$$x = \frac{1}{\omega^2}\frac{e}{m}E(t), \tag{3.53}$$

whence,

$$\varepsilon(\omega) = 1 - \frac{\omega_\mathrm{p}^2}{\omega^2}, \quad \omega_\mathrm{p}^2 = \frac{Ne^2}{m\varepsilon_0}, \tag{3.54}$$

where ω_p has the dimension of $[\mathrm{time}]^{-1}$ and is called the *plasma frequency*.

The plot of function (3.54) is presented in Figure 3.9a. Note, the dielectric function $\varepsilon(\omega) < 1$ everywhere and it takes negative values $\varepsilon(\omega) < 0$ for $\omega < \omega_\mathrm{p}$. This means that there is no plane electromagnetic wave in the electron gas for $\omega < \omega_\mathrm{p}$. For $\omega > \omega_\mathrm{p}$ there are electromagnetic waves. The principal peculiarity is propagation of electromagnetic waves under the condition $\varepsilon(\omega) < 1$. Recalling relations for phase velocity $v = \omega/k$ and $v = c/n$, and $n = \sqrt{\varepsilon}$, one may expect it not to be possible since the superluminal speed $v = c/\sqrt{\varepsilon} > c$ seems to become feasible. However this is not the case. To reveal the properties of electromagnetic waves in a plasma one should thoroughly account for the specific $\varepsilon(\omega)$ function in the range $0 < \varepsilon(\omega) < 1$. Indeed, the known relation,

$$\omega = ck/\sqrt{\varepsilon}, \tag{3.55}$$

with substitution of expression (3.54) for ε, gives $\omega(k)$ in the form,

$$\omega^2 = c^2k^2 + \omega_\mathrm{p}^2. \tag{3.56}$$

Table 3.4. Reflection coefficient (per cent) for a number of common metals at various wavelengths [13]

Metal	Wavelength (nm) 251	305	357	500	600	700	1000	5000
Aluminium	80	–	84	88	89	87	93	94
Copper	26	25	27	44	72	83	90	98
Steel	38	44	50	56	57	58	63	90
Silver	34	9	75	91	93	95	97	99
Nickel	38	44	49	61	65	69	72	94

This function is plotted in Fig. 3.9b along with the dispersion law in a vacuum $\omega = ck$ and dispersion law in a medium with frequency-independent refraction index $\omega = ck/n$. In the high frequency limit $\omega > \omega_p$, the dispersion law (3.56) tends to the vacuum one, whereas in the low frequency limit $\omega \to \omega_p$, the dispersion law obeys a quadratic dependence,

$$\omega = \omega_p + \frac{1}{2}\frac{c^2}{\omega_p}k^2. \tag{3.57}$$

One can see that the phase velocity, $v = \omega/k$ as well as the group velocity, $v_g = d\omega/dk$ always remain less than c everywhere in spite of $\sqrt{\varepsilon} < 1$.

Note, for $\omega < \omega_p$ an electromagnetic wave propagating in a dielectric medium, when meeting the border with a charged particle gas will be reflected back almost completely.

In metals, the free electron concentration is of the order of $N = 10^{22}$ cm^{-3} and the plasma frequency falls into the optical range. This is the principal reason for wide-band omnidirectional optical reflection of metal surfaces. The absolute values for a number of common metals are given in Table 3.4.

3.4 Propagation through a potential barrier: evanescent waves and tunneling

The peculiarity of this differential equation of Schrödinger consists in the fact that coefficient (E – U) can also become negative in some circumstances.

Max Planck, 1927

A potential step: evanescent waves and the skin effect

Consider a potential step with finite height U_0 and a particle energy $E < U_0$ (Fig. 3.10(a)). The wave function can now be written as,

$$\psi_1(x) = A\exp(ikx) + B\exp(-ikx) \quad \text{for } x < 0, k = \frac{1}{\hbar}\sqrt{2mE}$$

and

$$\psi_2(x) = C\exp(-\kappa x) \quad \text{for } x > 0, \kappa = \frac{1}{\hbar}\sqrt{2m(U_0 - E)}, \tag{3.58}$$

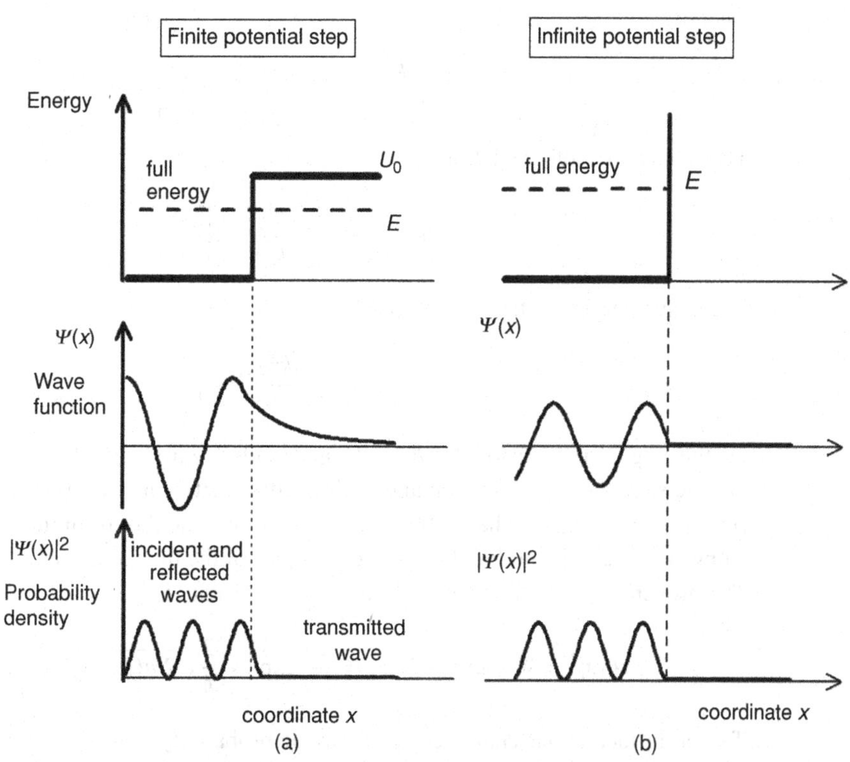

Fig. 3.10 **(a) A finite and (b) infinite potential steps. In case (a) the wave function of a particle whose energy is lower than U_0 exponentially decreases inside the barrier. The probability density oscillates in front of the step because of interference of the incident and reflected waves. There is a finite probability of finding a particle underneath the barrier behind the step. In case (b) the wave function $\psi(x)$ and the probability density $|\psi(x)|^2$ equal zero behind the step. In front of the step, interference of the incident and reflected waves give rise to oscillating $|\psi(x)|^2$.**

where $\psi_1(x)$ remains the same as in Section 3.2 whereas $\psi_2(x)$ now takes another form, since $E - U_0 < 0$ and the κ value unlike k is not wave number. There are incoming and reflected waves in front of the step described by the $\psi_1(x)$ function. We can take the numerical coefficient A to be unity for simplicity, since we are interested in the reflection coefficient r for amplitudes of wave functions which is $r = B/A$. The absolute values of all coefficients are defined by a normalization procedure. Unlike the case of $E > U_0$ (Fig. 3.1) there is no transmitted wave. Instead we have an evanescent wave described by an exponentially decreasing wave function underneath the step.

Using conditions of continuity of $\psi(x)$ and $\psi'(x)$ in $x = 0$ we arrive at,

$$1 + B = C, \quad 1 - B = \frac{i\kappa}{k}C,$$

whence,

$$B = \frac{k - i\kappa}{k + i\kappa} \quad \text{and} \quad C = \frac{2k}{k + i\kappa}.$$

The reflection coefficient R is,

$$R = |B|^2 = \left| \frac{k - i\kappa}{k + i\kappa} \right|^2 = 1. \tag{3.59}$$

The evanescent wave has the amplitude,

$$|C|^2 = \frac{4k^2}{k^2 + \kappa^2} = \frac{E}{U_0}, \tag{3.60}$$

i.e. the amplitude tends to 1 for $E \to U_0$ and tends to zero for $E/U_0 \to 0$. In the case of an infinite step, the wave function and probability density completely vanish behind the step (Fig. 3.10(b)). Thus we have 100% reflection of particle flux from the finite potential step along with the finite probability of finding a part of the flux beneath the step (Fig. 3.10(a)). The probability of finding a particle beneath the barrier is,

$$|\psi(x)|^2 = |C|^2 \exp(-2\kappa x) = \frac{E}{U_0} \exp\left[-\frac{2}{\hbar} \sqrt{2m(U_0 - E)} x \right]. \tag{3.61}$$

To get an idea about characteristic decay of probability beneath the barrier consider an electron with mass m_0 and energy $E = 1$ eV entering a potential step $U_0 = 2$ eV. Then,

$$|\psi(x)|^2 = 0.5 \exp(-10^{10} x) \quad \text{where } x \text{ is in meters,}$$

which means a 2.2×10^{-5} decrease in probability at a distance $x = 1$ nm beneath the step.

Note that regardless of the fact that κ in Eq. (3.58) has no meaning of wave number as it was in the case $E > U_0$, our definition Eq. (3.24) of the quantum mechanical relative refraction index formally still works. Indeed, now we have imaginary n_{QM},

$$n_{QM} = \sqrt{\frac{E - U_0}{E}} = i\sqrt{\frac{U_0 - E}{E}} = \frac{i\kappa}{k}, \tag{3.62}$$

and negative $n_{QM}^2 = -\kappa/k$. The reflection coefficient (3.59) can be written in the form,

$$R_{QM} = \frac{(1 - n_{QM})^2}{(1 + n_{QM})^2},$$

coinciding with Eq. (3.25).

Based on the considerations in Section 3.2 one can see that negative refraction index in quantum mechanics has an optical counterpart when electromagnetic wave propagation is considered at the dielectric–metal border under the condition $\omega < \omega_p$. The negative dielectric permittivity of a metal ε_2 determined by Eq. (3.54) gives rise to an evanescent

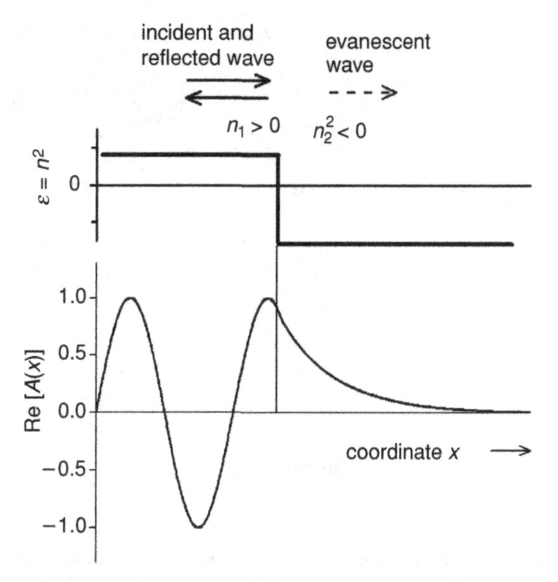

Fig. 3.11 Dielectric permittivity profile (top) and electric field amplitude (bottom) at a dielectric–metal interface. There are incident and reflected electromagnetic waves in front of the interface and an evanescent field inside the metal.

electromagnetic wave inside the metal in complete accordance with the above consideration of a quantum particle beneath a potential step (Fig. 3.11). The reader is requested to reproduce electromagnetic consideration themselves using functions like Eq. (3.58) for the electric field and notations,

$$k_{\mathrm{EM}} = \frac{\omega}{c}\sqrt{\varepsilon_1}, \quad \kappa_{\mathrm{EM}} = \frac{\omega}{c}\sqrt{-\varepsilon_2}. \tag{3.63}$$

By analogy with Eqs. (3.22) and (3.26) the reflection coefficient reads,

$$R = \left|\frac{\sqrt{\varepsilon_1}-\sqrt{\varepsilon_2}}{\sqrt{\varepsilon_1}+\sqrt{\varepsilon_2}}\right|^2 = \left|\frac{1-\sqrt{\varepsilon_2/\varepsilon_1}}{1+\sqrt{\varepsilon_2/\varepsilon_1}}\right|^2 = \left|\frac{1-in_{\mathrm{m}}/n_{\mathrm{d}}}{1+in_{\mathrm{m}}/n_{\mathrm{d}}}\right|^2 = 1, \tag{3.64}$$

where $\varepsilon_1 = n_{\mathrm{d}}^2$ is the permittivity of a dielectric medium whose index of refraction is n_{d} and the notation $\sqrt{\varepsilon_2} = in_{\mathrm{m}}$ is used to emphasize the imaginary refractive index of a metal.

The electric field decays inside the metal according to the formula,

$$A(x) = \frac{\varepsilon_1}{\varepsilon_1 - \varepsilon_2}\exp\left(-\frac{\omega}{c}\sqrt{-\varepsilon_2}x\right), \quad \varepsilon_1 > 0, \varepsilon_2 < 0,$$

or, with other notations,

$$A(x) = \frac{1}{1+(n_{\mathrm{m}}/n_{\mathrm{d}})^2}\exp(-\kappa x), \quad \kappa = \frac{\omega}{c}n_{\mathrm{m}}. \tag{3.65}$$

The values of κ and n_{m} for a number of common metals are listed in Table 3.5. The rapid exponential decay of the electromagnetic field amplitude inside a conductive medium is referred to as *the skin effect*. The surface layer where the field amplitude drops by a factor

Table 3.5. The values of n_m and κ for a number of common metals experimentally determined for wavelength 589.3 nm (adapted from Sivukhin [12])

Metal	n_m	κ, μm^{-1}
Aluminium	5.23	55.4
Copper	2.62	27.7
Gold	2.82	29.9
Magnesium	4.42	46.9
Mercury	4.41	46.7
Nickel	3.48	36.9
Silver	3.64	38.5
Sodium	2.61	24.6

of $1/e$ is called the *skin layer*. Its thickness equals $h_{skin} = 1/\kappa$. One can see from the values of κ in Table 3.5 that, in the visible typical κ values are of the order of 20 – 50 μm^{-1}. This means one order of magnitude decrease in field amplitude per 100 nm on average.

Finite barriers: tunneling

Gold, silver, and platinum are good conductors, and, yet, when formed into very thin plates, they allow light to pass through them. From the experiments which I have made on a piece of gold leaf, the resistance of which was determined by Mr. Hockin, it appears that its transparency is very much greater than is consistent with our theory.

James Clerk Maxwell, Treatise on Electricity and Magnetism (Oxford, 1873, sect. 800)

In the case of a rectangular potential barrier with height U_0 and electron energy E, lower than U_0, the solution of the steady-state Schrödinger equation can be found in the form,

$$\begin{aligned} &\psi_1(x) = A\exp(ikx) + B\exp(-ikx) \qquad \text{for } x < 0, k = \sqrt{2mE}/\hbar \\ &\psi_2(x) = C_1\exp(-\kappa x) + C_2\exp(\kappa x) \qquad \text{for } 0 < x < a, \kappa = \sqrt{2m(U_0 - E)}/\hbar \\ &\psi_3(x) = D\exp(ikx) \qquad \text{for } x > a, \end{aligned} \tag{3.66}$$

where A is defined from normalization and can be put as $A = 1$, and coefficients B, C, D are found from the continuity of the wave function and its first derivative in the points $x = 0, x = a$. There are incident and reflected waves in front of the barrier, an exponentially decaying function within the barrier and a transmitted wave behind the barrier (Fig. 3.12). The probability density oscillates in front of the barrier because of interference of the incident and reflected waves and takes a constant value outside the barrier. Omitting

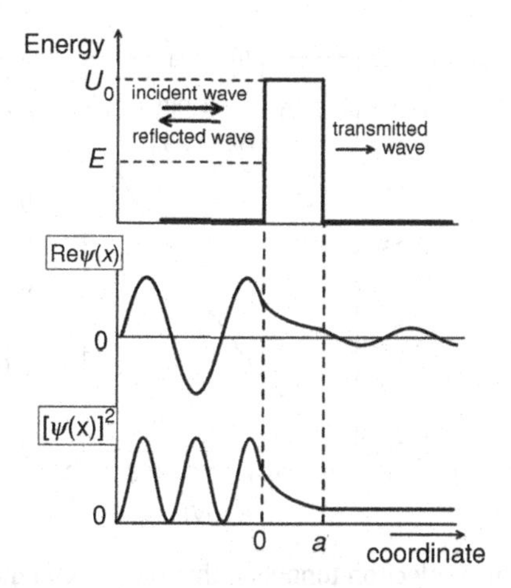

Fig. 3.12 A rectangular potential barrier (top), a particle wave function (middle) and probability density (bottom) for the case when the electron energy E is lower than the potential height U_0.

cumbersome calculations we write the final formula for the transmittance in terms of k and κ (see, e.g. [5]),

$$T_{\mathrm{QM}} = \left|\frac{D}{A}\right|^2 = \frac{4k^2\kappa^2}{4k^2\kappa^2 + (k^2+\kappa^2)^2 \sinh^2(\kappa a)}. \tag{3.67}$$

Here $\sinh(x) = (\mathrm{e}^x + \mathrm{e}^{-x})/2$ is the hyperbolic sine function. Recovering expressions for k and κ transmittance then reads,

$$T_{\mathrm{QM}} = \left|\frac{D}{A}\right|^2 = \left[1 + \frac{U_0^2}{4E(U_0-E)}\sinh^2(\kappa a)\right]^{-1} > 0, \tag{3.68}$$

and for reflectance,

$$R_{\mathrm{QM}} = 1 - T_{\mathrm{QM}} = \left[1 + \frac{4E(U_0-E)}{U_0^2}\sinh^{-2}(\kappa a)\right]^{-1} < 1 \tag{3.69}$$

for the particle flux. Penetration of a quantum particle through a barrier with finite height and width is called *tunneling*. The transmittance function for an electron versus its energy E and barrier width a is plotted for the barrier height $U_0 = 1$ eV to give an idea of the absolute transmittance values (Fig. 3.13). Transmittance grows almost linearly versus E and rapidly, almost exponentially, drops versus a. For $\kappa a \gg 1$, $\sinh^{-1}(x)$ reduces to $e^x/2$ and Eq. (3.68) can be written in a simpler form,

$$T_{\mathrm{QM}} \approx \frac{U_0^2}{4E(U_0-E)} \exp\left[-\frac{a}{\hbar}\sqrt{2m(U_0-E)}\right] \quad \text{for } \kappa a \gg 1. \tag{3.70}$$

For example, when $U_0 - E = 1$ eV and $a = 1$ nm the value of $\kappa a \approx 10$. Note, formally $\kappa = \sqrt{2m(U_0-E)}/\hbar$ resembles the wave number of a particle with kinetic energy $U_0 - E$, or in other words, $\kappa = 2\pi/\lambda$ where λ is the de Broglie wavelength of a particle with kinetic energy $U_0 - E$. This presentation helps to estimate the absolute value of κa as the ratio of

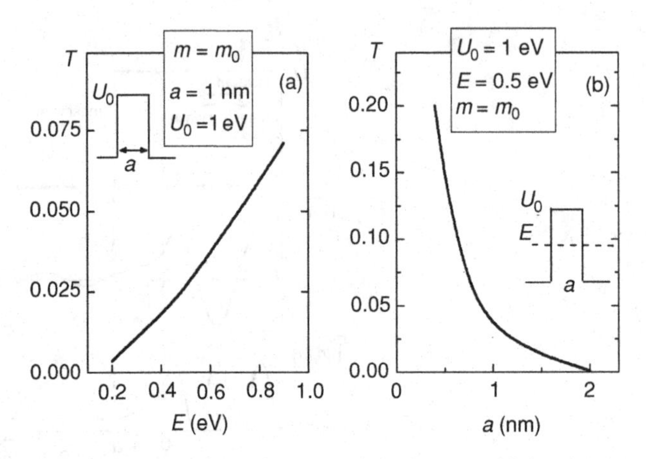

Fig. 3.13 **Probability of an electron tunneling through a rectangular barrier with height $U_0 = 1$ eV as a function of (a) electron energy E and (b) barrier width a.**

the barrier width and a fictitious particle wavelength with the same mass but with kinetic energy $U_0 - E$.

In the case where the barrier is not rectangular but has a complex shape, κa is replaced by the integral,

$$\kappa a \rightarrow \int_0^a \kappa(x)\mathrm{d}x,$$

and Eq. (3.70) reads,

$$T_{\mathrm{QM}} \approx \frac{U_0^2}{4E(U_0 - E)} \exp\left[-\frac{\sqrt{2m}}{\hbar} \int_0^a \sqrt{U(x) - E}\mathrm{d}x \right]. \tag{3.71}$$

Recalling the notation of the quantum mechanical refraction index n_{QM}, defined by Eq. (3.24), we can formally use this notation as shown by Eq. (3.62),

$$n_{\mathrm{QM}}^2 = (E - U_0)/E < 0, \tag{3.72}$$

and represent the transmittance given by Eq. (3.68) as,

$$T_{\mathrm{QM}} = \left[1 - \frac{1}{4}\left(1 - n_{\mathrm{QM}}^2\right)\left(1 - \frac{1}{n_{\mathrm{QM}}^2}\right)\sinh^2(\kappa a) \right]^{-1}, \quad \kappa = \frac{1}{\hbar}\sqrt{2m(U_0 - E)}. \tag{3.73}$$

Let us now use the results of Section 3.3 to look for an optical counterpart of tunneling. Notably, Eq. (3.72),

$$n_{\mathrm{QM}}^2 = \frac{E - U_0}{E} = 1 - \frac{U_0}{E},$$

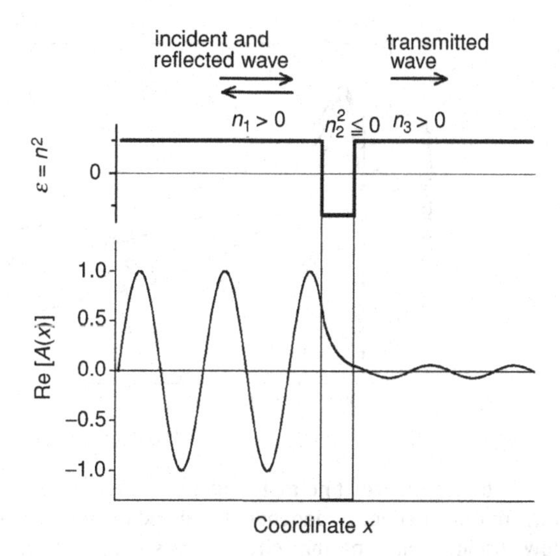

Fig. 3.14 **Propagation of an electromagnetic wave through a rectangular well of dielectric permittivity ε, reaching negative values in the finite interval of x. Such a well develops if a metal film is embedded in a dielectric environment. Evanescent waves give rise to the transmitted plane wave behind the film.**

has the same form as Eq. (3.54) for the dielectric permittivity of a gas of charged particles, if we consider $\sqrt{E}$ and $\sqrt{U_0}$ as the quantum mechanical counterparts of the wave frequency ω (property of an electromagnetic wave) and plasma frequency ω_p (property of the barrier). It is not surprising that tunneling of a quantum particle through a potential barrier resembles the principal features of light propagation through a metal film (Fig. 3.14). An electric field obeys the same dependencies in dielectric and in metal as a particle wave function in Eq. (3.66) but with substituted k_{EM}, κ_{EM} according to Eq. (3.63). Instead of cumbersome calculations we can write the transmittance directly based on the quantum mechanical formula (3.67) using the above mentioned substitutions k_{EM}, κ_{EM} and arrive at,

$$T_{EM} = \left[1 + \frac{\left(k_{EM}^2 + \kappa_{EM}^2\right)^2}{4k_{EM}^2\kappa_{EM}^2}\sinh^2\left(\kappa_{EM}a\right)\right]^{-1} = \left[1 - \frac{\left(\varepsilon_1^2 - \varepsilon_2^2\right)^2}{4\varepsilon_1^2\varepsilon_2^2}\sinh^2\left(\kappa_{EM}a\right)\right]^{-1}. \quad (3.74)$$

Using the notation of the relative refractive index n_{EM} electromagnetic transmittance is,

$$T_{EM} = \left[1 - \frac{1}{4}\left(1 - n_{EM}^2\right)\left(1 - \frac{1}{n_{EM}^2}\right)\sinh^2\left(\kappa_{EM}a\right)\right]^{-1}, \quad (3.75)$$

where,

$$\kappa_{EM} = \frac{\omega}{c}\sqrt{-\varepsilon_2}, \quad n_{EM}^2 = \frac{\varepsilon_1}{\varepsilon_2}, \quad \varepsilon_2 < 0 < \varepsilon_1.$$

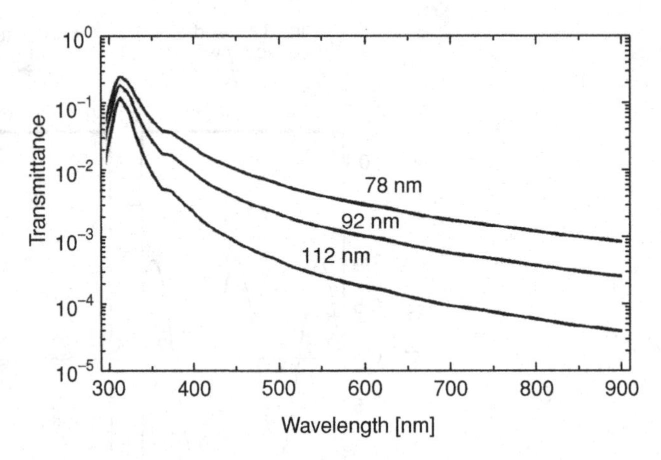

Fig. 3.15 **Calculated transmittance at normal incidence of 78 nm, 92 nm and 114 nm thick Ag films on glass with the realistic frequency-dependent real and imaginary part of the Ag dielectric function. Note the low absolute values of transmittance, its strong thickness dependence and its increase with decreasing wavelength. Reprinted with permission from [16]. Copyright 1998 AIP.**

Similar to quantum mechanics, for $\kappa_{\mathrm{EM}}a \gg 1$ (i.e. for common metals in the visible range for $a > 100$ nm as is seen from Table 3.5) Eq. (3.75) reduces to the exponential law,

$$T_{\mathrm{EM}} \approx \frac{1}{4}\left(1 - n_{\mathrm{EM}}^2\right)\left(1 - \frac{1}{n_{\mathrm{EM}}^2}\right)\exp\left(-\kappa_{\mathrm{EM}}a\right), \quad \text{for } \kappa_{\mathrm{EM}}a \gg 1. \tag{3.76}$$

Therefore, propagation of an electromagnetic wave through a metal film resembles the principal features of quantum mechanical tunneling. Because of the negative dielectric function for frequencies lower than the plasma frequency, the index of refraction becomes imaginary. This corresponds to the imaginary wave number inherent in quantum mechanics for a particle energy lower than the potential barrier height. Note that, at the same time, for frequencies higher than the plasma frequency the refractive index is real, though less than 1. For $\omega > \omega_{\mathrm{p}}$, metals support propagation of electromagnetic waves. At $\omega = \omega_{\mathrm{p}}$ the dielectric permittivity and refraction index are equal to zero. This is the case of $E = U_0$. There is finite transmittance in these cases, both for the quantum and the optical problem. Interestingly, wave function and electric field exhibit in these cases neither oscillating (found for $E > U_0$ and its optical counterpart), nor evanescing (inherent in the tunneling case). Instead, both wave function and electric field exhibit a linear dependence on coordinate. Indeed, for these cases the Helmholtz and Schrödinger equations reduce to,

$$\nabla^2\psi(x) = 0, \quad \nabla^2 A(x) = 0,$$

with the solutions $\psi(x) = \alpha_1 x + \alpha_2$, $A(x) = \beta_1 x + \beta_2$ (Problem 7).

In Figure 3.15 the calculated transmittance spectra of silver films are presented, obtained with the real and imaginary parts of the refraction index taken into account. The figure represents basic features evaluated on the basis of a very simple model of an electron gas in a metal. It is clearly seen that there are small but finite transmittance values for the films with thickness of the order of 100 nm, strong dependence of transmittance on thickness, as well as

Table 3.6. Wave optics versus quantum mechanics: propagation through a potential/refraction step, well and barrier

Property	Quantum mechanics	Wave optics
Function	Wave function $\psi(x)$	Electric field $A(x)$
Basic equation	$\psi''(x) + \frac{2m}{\hbar^2}[E - U(x)]\psi(x) = 0$	$A''(x) + \frac{\omega^2}{c^2}\varepsilon(x)A(x) = 0$
Medium parameter*	Potential $U(x)$	Dielectric function $\varepsilon(x)$
		Refraction index $n = \sqrt{\varepsilon}$ for non-absorbing media
Evanescence and tunneling	$E < U(x), \quad \psi(x) = C\exp(-\kappa_{QM}x)$	$\varepsilon_2(x) < 0, \quad A(x) = C\exp(-\kappa_{EM}x)$
Decay factor	$\kappa_{QM} = \sqrt{2m(U_0 - E)}/\hbar, \quad U_0 > E$	$\kappa_{EM} = \frac{\omega}{c}\sqrt{-\varepsilon_2}, \quad \varepsilon_2 < 0$
Relative refraction index n_{QM}	$n_{QM}^2 = \frac{E - U_0}{E} = -\frac{\kappa^2}{k^2} < 0$	$n_{EM}^2 = \frac{\varepsilon_1}{\varepsilon_2} = -\frac{k^2}{\kappa^2} < 0, \quad \varepsilon_2 < 0 < \varepsilon_1$
Transmittance, accurate expression	$T_{QM} = \left[1 - \frac{1}{4}(1 - n_{QM}^2)\left(1 - \frac{1}{n_{QM}^2}\right)\sinh^2(\kappa_{QM}a)\right]^{-1}$	$T_{EM} = \left[1 - \frac{1}{4}(1 - n_{EM}^2)\left(1 - \frac{1}{n_{EM}^2}\right)\sinh^2(\kappa_{EM}a)\right]^{-1}$
Transmittance, for $\kappa_{QM}a \gg 1, \quad \kappa_{EM}a \gg 1$	$T_{QM} \approx \frac{1}{4}(1 - n_{QM}^2)\left(1 - \frac{1}{n_{QM}^2}\right)\exp(-\kappa_{QM}a)$	$T_{EM} \approx \frac{1}{4}(1 - n_{EM}^2)\left(1 - \frac{1}{n_{EM}^2}\right)\exp(-\kappa_{EM}a)$

* Upward potential step in quantum mechanics (QM) corresponds to downward refraction step in optics, potential barrier (well) corresponds to refraction well (barrier). This asymmetry arises from the linear speed versus wave number relation in QM and reciprocal speed versus wave number relation in optics. This asymmetry has no effect on formulas for T because these contain functions of n which are invariant with respect to substitution $n \to 1/n$.

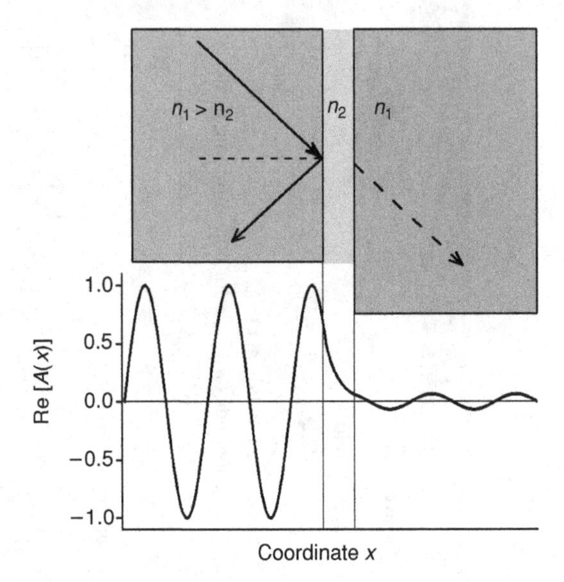

Fig. 3.16 **Frustrated total reflection in optics. Evanescent wave behind the high-refraction slab generates the transmitted plane wave provided that the thin (as compared to wavelength) low-refraction layer is followed by the high refraction medium.**

a pronounced tendency for higher transparency for shorter wavelengths. The latter is because of close proximity to the plasma frequency (from the negative range of permittivity). For silver the plasma frequency corresponds to 320 nm. In alkali metals, high transmission for $\omega > \omega_p$ is well documented and corresponds to the ultraviolet spectral range.

Table 3.6 summarizes the principal notations, formulas and phenomena typical for tunneling in quantum physics and in optics.

There is yet another familiar case in wave optics where an evanescent wave develops and optical tunneling occurs. This is the case of *frustrated total reflection*. If an electromagnetic wave propagates in a medium with refractive index n_1 bordering another medium with lower refraction index $n_2 < n_1$, then total reflection occurs when the angle of incidence (with respect to the normal to the border) exceeds the critical value defined by the relation,

$$\sin \alpha_{\text{crit}} = \frac{n_2}{n_1}. \tag{3.77}$$

In this case the electromagnetic field evanesces inside the low-refraction medium similar to the case of a metal (Fig. 3.16). Figure 3.17 shows an experimental visualization of evanescent wave first proposed by L. I. Mandelstam in 1914. If the low-refraction medium offers a short enough path and then is followed by a high-refraction medium, then the evanescent wave recovers into a plane electromagnetic wave resembling another optical analog of tunneling.

If we compare this optical case with a quantum potential barrier in more detail, one can see that instead of a particle energy versus a potential barrier height we should consider the angle of incidence. In fact, for $\alpha < \alpha_{\text{crit}}$ we have normal propagation of a wave throughout a triple "sandwich" of three media as it holds for a quantum particle in the case that its energy

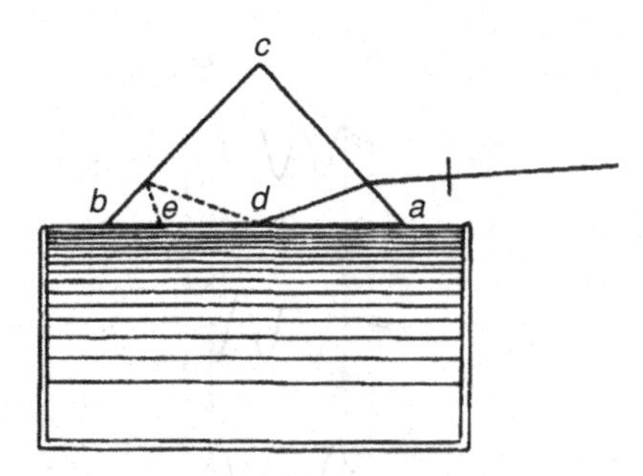

Fig. 3.17 The elegant performance providing visualization of an evanescent wave under the condition of frustrated total reflection. A glass prism borders a solution of fluorescent dye (L. Mandelstam, 1914 [17])

exceeds the potential barrier height. For $\alpha > \alpha_{\mathrm{crit}}$, evanescence and tunneling occur as in quantum mechanics when a particle energy is lower than the potential barrier height. The intermediate case $\alpha = \alpha_{\mathrm{crit}}$ versus $E = U_0$ is not so straightforward and remains for detailed consideration elsewhere. In quantum mechanics the case $E = U_0$ gives finite transmittance (see Fig. 3.5 and Eq. 3.36). In the optical case, at $\alpha = \alpha_{\mathrm{crit}}$ the surface wave develops at the interface. Here polarization of incident light has to be thoroughly accounted for, contrary to the one-dimensional problem where polarization is not important.

Potential and refraction barriers/steps compared

Again, as have been already discussed in Section 3.2, we highlight that pictorial representation of quantum particles and electromagnetic waves traveling in a tunneling regime can be made to look alike. Strictly tracing the mathematical identity of Eqs. (3.3) and (3.4) in the very beginning of this chapter one can see it is the value $E - U(x)$ for a quantum particle to be superimposed with $n^2(x) = \varepsilon(x)$ for electromagnetic waves. Plotting these values instead of the more conventional presentation given in Figures 3.10 and 3.12 results in complete merging with the electromagnetic counterparts presented in Figures 3.11 and 3.14.

3.5 Resonant tunneling in quantum mechanics and in optics

In this section we consider tunneling of a quantum particle in a potential with two barriers (Fig. 3.18) and its optical counterpart. We restrict consideration to the case of a symmetrical potential with identical barriers, for simplicity and clarity. Based on the previous analogies evaluated for quantum particles and electromagnetic waves an evident optical counterpart is nothing else but a well-known Fabry–Perot interferometer consisting of a couple of parallel identical mirrors (Fig 3.19). Mirrors can be made as elementary metal films or composite

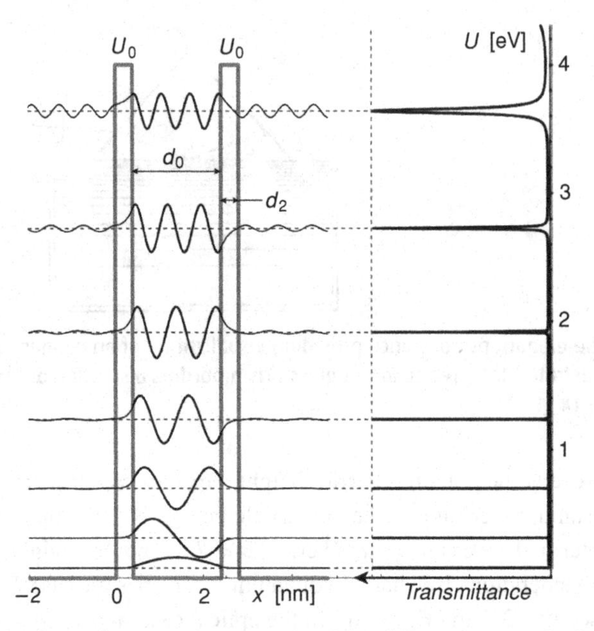

Fig. 3.18 **Accurate numerical solution for a quantum particle whose full energy outside the barriers, E is lower than the height of the two potential barriers, U_0 with thickness d_2 and spacing d_0. Parameter values are $d_0 = 2$ nm, $d_2 = 0.4$ nm, $U_0 = 4$ eV. The left panel shows the particle wave function corresponding to reflectionless propagation. The right panel shows the transmittance on the same energy axis [18]. The particle mass is equal to the electron mass.**

periodic multilayer dielectric structures. The latter offer superior reflection with negligible losses. These so-called Bragg reflectors, more fashionably known as "one-dimensional photonic crystals" will be the subject for close consideration in Chapter 7. Keeping ourselves within the framework of barriers, layers, slabs and wells considered to this point, we imply that ideal reflecting layers are hypothetical lossless metal mirrors with reflection coefficient r close to 1 and transmission coefficent t close to 0, arranged parallel to each other (Fig. 3.22).

When a quantum particle moves in the vicinity of a couple of potential barriers there is a non-trivial effect referred to as *resonant tunneling.* For certain energies a particle gains unit probability to pass through the two barriers without reflection. The key notion in understanding this phenomenon is to correctly account for all reflected and transmitted waves, as was done for the more simple geometries of barriers in Sections 3.2 and 3.4. This is solvable but rather cumbersome. The following intuitive consideration appears to be instructive. Recalling propagation of a particle over a barrier or well (Figures 3.5 and 3.6 in Section 3.2) along with its optical counterpart (Figure 3.7 and Eqs. (3.40) – (3.42) in Section 3.2) we can use partial reflection and transmission coefficients r, t (Eqs. (3.20) and (3.21)) derived for each of the two barriers and count the total particle flux as we did for electromagnetic waves when deriving expressions (3.40) and (3.41). This simplified approach is used in many textbooks on wave optics to get the formula for an optical interferometer. It is physically correct for the electromagnetic problem and based

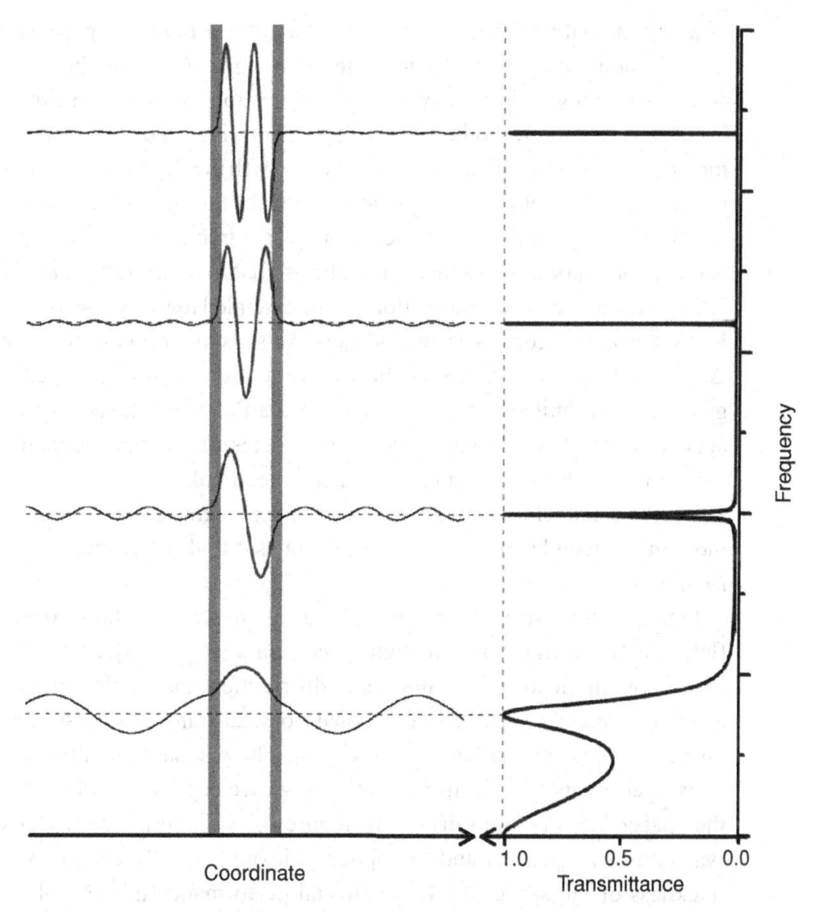

Fig. 3.19 **Accurate numerical solutions for the electric field amplitude (left) and transmission spectrum (right) of a Fabry–Perot resonator consisting of two metal films with thickness d_2, refraction index $i\chi_2$, separated by a dielectric spacer of thickness d_0 and refraction index n_0, coinciding with that of the ambient environment. Parameters used in calculations are, $d_0 = 200$ nm, $n_0 = 1, d_2 = 40$ nm, $\chi_2 = 2$ [18].**

on elaborate quantum mechanical and electromagnetic wave analogies, we are in a position to apply the same approach to a quantum mechanical problem.

The net result of such a consideration is the formula coinciding with Eq. (3.45) which unambiguously predicts resonant propagation for a quantum particle if the integer number of its de Broglie half-wavelength fits the well width between the barriers. In Figure 3.18 correct numerical solutions are presented for a particle wave function and transmittance.

Several physically meaningful features are clearly observed in Figure 3.18. First, for resonant energies the wave function amplitude is noticeably higher between the barriers than outside. Second, the difference is bigger for lower particle energies (deeper states in the well). Third, resonance transmission peaks widen with growing energies. All these features can be consistently discussed in terms of a certain Q-factor to be ascribed to a pair of potential barriers separated by a well. A finite Q-factor results in "accumulation" of a particle within the well, i.e. the probability of finding a particle in the well is higher than of finding it outside. Deeper states gain higher Q-factors because of the more perfect confinement (higher barriers with higher reflection and lower transmission coefficient for a barrier). Accordingly, higher Qs manifest themselves in sharper transmission spectra. Thus transmission peak width correlates perfectly with the degree of "accumulation" of a particle inside the space between the barriers. A reader familiar with radiophysics is expected to recall an analogy with storing energy in an LC-circuit and correlation between its Q-factor and the corresponding resonant curve of oscillation amplitude versus frequency.

Further qualitative consideration can be extended towards discussion of the finite particle lifetime in the space between the barriers. This results from the known energy–time relation $\Delta E \Delta t \approx \hbar$ where ΔE should be treated as the transmission peak width, whereas Δt gives rise to a finite particle "lifetime" inside the well. The above Q-factor consideration appears instructive in understanding the basics of resonant tunneling although it is not comprehensively adopted and elaborated in textbooks.

Resonant tunneling of electrons in complex potential barriers constitutes a solid field in modern semiconductor physics forming an essential conceptual contribution to *nanoelectronics* [19].

Figure 3.19 presents the results of an accurate solution showing electric field amplitude (left panel) and transmission (right panel) on a single frequency axis. A clear analogy is seen between electric field amplitude distribution and particle wavefunction (Fig. 3.18), as well as analogy between the resonant optical transmission spectrum and that for resonance quantum tunneling in Figure 3.18. The resonant condition remains the same, i.e. an integer number of electromagnetic half-wavelengths should fit the optical thickness of the spacer between the mirrors. In Figure 3.19 the refractive index of spacing equals 1 (vacuum or air spacer) and the optical thickness strictly coincides with the geometrical thickness of the spacer d_0. Experimental performance of Fabry–Perot interferometers is used commercially to produce narrow-band optical transmission filters. However because of the inevitable losses in metal films a typical transmission maximum does not exceed 50% (Fig. 3.20).

There is also a difference seen in Figure 3.19 as compared to Figure 3.18. In quantum mechanics rising particle energy (shortening of its de Broglie wavelength) results in smoother (wider) transmission peaks, whereas in optics rising frequency (shortening of electromagnetic wavelength) results in sharper (narrower) transmission peaks. This difference results from dependence of the tunneling probability for an individual barrier on its relative width and height. In the quantum mechanical problem, barrier width remains constant with energy whereas relative barrier height $(U_0 - E)/E$ reduces when particle energy rises. Therefore a barrier becomes more and more transparent. In optics, barrier height in terms of χ_2 and n_0 values remains constant, whereas relative barrier width d_2/λ infinitely increases for shorter λ. In the opposite limit for $\lambda \gg d_2$ the relative thickness of the barrier goes down and transmission approaches 1 for $\omega \to 0$. Thus in quantum mechanics Figure 3.18 gives transmissions for different barrier heights, whereas in optics Figure 3.19 gives transmissions for different barrier widths.

To complete the consideration of resonant tunneling in quantum mechanics and optics there now follows a rigorous mathematical justification for the above qualitative discussion.

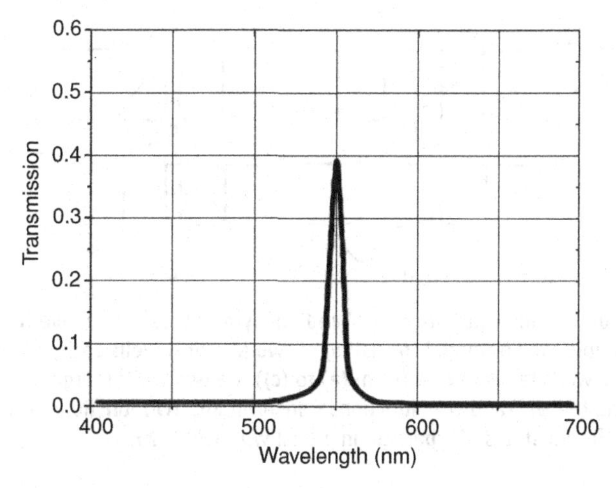

Fig. 3.20 **Experimentally observed resonant transmission spectrum for a commercial Fabry–Perot interferometer with metal mirrors. Maximal transmission is noticeably less than 1 because of losses in the metal films.**

Amplitude reflection and transmission coefficients for a step formed by medium 1 and 2 in optics (EM) and in quantum mechanics (QM) are given by,

$$r_{12}^{(EM)} = \frac{k_1^{(EM)} - k_2^{(EM)}}{k_1^{(EM)} + k_2^{(EM)}}, \quad t_{12}^{(EM)} = \frac{2k_1^{(EM)}}{k_1^{(EM)} + k_2^{(EM)}},$$
$$r_{12}^{(QM)} = \frac{k_1^{(QM)} - k_2^{(QM)}}{k_1^{(QM)} + k_2^{(QM)}}, \quad t_{12}^{(QM)} = \frac{2k_1^{(QM)}}{k_1^{(QM)} + k_2^{(QM)}}. \tag{3.78}$$

If layer 2 is embedded in an ambient dielectric environment whose parameters are denoted by subscript "0" amplitude reflection and transmission coefficients in optics are,

$$t_{020}^{\text{diel}} = \frac{t_{02}t_{20}\exp[ik_2d_2]}{1 - r_{20}^2\exp[2ik_2d_2]}, \quad r_{020}^{\text{diel}} = r_{02} + \frac{t_{02}r_{20}t_{20}\exp[2ik_2d_2]}{1 - r_{20}^2\exp[2ik_2d_2]}, \tag{3.79}$$

if layer 2 is dielectric (denoted by superscript "diel") and,

$$t_{020}^{\text{met}} = \frac{t_{02}t_{20}\exp[-\kappa_2d_2]}{1 - r_{20}^2\exp[-2\kappa_2d_2]}, \quad r_{020}^{\text{met}} = r_{02} + \frac{t_{02}r_{20}t_{20}\exp[-2\kappa_2d_2]}{1 - r_{20}^2\exp[-2\kappa_2d_2]} \tag{3.80}$$

if layer 2 is metal (denoted by superscript "met"). Based on the identity of expressions (3.78) for electromagnetic and quantum mechanical problems, expressions (3.79) and (3.80) are valid for quantum mechanical problems as well. These formulas can be extensively used in recurrent and matrix-based calculation schemes to calculate transmission and reflection for complex structures both in electromagnetism and in quantum mechanics. In particular they have been applied to the examination of electromagnetic wave propagation in fractal multilayer structures with the number of individual layers approaching 1000 [20, 21]. Symmetrical double-barrier structures in Figures 3.18 and 3.19 are characterized by the

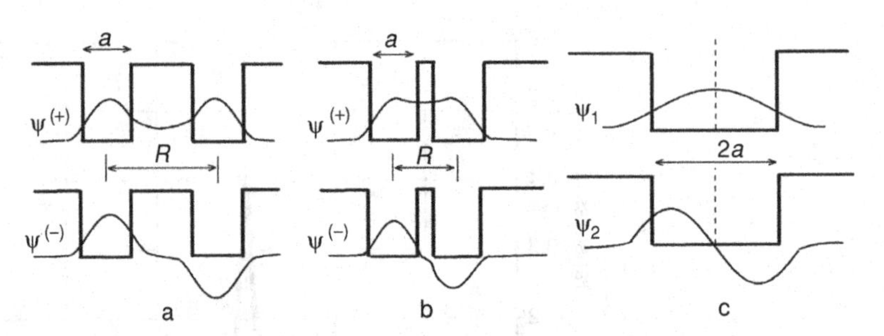

Fig. 3.21 Evolution of symmetrical $\psi^{(+)}$ and antisymmetrical $\psi^{(-)}$ wave functions of a particle in a double-well potential consisting of two identical wells each of width a. As the distance between the wells tends to zero (from (a) to (c)) the two wells merge into a single well of double width $2a$ and the $\psi^{(+)}$ and $\psi^{(-)}$ functions transform into two functions ψ_1 and ψ_2, corresponding to the two different states of a particle in a well with width $2a$.

coefficients,

$$t_{02020} = \frac{t_{020}^2 \exp[ik_0 d_0]}{1 - r_{020}^2 \exp[2ik_0 d_0]},$$
$$r_{02020} = r_{020} + \frac{t_{020} r_{020} t_{020} \exp[2ik_0 d_0]}{1 - r_{020}^2 \exp[2ik_0 d_0]}. \tag{3.81}$$

These expressions can be further simplified and in particular for transmission, one can finally arrive at a formula coinciding with Eq. (3.42).

3.6 Multiple wells and barriers: spectral splitting

Consider a pair of identical wells of width a separated by a barrier with finite height and thickness (Fig. 3.21). We shall trace evolution of a particle energy spectrum and wave function when the barrier height and width both tend to zero.

If the distance between wells is large compared to the well width ($R \gg a$), a particle wave function is close to zero between the wells. Solution of the Schröedinger equation in this case nearly coincides with the wave function for an isolated well ψ_1, the only difference being that $|\psi_1|^2$ is halved because of normalization (a particle with equal probability can be found in either of the two wells). This is shown in the upper panel of Figure 3.21a. However, one more solution of the Schrödinger equation exists, for which ψ has a different sign in one of the two wells (lower panel in Figure 3.21(a)). Energy values for the two wave functions coincide. The function $\psi^{(+)}$ is symmetric whereas the function $\psi^{(-)}$ is referred to as an asymmetric wave function. When the interwell distance gets smaller $\psi^{(+)}$ and $\psi^{(-)}$ change (Fig. 3.21(b)). At close distance, the $\psi^{(+)}$-state has lower energy compared with the $\psi^{(-)}$-state. In the limit, functions $\psi^{(+)}$ and $\psi^{(-)}$convert into functions ψ_1 and ψ_2 of the ground and the first excited states, respectively of a particle in a well with width equal to $2a$ (Fig. 3.21(c)). For sufficiently high potential walls, particle lower states can be approximated by the formula derived for a well with

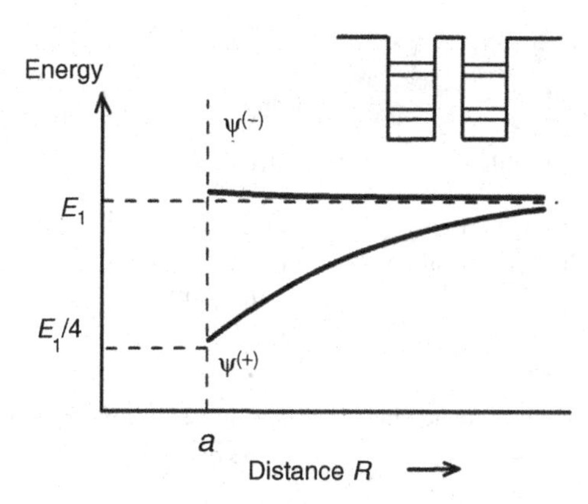

Fig. 3.22 Splitting of the lower state in a pair of rectangular wells.

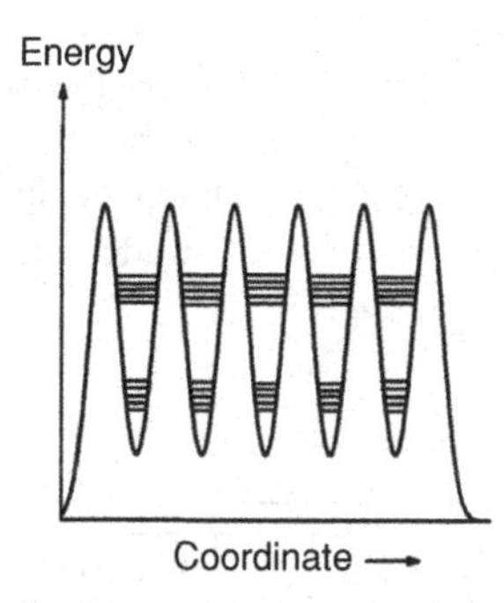

Fig. 3.23 Splitting of energy levels in a multiwell potential consisting of identical wells/barriers.

infinite barriers (see Eq. (2.73)), i.e. $E_n = E_1 n^2$, $n = 1, 2, 3, \ldots$, $E_1 = \pi^2 \hbar^2/(2ma^2)$. Comparing the energy set for a well width $a\,(E_1, 4E_1, 9E_1, \ldots)$ with that for a well width $2a\,(E_1/4,\ E_1,\ 9E_1/4,\ 4E_1,\ 25E_1/4,\ 9E_1, \ldots)$ one can see that the double-width well has the same set of energies as the original well along with additional levels lying lower than the original ones. Exact solutions show that every level (twice degenerate) in individual wells splits into two non-degenerate levels continuously upon $R \to a$. For the lowest states this is shown in Figure 3.22. Such splitting occurs in any double-well potential, not only in rectangular wells.

In a set of N identical multiple wells separated by finite barriers N-fold splitting of every energy level occurs (Figure 3.23). For very large N one can speak about evolution of a discrete set of energy levels into energy bands. This consideration offers a reasonable hint to the evolution of electron states in solids as compared to isolated atoms.

All the above splitting phenomena do have direct optical counterparts. Multiple splitting of an optical transmission spectrum is inherent in coupled optical cavities, i.e. in multicavity interferometers. These may be constructed as a set of dielectric spacers coupled via semi-transparent metal mirrors. However inevitable losses in metal films inhibit efficient multiple interference (i.e. make the Q-factor of the system lower) and resonant peaks smear readily. Purely dielectric structures are much more efficient in every case where multiple

Table 3.7. Analogies in the properties of electrons and electromagnetic waves propagating in complex media

Potential profile	Electron	Electromagnetic wave
Semi-infinite barrier	Transmission/reflection	Transmission/reflection
Well with finite width	Transmission/reflection *over* the well	Transmission, reflection and Fabry–Perot modes in an air slit between the dielectric plates
Barrier with finite height	Transmission/reflection *over* the barrier	Transmission, reflection and Fabry–Perot modes of a dielectric plate or film in air
	Tunneling *under* the barrier	Transparency of thin metal films Frustrated total reflection
Well between two barriers	Resonant tunneling *under* barriers	Transparency of a Fabry–Perot interferometer
A sequence of identical barriers/wells	Multiple splitting of the steady-state energy levels	Multiple splitting of the resonant transmission bands in coupled microcavities

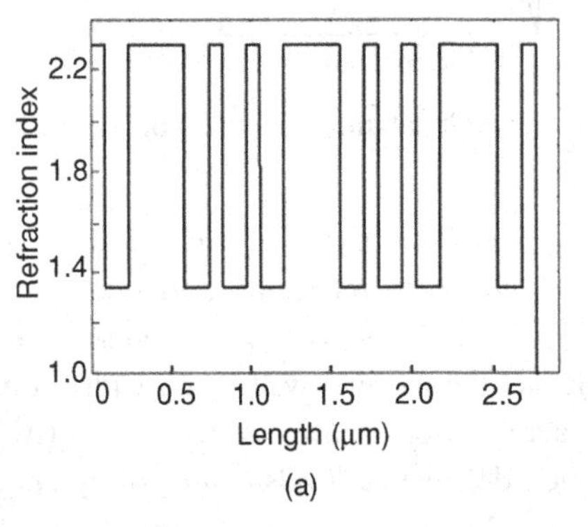

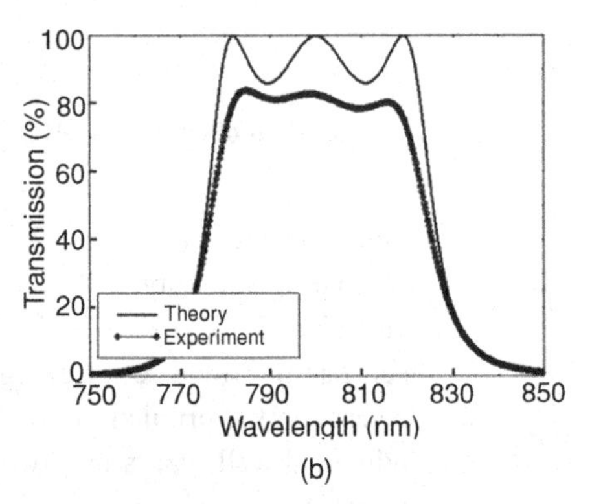

Fig. 3.24 **A three-cavity dielectric structure and its transmission spectrum [22]. (a) Refraction index profile of an experimentally developed multilayer film. Wider portions form cavities whereas alternating thinner layers form dielectric reflectors. (b) Calculated and measured central part of a transmission spectrum.**

interference is important. Coupled optical cavities are organized as complex multilayer structures where several subwavelength films form Bragg reflectors (propagation of light in periodic structures will be discussed in detail in Chapter 7). In Figure 3.24 a three-cavity structure is presented along with calculated and experimentally observed three-fold transmission bands.

It is reasonable to summarize the whole set of optical and quantum phenomena that can be considered as counterparts (Table 3.7).

Louis de Broglie (1892–1987)

Erwin Schrödinger (1897–1961)

3.7 Historical comments

The laws of the new mechanics are found simply by retaining throughout the analogy of mechanics with optics.

Max Planck, 1927 [23]

In conclusion to this chapter we recount a few reminiscences from the history of physics indicating the role of optics in the elaboration of quantum principles as well as evaluation of early optical counterparts among newly discovered quantum phenomena.

There are many indications pointing at wave optics as the basic contribution to the formulation of the principal ideas of wave mechanics. The Max Planck quotation in the epigraph of this section is clear confirmation of this statement. By the beginning of the dawn of wave mechanics, an analogy between classical mechanics and geometrical optics had already been thoroughly established. The Hamilton principle in mechanics correlates with the Fermat's principle in optics as well as the eikonal equation in optics remarkably correlating with the Hamilton–Yakobi equation in mechanics. Thus optics contained two parts, geometrical optics and wave optics, the former correlating with classical mechanics, whereas the latter had as yet no mechanical counterpart. Louis de Broglie and Erwin Schrödinger recognized a way to search for an adequate mechanical counterpart with respect to wave optics, as a way towards new mechanics capable of explaining atomic phenomena. De Broglie wrote that the new dynamics of material particles relates to the former dynamics as wave optics relates to geometrical optics [24]. Schrödinger wrote specifically about this route in lectures on wave mechanics and in his Nobel lecture [25]. He pointed out that, since the Fermat principle in geometrical optics represents the consequences of wave optics, it is reasonable to suggest that the Hamilton principle in classical mechanics possibly points to more general wave-like mechanical laws. Schrödinger did emphasize that the original hint, at a wave origin for the Hamilton principle had been suggested by Hamilton himself, but that this had been neglected for many decades. Schrödinger believed it was necessary to reveal that hint, and to build self-consistent wave mechanics as optics had been developed with the wave nature of mass-points and with the Hamilton principle in action as the ultimate short-wave limit. To emphasize this symmetry in mechanics and in optics, Schrödinger called classical mass-point mechanics "geometrical mechanics" by analogy to geometrical optics.

Max Planck (1858–1947)

William Hamilton (1805–1865)

Surprisingly enough, this organic consistency of optics and mechanics at the very beginning of the quantum era in physics, as well as the guiding role of optics in the elaboration of its mechanical counterpart, seem to have been forgotten and are regretfully not included in modern textbooks on quantum mechanics.

It was reasonable that the first results and the first predictions of the newborn wave mechanics were subjected to a thorough analysis in terms of possible optical counterparts. For example, the smooth decay of a particle wave function inside a potential step had been studied by Max Planck in 1929 as the wave mechanical analog of electromagnetic wave propagation under conditions of total reflection when the field evanesces smoothly behind the border of the two media [26]. The tunneling effect, first predicted by L. Mandelstam and M. Leontovich in 1928 [27], was used by W. Heisenberg in 1932 as the quantum mechanical counterpart to electromagnetic wave propagation through a metal film [28]. Much later, in 1949, A. Sommerfeld considered frustrated total reflection as a further analog to the quantum tunneling effect [29] and introduced the quantum mechanical analog of the optical refractive index [8].

In this connection, the curious situation in the history of physics related to the tunneling effect and frustrated total reflection as its optical counterpart is noteworthy. As was mentioned, the very first hint of the quantum tunneling effect was published by Mandelstam and Leontovich in 1928 [27]. In this paper, the possibility of a particle wave function percolating through a classically forbidden area had been predicted (Figure 3.25). Remarkably, approximately a decade before L. I. Mandelstam extensively examined frustrated total reflection in optics ([17] and Figure 3.17) as well as the propagation of evanescent waves

Arnold Sommerfeld (1868–1951)

L. I. Mandelstam (1879–1944)

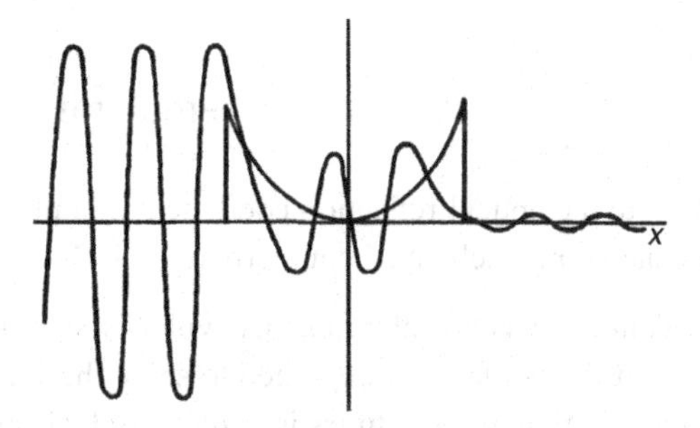

Fig. 3.25 The wave function of a quantum particle in a double-barrier structure representing a precursor to the quantum tunneling effect in 1928 by L. Mandelstam and M. Leontovich [27].

in radiophysics for the purpose of applications in signal transfer. Based on his experience and taking into account the general route to wave mechanics through wave optics, it looks plausible that the pioneering paper [27] had been inspired by the optical analogy. However no documented reference has been found so far to prove this and it remains open for further investigation in the history of quantum mechanics.

Quantum tunneling phenomena have found wide application in electronics. Leo Esaki, the inventor of the tunnel diode [30] received the Nobel Prize in 1973 for this achievement. Remarkably in his Nobel lecture he highlighted resonant tunneling through a double-barrier potential as an intriguing effect, and mentioned that a Fabry–Perot interferometer probably constituted the optical analog to this effect. He said:

"This effect is quite intriguing because the transmission coefficient (or the attenuation factor) for two barriers is usually thought of as the product of two transmission coefficients, one for each barrier, resulting in a very small value for overall transmission. The situation, however, is somewhat analogous to the Fabry–Perot-type interference filter in optics. The high transmissivity arises because, for certain wavelengths, the reflected waves from inside interfere destructively with the incident waves, so that only a transmitted wave remains." [31].

Different issues related to optical analogs to quantum mechanical phenomena were later discussed by several authors [9, 11, 32, 33, 34]. Many further analogs will be discussed in the forthcoming chapters in this book. In the context of the history of science, it is noteworthy that not only quantum mechanics consumed certain ideas and concepts from wave optics. A few decades later, in the late twentieth century, many results of quantum theory of complex structures have been transferred to wave optics. In the dawn of quantum theory the laws of new mechanics were searched to a large extent by the transfer of wave -optical concepts to mechanics. Nowadays, the reverse process is noticeably pronounced. Nanophotonics has been developed to a large extent by the systematic transfer of quantum mechanical results to wave optics. The very field of so-called photonic crystals and photonic solids has been elaborated by means of the direct transfer of results and conclusions from the electron theory of solids to electromagnetism, and the search for optical counterparts to a number of quantum phenomena in complex structures exhibiting light confinement and multiple interference events. These optical phenomena today form a significant contribution to the constitution of nanophotonics as a well-defined field in science and technology.

Problems

1. Calculate the optical reflection coefficient of a plane slab and demonstrate coincidence with the quantum mechanical counterpart.

2. In quantum mechanical problems typically continuity of the wave function and its derivative at the borders are exploited to solve the wave equation, whereas in wave optics sequential addition of amplitudes is widely used. Using a couple of simple problems show that both techniques give the same result.

3. Consider transmission of light by a plane dielectric lossless slab, with reflection at the boundaries in terms of additive calculation of light intensities of multiply reflected beams rather than amplitudes. Compare the results with the case where amplitudes are additively calculated with respect to the resulting transmission coefficient and intensity of light inside the slab. Explain the difference.

4. Evaluate the refractive index ratio for the case presented in Figure 3.7.

5. A higher layer of atmosphere consists of ions with rather low concentration. Its plasma frequency falls into the radiofrequency range. Reflection of radiowaves from this layer enables distant radio communication in the short-wave range ($\lambda = 10^1 - 10^2$ m). Estimate the concentration of ions taking into account that efficient reflection occurs for radiowaves with wavelength of the order of 10 m, and assuming the average charge of ions $Z = 10$ and their average mass $m = 10^4 m_0$.

6. Evaluate the tunneling probability through the same barrier for a proton in comparison with an electron.

7. Evaluate the wave function and electric field profile for the cases $E - U_0 = 0,\ \varepsilon = 0$ in the tunneling problem.

8. Based on the complete numerical solutions presented in Figure 3.7 for an optical problem try to restore the wave functions for its quantum counterparts, given in Figures 3.5 and 3.6.

9. Try to formulate a definition for the Q-factor of a double-barrier potential in wave mechanics.

10. Try to draw a sketch of the transmission spectrum for an electron in the case of a multi-well potential shown in Figure 3.23.

11. Metal film offers a possibility of tunneling for electromagnetic waves along with energy losses. Tunneling in optics was shown to resemble the quantum mechanical counterpart. Find out a quantum mechanical counterpart for electromagnetic energy losses.

12. Try to recall and analyze resonant splitting phenomena in mechanics (pendulums) and in radiophysics (LC-circuits).

References

[1] L. I. Schiff. *Quantum Mechanics* (New York: McGraw-Hill, 1968).
[2] E. H. Wichman. *Quantum Physics* (New York: McGraw-Hill, 1967).
[3] S. Flügge. *Practical Quantum Mechanics I* (Berlin: Springer, 1971).
[4] A. S. Davydov. *Quantum Mechanics* (New York: Pergamon, 1965).
[5] L. D. Landau and E. M. Lifshitz. *Quantum Mechanics* (Moscow: Nauka, 1989).
[6] S. Brandt and H. D. Dahmen. *The Picture Book of Quantum Mechanics* (New York: Wiley & Sons, 1985).
[7] S. Brandt and H. D. Dahmen. *Quantum Mechanics on the Personal Computer* (Berlin: Springer, 1990).
[8] A. Sommerfeld. *Atombau und Spektrallinien* (Brawnschweig: F. Vieweg & Sohn, 1951).
[9] D. I. Blokhintsev. *Principles of Quantum Mechanics* (Moscow: Nauka, 1976) (in Russian).
[10] M. Born and E. Wolf. *Principles of Optics* (New York: MacMillan, 1964).
[11] A. Yariv and P. Yeh. *Optical Waves in Crystals* (New York: John Wiley and Sons, 1964).
[12] D. V. Sivukhin. *Optics* (Moscow, Nauka, 1980) (in Russian).
[13] S. A. Akhmanov and S. Yu. Nikitin. *Physical Optics* (Moscow: Moscow State University Press 2004) (in Russian).
[14] D. N. Chigrin, A. V. Lavrinenko, D. A. Yarotsky and S. V. Gaponenko. Observation of total omnidirectional reflection from a one-dimensional dielectric lattice. *Appl. Phys. A*, **68** (1999), 25–28.
[15] V. N. Bogomolov, S. V. Gaponenko, I. N. Germanenko, A. M. Kapitonov, E. P. Petrov, N. V. Gaponenko, A. V. Prokofiev, A. N. Ponyavina, N. I. Silvanovich and S. M. Samoilovich. Photonic band gap phenomenon and optical properties of artificial opals. *Phys. Rev. E*, **55** (1997), 7619–7625.
[16] M. J. Bloemer and M. Scalora. Transmissive properties of Ag/MgF_2 photonic bandgaps. *Appl. Phys. Lett.*, **72** (1998), 1677–1679.

[17] L. I. Mandelstam. Strahlung einer Lichtquell, die sich sehr nahe an der Trennungsfläche zweier durchsichtiger Medien befindet. (Radiation of a light source placed in the vicinity of the border between two transparent media). *Z. Physik,* **15** (1914), 220–225.

[18] S. V. Gaponenko, S. V. Zhukovskii and V. N. Khilmanovich. *Optical Analogies to Quantum Phenomena* (Minsk: Belarussian State University, 2009) (in Russian).

[19] V. V. Mitin, V. A. Kochelap and M. A. Stroscio. *Introduction to Nanoelectronics* (Cambridge: Cambridge University Press, 2008).

[20] S. V. Zhukovskii, A. V. Lavrinenko, K. S. Sandomirskii and S. V. Gaponenko. Propagation of waves in non-periodic deterministic structures: Scaling properties of an optical Cantor filter. *Physical Review E*, **65** (2002), 036621.

[21] S. V. Zhukovsky, A. V. Lavrinenko and S. V. Gaponenko. Spectral scalability as a result of geometrical self-similarity of fractal multilayers. *Europhysics Letters,* **66** (2004), 455–461.

[22] A. Belardini, A. Bosco, G. Leahu, M. Centini, E. Fazio, C. Sibilia, M. Bertolotti, S. V. Zhukovsky and S. V. Gaponenko. Femtosecond pulses chirping compensation by using one-dimensional compact multiple-defect photonic crystals. *Appl. Phys. Lett.*, **89** (2006), p.031111.

[23] M. Planck. *Theory of Light* (London: Macmillan, 1932) (translation from the original German edition published in 1927).

[24] L. de Broglie. Quanta de lumière, diffraction et interférences. *Comp. Rend.,* **177** (1923), 548–553.

[25] E. Schrödinger. The Fundamental Idea of Wave Mechanics. Nobel Lecture (1933). *Nobel Lectures, Physics 1922–1941* (Amsterdam: Elsevier, 1965).

[26] M. Planck. *Lecture at the University of Leipzig* (Leipzig: J. B. Barth Verlag, 1929).

[27] L. Mandelstam and M. Leontovich. Zur Theorie der Schrödingerschen Gleichung. *Z. Phys.,* **47** (1928), 131–138.

[28] W. Heisenberg, Addition to the Russian edition of the book "Physical Principles of Quantum Theory" (Leningrad–Moscow: State publishing house on theoretical physics, 1932) (in Russian).

[29] A. Sommerfeld. *Optics* (New York: Academic Press, 1954).

[30] R. Tsu and L. Esaki. Tunneling in a finite superlattice. *Appl. Phys. Lett.*, **22** (1973), 562–564.

[31] L. Esaki. Long journey into tunneling. Nobel Lecture (1973). *Nobel Lectures, Physics 1971–1980*, Ed. S. Lundqvist (Singapore: World Scientific, 1992).

[32] J. H. Davies. *The Physics of Low-Dimensional Semiconductors* (Cambridge: Cambridge University Press, 1998).

[33] D. Dragoman and M. Dragoman. *Quantum Classical Analogies* (Berlin: Springer, 2004).

[34] D. Kossel. Analogies between thin film optics and electron-band theory of solids. *J. Opt. Soc. Amer.*, **56** (1966), 1434–1440.

4 Electrons in periodic structures and quantum confinement effects

The optical properties of matter result from the quantum transitions electrons perform in the course of light–matter interactions. This chapter contains a brief introduction to the electron theory of solids in terms of quantum particle properties in a periodic potential, the concept of quasiparticles and band structure. Therefore, this chapter is important for the understanding of the basic optical properties of crystals including crystalline semiconductors and dielectrics. It is also important to understand the essence of quantum confinement phenomena in crystalline solids. Moreover, it forms a precursor to the concept of photonic crystals, the subject of Chapter 7. Basic notions of the electron theory of solids along with an introduction to the realm of quantum confinement phenomena in low-dimensional systems will be considered. All issues are discussed in terms of key problem formulations and principal results overview. For more detail on the electron theory of solids and the optical properties of semiconductors the reader is referred to topical textbooks [1–7]. Quantum confinement in semiconductor nanostructures is discussed in textbooks [5–7]. For close consideration of quantum confinement effects on the electronic and optical properties of quantum wells, quantum wires and quantum dots the books [8–13] are recommended.

4.1 Bloch waves

Consider a one-dimensional Schrödinger equation for a particle with mass m,

$$-\frac{\hbar^2}{2m}\frac{\mathrm{d}^2}{\mathrm{d}x^2}\psi(x) + U(x)\psi(x) = E\psi(x), \tag{4.1}$$

with a periodic potential,

$$U(x) = U(x + a). \tag{4.2}$$

Note, Eq. (4.2) leads to a condition,

$$U(x) = U(x + na), \tag{4.3}$$

where n is an integer number. Thus we have translational symmetry of potential, i.e. invariance of U with respect to coordinate shift over an integer number of periods. To give a primary insight into the properties of wave functions satisfying the Schrödinger equation with a periodic potential, consider the replacement, $x \to x + a$. Then we arrive at equation,

$$-\frac{\hbar^2}{2m}\frac{\mathrm{d}^2}{\mathrm{d}x^2}\psi(x + a) + U(x)\psi(x + a) = E\psi(x + a), \tag{4.4}$$

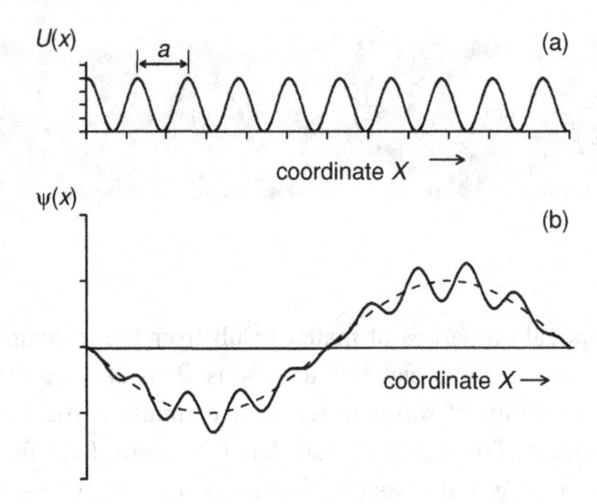

Fig. 4.1 (a) A periodic potential and (b) typical wave function resembling a plane wave with amplitude modulated periodically in accordance with the periodicity of the potential.

where periodicity of the potential (4.2) has been implied. One can see that $\psi(x)$ and $\psi(x+a)$ do satisfy the same second-order differential equation. This means that either $\psi(x)$ is periodic, or at least it differs from a periodic function by a complex coefficient whose square of modulus equals 1. In fact, the *Floquet* theorem derived in 1883 [14] states that the solution of Eq. (4.1) with a periodic potential reads,

$$\psi(x) = e^{ikx} u_k(x), \; u_k(x) = u_k(x+a), \tag{4.5}$$

i.e. the wave function is a plane wave (e^{ikx}) with amplitude modulated in accordance with the periodicity of the potential $U(x)$. The "k" subscript for u_k means the function u_k is different for different wavenumbers.

An important property of the wave function (4.5) is its periodicity with respect to wave number. In fact, substituting $x \to x+a$ and $k \to k + 2\pi/a$ returns $\psi(x)$ to its original value.

In a three-dimensional case the periodic potential reads,

$$U(\mathbf{r}) = U(\mathbf{r} + \mathbf{T}), \tag{4.6}$$

where,

$$\mathbf{T} = n_1 \mathbf{a}_1 + n_2 \mathbf{a}_2 + n_3 \mathbf{a}_3 \tag{4.7}$$

is a translation vector, and $\mathbf{a}_i$ are elementary translation vectors defined as,

$$\mathbf{a}_1 = a_1 \mathbf{i}\,,\; \mathbf{a}_2 = a_2 \mathbf{j}\,,\; \mathbf{a}_3 = a_3 \mathbf{l}, \tag{4.8}$$

with a_1, a_2, a_3 being the periods and $\mathbf{i}, \mathbf{j}, \mathbf{l}$ the unit vectors in the x, y, z directions, respectively. The solution of a three-dimensional Schrödinger equation,

$$-\frac{\hbar^2}{2m}\nabla^2 \psi(\mathbf{r}) + U(\mathbf{r})\psi(\mathbf{r}) = E\psi(\mathbf{r}) \tag{4.9}$$

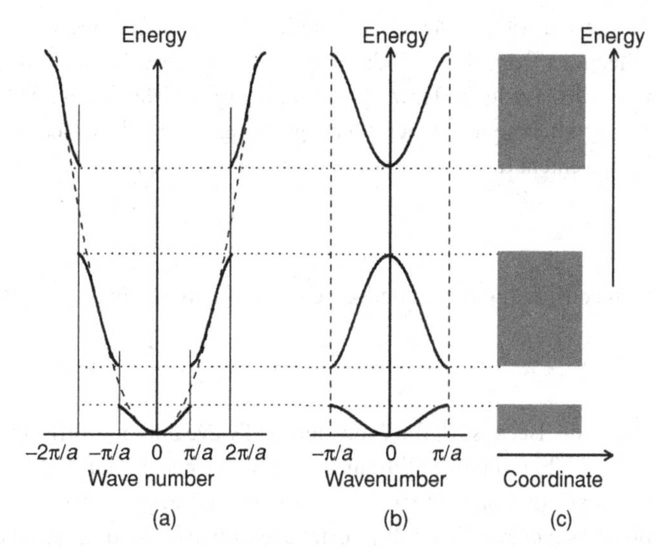

Fig. 4.2 **Original (a) and reduced (b) presentation of the dispersion law of a particle in a one-dimensional periodic potential, and the corresponding energy bands in space (c). The dashed line in (a) is the parabolic dispersion law for a particle in free space.**

in the case of a periodic potential has the form,

$$\psi(\mathbf{r}) = e^{i\mathbf{k}\cdot\mathbf{r}}u_k(\mathbf{r}),\ u_k(\mathbf{r}) = u_k(\mathbf{r}+\mathbf{T}). \tag{4.10}$$

This statement is known as the *Bloch theorem* since the three-dimensional version of the Floquet theorem was first considered by F. Bloch [15]. His paper was published at an early period in quantum mechanics and systematically addressed the properties of the Schrödinger equation for electrons in a crystal lattice. Therefore solutions of Eq. (4.9) (and (4.1)) in solid-state physics are referred to as *Bloch waves*.

In what follows we restrict consideration to a one-dimensional case only; the three-dimensional case will be discussed in Section 4.2. Periodicity of $u_k(x)$ in the Bloch function (4.5) along with the periodicity of the phase coefficient e^{ikx} with respect to kx with period 2π gives rise to an important property of particles in a periodic potential, namely that wavenumbers differing in integer number of $2\pi/a$, i.e.

$$k_1 - k_2 = \frac{2\pi}{a}n, \quad n = \pm1, \pm2, \pm3, \ldots,$$

appear to be equivalent. This is a direct consequence of the translational symmetry of the space. Therefore, the whole multitude of k values consists of equivalent intervals each with width $2\pi/a$. Every interval contains the full set of nonequivalent k values and is called the *Brillouin zone* to acknowledge the principal contribution to this notion by L. Brillouin. The energy spectrum and the dispersion curve differ from those of a free particle (Fig. 4.2(a)). The dispersion curve has discontinuities at points

$$k_n = \frac{\pi}{a}n; \quad n = \pm1, \pm2, \pm3, \ldots. \tag{4.11}$$

At these k values the wave function is a *standing wave* which arises due to multiple reflections from the periodic structure. For every k_n satisfying Eqs. (4.12), two standing waves exist with different potential energies. This leads to the emergence of forbidden energy intervals for which no propagating wave exists. Because of these discontinuities it is convenient to consider Brillouin zones starting from the first one around $k = 0$, i.e.

$$-\frac{\pi}{a} < k < \frac{\pi}{a},$$

then considering the second zone consisting of the two symmetrical equal intervals,

$$-\frac{2\pi}{a} < k < -\frac{\pi}{a}, \quad \frac{\pi}{a} < k < \frac{2\pi}{a},$$

and so on. Because of the equivalence of wave numbers differing in integer number of $2\pi/a$ it is possible to move all branches of the dispersion curve towards the first Brillouin zone by means of a shift along the k-axis by the integer number $2\pi/a$. Therefore, the original dispersion curve (Fig. 4.2(a)) can be modified to yield the reduced zone scheme (Fig. 4.2b). The whole wealth of particle dynamics in a periodic potential can be treated in terms of events within the first Brillouin zone, the energy being a multivalue function of the wave number. Presentation of the dispersion law in Fig. 4.2(b) is referred to as the *band structure*.

The value,

$$p = \hbar k, \tag{4.12}$$

is called the "quasi-momentum". It differs from the momentum by a specific conservation law. It conserves with an accuracy of $2\pi\hbar/a$, which is, again, a direct consequence of the translational symmetry of space. The quasi-momentum conservation law is to be considered in line with the known conservation laws, namely momentum conservation (resulting from space homogeneity), energy conservation (resulting from time homogeneity) and circular momentum conservation (resulting from space isotropy), and all these conservation laws agree with the *Noether theorem*. This states that every type of space and time symmetry generates a certain conservation law.

The relation (4.12) gives rise to Brillouin zones for momentum. For Figure 4.2(b) the first Brillouin zone for momentum is the interval,

$$-\hbar\frac{\pi}{a} < p < \hbar\frac{\pi}{a}.$$

Standing waves at points in Eq. (4.11) result in zero values of the derivative,

$$\left.\frac{\mathrm{d}E(p)}{\mathrm{d}p(k)}\right|_{k_n} = 0, \tag{4.13}$$

i.e. in the center and at the edges of the first Brillouin zone. Therefore the expansion of the $E(k)$ function near a given extremum point $E_0(k_0)$,

$$E(k) = E_0 + (k - k_0)\left.\frac{\mathrm{d}E}{\mathrm{d}k}\right|_{k=k_0} + \frac{1}{2}(k - k_0)^2\left.\frac{\mathrm{d}^2E}{\mathrm{d}k^2}\right|_{k=k_0} + \cdots, \tag{4.14}$$

can be reduced to a parabolic $E(k)$ law,

$$E(k) = \frac{1}{2}k^2 \frac{d^2 E}{dk^2}\bigg|_{k=0}, \tag{4.15}$$

by putting $E_0 = 0,\ k_0 = 0$ and omitting the higher-order derivatives in the raw. In turn, the parabolic dispersion law means that the *effective mass* of the particle under consideration can formally be introduced in the vicinity of every extremum of $E(k)$ as,

$$\frac{1}{m^*} = \frac{1}{\hbar^2}\frac{d^2 E}{dk^2} \equiv \frac{d^2 E}{dp^2} = \text{const.} \tag{4.16}$$

Note, from the relation $E = p^2/2m = \hbar^2 k^2/2m$, for a free particle we have everywhere,

$$\frac{1}{\hbar^2}\frac{d^2 E}{dk^2} = \frac{d^2 E}{dp^2} \equiv m^{-1}.$$

The effective mass (4.16) determines the reaction of a particle to the external force, $\mathbf{F}$, via a relation,

$$m^*\mathbf{a} = \mathbf{F}, \tag{4.17}$$

where $\mathbf{a}$ is the acceleration. Equation (4.17) coincides formally with Newton's second law.

Comparing Figures 4.2(a) and (b), one can see that, for example, in the vicinity of the $k = 0$ point, the effective mass is noticeably smaller than the intrinsic inertial mass of a particle. This is evident, because the curvature of the $E(k)$ function, which is equal to the second derivative, is larger in case (b) near the $k = 0$ point than in case (a), shown by a dashed line. Therefore, a particle in a periodic potential can sometimes be "lighter" than in free space. Sometimes, however, it can be "heavier". Moreover, it can even possess a negative mass. This corresponds to the positive curvature of the $E(k)$ dependence in the vicinity of the maximum. The negative effective mass is not an artifact but an important property peculiar to a particle which interacts simultaneously with a background periodic potential and with an additional perturbative potential. The negative mass means the momentum of a particle decreases in the presence of an extra potential. This results from reflection from the periodic potential steps/wells. The difference in momentum does not vanish but is transferred to the material system responsible for the periodic potential, for example, the ion lattice of the crystal.

In the vicinity of every extremum, the Schrödinger equation with periodic potential (4.10) reduces to the equation,

$$-\frac{\hbar^2}{2m^*}\frac{d^2\psi(x)}{dx^2} = E\psi(x), \tag{4.18}$$

which describes the *free motion* of a particle with effective rather than original mass.

To summarize the properties of a particle in a periodic potential, we outline a few principal results. First, a particle is described by a plane wave modulated by a period of the potential. Second, the particle state is characterized by the quasi-momentum. The latter has a set of equivalent intervals, the Brillouin zones, each containing the complete multitude of non-equivalent values. Third, the energy spectrum consists of wide continuous bands separated from each other by forbidden gaps. As a plane wave, a particle in a periodic potential exhibits quasi-free motion without an acceleration. With respect to the external force, the particle's behavior is described in terms of the effective mass. The latter is, basically, a complicated function of energy, but can be considered a constant in the vicinity of a given extremum of the $E(k)$ curve. Generally, the renormalization of mass is nothing else but a result of the interaction of a particle with a given type of periodic potential.

For a step-wise periodic potential the one-dimensional Schrödinger equation can be solved explicitly. This problem was considered for the first time in 1931 by R. de Kronig and W. G. Penney [16] and is referred to as the Kronig–Penney model.

4.2 Reciprocal space and Brillouin zones

For Brillouin zones to be discussed in the three-dimensional case, one more important concept is to be introduced. It is the *reciprocal crystal lattice*. The primitive vectors $\mathbf{b}_1, \mathbf{b}_2, \mathbf{b}_3$ of the reciprocal lattice for a given original lattice with primitive vectors $\mathbf{a}_1, \mathbf{a}_2, \mathbf{a}_3$ are defined as follows:

$$\mathbf{b}_1 = \frac{2\pi}{V_0}[\mathbf{a}_2 \times \mathbf{a}_3], \quad \mathbf{b}_2 = \frac{2\pi}{V_0}[\mathbf{a}_1 \times \mathbf{a}_3], \quad \mathbf{b}_3 = \frac{2\pi}{V_0}[\mathbf{a}_2 \times \mathbf{a}_1], \tag{4.19}$$

where $V_0 = \mathbf{a}_1 \cdot [\mathbf{a}_2 \times \mathbf{a}_3]$ is the volume of the elementary cell of the original lattice. A parallelepiped formed by the three vectors $\mathbf{b}_1, \mathbf{b}_2, \mathbf{b}_3$ defines the elementary cell of the reciprocal lattice. Its volume is,

$$V_b = \mathbf{b}_1 \cdot [\mathbf{b}_2 \times \mathbf{b}_3] = 8\pi^3 / V_0. \tag{4.20}$$

According to the definition (4.19) every primitive vector of the reciprocal lattice is normal to each of the couple of original basic vectors whose number differs from the number of the reciprocal lattice vector, i.e. $\mathbf{b}_i \perp \mathbf{a}_j, \mathbf{a}_l$, where $i, j, l = 1, 2, 3$. For cubic and hexagonal lattices the primitive vectors are all orthogonal. Therefore the relevant reciprocal lattices also have orthogonal primitive vectors. For a cubic lattice $a_1 = a_2 = a_3 = a$ and $b_1 = b_2 = b_3 = 2\pi/a$ hold.

In the previous section we saw that, in a one-dimensional periodic potential, wave numbers differing by the value $2n\pi/a$ are equivalent. In a three-dimensional periodic potential those *wave vectors are equivalent whose difference equals a reciprocal lattice vector.* That is translational symmetry of the original lattice gives rise to the translational symmetry of wave vectors in the space of the reciprocal lattice. Unlike the one-dimensional case, Brillouin zones in two- and three-dimensional space are no longer segments, but rather polygons and polyhedrons, respectively.

Brillouin zones for two and three dimensions are composed as follows. The origin of the coordinates is to be chosen at a node of the reciprocal lattice. Then the origin point is connected with segments to the nearest nodes. The first Brillouin zone will then be

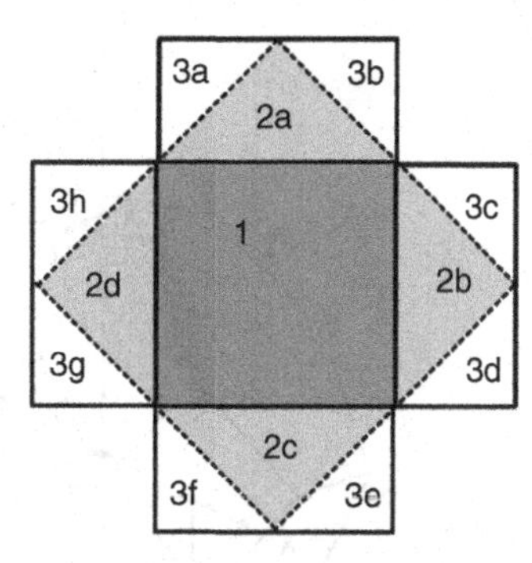

Fig. 4.3 The first three Brillouin zones (BZ) of a simple two-dimensional square lattice. The first BZ is a square with the side equal to $2\pi/a$. The second BZ consists of four triangles. The third BZ consists of eight triangles. Every BZ has a square equal to $(2\pi/a)^2$.

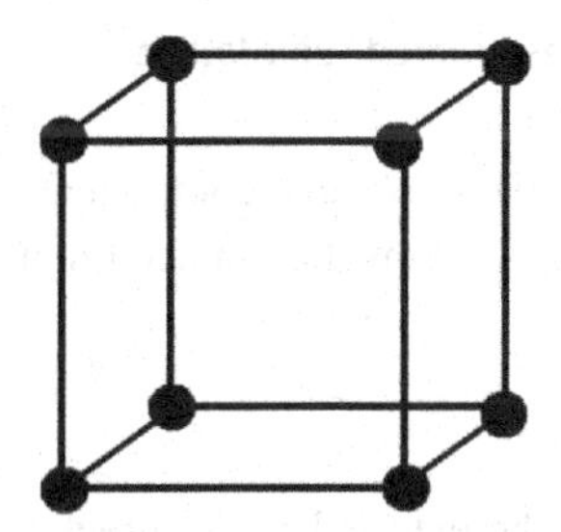
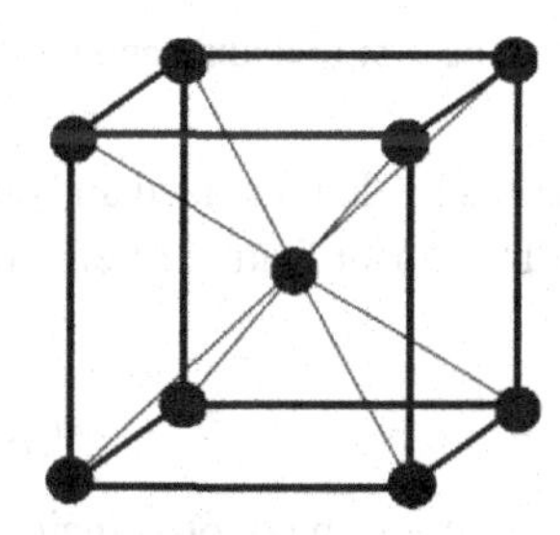
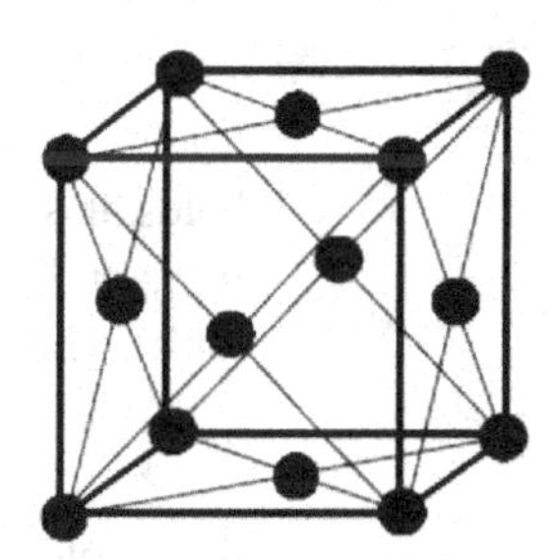

Fig. 4.4 Elementary cells of cubic lattices (from left to right): simple, body-centered, and face-centered.

confined by the planes orthogonal to these segments and crossing their centers. Further, segments to the next-nearest nodes are the next to be drawn. Orthogonal planes are built through their centers. The second Brillouin zone will be confined from the outside by these planes and from the inside by the first zone. Figure 4.3 represents the first three Brillouin zones of a two-dimensional square lattice. Figure 4.4 shows elementary cells of the simple cubic, body-centered cubic, and face-centered cubic lattices. The first Brillouin zone for the simple cubic lattice has cubic shape, whereas the other lattices have the first Brillouin zones in the form of complex polyhedrons.

Electron states in crystals are characterized using points and lines in the first Brillouin zone for which non-reducible representation groups occur. These points and lines are labeled using Latin and Greek capitals (Fig. 4.5). The center of the first Brillouin zone ($k = 0$) is labeled as Γ-point.

In a 3-dimensional periodic potential, the effective mass defined as,

$$m_{ij}^{*-1} = \frac{1}{\hbar^2}\frac{\partial^2 E}{\partial k_i \partial k_j}, \tag{4.21}$$

is the second-order tensor. Note that since,

$$m_{ij}^* = m_{ji}^*$$

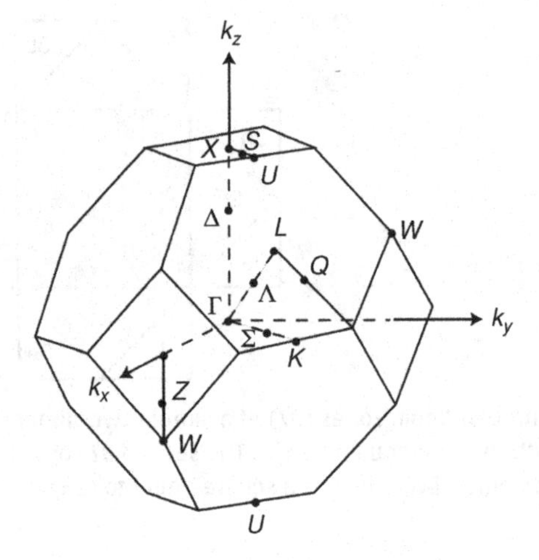

Fig. 4.5 The first Brillouin zone for the face-centered cubic lattice.

holds, this tensor can be diagonalized and characterized by three components only, namely m^*_{xx}, m^*_{yy}, m^*_{zz}. The tensor feature of effective mass means that, in accordance with Newton's second law,

$$m^*_{ij}\mathbf{a} = \mathbf{F}, \tag{4.22}$$

the acceleration direction may not coincide with the external force direction. The explanation of this effect can be found recalling that a particle experiences the impacts of the periodic potential and external force simultaneously.

4.3 Electron band structure in solids

The dispersion law for the first Brillouin zone of a particle in a periodic potential shown in Figure 4.2 provides an idea of the band structure for a hypothetical one-dimensional crystal. In a crystal the periodic potential for an electron arises from periodic displacement of ions which in turn can be understood, for example, in terms of close-packed balls comprising cubic lattices, shown in Figure 4.4. This is the case for elementary solids, like, e.g. Si, C, Ge and others. For binary solids, not only cubic, but more complex lattices occur. In particular for certain binary semiconductors, e.g. CdS, CdSe, a hexagonal lattice forms with anisotropic displacement of atoms along one axis (usually labeled as the *c*-axis).

The electronic properties of solids are determined by occupation of the bands and by the absolute values of the forbidden gap between the completely occupied and the partly unoccupied, or free bands. If a crystal has a partly occupied band, it exhibits *metallic* properties, because electrons in this band provide electrical conductivity. If all the bands at $T = 0$ are either occupied or completely free, material will show *dielectric* properties. Electrons within the occupied band cannot provide any conductivity because of *Pauli's exclusion principle*: only one electron may occupy any given state. Therefore, under an

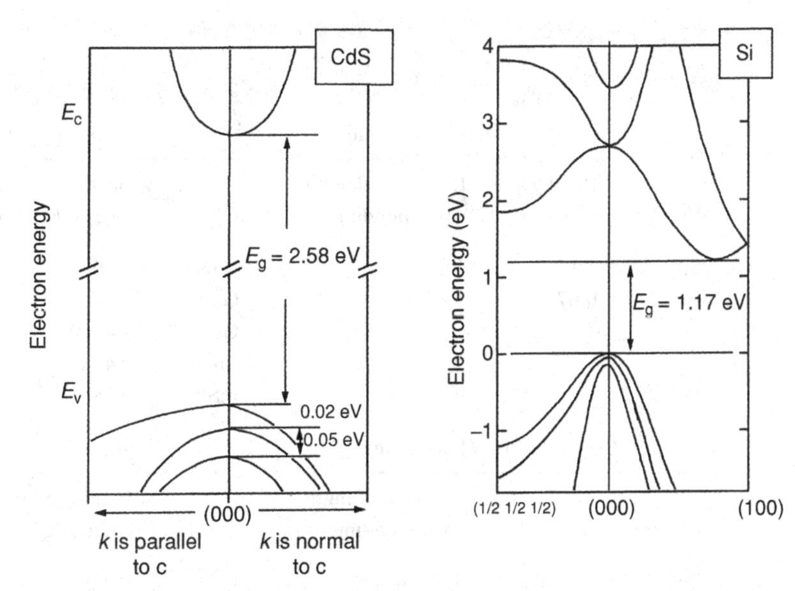

Fig. 4.6 **Band structures of two representative semiconductors, CdS and Si. Adapted from [1]. In CdS the top of the valence band and the bottom of the conduction band correspond to the same wave number, i.e., CdS is a "direct-gap" semiconductor. In Si the extrema of the conduction and the valence band correspond to the different wave numbers, i.e. Si is an "indirect-gap" semiconductor.**

external field an electron in the completely occupied band cannot change its energy because all neighboring states are already filled. The highest occupied band is usually referred to as the "*valence band*" and the lowest unoccupied band is called the "*conduction band*". The difference between the top of the valence band, E_v, and the bottom of the conduction band, E_c, is called the *band gap energy*, E_g:

$$E_g = E_c - E_v. \tag{4.23}$$

Depending on the absolute E_g value, solids that show dielectric properties (i.e., zero conductivity) at $T = 0$, are classified into *dielectrics* and *semiconductors*. If E_g is less than 3–4 eV, the conduction band has a non-negligible population at elevated temperatures, and these types of solids are called "*semiconductors*."

The dispersion curve $E(k)$ and the band structure for real crystals is rather complicated. The effective mass cannot be considered as a constant and in a number of cases can be described as a second-rank tensor. However, in a lot of practically important cases, the events within the close vicinity of E_c and E_v are most important and can be described using an approximation of the constant effective mass, but are sometimes different for different directions. The band structures of the two representative semiconductors, cadmium sulfide and silicon, are given in Figure 4.6. For CdS crystals the minimal gap between E_c and E_v occurs at the same k value. Crystals of this type are called "*direct-gap semiconductors.*" For the Si crystal, the minimal energy gap corresponds to the different k values for E_c and E_v. These types of crystals are usually referred to as "*indirect-gap semiconductors.*" The band gap energies of the most common semiconductors are given in Table 4.1.

Table 4.1. Band gap energy and relevant wavelength for semiconductor and dielectric crystals [17]

Crystals of group IV elements			Crystals of III–V compounds		
Crystal	Band gap energy E_g	Light wavelength corresponding to E_g	Crystal	Band gap energy E_g	Light wavelength corresponding to E_g
Si*	1.14 eV	1.1 μm	GaN	3.50 eV	354 nm
Ge*	0.67 eV	1.85 μm	GaP*	2.26 eV	550 nm
			GaAs	1.43 eV	870 nm
			InAs	0.42 eV	2.95 μm
			InSb	0.18 eV	6.9 μm
Crystals of II–VI compounds			**Crystals of I–VII compounds**		
Crystal	Band gap energy E_g	Light wavelength corresponding to E_g	Crystal	Band gap energy E_g	Light wavelength corresponding to E_g
ZnS	3.68 eV	337 nm	LiF	12 eV	100 nm
ZnSe	2.80 eV	440 nm	NaCl	>4 eV	<300 nm
ZnTe	2.25 eV	550 nm	CuCl	3.2 eV	390 nm
CdS	2.58 eV	480 nm	CuBr	2.9 eV	420 nm
CdSe	1.84 eV	670 nm	**Crystals of IV–VI compounds**		
CdTe	1.6 eV	770 nm	PbS	0.41 eV	3.0 μm
HgTe	0.15 eV	8.2 μm	PbSe	0.28 eV	4.4 μm

* Indirect-gap materials

At zero temperature semiconductor and dielectric materials may have the higher occupied band (valence band) completely filled with electrons and the lower unoccupied band (conduction band) with no electrons at all. For the conduction band to contain no electrons the material should consist either of group IV elements (carbon, germanium, silicon) or of a pair of elements I–VII, II–VI, and III–V groups to complete eight electrons in the valence band. These types of solids are presented as separate groups in Table 4.1 in accordance with the decreasing band gap within each group. Note that the band gap decreases within every group with increasing element numbers, i.e. the number of electronic shells. Consider for example band gaps in the rows **CdS**→**Cd**Se→**Cd**Te or Zn**Te**→Cd**Te**→Hg**Te**. This is because a higher number of electron shells results in screening of the Coulomb potential in the lattice. There is also a regularity with respect to the horizontal position of elements in the Periodic table. For the same total number of shells, the band gap typically rises in the rows III–V→II–VI→I–VII. The reason is stronger polarity of the lattice structure with growing charge difference. For a smaller charge difference bonding is close to covalent, whereas for a larger charge difference it is strongly ionic. The reader is encouraged to trace this regularity for themselves (Problem 1). Note, the whole range of the band gap value is two orders of magnitude.

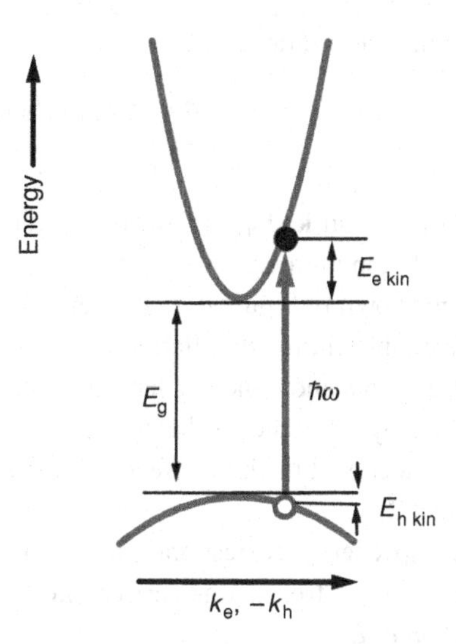

Fig. 4.7 A process of photon absorption resulting in creation of one electron-hole pair. In a diagram including dispersion curves for conduction and valence bands this event can be shown as a vertical transition exhibiting simultaneous energy and momentum conservation.

4.4 Quasiparticles: holes, excitons, polaritons

An electron in the conduction band of a crystal, as we have seen in previous sections, can be described as a particle with charge $-e$, spin $^1/_2$, mass m_e^* (basically variable rather than constant), and quasi-momentum $\hbar k$, with a specific conservation law. One can see that among the above mentioned parameters, only charge and spin remain the same for an electron in a vacuum and in a crystal. Therefore, when speaking about an electron in the conduction band, one implies a particle whose properties result from interactions in a many-body system consisting of a large number of positive nuclei and negative electrons. It is the standard approach in the theory of many-body systems to replace consideration of the large number of interacting particles by the small number of non-interacting *quasiparticles*. These quasiparticles are described as elementary excitations of the system consisting of a number of real particles. Within the framework of this consideration, an electron in the conduction band is the primary elementary excitation of the electron subsystem of a crystal. The further elementary excitation is a "*hole*", which is a quasiparticle relevant to an ensemble of electrons in the valence band from which one electron is removed (e.g., to the conduction band). This excitation is characterized by the positive charge $+e$, spin $^1/_2$, effective mass m_h^*, and the quasi-momentum. In this presentation, the kinetic energy of the hole has an opposite sign compared to that of the electron kinetic energy.

Using concepts of elementary excitation, we can consider the ground state of a crystal as a *vacuum state* (neither an electron in the conduction band nor a hole in the valence band exists), and the first excited state (one electron in the conduction band and one hole in the valence band) in terms of creation of one electron–hole-pair (e–h-pair). A transition from the ground to the first excited state occurs due to photon absorption (Fig. 4.7) with energy and momentum conservation,

$$\hbar\omega = E_g + E_{e\,kin} + E_{h\,kin}$$
$$\hbar\mathbf{k}_{phot} = \hbar\mathbf{k}_e + \hbar\mathbf{k}_h. \tag{4.24}$$

Here $E_{e\,kin}$ ($E_{h\,kin}$) and $\mathbf{k}_e(\mathbf{k}_h)$ are electron (hole) kinetic energy and wave vector, respectively. As the photon momentum is negligibly small, we simply have the vertical transition in the diagram shown in Figure 4.7. The reverse process, i.e., a downward radiative transition which is equivalent to annihilation of an e–h-pair and creation of a photon is possible as well. These events and concepts have a lot in common with real vacuum, electrons and positrons. The only difference is that the positron mass is exactly equal to the electron mass m_0, whereas in a crystal the hole effective mass m_h^* is usually larger than the electron mass m_e^* (see Table 4.2).

The band gap energy corresponds to the minimal energy which is sufficient for the creation of one pair of free charge carriers, i.e., electron and hole. *This statement can serve as the definition of E_g.*

A description based on non-interacting electrons and holes as the only elementary excitations corresponds to the so-called single-particle presentation. In reality, electrons and holes as charged particles do interact via Coulomb potential and form an extra quasiparticle, which corresponds to the hydrogen-like bound state of an electron–hole pair and denoted as "*exciton.*" Interacting holes and electrons can be described by a Hamiltonian,

$$H = -\frac{\hbar^2}{2m_e^*}\nabla_e^2 - \frac{\hbar^2}{2m_h^*}\nabla_h^2 + \frac{e^2}{\varepsilon\,|\mathbf{r}_e - \mathbf{r}_h|}. \tag{4.25}$$

This is the same as the Hamiltonian (2.104) of the hydrogen atom with m_e^* and m_h^* instead of a free electron mass m_0 and a proton mass $m_p = 1836\,m_0$, respectively and with the dielectric constant of the crystal $\varepsilon > 1$. Therefore, similarly to the hydrogen atom (Eq. (2.109)), an exciton is characterized by the *exciton Bohr radius* (in SI units),

$$a_B^* = 4\pi\varepsilon_0\frac{\varepsilon\hbar^2}{\mu_{eh}e^2} = \varepsilon\frac{\mu_H}{\mu_{eh}} \times 0.053\ \text{nm} \approx \varepsilon\frac{m_0}{\mu_{eh}} \times 0.053\ \text{nm} \tag{4.26}$$

where μ_H is the electron–proton reduced mass defined as,

$$\mu_H^{-1} = m_0^{-1} + m_{proton}^{-1} \approx m_0^{-1}, \tag{4.27}$$

which to great accuracy equals the electron mass m_0 because $m_0 \ll m_{proton}$, and μ_{eh} is the electron–hole reduced mass,

$$\mu_{eh}^{-1} = m_e^{*-1} + m_h^{*-1}. \tag{4.28}$$

Also, similarly to a hydrogen atom, the *exciton Rydberg energy* can be written as,

$$\text{Ry}^* = \frac{e^2}{2\varepsilon a_B^*} = \frac{1}{4\pi\varepsilon_0}\frac{\mu_{eh}e^4}{2\varepsilon^2\hbar^2} = \frac{\mu_{eh}}{\mu_H}\frac{1}{\varepsilon^2} \times 13.60\ \text{eV} \approx \frac{\mu_{eh}}{m_0}\frac{1}{\varepsilon^2} \times 13.60\ \text{eV}. \tag{4.29}$$

Table 4.2. Electron, hole and exciton parameters [17]

	Exciton Rydberg Ry* (meV)	Electron effective mass m_e/m_0	Hole effective mass m_h/m_0	Exciton Bohr radius a_B^* (nm)
Ge	4.1	⊥0.19 ‖0.92	0.54 (hh) 0.15 (lh)	24.3
Si	15	⊥0.081 ‖1.6	0.3 (hh) 0.043 (lh)	4.3
GaAs	4.6	0.066	0.47 (hh) 0.07 (lh)	12.5
CdTe	10	0.1	0.4	7.5
CdSe	16	0.13	⊥0.45 ‖1.1	4.9
CdS	29	0.14	⊥0.7 ‖2.5	2.8
ZnSe	19	0.15	0.8 (hh) 0.145 (lh)	3.8
CuBr	108	0.25	1.4 (hh)	1.2
CuCl	190	0.4	2.4 (hh)	0.7
GaN	28	0.17	0.3 (lh) 1.4 (hh)	2.1
PbS	2.3	⊥0.080 ‖0.105	⊥0.075 ‖0.105	18
PbSe	2.05	⊥0.040 ‖0.070	⊥0.034 ‖0.068	46

hh and lh stand for heavy and light hole, respectively;
⊥(‖) means transverse (longitudinal) values.

The reduced electron–hole mass is smaller than the electron mass m_0, and the *dielectric constant* ε is several times larger than that of a vacuum. This is why the exciton Bohr radius is significantly larger and the exciton Rydberg energy is significantly smaller than the relevant values for the hydrogen atom. Absolute values of a_B^* for the common semiconductors range between 10–100 Å, and the exciton Rydberg energy takes values approximately from 1 meV to 100 meV (Table 4.2).

An exciton exhibits the translational center-of-mass motion as a single uncharged particle with mass $M = m_e^* + m_h^*$. The dispersion relation can be expressed as,

$$E_n(\mathbf{K}) = E_g - \frac{\mathrm{Ry}^*}{n^2} + \frac{\hbar^2 \mathbf{K}^2}{2M}, \quad n = 1, 2, 3, \ldots, \tag{4.30}$$

where **K** is the exciton wave vector. Equation (4.30) includes the hydrogen-like set of energy levels, the kinetic energy of the translational motion and the band gap energy. The exciton

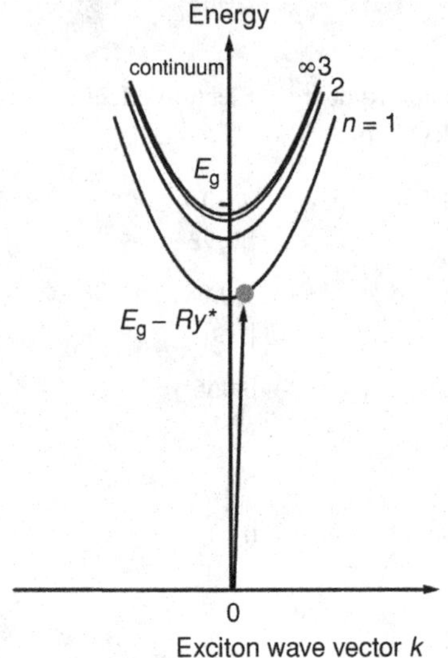

Fig. 4.8 **Dispersion curves of an exciton and the optical transition corresponding to photon absorption and exciton creation. Dispersion curves correspond to the hydrogen-like set of energies $E_n = E_g - \mathrm{Ry}^*/n^2$ at $k = 0$ and a parabolic $E(k)$ dependence for every E_n, describing the translational center-of mass motion. For $E > E_g$, the exciton spectrum overlaps with the continuum of unbound electron–hole states. Exciton creation can be presented as intercrossing of the exciton and the photon dispersion curves corresponding to simultaneous energy and momentum conservation. The photon dispersion curve is a straight line in agreement with the formula $E = pc$.**

energy spectrum consists of subbands (Fig. 4.8) which converge to the dissociation edge corresponding to the free electron–hole-pair. Similarly to the free e–h-pair, an exciton can be created by photon absorption. Taking into account that a photon has a negligibly small momentum, exciton creation corresponds to a discrete set of energies,

$$E_n = E_g - \frac{\mathrm{Ry}^*}{n^2}, \quad n = 1, 2, 3, \ldots. \tag{4.31}$$

An exciton gas can be described as a gas of bosons with an energy distribution function obeying the *Bose–Einstein statistics*, i.e. there is no limit to the number of excitons occupying a single state. For a given temperature T, the concentration of excitons n_{exc} and of the free electrons and holes $n = n_e = n_h$, are related via the *ionization equilibrium equation* known as the *Saha equation*:

$$n_{\mathrm{exc}} = n^2 \left(\frac{2\pi \hbar^2}{k_B T} \frac{1}{\mu_{\mathrm{eh}}} \right)^{3/2} \exp \frac{\mathrm{Ry}^*}{k_B T}. \tag{4.32}$$

For $k_B T \gg \mathrm{Ry}^*$ most of the excitons are ionized and the properties of the electron subsystem of the crystal are determined by the gas of free electrons and holes. At $k_B T \leq \mathrm{Ry}^*$ a significant number of electron – hole-pairs exist in the bound state.

As a result of creation of excitons and free electron – hole-pairs, the absorption spectrum of direct-gap semiconductor monocrystals contains a pronounced resonance peak at energy

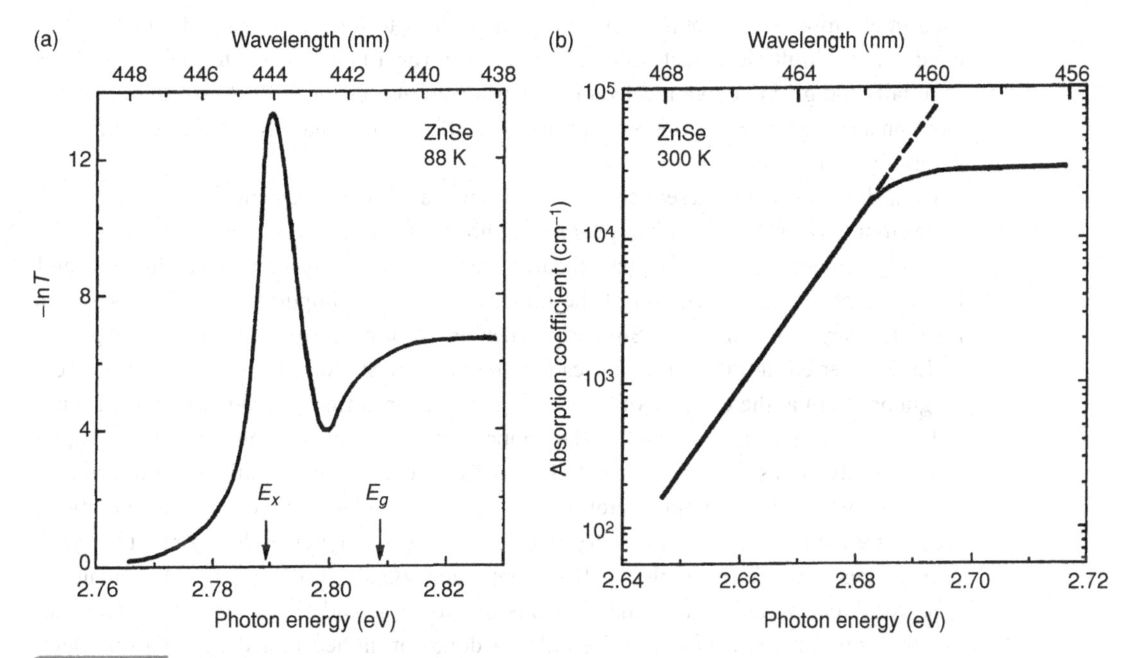

Fig. 4.9 **Absorption spectrum of a ZnSe single crystal near the fundamental absorption edge at a temperature equal to 88 K (a) and 300 K (b) [18]. In panel (a) $-\ln T = -\ln(I_{\text{out}}/I_{\text{in}})$ where T is the optical transmission coefficient, $I_{\text{out}}(I_{\text{in}})$ is output (input) light intensity.**

$\hbar\omega = E_g - \text{Ry}^*$, a set of smaller peaks at energies E_n (Eq. 4.31), and a smooth continuous absorption for $\hbar\omega \geq E_g$. For $\hbar\omega \geq E_g$ the absorption coefficient increases monotonically, the dependence of absorption coefficient versus frequency following the density of states dependence $D(E) = \text{const}\sqrt{E - E_g}$ as a first approximation. In reality, absorption is enhanced because of the electron–hole Coulomb interaction. In Figure 4.9 an experimentally observed absorption spectrum for a ZnSe single crystal film is presented. Zinc selenide possesses band-gap energy $E_g = 2.809$ eV at $T = 80$ K and 2.67 eV at $T = 300$ K, the exciton Rydberg energy, $\text{Ry}^* = 18$ meV. At $k_B T \gg \text{Ry}^*$ the spectrum contains a pronounced peak for exciton absorption corresponding to $n = 1$, whereas the higher sub-bands are smeared due to thermal broadening. At $k_B T > \text{Ry}^*$ the exciton band is not pronounced, however enhancement of absorption at $\hbar\omega < \text{E}_g$ due to electron–hole Coulomb interaction occurs. The long-wave absorption tail shows exponential dependence of the absorption coefficient on the photon energy (the *Urbach rule*) and corresponds to a straight line on a semilogarithmic scale.

4.5 Defect states and Anderson localization

In semiconductors and dielectrics substitution of intrinsic atoms by impurity atoms results in substitution of the corresponding intrinsic states of electrons by impurity states. These states may have energy inside the band gap. Doping of semiconductors is systematically used in electronics to increase conductivity as well as to form electron and hole types of conductivity. Multiple defect states of an electron (or a hole) evolve to defect bands or even tails and give rise to a localization–delocalization transition which has an analog for electromagnetic waves. In this section disorder effects on quantum particle behavior will be briefly overviewed.

Consider, for example, a semiconductor crystal of a group IV element, say Si. Every atom forms four covalent bonds with its nearest neighbors. If one atom is replaced by an atom of a group V element, e.g. As, P, Sb, then the fifth valent electron can enter the conduction band by means of thermal ionization of the impurity atom. The impurity atom then becomes a single charged positive ion. Similarly, when an atom of a group III element (e.g., Ga or In) is inserted into the lattice instead of the original Si atom, bonds with four nearest neighbors form at the expense of "borrowing" an electron from the valence band. Then a hole appears in the valence band an the impurity atom becomes a single charged negative ion. Impurity atoms which give rise to higher free electron concentrations are referred to as *donors* whereas the impurity atoms which give rise to higher free hole concentrations are referred to as *acceptors*. In binary semiconductors, two types of donors and two types of acceptors exist. For example, in III–V compounds (GaAs, InP, GaN, InSb) elements of groups IV and VI are donors and elements of groups II and IV are acceptors. One can see the same impurity atom may be either a donor or an acceptor depending on which element it substitutes. For example, Si atoms in GaAs are donors when they substitute Ga and acceptors when they substitute As.

Simple quantitative consideration of impurity electron states can be performed within the approximations of isotropic electron effective mass m_{e}^{*}, dielectric permittivity ε and the Coulomb potential of a single charged donor atom. Since an impurity atom mass is 3–4 orders of magnitude higher than an electron effective mass, an impurity center can be considered as immobile. Then the Schrödinger equation for an electron in a Coulomb potential of an impurity ion formally coincides with the equation for an electron in a Coulomb potential, considered in Section 2.6, with replacements $m \to m_{\mathrm{e}}^{*}$ and $e^2/r \to e^2/\varepsilon r$. Therefore we can write an expression for the energy spectrum,

$$E = -\frac{E_{\mathrm{d}}}{n^2}, \quad n = 1, 2, 3, \ldots \tag{4.33}$$

where donor ionization energy E_{d} reads (in SI units),

$$E_{\mathrm{d}} = \frac{1}{2\pi\varepsilon_0} \frac{e^4 m_{\mathrm{e}}^{*}}{2\varepsilon^2 \hbar^2}. \tag{4.34}$$

The donor Bohr radius, by analogy with Eq. (2.101), reads (in SI units),

$$a_{\mathrm{d}} = 4\pi\varepsilon_0 \frac{\varepsilon \hbar^2}{m_{\mathrm{e}}^{*} e^2}. \tag{4.35}$$

Perturbation of the intrinsic electron energy spectrum in a periodic lattice by a single donor impurity is presented in Figure 4.10. For an acceptor impurity similar consideration can be applied to give the acceptor ionization energy E_{a} and acceptor Bohr radius,

$$E_{\mathrm{a}} = \frac{1}{2\pi\varepsilon_0} \frac{e^4 m_h^{*}}{2\varepsilon^2 \hbar^2}, \quad a_{\mathrm{h}} = 4\pi\varepsilon_0 \frac{\varepsilon \hbar^2}{m_{\mathrm{h}}^{*} e^2}. \tag{4.36}$$

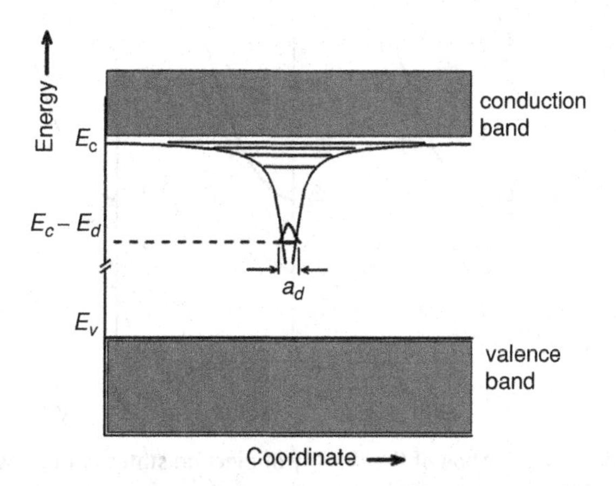

Fig. 4.10 **The Coulomb potential of a donor impurity superimposed with conduction and valence bands, showing the electron energy spectrum and wave function of the lowest impurity state.** $E_c - E_d \ll E_g$, $a_d > a_L$.

Consider the evolution of impurity states with increasing impurity concentration N. Based on general quantum mechanical considerations one can foresee the impurity states will modify as the average distance between impurity atoms is no longer negligible as compared with the impurity Bohr radius. Along with the impurity Bohr radius another length parameter, namely the concentration-dependent *Debye screening radius* r_D becomes important. This depends on the free charge carrier concentration n_e and reads,

$$r_D = \left(\frac{\varepsilon_0 \varepsilon \kappa_B T}{4\pi e^2 n_e}\right)^{1/2}. \tag{4.37}$$

Provided that the following two inequalities hold,

$$N^{-1/3} \gg r_D, \tag{4.38}$$

$$N^{-1/3} \gg a_d, \tag{4.39}$$

the above consideration of impurity states in terms of isolated hydrogen-like ones is justified. Hereafter we consider electrons and donor impurities for clarity. When donors are the only source of free electrons, the inequality $r_D(N) > a_d$ holds. Therefore with increasing N, first the inequality (4.38) becomes invalid and then, at higher N, the inequality (4.39) also becomes invalid. Consider modification of the electron states (Fig. 4.11). First, when the average distance between impurity sites $N^{-1/3}$ becomes comparable to r_D, the energy level of every impurity atom will be perturbed by the neighbor atoms. Chaotic displacement of impurities will give rise to random deviation of every state with respect to that given by Eq. (4.34). Therefore the electron density of states $D(E)$ can be described in terms of the Gaussian distribution function,

$$D(E) = D_0 \exp\left[-\frac{(E - E_0)^2}{(\Delta E)^2}\right], \tag{4.40}$$

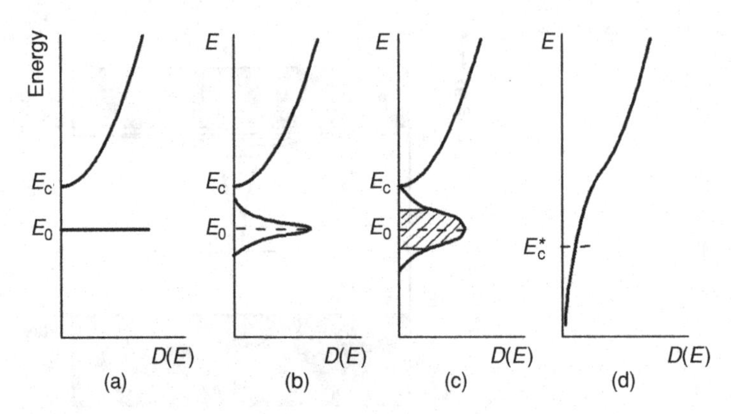

Fig. 4.11 **Gradual modification of the density of electron states with growing density N of a donor impurity. (a) Low doping level, no broadening of impurity states. (b) Intermediate doping. Formation of the impurity band due to classic broadening. (c) High doping level. Formation of delocalized states due to quantum broadening. (d) Very high doping. Impurity band merges with conduction band to form a continuous range of electron states far below E_c. Mobility edge E_c^* separates localized and delocalized states.**

where parameter D_0 can be found from the condition of conservation of the total density of impurity states,

$$\int D(E)\mathrm{d}E = N.$$

Therefore, identical impurity energy levels (Fig. 4.11a) evolve to the impurity band with width ΔE (Fig. 4.11b). The above broadening resulting from chaotic displacement of impurity atoms rather than from wave properties of electrons is referred to as *classical broadening*.

At higher concentrations when the inequality (4.39) becomes invalid the electron energy states split because of the wave properties of electrons, as was discussed for a quantum particle in a double-well potential in Section 3.6. In the case of regular displacement of impurity atoms in the lattice every electron energy level will split into N levels and electrons will delocalize, with the electron wave function extended over the whole crystal. Because of the random displacement of impurities even at higher concentration there are still localized electron states coexisting with delocalized ones (Fig. 4.11(c)). Finally the impurity band merges with the conduction band with the threshold energy referred to as the mobility edge E_c^*.

An electron with energy higher than the mobility edge is completely delocalized. The density of states function obeys a form of exponential tail (Fig. 4.11(d)). In Figure 4.12 this case is considered in an energy–coordinate presentation superimposed with the density of states function. There are two cases possible for the relation between the mobility edge and the Fermi level E_F. If $E_F < E_c^*$ (Fig. 4.12(a)) then at $T = 0$ all electrons will be in localized states. If $E_F > E_c^*$ then delocalized electrons exist even at zero temperature. The first case corresponds to zero conductivity whereas the second case corresponds to high conductivity. A transition from isolating to conductive properties of electrons in random potentials is

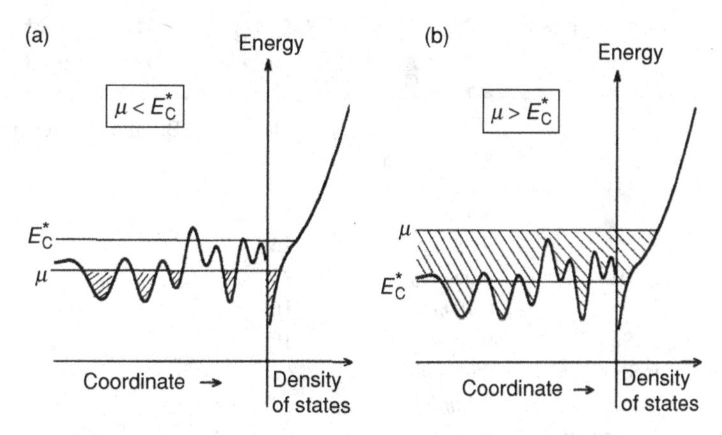

Fig. 4.12 **A sketch of electron energy and density of states in a random potential for the case (a) of the Anderson dielectric and (b) of metallic conductivity.**

called an *Anderson transition*, after P. Anderson who was the first to outline the absence of electron diffusion in certain random potentials in 1958 [19]. Localization of an electron in a random potential can be intuitively understood in terms of the *Ioffe–Regel* criterion which considers electron mean free path ℓ versus its de Broglie wavelength λ, namely,

$$k\ell < 1, \quad \text{or} \quad \ell < \lambda/2\pi. \tag{4.41}$$

The Ioffe–Regel criterion states that an electron becomes localized when its mean free path is less than its de Broglie wavelength by a factor of 2π. The Anderson transition is a subject of extensive theoretical modeling with respect to space dimensionality and randomness type. Importantly, for the one-dimensional problem, a random potential always results in localized states and only for two- and three-dimensional problems does space delocalization occur.

4.6 Quantum confinement effects in solids

Consider an electron's de Broglie wavelength λ_e (Eq. 2.8), assuming the effective mass inherent in real crystals and using the room temperature $k_B T = 26.7$ meV value as a measure for the electron kinetic energy, i.e.

$$\lambda_e = \frac{h}{\sqrt{2m_e^* k_B T}}. \tag{4.42}$$

Note that the $k_B T$ value gives the mean energy of a particle per one degree of freedom. For an electron in a vacuum Eq. 4.42 gives $\lambda_e = 7.3$ nm. For an electron in various materials the λ_e values are given in Table 4.3, along with the lattice constants a_L for the same crystals. One can see λ_e is more than one order of magnitude bigger than a_L.

Therefore it is possible to perform spatial confinement of an electron (as well as hole and exciton) motion in one, two or all three directions while keeping the crystal lattice structure

Table 4.3. Lattice constants and electron de Broglie wavelengths at room temperature for different semiconductor crystals

Material	Electron mass	Electron de Broglie wavelength λ_e (nm)	Lattice constant a_L (nm)
SiC	$0.41\,m_0$	11	0.308
Si	$0.08\,m_0$	26	0.543
GaAs	$0.067\,m_0$	28	0.564
Ge	$0.19\,m_0$	16	0.564
ZnSe	$0.15\,m_0$	19	0.567
InSb	$0.014\,m_0$	62	0.647
CdTe	$0.1\,m_0$	23	0.647
In vacuum	m_0	7.3	

non-perturbed or rather slightly perturbed. In a thin film whose thickness is less than the electron de Broglie wavelength the electron motion will be confined in the direction normal to the film plane, remaining infinite in the other two dimensions. Therefore, in a thin film two-dimensional electron motion and a two-dimensional electron gas can be performed. Thin film is hard to grow free standing, typically it is grown on a crystalline substrate or between the two crystalline layers of different material to provide potential barriers for the electrons. Such a film is referred to as a *quantum well*.

The motion of quasiparticles can also be confined in two directions if a crystal has a needle-like or rod-like shape with cross-sectional size smaller than λ_e. Such structures can be grown as isolated species or again, can be developed on top of different materials. These are usually referred to as *quantum wires*.

Finally, crystalline nanoparticles (nanocrystals) can offer confinement for electron motion in all three directions provided that their size is no more than a few nanometers. These are referred to as *quantum dots*. Nanocrystals can be grown in solutions and polymers, in glasses and on top of a crystalline substrate.

For an electron in a quantum well with thickness L_z and infinite potential barriers the wave function can be written as,

$$\psi(\mathbf{r}) = \sqrt{\frac{2}{L_x L_y L_z}} \sin(k_n z) \exp(i\mathbf{k}_{xy} \cdot \mathbf{r}), \tag{4.43}$$

where the wave number,

$$k_n = \frac{\pi}{L_z} n, \quad n = 1, 2, 3, \ldots \tag{4.44}$$

takes a discrete value determined by the well width, whereas $\mathbf{k}_{xy}$ takes continuous values and corresponds to infinite in-plane motion of an electron. Accordingly, the electron energy is the sum of the kinetic energy relevant to infinite motion $E = \hbar^2 k_{xy}^2 / 2m_e^*$ and the discrete set of energy values $E_n = \hbar^2 k_n^2 / 2m_e^*$ relevant to quantized states of an electron, i.e.

$$E = \frac{\hbar^2 k_{xy}^2}{2m_e^*} + \frac{\pi^2 \hbar^2}{2m_e^* L_z^2} n^2, \quad n = 1, 2, 3, \ldots. \tag{4.45}$$

For a quantum wire with a rectangular cross-section which allows infinite motion in the x-direction, with infinite potential barriers in the y- and z-directions, the wave function can be written as,

$$\psi(\mathbf{r}) = 2\sqrt{\frac{1}{L_x L_y L_z}} \sin\left(k_n^{(z)} z\right) \sin\left(k_m^{(y)} y\right) \exp(i k_x x), \tag{4.46}$$

where the wave number k_x takes continuous values whereas $k_n^{(z)}$, $k_m^{(y)}$ are discrete:

$$k_n^{(z)} = \frac{\pi}{L_z} n, \quad n = 1, 2, 3, \ldots, \qquad k_m^{(y)} = \frac{\pi}{L_y} m, \quad m = 1, 2, 3, \ldots. \tag{4.47}$$

The energy spectrum consists of a continuous term for motion along the x-axis and the two discrete sets resulting from confinement in the y- and z-directions,

$$E = \frac{\hbar^2 k_x^2}{2m_e^*} + \frac{\pi^2 \hbar^2}{2m_e^* L_z^2} n^2 + \frac{\pi^2 \hbar^2}{2m_e^* L_y^2} m^2, \quad n = 1, 2, 3, \ldots, \; m = 1, 2, 3, \ldots. \tag{4.48}$$

For a cubic quantum dot with infinite potential barriers only a discrete set of completely localized states is allowed. The wave function reads,

$$\psi(\mathbf{r}) = 2\sqrt{\frac{2}{L_x L_y L_z}} \sin\left(k_n^{(z)} z\right) \sin\left(k_m^{(y)} y\right) \sin\left(k_\ell^{(x)} x\right), \quad n, m, \ell = 1, 2, 3, \ldots \tag{4.49}$$

The energy spectrum of an electron in a cubic box with $L_x = L_y = L_z = L$ reads,

$$E = \frac{\pi^2 \hbar^2}{2m_e^* L^2} (n^2 + m^2 + \ell^2), \quad n, m, \ell = 1, 2, 3, \ldots. \tag{4.50}$$

The energy spectrum of an electron in a spherical box will be considered in detail in Chapter 5.

4.7 Density of states for different dimensionalities

We need now to recall Section 2.2 and count standing waves (normal) modes in 1- and 2-dimensional boxes with size L. We shall use notations $1d$ and $2d$, respectively. For a $1d$-box there is space between neighboring k values equal to π/L. To count the number of modes in the range from k to $k + \mathrm{d}k$ in unit volume one has to divide dx by π/L and by L. This gives a mode number $\mathrm{d}k/\pi$ and mode density,

$$D_1(k) = 1/\pi. \tag{4.51}$$

In the $2d$-case, an elementary "cell" per one mode will become $(\pi/L)^2$. Now instead of dk we need to take the square of the elementary positive layer segment in k-space confined between the two circumferences with radii k, $k + \mathrm{d}k$ (Fig. 4.13). Its square reads $\pi k \mathrm{d}k/2$. Dividing this value by the size of a unit cell $(\pi/L)^2$ and the volume L^2 we count the number of modes $(k/\pi)\mathrm{d}k$, whence the mode density per unit k interval reads,

$$D_2(k) = \frac{k}{2\pi}. \tag{4.52}$$

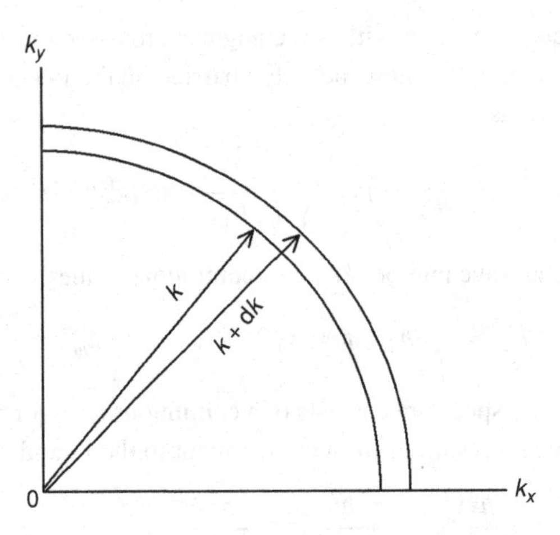

Fig. 4.13 An elementary layer in 2d k-space.

Now mode densities (4.51) and (4.52) are ascribed to electron state densities. Using known relations (2.24) we arrive at the densities of electron states on the energy scale as follows,

$$D_1(E) = D_1(k)\frac{\mathrm{d}k}{\mathrm{d}E} = \frac{\sqrt{2}m^{1/2}E^{-1/2}}{h}$$
$$D_2(E) = D_2(k)\frac{\mathrm{d}k}{\mathrm{d}E} = \frac{2\pi m}{h^2}. \tag{4.53}$$

One can see a clear correlation between the density of states functions and the dimensionality of space d. Based on Eqs. (4.53) combined with their $3d$ analog Eq. (2.26), one can see the general relation holds,

$$D_d(E) = \mathrm{const}\frac{m^{\frac{d}{2}}E^{\frac{d}{2}-1}}{h^d}. \tag{4.54}$$

Formulas (4.51) and (4.52) are general and valid for all waves. Formulas (4.53) and (4.54) are valid for all types of quantum particles with non-zero rest mass m. When electrons (or holes in solids) are considered, a factor of 2 is added to account for different possible spin orientations.

4.8 Quantum wells, quantum wires and quantum dots

In Figure 4.14 the density of states functions are compared for spaces with dimensionalities $d = 3, 2, 1$. In Figure 4.15 the density of states functions are plotted for a quantum well (left) and a quantum wire (center). For wells and wires the densities of states are determined by energy quantization and by functions (4.54). For quantum dots an expression like

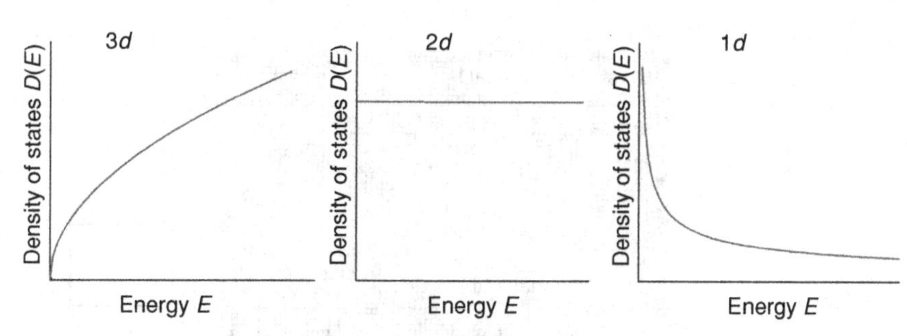

Fig. 4.14 Density of states functions for a quantum particle with non-zero rest mass in spaces with different dimensionalities.

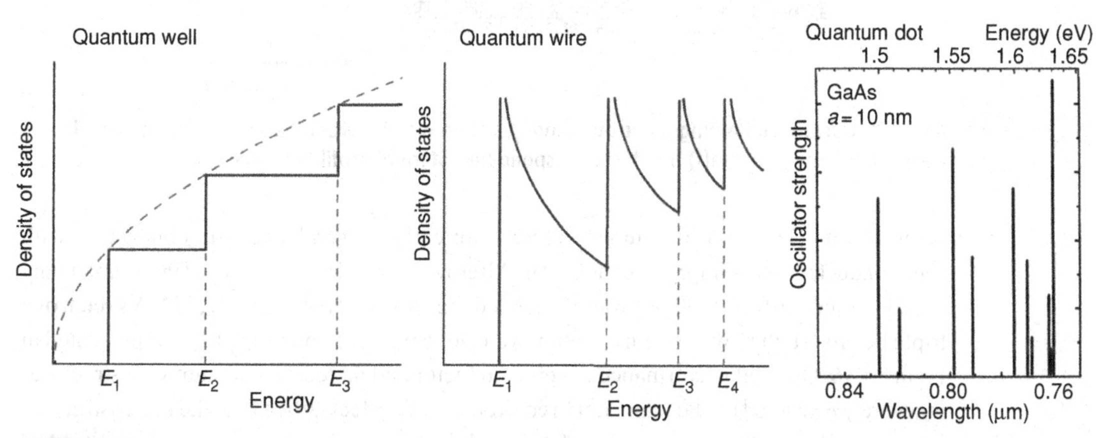

Fig. 4.15 Electron density of states in a quantum well, in a quantum wire and calculated probabilities for optical transitions in a realistic GaAs quantum dot with size 10 nm. Data for GaAs dots are taken from Ref. [20].

Eq. (4.54) can not be derived. The density of states function in a quantum dot condenses into a delta-like sharp peak for every energy state allowed in a quantum box. To give an idea of the optical properties of quantum dots, an example is given of oscillator strengths for optical transitions in a GaAs quantum dot in Fig. 4.15 (right). Relative oscillator strength is influenced by the wave functions of quantized electrons and holes, as well as their overlap within the dot.

Quantum wells and quantum well superlattices

Quantum wells can be fabricated by means of epitaxial growth of a multilayer semiconductor structure in which a material with a narrower band gap is "buried" into another material with wider band gap. Notably, this technique is essentially based upon the heterostructure concept, i.e. the structure which exhibits a continuous crystal lattice but discontinuous chemical composition. It is possible only provided that the crystal lattices of the adjacent materials feature equal or very close crystal lattices, the same lattice symmetry and chemical

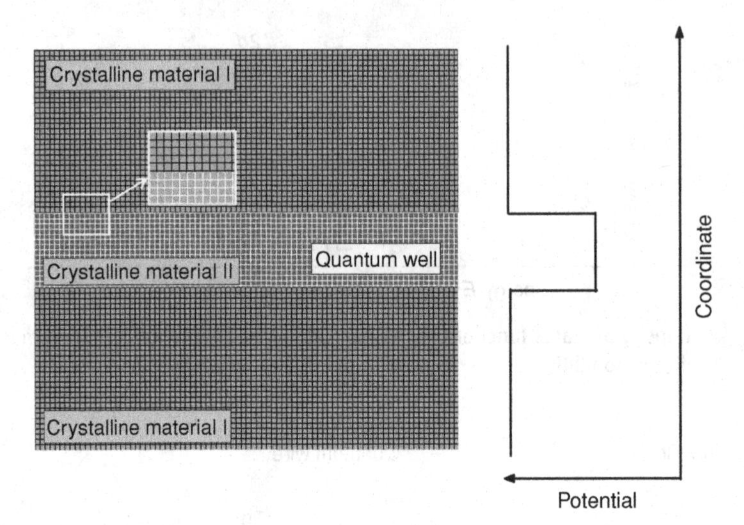

Fig. 4.16 **A heterostructure containing a narrow-band quantum well between two crystalline materials with wider band gaps (left) and the corresponding potential profile for an electron.**

compatibility. The idea of heterostructures fabrication was advanced in connection with semiconductor lasers improvement by Zh. Alferov and H. Kroemer in the 1960s, and many decades afterwards in 2000 this was acknowledged with a Nobel Prize [21, 22]. A sketch of a triple heterostructure with a thin quantum well along with the potential profile is presented in Figure 4.16. Optimal performance of semiconductor and dielectric quantum well structures is severely restricted by the mandatory requirement of perfect lattice fit at the heterostructure interface. The representative pairs of materials meeting this requirement are e.g. GaAs–$GaAl_xAs_{1-x}$, ZnS–ZnS_xSe_{1-x}, Ge–Ge_xSi_{1-x}, CdSe–$CdSe_xTe_{1-x}$. Not only single quantum wells but also sequential equally spaced multiple quantum wells can be developed with discrete electron and hole energy levels evolving to minibands. These types of structures are referred to as quantum well *superlattices*. Formation of minibands occurs similarly to energy level splitting in multiple potential wells (Fig. 3.23) and similarly to band structure formation for a quantum particle in a one-dimensional periodic potential.

Quantum wells feature a number of amazing properties as compared to the parent three-dimensional bulk crystals. The Schrödinger equation with Coulomb potential for two-dimensional space gives rise to a different solution compared to the 3*d*-case. Instead of an energy spectrum given by Eq. (2.108), another expression comes up,

$$E_n = -\frac{\mathrm{Ry}}{\left(n+\frac{1}{2}\right)^2} \quad \text{with } n = 0, 1, 2, 3, \ldots, \tag{4.55}$$

and the lowest exciton state now gains energy $E_g^* - 4\mathrm{Ry}^*$ where E_g^* stands for the band gap energy modified by electron and hole confinement. A four-fold increase in exciton binding energy offers a strong manifestation of excitonic effects for higher temperatures.

Figure 4.17 shows the behavior of exciton binding energy from the 3*d*- to 2*d*-value as a quantum well width tends to a negligibly small value. It corresponds to infinite barriers,

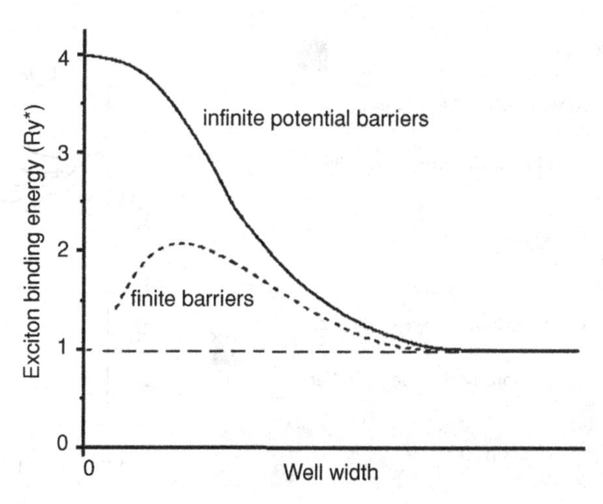

Fig. 4.17 **Evolution of the exciton binding energy for a quantum well width tending to zero. For the case of unrealistic infinite potential barriers confining electron and hole motion, exciton binding energy tends to 4Ry*, predicted for the pure two-dimensional Schrödinger problem with Coulomb potential. For realistic finite barriers inherent in heterostructures, the exciton binding energy increases with decreasing well width, but eventually drops back to the Ry* value at vanishing well width.**

which is impossible to perform experimentally. In the more realistic case of finite barriers at the interfaces, the binding energy at first rises up but then falls down. It happens because for finite barriers the limiting case of zero width corresponds to *3d-excitons* in the ambient material. A notable increase in exciton binding energy for two-dimensional semiconductors was predicted for the first time in 1966 by M. Shinada and S. Sugano when considering exciton properties in extremely anisotropic crystals [23].

Another amazing property of quantum wells is the strong sensitivity of the absorption spectrum to an external electric field. This was discovered in the 1980s by D. A. B. Miller and co-workers for GaAs–$GaAl_xAs_{1-x}$ quantum well structures [24]. Recently it has been demonstrated for the Ge–Si quantum well structure [25], (Fig. 4.18). Noteworthy is the challenging issue of making use of advanced Si–Ge based technology for optical communication and optical processing circuitry. This technology offers easy and cheap integration of newly developed components because of technological compatibility with CMOS electronic circuitry as well as with commercial technological equipment. The Si–Ge paradigm keeps CMOS compatibility but allows us to go from electro-optically inefficient silicon to Ge, whose direct gap transitions much more readily fall in the vicinity of 1.5 μm photon wavelength. Germanium quantum wells look to be a very promising candidate for such an application.

A breakthrough in the problem has been proposed based on exploitation of the direct gap optical transitions in Ge in the spectral range close to the desirable optical communication domain. Strong modulation of Ge absorption in quantum wells has been reported [25, 26] and the first evidence of a feasible optical modulator based on electric field control of optical cavity finesse has been demonstrated [27]. The electric field effect on intrinsic band-to-band

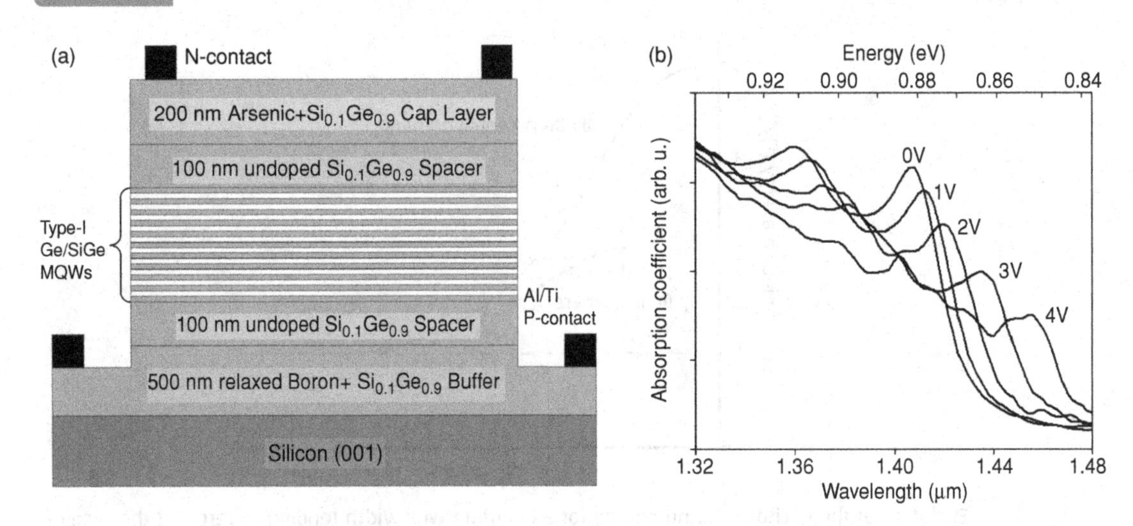

Fig. 4.18 **(a) Electro-absorptive multiple quantum well structure and (b) its absorption spectra at various applied voltages. Reprinted with permission from [25]. Copyright 2005 Nature Publ. Group.**

transitions in Ge quantum wells far from the indirect absorption onset (0.7 eV) changes absorption significantly in the spectral range 1340–1460 nm (Fig. 4.18). This pioneering observation means Ge multiple quantum well structures are considered to be a promising active material for compact CMOS-compatible electro-optical modulators.

For spectral shift of the response towards the desirable 1550 nm (1540–1560 nm is the operating range of Er amplifiers in optical communication networks), modified quantum well stoichiometry has been proposed, namely $Ge_{0.9925}Si_{0.0075}$ [26]. However, change in stoichiometry gives rise to strain which is not desirable. Pure Ge wells, even if combined with a Fabry–Perot cavity exhibit an electro-optical response somewhat outside the desirable range, namely 1440–1460 nm [27]. Most probably, electro-refractive rather than electro-absorptive properties will be exploited. A change in absorption edge will always be accompanied by a change in refraction index at the red-side with respect to absorption onset in accordance with the Kramers–Kronig relations. Therefore long-wave operation becames feasible. Modulation of the refraction index can be transformed into transmission modulation by means of interferometers or more complex photonic circuitry solutions like those to be described in Chapter 9.

Semiconductor quantum wells have found extensive application in injection lasers. Electron-hole gases of lower dimensionalities feature a narrow spread in velocities at finite temperatures because of the specific density of states functions (Fig. 4.14). This in turn lifts the threshold current dependence on temperature. The latter is the crucial injection laser parameter. Heating of the active layer upon operation gives rise to a higher threshold current, i.e. undesirable decrease in efficiency of laser operation. It comes from a higher spread of charge carriers over energy. For lower dimensionality this spread is inhibited because of the lower number of states available for higher energies. Applications of quantum wells in lasers are discussed in more detail in [21, 28, 29].

In past decades, in addition to traditional semiconductor lasers, multiple quantum wells have been suggested as active media for cascade lasers capable of generating terahertz radiation (wavelengths of the order of 0.1–1 mm). This frequency range does not have reliable, efficient and compact coherent sources as optical and infrared ranges do. However the range is rather demanding for the purpose of security issues (terahertz tomography can readily replace x-ray scanning of luggage and personal belongings at airports etc.) as well as for multiple applications in spectral analysis of gases and atmospheric pollution, where traditional infrared spectroscopy is not efficient because of spectral overlapping of characteristic absorption lines. The very idea of using a semiconductor superlattice for terahertz generation dates back to the pioneering work by R. Kazarinov and R. Suris in 1971 [30]. Since the first demonstration of quantum cascade laser operation in 2001 [31], milliwatt power at an operating temperature of about 150 K in the spectral range of 1–5 THz have been developed [32].

Quantum wires and nanorods

An ideal one-dimensional conductor, a quantum wire, exhibits a remarkable property. In the case of purely ballistic electron propagation, i.e. in the absence of electron scatterers, its conductivity G is defined by two fundamental constants, namely,

$$G = \frac{e^2}{\pi \hbar}. \tag{4.56}$$

The origin of this amazing relation is as follows. Conductivity is proportional to electron velocity but an electron with higher velocity has lower density of states available, in inverse proportion to the velocity, as follows from Eq. (4.54). The product of velocity and density of states obeys therefore a constant value resulting in constant conductivity. The proof of this universal law was proposed by Landauer in 1970 [33] and can be found in many books related to nanoelectronics [34–36]. The G value defined by Eq. (4.56) is called the *conductivity quantum.* Its inverse value has the dimensions of resistance and is equal to $G^{-1} \approx 13$ kOhm.

Quantum wires are of prime importance in nanoelectronics. Quantum wires have not found extensive applications in optical engineering. A few examples of needle-like crystal structures with a high aspect ratio can be found among minerals. For example, *asbestos* consists of needle-like nanometer-sized crystals.

In very recent years reliable chemical routes for synthesis of elongated semiconductor nanocrystals with high aspect ratio have been developed [37–41]. These nanocrystals are known as "*nanorods*." Nanorods of II–VI (e.g. CdSe) compounds can be chemically synthesized with pronounced match- or needle-like geometry and size-dependent optical properties (Figs. 4.19 and 4.20).

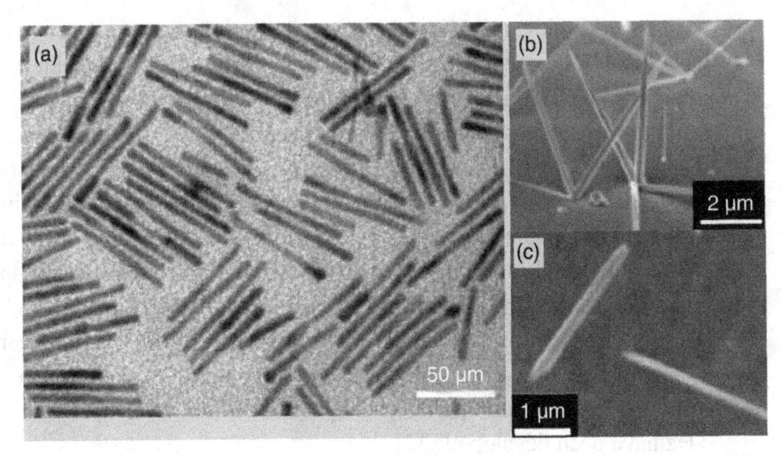

Fig. 4.19 **Images of semiconductor nanorods obtained by means of electron microscopy.** ***Courtesy of L. Manna.***

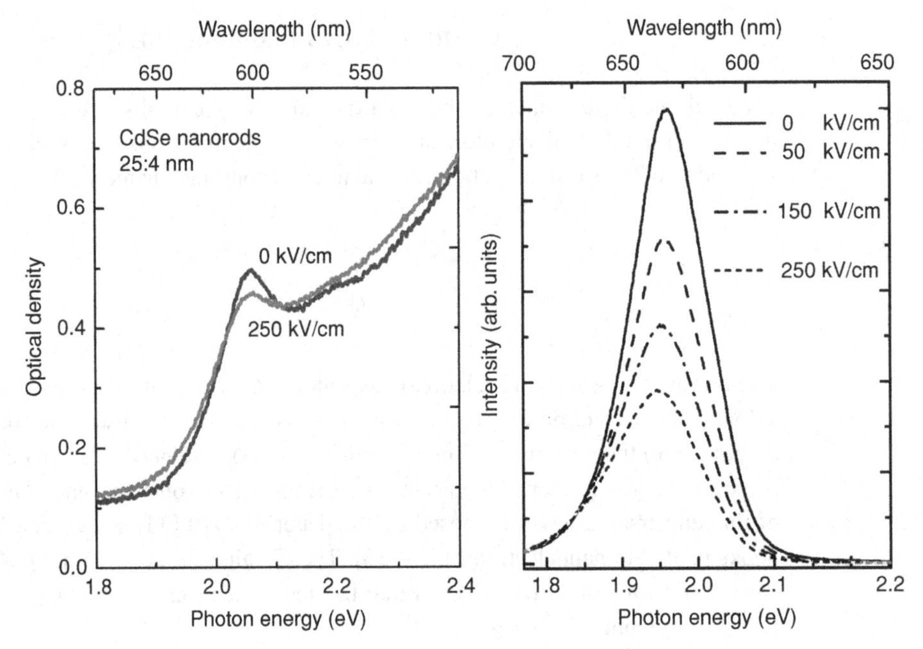

Fig. 4.20 **Optical properties of CdSe nanorods with length 25 nm and diameter 4 nm in a polymer matrix. Left: Absorption spectrum with and without an external electric field applied. Right: Photoluminescence emission spectra of nanorods for various applied voltages.**

Quantum dots

Quantum dots in the form of semiconductor nanocrystals in various matrices, or semiconductor pyramids and islands on crystalline substrates have become the subject of extensive theoretical and experimental investigations and a number of applications. Lasers, light-emitting devices and other photonic components have been proposed. The properties of quantum dots have been reviewed in a number of papers and books [11, 12, 42–44]. Quantum dots will be the subject of extensive discussion in Chapter 5.

Problems

1. Consider modification of the band gap energy for a number of semiconductor and dielectric compounds (Table 4.1.) within the series: CdTe→CdSe→CdS; ZnTe→ZnSe→ZnS; HgTe→CdTe→ZnTe; InSb→InAs; InAs →GaAs.

Try to predict band gaps for HgS, HgSe, PbTe and CuI based on evaluated correlations.

2. Consider and summarize similarities and differences between excitons and a hydrogen atom.

3. Consider expressions for the Bohr radius and the ionization energy for an impurity atom versus those for excitons and explain similarities and differences. Evaluate and compare absolute values.

4. Find the factor of h^{-d} (where h is the Planck constant and d is the space dimensionality) in the expression for the electron density of states for various dimensionalities and explain, based on Planck's ideas on an elementary cell in phase space. See Section 2.4 for detail.

5. Try to solve the one-dimensional Schrödinger equation with Coulomb-like potential $U = -e^2/x$ and evaluate the $1d$-exciton spectrum.

6. Evaluate the dimensions of the conductivity quantum (Eq. (4.56)) and show that it is actually Ohm^{-1}.

7. Estimate the quantum confinement effect on the optical absorption spectrum of CdSe nanorods in Figure 4.20, comparing with the band gap energy of the parent bulk crystal.

8. Compare the electric field effect on absorption spectra of quantum wells and nanorods. Evaluate the difference.

References

[1] J. S. Blakemore. *Solid State Physics* (Cambridge: Cambridge University Press, 1985).
[2] Ch. Kittel. *Introduction to Solid State Physics* (New York: Wiley, 1976).
[3] A. S. Davydov. *Solid State Theory* (Moscow: Nauka, 1976).
[4] V. P. Gribkovskii. *Theory of Light Absorption and Emission in Semiconductors* (Minsk: Nauka i Tekhnika, 1975) – in Russian.
[5] C. Klingshirn. *Semiconductor Optics* (Berlin: Springer, 1995).
[6] N. Peyghambarian, S. W. Koch and A. Mysyrowicz. *Introduction to Semiconductor Optics* (Englewood Cliffs: Prentice Hall, 1993).

[7] P. Y. Yu and M. Cardona. *Fundamentals of Semiconductor Optics* (Berlin: Springer, 1996).
[8] G. Bastard. *Wave Mechanics Applied to Semiconductor Heterostructures* (Paris: Les Editions de Physique, 1988).
[9] C. Weissbuch and B. Vinter. *Quantum Semiconductor Structures* (New York: Academic Press, 1991).
[10] H. T. Grahn. *Semiconductor Superlattices* (Singapore: World Scientific, 1995).
[11] U. Woggon. *Optical Properties of Semiconductor Quantum Dots* (Berlin: Springer, 1997).
[12] S. V. Gaponenko. *Optical Properties of Semiconductor Nanocrystals* (Cambridge: Cambridge University Press, 1998).
[13] V. V. Mitin, V. A. Kochelap and M. A. Stroscio. *Quantum Heterostructures* (Cambridge: Cambridge University Press, 1999).
[14] G. Floquet. Sur les équation différentielles linéaries à coefficients périodiques. *Ann. Ecole. Norm. Sup.*, **12** (1883), 47–88.
[15] F. Bloch. Uber die quantenmechanik der electronen in kristallgittern. *Zs. Physik,* **52** (1928), 555–600.
[16] R. de Kronig and W. G. Penney. Quantum mechanics of electrons in crystal lattices. *Proc. Royal Soc. London*, **130A** (1931), 499–512.
[17] O. Madelung. *Semiconductor Data Handbook* (Berlin: Springer, 2004).
[18] V. P. Gribkovskii, L. G. Zimin, S. V. Gaponenko, I. E. Malinovskii, P. I. Kuznetsov and G. G. Yakushcheva. Optical absorption near excitonic resonance of MOCVD-grown ZnSe single crystals. *Phys. Stat. Sol. (b),* **158** (1990), 359–366.
[19] P. W. Anderson. Absence of electron diffusion in certain random lattices. *Phys. Rev.*, **109** (1958), 1492–1505.
[20] J. L. Pan. Oscillator strengths for optical dipole interband transitions in semiconductor quantum dots. *Phys. Rev. B,* **46** (1992), 4009–4019.
[21] Zh. I. Alferov. The double heterostructure: concept and its applications in physics, electronics and technology. *Nobel Lecture* (Nobel Foundation, 2000). In *Nobel Lectures in Physics 1996–2000*, edited by G. Ekspong (Singapore: World Scientific, 2002), pp. 413–441.
[22] H. Kroemer. *Nobel Lecture* (Nobel Foundation, 2000). In *Nobel Lectures in Physics 1996–2000*, edited by G. Ekspong (Singapore: World Scientific, 2002), pp. 442–469.
[23] M. Shinada and S. Sugano. Interband optical transitions in extremely anisotropic semiconductors. I. Bound and unbound exciton absorption. *J. Phys. Soc. Japan,* **21** (1966), 1936–1946.
[24] D. A. B. Miller, D. S. Chemla, T. C. Damen, A. C. Gossard, W. Wiegmann, T. H. Wood and C. A. Burrus. Electric field dependence of optical absorption near the bandgap of quantum well structures. *Phys. Rev. B*, **32** (1985), 1043–1060.
[25] Y.-H. Kuo, Y. Lee, Y. Ge, S. Ren, J. E. Roth, T. I. Kamins, D. A. B. Miller and J. S. Harris. Strong quantum-confined Stark effect in germanium quantum-well structures on silicon. *Nature*, **437** (2005), 1334–1336.

[26] J. Liu, D. Pan, S. Jongthammanurak, K. Wada, L. C. Kimerling and J. Michel. Design of monolithically integrated GeSi electro-absorption modulators and photodetectors on an SOI platform. *Opt. Express*, **15** (2007), 623–628.

[27] J. E. Roth, O. Fidaner, R. K. Schaevitz, Yu-Hsuan Kuo, Th. I. Kamins, J. S. Harris and D. A. B. Miller. Optical modulator on silicon employing germanium quantum wells. *Optics Express,* **15** (2007), 5851–5860.

[28] P. S. Zory (Ed.). *Quantum Well Lasers* (New York: Academic Press, 1993).

[29] W. W. Chow and S. W. Koch. *Semiconductor-Laser Fundamentals* (Berlin: Springer, 1999).

[30] R. F. Kazarinov and R. A. Suris. Possibility of the amplification of electromagnetic waves in a semiconductor with a superlattice. *Sov. Phys. Semiconductors,* **5** (1971), 707–709.

[31] R. Kehler *et al.* Terahertz semiconductor-heterostructure laser. *Nature*, **417** (2002), 156–159.

[32] B. S. Williams. Terahertz quantum-cascade lasers. *Nature Photonics,* **1** (2007), 517–527.

[33] R. Landauer. Electrical resistance of disordered one-dimensional lattices. *Phil. Mag.*, **21** (1970) 863–872.

[34] J. H. Davies. *The Physics of Low-Dimensional Semiconductors* (Cambridge: Cambridge University Press, 1998).

[35] Y. Imry. *Introduction to Mesoscopic Physics* (Oxford: Oxford University Press, 2002).

[36] V. V. Mitin, V. A. Kochelap and M. A. Stroscio. *Introduction to Nanoelectronics* (Cambridge: Cambridge University Press, 2007).

[37] D. V. Talapin, E. V. Shevchenko, C. B. Murray, A. Kornowski, S. Förster and H. Weller. CdSe and CdSe/CdS nanorod solids. *J. Am. Chem. Soc.*, **126** (2004), 12984–12988.

[38] N. Le Thomas, E. Herz, O. Schöps, U. Woggon and M. V. Artemyev. Exciton fine structure in single CdSe nanorods. *Phys. Rev. Lett.*, **94** (2005), 016803.

[39] A. Fasoli, A. Colli, S. Kudera, L. Manna, S. Hofmann, C. Ducati, J. Robertson and A. C. Ferrari. Catalytic and seeded shape-selective synthesis of II–VI semiconductor nanowires. *Physica E,* **37** (2007), 138–141.

[40] L. I. Gurinovich, A. A. Lutich, A. P. Stupak, S. Ya. Prislopsky, E. K. Rusakov, M. V. Artemyev, S. V. Gaponenko and H. V. Demir. Luminescence of cadmium selenide quantum dots and nanorods in the external electric field. *Semiconductors,* **43** (2009), 1045–1053.

[41] A. A. Lutich, L. Manna, V. A. Sokol, S. V. Volchek and S. V. Gaponenko. Alignment of semiconductor nanorods using nanoporous alumina templates. *Physica Status Solidis RRL* **3** (2009), 151–153.

[42] U. Woggon and S. V. Gaponenko. Excitons in quantum dots. *Phys. Stat. Sol. (b)*, **189** (1995), 286–343.

[43] D. Bimberg, N. N. Ledentsov and M. Grundmann. *Quantum Dot Heterostructures.* (Chichester: John Wiley and Sons, 1999).

[44] S. V. Gaponenko, H. Kalt and U. Woggon. *Semiconductor Quantum Structures. Part 2. Optical Properties* (Berlin: Springer Verlag, 2004).

5 Semiconductor nanocrystals (quantum dots)

From the standpoint of a solid-state physicist, nanocrystals are nothing else but a kind of low-dimensional structure complementary to quantum wells (two-dimensional structures) and quantum wires (one-dimensional structures). However, nanocrystals have a number of specific features that are not inherent in the two- and one-dimensional structures. Quantum wells and quantum wires still possess a translational symmetry in two or one dimensions, and a statistically large number of electronic excitations can be created. In nanocrystals, the translational symmetry is totally broken and only a finite number of electrons and holes can be created within the same nanocrystal. Therefore, the concepts of the electron–hole gas and quasi-momentum fail in nanocrystals.

From the viewpoint of molecular physics, nanocrystals can be considered as a kind of large molecule. Similar to molecular ensembles, nanocrystals dispersed in a transparent host environment (liquid or solid) exhibit a variety of guest–host phenomena known for molecular structures. Moreover, every nanocrystal ensemble has inhomogeneously broadened absorption and emission spectra due to distribution of sizes, defect concentration, shape fluctuations, environmental inhomogeneities and other features. Therefore, the most efficient way to examine the properties of a single nanocrystal which are smeared by inhomogeneous broadening is to use selective techniques, including single nanoparticle luminescence spectroscopy. The present chapter summarizes the principal quantum confinement effects on absorption and emission of light, recombination dynamics and many-body phenomena in semiconductor nanocrystals, as well as primary application issues. Since pioneering papers on quantum confinement effects in semiconductor nanocrystals by Ekimov, and co-workers [1, 2], Efros and Efros [3], and Brus [4, 5], great progress has been achieved in the field due to extensive studies performed by thousands of researchers across the world. The theory and photophysics of semiconductor nanocrystals have been described thoroughly in a number of books [6–12] and reviews [13–20]. In addition to these publications, we shall consider a number of possible applications based on the most recent advances.

5.1 From atom to crystal

Let us discuss the size ranges for semiconductor nanocrystals related to different models and approaches which provide an adequate description of their optical properties (Fig. 5.1). The characteristic length parameters involved in such a classification are the crystal lattice constant, a_L, the exciton Bohr radius, a_B^*, and the photon wavelength λ corresponding to the lowest optical transition.

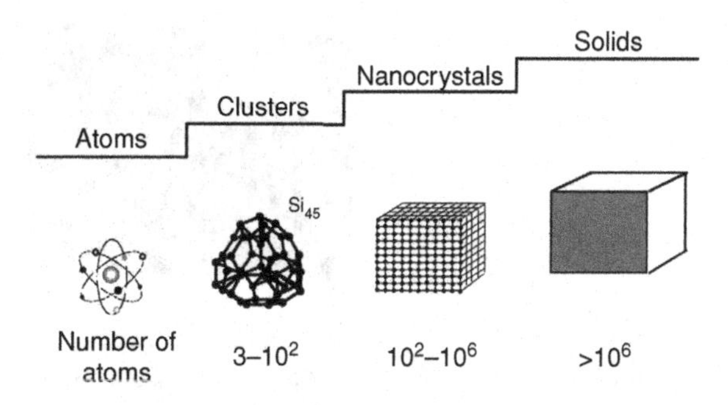

Fig. 5.1 **Evolution of matter from a single atom to bulk solid.**

First of all, if the size a of a nanocrystal is comparable to a_L, i.e. close to 1 nanometer, an adequate description can be provided only in terms of the quantum-chemical (molecular) approach with the specific number of atoms and configuration taken into account. This size interval can be classified as a *cluster range*. The number of atoms in a cluster ranges from several to a few hundreds. The main distinctive feature of clusters is the absence of a monotonic dependence of the optical transition energies and probabilities versus number of atoms. The size as a characteristic of clusters is by no means justified. A subrange can be outlined corresponding to surface clusters in which every atom belongs to the surface. This type of cluster features enhanced chemical activity.

In the case when $a_L \ll a \ll \lambda$ (the signs $\ll$ and $\gg$ imply here at least a few times), a nanocrystal can be treated as a quantum box for electron excitations. The solid-state approach in terms of the effective mass approximation predicts a monotonic evolution of the optical properties of nanocrystals with size towards the properties of the parent bulk single crystals. The exciton Bohr radius value, a_B^*, divides this range into two subranges, $a \ll a_B^*$ and $a \gg a_B^*$, providing an interpretation of size-dependent properties in terms of either electron and hole, or a hydrogen-like exciton confined motion. A number of analytical relations describing size-dependent properties of nanocrystals have been derived for these cases. Generally, throughout this range, the concept of an exciton as an interacting electron–hole pair is relevant, and numerical calculations provide evidence of monotonic optical transition energies and probabilities behavior without any discontinuity around $a = a_B^*$. Remarkable is the fact that for smaller sizes the results provided by the solid-state and the molecular approaches do converge. This range can be classified as the *quantum dot range.* The very notation "quantum dot" implies modeling and interpretation of electronic and optical properties of nanocrystals in terms of electron confinement phenomena. The number of atoms for this range is approximately from a couple of hundreds to one million. A subrange can be identified within this range for smaller sizes where shell-like structure can be assigned to a nanocrystal in terms that every additional atomic shell discretely shifts optical spectra and electron energies. Pronounced crystallographic structures can be identified for particles as small as 3 nm (Fig. 5.2) whose size counts 5–6 lattice periods.

As the size of a nanocrystal reaches a value of the order of 100 nm, it becomes comparable to the photon wavelength relevant to resonant optical transitions. This circumstance brings

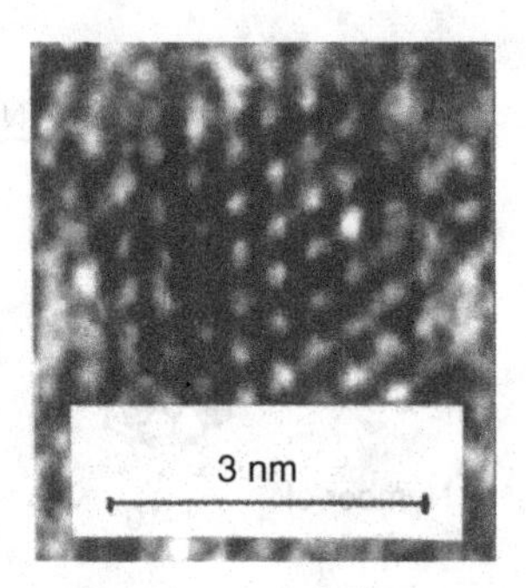

Fig. 5.2 Scanning electron microscope image of a single CdTe nanocrystal [21].

about a number of important aspects. First of all, light scattering should be explicitly considered. Second, a concept of *photon confinement* should be introduced. A nanocrystal in this case should be treated as a *microcavity* possessing a definite number of electromagnetic modes. Light absorption and emission in this case will be strongly influenced by *exciton–photon coupling*. In the bulk crystal exciton–photon coupling is described in terms of the new quasiparticle, a *polariton*. The concept of a polariton leads to a description of light propagation in terms of polariton creation and annihilation at the front and the rear boundary of the sample, respectively, whereas photon absorption is interpreted in terms of polariton scattering by lattice imperfections (phonons and impurities). This size range of nanocrystals is less investigated. It can be referred to as the *polaritonic range*. A number of interesting phenomena can be foreseen for this range due to exciton–photon coupling under the conditions of photon confinement.

5.2 Particle-in-a-box theory of electron–hole states

The principal quantum confinement effects on electron and hole states in semiconductor nanocrystals and relevant optical transitions can be revealed within the framework of the effective-mass consideration, using the "particle-in-a-box" approach. For a spherical potential box with infinite potential, and electrons and holes with isotropic effective masses, clear physical results and elegant analytical expressions can be derived for the two limiting cases, the so-called "weak confinement" and "strong confinement" limits, proposed by Efros and Efros [3].

Weak confinement regime

In larger quantum dots, when the dot radius, a, is small but still a few times larger than the exciton Bohr radius, a_B^*, quantization of the exciton center-of-mass motion occurs. Starting from the dispersion law of an exciton in a crystal (Chapter 4, Eq. (4.30)), we have to replace the kinetic energy of a free exciton by a solution relevant to a particle in a spherical box (see Chapter 2, Eq. (2.98)). The energy of an exciton is then expressed in the form,

$$E_{nml} = E_g - \frac{\mathrm{Ry}^*}{n^2} + \frac{\hbar^2 \chi_{ml}^2}{2Ma^2}, \quad n, m, l = 1, 2, 3, \ldots \tag{5.1}$$

with the roots of the Bessel function χ_{ml} (Table 2.1 and Fig. 2.7). Here $M = m_e^* + m_h^*$ is the exciton mass which equals the sum of the electron m_e^* and hole m_h^* effective masses. The reader is requested to reveal the generic relation between Eq. (5.1) and Eq. (2.98) by themselves (Problem 1). One can see an exciton in a spherical quantum dot is characterized by the quantum number n describing its *internal* states arising from the Coulomb electron–hole interaction (1S; 2S, 2P; 3S, 3P, 3D; ...), and by the two additional numbers, m and l, describing the states connected with the center-of-mass motion in the presence of the *external* potential barrier featuring spherical symmetry (1s, 1p, 1d..., 2s, 2p, 2d..., etc.). Here S(s), P(p), D(d), F(f), ... letters label states with $l = 0, 1, 2, 3, \ldots,$ respectively as it is adopted in atomic spectroscopy. To distinguish the "internal" and the "external" states, we shall use capital letters for the former and lower case for the latter.

For the lowest 1S1s state ($n = 1$, $m = 1$, $l = 0$) the energy is expressed as,

$$E_{1S1s} = E_g - \mathrm{Ry}^* + \frac{\pi^2 \hbar^2}{2Ma^2}, \tag{5.2}$$

or in another way,

$$E_{1S1s} = E_g - \mathrm{Ry}^* \left[1 - \frac{\mu}{M} \left(\frac{\pi a_B^*}{a} \right)^2 \right], \tag{5.3}$$

where μ is the electron–hole reduced mass $\mu = m_e^* m_h^* / (m_e^* + m_h^*)$. In Eqs. (5.2) and (5.3) the value $\chi_{10} = \pi$ was used. Therefore the first exciton resonance in a spherical quantum dot experiences a high-energy shift by the value,

$$\Delta E_{1S1s} = \frac{\mu}{M} \left(\frac{\pi a_B^*}{a} \right)^2 \mathrm{Ry}^*. \tag{5.4}$$

Note this value remains small as compared to Ry* since we consider the case $a \gg a_B^*$. This is the quantitative justification of the term "weak confinement".

Taking into account that photon absorption can create an exciton with zero angular momentum only, the absorption spectrum will consist of a number of lines corresponding to states with $l = 0$. Then the absorption spectrum can be derived from Eq. (5.1) with $\chi_{m0} = \pi m$, (see Table 2.1) i.e.

$$E_{nm} = E_g - \frac{\mathrm{Ry}^*}{n^2} + \frac{\hbar^2 \pi^2}{2Ma^2} m^2, \quad n, m = 1, 2, 3, \ldots. \tag{5.5}$$

A weak confinement regime is feasible in wide-band semiconductors of I–VII compounds (copper halides) featuring a small exciton Bohr radius and large exciton Rydberg energy (see Table 4.2). Such nanocrystals are incorporated in certain commercial photochromic glasses [1]. Typical size-dependent optical spectra for this case are shown in Figure 5.3.

Copper chloride (CuCl) monocrystals have two exciton bands (Z_3 and Z_{12}) related to different valence band branches and a flat interband absorption for $\hbar\omega \geq E_g$. The exciton Rydberg energy is $\mathrm{Ry}^* = 200$ meV, and the exciton Bohr radius is $a_B^* = 0.7$ nm. Large

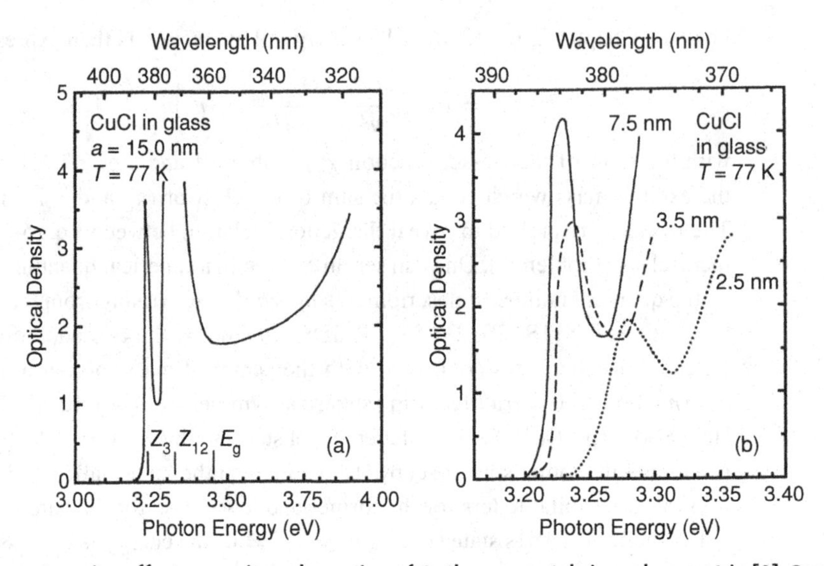

Fig. 5.3 **The quantum size effect on exciton absorption of CuCl nanocrystals in a glass matrix [8]. Panel (a) shows an absorption spectrum of larger nanocrystals ($a = 15$ nm) in a wide spectral range. The sharp increase in absorption at $\lambda < 340$ nm is due to glass matrix absorption. Panel (b) shows in detail the Z_3-exciton absorption band for $a = 7.5, 3.5, 2.5$ nm.**

nanocrystals exhibit basically the same features as bulk crystals (Fig. 5.3a). Note, the sharp increase in absorption at $\lambda < 340$ nm is due to glass matrix absorption. With decreasing size, the exciton band shows a monotonic blue shift (Fig. 5.3b). As the absolute value of the shift remains considerably smaller than the exciton Rydberg energy, $\text{Ry}^* = 200$ meV, this is a typical example of the quantum-size effect in a weak confinement limit. Note the complex shape of the exciton band in Figure 5.3b. It was shown to be inhomogeneously broadened because of the size dispersion for a statistically large number of nanocrystals contributing to the absorption spectrum in the experiment.

The experimental dependence of the first absorption maximum on particle mean size $\bar{a}$ (Fig. 5.4) features $\bar{a}^{-2}$ behavior which is typical of a "particle-in-a-box" model. The inevitable size distribution of nanocrystals results in a modified $E(\bar{a})$ dependence differing from the exact analytical solution for a single particle in a box by a numerical coefficient A to be added to Eq. (5.4) and to the third term in Eq. (5.5). The coefficient is specific for every distribution function. In the case of the asymptotic Lifshitz–Slyozov distribution which is characterisctic of diffusion-limited aggregation in a glass matrix, $A = 0.67$ holds.

Strong confinement limit

The strong confinement limit corresponds to the condition $a \ll a_B^*$ in which the sign $\ll$ in a real experimental situation means "several times smaller". That means that the confined electron and hole have no bound state corresponding to the hydrogen-like exciton. Supposing non-interacting between an electron and a hole as a reasonable starting approximation,

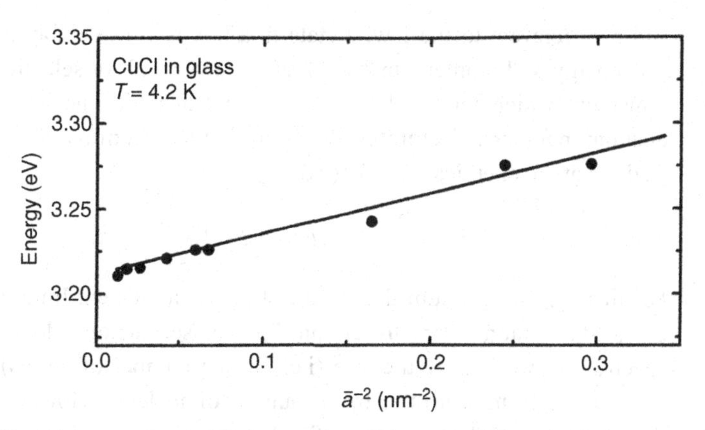

Fig. 5.4 The size dependence of the Z_3-exciton band for CuCl nanocrystals in a glass matrix [8, 22]. The dots are experimental data, the curve corresponds to Eq. (5.5) with an additional coefficient for the third term accounting for the size distribution function.

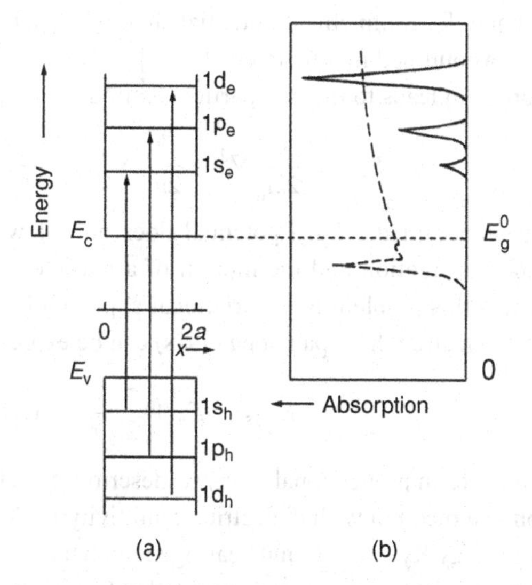

Fig. 5.5 The optical properties of an ideal spherical quantum dot with isotropic scalar effective masses of a non-interacting electron and hole. (a) A sketch of electron–hole energy levels, optical transitions and (b) reduction of original bulk absorption spectrum (dashed curve) to discrete bands (solid curves).

we consider quantization of an electron and a hole motion separately and apply the results of Section 2.6 (Eq. 2.98) independently to an electron and a hole. The "free" electron and hole in a spherical potential box have the energy spectra,

$$E^{\mathrm{e}}_{ml} = E_{\mathrm{g}} + \frac{\hbar^2 \chi^2_{ml}}{2m^*_{\mathrm{e}} a^2}, \; E^{\mathrm{h}}_{ml} = -\frac{\hbar^2 \chi^2_{ml}}{2m^*_{\mathrm{h}} a^2}, \tag{5.6}$$

if we consider the top of the valence band E_{V} as the origin of the energy scale, i.e. $E_{\mathrm{V}} = 0$. These energy levels are shown in Figure 5.5a. The zero-point kinetic energy of the electron and hole relevant to the lowest state in a box is considerably larger than the Ry* value. The energy and momentum conservation laws result in selection rules that allow optical transitions which couple electron and hole states with the same principal n and orbital l quantum numbers. Therefore, the absorption spectrum reduces to a set of discrete bands peaking at the energies (Fig. 5.5(b)),

$$E_{nl} = E_g + \frac{\hbar^2}{2\mu a^2}\chi_{nl}^2. \tag{5.7}$$

For this reason, quantum dots in the strong confinement limit are sometimes referred to as "artificial atoms" or "hyperatoms", because quantum dots exhibit a discrete optical spectrum controlled by the size (i.e., by the number of atoms), whereas an atom has a discrete spectrum controlled by the number of nucleons. However, one should bear in mind that an electron and a hole are confined in space comparable to, or even more compact than, spatial extension of the exciton ground state in the ideal infinite parent crystal. Therefore, an independent treatment of the electron and hole motion is by no means justified, and the problem including the two-particle Hamiltonian with the two kinetic energy terms, Coulomb potential and the confinement potential should be examined. This important aspect of the problem was outlined first by Brus [23, 24].

The problem leads to the two-particle Schrödinger equation with Hamiltonian,

$$\mathbf{H} = -\frac{\hbar^2}{2m_e^*}\nabla_e^2 - \frac{\hbar^2}{2m_h^*}\nabla_h^2 - \frac{e^2}{\varepsilon\left|\mathbf{r}_e - \mathbf{r}_h\right|} + U(r). \tag{5.8}$$

The appearance of the $U(r)$ potential does not allow independent consideration of the center-of-mass motion and the motion of a particle with reduced mass. Several authors have treated this problem by a variational approach [23–27] and found that the energy of the ground electron–hole pair state (1s1s) can be expressed in a form,

$$E_{1s1s} = E_g + \frac{\pi^2\hbar^2}{2\mu a^2} - 1.786\frac{e^2}{\varepsilon a}, \tag{5.9}$$

in which the term proportional to $e^2/\varepsilon a$ describes the effective Coulomb electron–hole interaction in a medium with dielectric permittivity ε. Comparing this term with the exciton Rydberg energy $\mathrm{Ry}^* = \frac{e^2}{2\varepsilon a_B^*}$ and bearing in mind that our consideration is still for the strong confinement limit ($a \ll a_B^*$), one can see that Coulomb interaction by no means vanishes in small quantum dots. Moreover, the Coulomb-term contribution to the ground-state energy is even greater in its absolute value than in the bulk monocrystal. This is a principal difference of quantum dots as compared to crystals, quantum wells and quantum wires, where the Coulomb energy of a free electron–hole pair is zero. Therefore, an elementary excitation in a quantum dot can be classified as an *exciton* with a notation "exciton in quantum dot" [7, 8, 13]. Within the framework of this agreement we shall use the term "exciton" throughout this chapter even when the relevant state does not obey the hydrogen-like model.

The exciton lowest state energy measured as a deviation from the bulk band gap energy E_g in the strong confinement limit can be written in a more general way as a raw expansion,

$$E_{1s1s} - E_g = \left(\frac{a_B^*}{a}\right)^2 \mathrm{Ry}^* \left[A_1 + \frac{a}{a_B^*}A_2 + \left(\frac{a}{a_B^*}\right)^2 A_3 + \cdots\right], \tag{5.10}$$

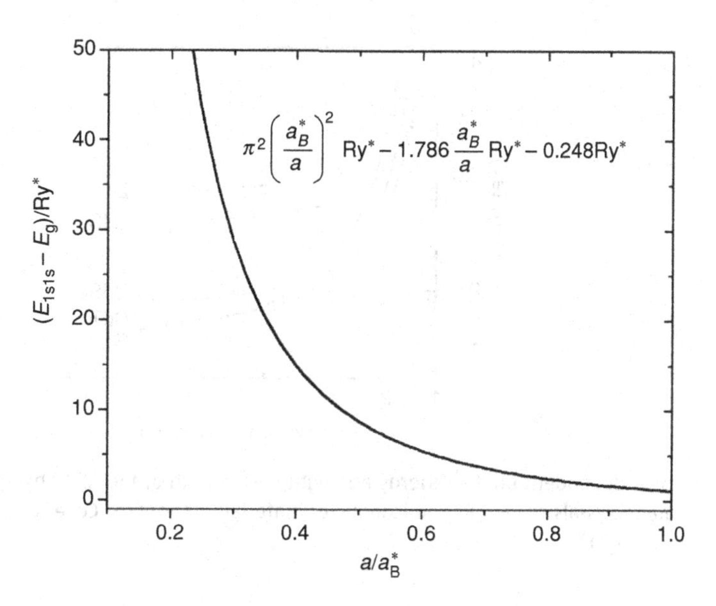

Fig. 5.6 **The universal law in terms of dimensionless energy and radius units describing the size dependence of the first optical resonance in a quantum dot (Eq. 5.11).**

with the small parameter $a/a_B \ll 1$. The first coefficient A_1 for various states is described by the roots of the Bessel function (see Table 2.1). The second coefficient A_2 corresponds to the Coulomb term and takes the following values: $A_2 = -1.786$ for the 1s1s-state, $A_2 = -1.884$ for the 1p1p-state, and values between -1.6 and -1.8 for other configurations [26]. The A_3 coefficient for the 1s1s-state was found to be $A_3 = -0.248$ [25]. Summarizing the findings relevant to the ground state, we can write the energy of the first absorption peak as follows,

$$E_{1s1s} = E_g + \pi^2\left(\frac{a_B^*}{a}\right)^2 Ry^* - 1.786\frac{a_B^*}{a}Ry^* - 0.248Ry^*. \tag{5.11}$$

Generally terms in the right part of Eq. (5.11) successively reduce in absolute value, though for small nanocrystals of narrow band semiconductors with small electron effective mass (PbSe, PbTe, PbS, HgTe, HgSe, HgS, GaAs, InSb) the second term describing the sum of the kinetic energies of an electron and a hole can be comparable to, or even greater than the original band gap energy E_g of the parent bulk crystal. Presentation of the optical absorption shift energy E_{1s1s} in terms of a dimensionless dot radius a/a_B^* manifests as the universal material-independent law for size-dependent optical absorption if the photon energy is measured in dimensionless units E_{1s1s}/Ry^*. This law is plotted in Figure 5.6. It appears to be rather instructive in many cases for the prompt estimation of the quantum confinement effect on optical absorption and color of nanocrystalline materials.

The size dependence of the E_{1s1s} energy is plotted in Figure 5.7 for a number of semiconductor materials. One can see, materials with a narrow band gap (GaAs, CdTe, CdSe) promise wide-band tunability of the absorption edge through hundreds of nanometers i.e. the width of the whole visible range of the electromagnetic wave spectrum. The small size

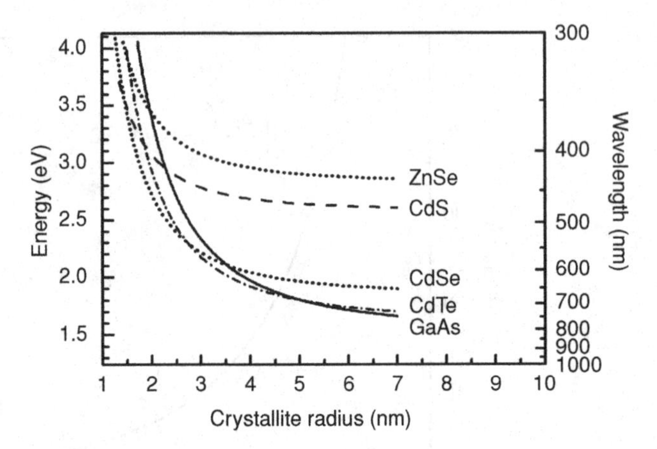

Fig. 5.7 **Size-dependent photon energy and light wavelength of the first absorption resonance for nanocrystals of various semiconductor materials (ZnSe, CdS, CdSe, CdTe, GaAs) calculated using Eq. (5.11).**

cut-off ($a = 1.5$ nm) in the plots corresponds approximately to the reasonable applicability of the "particle-in-a-box" model.

It should be noted that the above consideration in terms of either exciton center-of-mass motion quantization, or electron–hole motion quantization does not mean any fundamental physical effect or discontinuity when the size of a dot moves around $a = a_B^*$. A smooth evolution of quantum dot properties from crystal-like to cluster-like features occurs and this can be successfully proven within the framework of the effective-mass approximation by means of explicit numerical solution of the Schrödinger problem with Hamiltonian (5.8).

Further steps from the "particle-in-a-box" models toward real semiconductor nanocrystals take into account a complicated *valence band structure*, *surface polarization* effects and *finite barrier* effects. The valence band structure of real semiconductors brings about a multitude of hole states and modifies the selection rules for dipole-allowed transitions. Surface polarization effects arise from a dielectric medium with a dielectric constant ε_2, normally being less than that of the semiconductor nanocrystal ε_1. The finite barrier height results in a general energy lowering than can be foreseen on the basis of elementary quantum mechanics.

Generally, even though straightforward quantum mechnical calculation is not applicable for extremely small crystallites and clusters, it gives very clear and instructive results on size-dependent optical properties of matter.

5.3 Quantum chemical theory

Real nanometer-sized semiconductor crystallites, if treated correctly, should be described like large molecules. This means that a particular number of atoms and specific spatial configurations should be involved rather than the size. The importance of such an approach

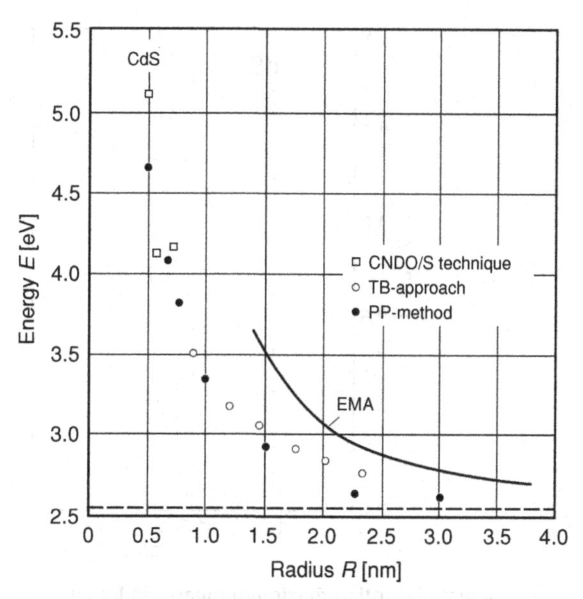

Fig. 5.8 **Calculated energy of the first allowed optical transition versus crystallite radius for CdS clusters/dots [8,12]. (□) CNDO/S-technique; (○) tight-binding approach; (•) pseudopotential method; solid line – particle-in-a-box model (Eq. (5.11)). The dashed line shows the energy of the first exciton resonance in the bulk crystal.**

becomes crucial for smaller crystallites consisting of less than 100 atoms. In this case, the properties of semiconductor particles should be deduced from the properties of individual atoms rather than crystals. Therefore, we have to deal with specific types of clusters which can be examined using molecular quantum mechanics, often referred to as *quantum chemistry*. Quantum-chemical consideration provides an opportunity to reveal the development of crystal-like properties starting from the atomic and molecular level. Quantum-chemical calculations for semiconductor clusters have been performed by many groups using various techniques. Unlike the "particle-in-a-box" approach, quantum-chemical consideration does not lead to any unified dependence and each specific cluster configuration and composition has to be examined in detail. In Figure 5.8 a summary is presented of data for CdS-based clusters [28–30] and dots evaluated by means of both the solid-state and molecular approaches. One can see a clear agreement of the quantum-chemical and pseudopotential calculations for smaller sizes and a convergence of the results from effective mass approximation (EMA) and the numerical calculations for larger sizes. A discrepancy between the EMA results and the data provided by means of other techniques for smaller particles can be reduced if a finite potential barrier is introduced in the "particle-in-a-box" problem.

Figure 5.9 shows a sketch of the data reported for CdS nanocrystals and clusters in various matrices [31–34]. Experimental data demonstrate reasonable convergence of results obtained by different groups as well as satisfactory agreement of experimental findings with theoretical predictions. When comparing experimental data with theory one should bear in mind guest–host interface effects, size distribution and the resulting inhomogeneous broadening of the absorption spectrum as well as certain problems in the precise determination

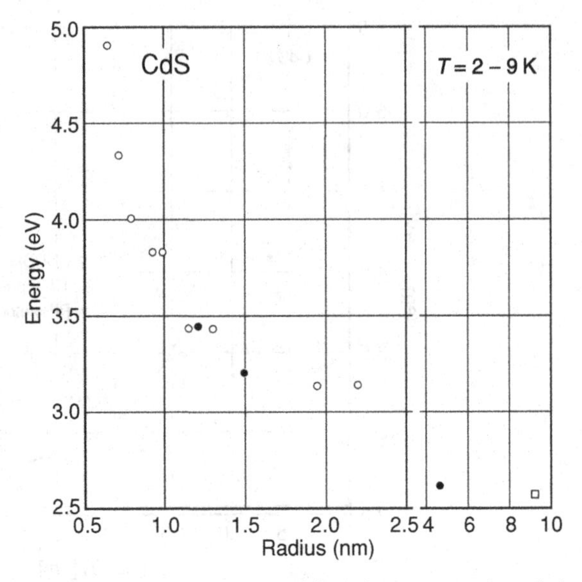

Fig. 5.9 **Energy of the first absorption maximum measured for CdS nanocrystals and clusters at temperature 2–9 K in different matrices. Adapted from [12].**

of the mean nanocrystal/cluster size, including certain divergence of size data obtained by different techniques.

5.4 Synthesis of nanocrystals

There are several approaches to the synthesis of nanocrystals based on different techniques. In what follows these approaches are discussed at the introductory level. For a more thorough description [8, 9, 12, 15] are recommended.

Diffusion-controlled growth in glass matrices

Synthesis of nanometer sized crystallites in a glass matrix by means of diffusion-controlled growth is based on commercial technologies developed for fabrication of color cut-off filters and photochromic glasses. Color cut-off filters produced by "Corning" (USA), "Schott" (Germany), "Rubin" (Russia), and "Hoya" (Japan) are nothing but glasses containing nanometer-sized crystallites of solid solutions of II–VI compounds (mainly CdS_xSe_{1-x} and possibly $Zn_yCd_{1-y}S_xSe_{1-x}$). Empirical methods for the diffusion-controlled growth of semiconductor nanocrystals in a glass matrix have been known over many decades. Commercial photochromic glasses developed much later contain nanocrystals of I–VII compounds (e.g. CuCl, CuBr, AgBr). Typically, silicate or borosilicate matrices are used with

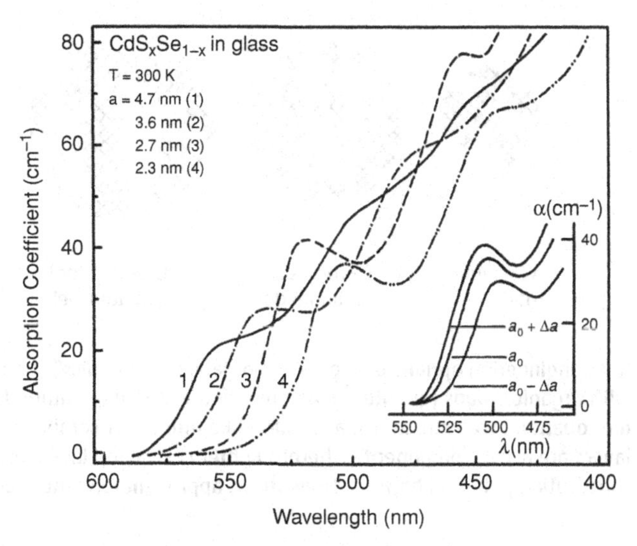

Fig. 5.10 **Absorption spectra of CdS_xSe_{1-x} nanocrystals in a glass matrix at room temperature [35]. Insert shows absorption spectra for small deviations of mean size around $\overline{a}_0 = 2.3$ nm. The spectra show a clear blue shift with decreasing mean radius $\overline{a}$ due to the quantum confinement effect. The absolute values of the shift is one order of magnitude larger than the exciton Rydberg energy which is indicative of the strong confinement regime. For smaller $\overline{a}$ values the shift of the first band with growing $\overline{a}$ systematically results in an increase in the total absorption (see data for $\overline{a} = 2.3, 2.7$ and the insert) which is indicative of the normal growth stage. At this stage, mean size growth occurs due to increase in the total volume of the semiconductor phase in the matrix. Therefore, absorption spectra at this stage do not overlap.**

the absorption onset near 4 eV (about 300 nm), thus allowing optical transmission of the semiconductor inclusions to be studied (and utilized) over the whole visible range.

Growth of crystallites occurs due to phase transition in a supersaturated viscous solution. The process is controlled by diffusion of ions dissolved in the matrix and can be performed in the temperature range $T_{\rm glass} < T < T_{\rm melt}$, where $T_{\rm glass}$ is the temperature of the glass transition and $T_{\rm melt}$ is the melting temperature of the matrix. Typically, growth temperatures range between 550 °C and 700 °C depending on the desired size of the crystallites and the matrix composition.

Diffusion-controlled growth from a supersaturated solution can be described in terms of *nucleation, normal growth* and *competitive growth* stages. At the first stage small nuclei are formed. At the second stage, crystallites exhibit a monotonic growth due to atoms jumping across the nucleus–matrix interface. At this stage the supersaturation decreases with time and the total volume of the semiconductor phase monotonically increases. Finally, when the crystallites are large enough and the degree of supersaturation is negligible (i.e. all ions are already incorporated in crystallites) surface tension plays the main role and the growth dynamics features diffusive mass transfer from smaller particles to larger ones. This stage is commonly referred to as "Ostwald ripening", "competitive growth", "diffusion-limited aggregation" or "coalescence". Examples of size-dependent optical properties of laboratory samples grown on the basis of commercial technology are presented in Figure 5.3 (copper halide crystallites) and in Figure 5.10 (cadmium halcogenide solid solutions).

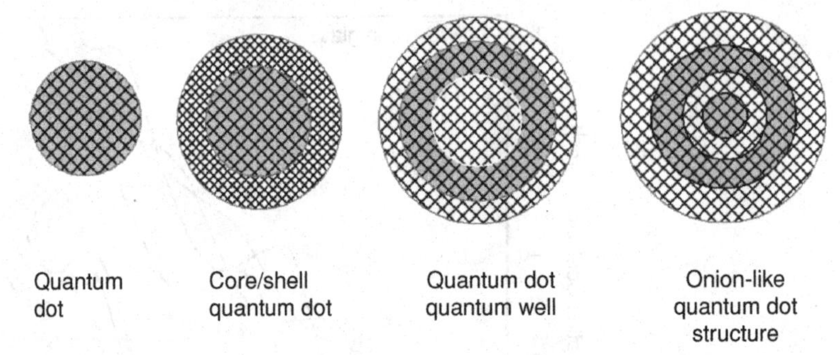

Fig. 5.11 **Nanoengineering options offered by "inorganics-in-organics" chemistry: nanocrystals capped with organic groups (quantum dots), binary core/shell quantum dot structures, ternary quantum dot/quantum well structures, and onion-like multilayer composite systems. Darker circles and layers are active components whereas lighter circles and layers are wider-band-gap materials constituting potential barriers. Lines show, approximately, the atomic planes.**

Semiconductor-doped glasses are widely used as optical commercial filters, laser shutters and can be used as optical modulators and other components of laser systems. The evident advantage is high durability and reliability. However, typically, their luminescence quantum yield is rather low and therefore glasses are not considered in the context of light-emitting applications. Most probably guest–host effects, limited surface control options, a wide size distribution and photoinduced processes constitute drawbacks which prevent development of light-emitting nanocrystalline glassy materials. Strain effects should also be taken into account as undesirable guest–host phenomena because of the different thermal expansion coefficients of matrix and crystallites.

Colloidal nanocrystals in solutions and polymers

This approach can be classified as "inorganics-in-organics" nanotechnology. It offers actual molecular-scale flexible nanoengineering opportunities and promises novel luminophores and fluorescent labels.

Nanometer-sized II–VI crystallites can be developed in an organic environment using a variety of techniques based on organometallic and polymer chemistry. Basic features of structures fabricated in this way can be summarized as follows. A relatively low precipitation temperature (usually not exceeding 200 °C) is favorable to minimize the number of lattice defects. The possibility of capping the crystallite surface with organic groups provides a way to control the surface states. It is possible to obtain isolated clusters or to disperse them in a very thin film. Under certain conditions, an extremely narrow size distribution of clusters can be obtained. Bawendi and co-workers have proposed a basic approach to the synthesis of high-quality luminescent II–VI nanocrystallites in an organic environment which has been followed by many researchers across the world, and has even been introduced into the realm of commercial production [36]. In Figure 5.11 the flexibility of this approach is

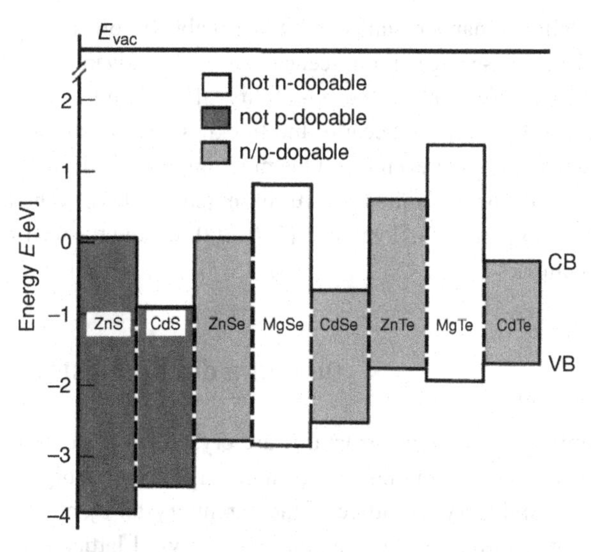

Fig. 5.12 **Band gap positions for several II–VI compounds [12, 41].**

demonstrated. First, colloidal nanocrystalline particles can be developed, typically capped with organic groups over the surface. Second, core/shell structures consisting of a narrow band-gap nanocrystalline core capped with a wide-band-gap semiconductor shell are also feasible. The core crystallite appears to be well isolated from the matrix by chemically "friendly" capping forming a potential barrier of approximately 1 eV. The core/shell system is most popular in research and potential applications related to molecular scale photonic technologies including luminophores, fluorescent labels, biochips and biosensors.

A further example of colloidal nanoengineering is a ternary structure that comprises an active narrow-band-gap shell surrounded by a wide-band-gap core and outer wide-band-gap shell. It can be classified as a "quantum dot/quantum well" since the active middle layer is pulled over a spherical core [37]. Electron–hole states and the resulting optical properties of such a structure are formed by confinement of electrons and holes in a quasi-two-dimensional middle layer and are described in terms of a quantum well with spherical curvature. The spherical dot model described in Section 5.1 is not applicable. Theoretical results for such structures are presented in [20, 38]. Finally, a quantum dot can be combined with a single or multiple concentric spherical layer(s) separated by wide-band-gap shells [39]. Many II–VI nanocrystalline colloidal structures, mainly Cd halcogenides, have been widely developed and examined extensively [7, 8, 12–18, 36] with a mean crystallite radius ranging from 1 to 10 nm. Synthesis of III–V nanocrystals by means of colloidal chemistry can also be performed. Additionally, new techniques in the synthesis of semiconductor nanocrystals are being described. An example is using an electric discharge in water to obtain luminescent ZnO nanocrystals [40]. Figure 5.12 shows the relative positions of band gaps for several representative II–VI compounds to illustrate options for material combinations to form potential barriers in the heterostructures shown in Figure 5.11.

Colloidal nanocrystals exhibit sharp absorption spectra with clearly pronounced bands and an intense edge luminescence. An evident advantage of structures like "semiconductor-in-an-organic-film" is the possibility of applying a strong electric field when studying electric field effects because the thickness of the structure, unlike glasses, can be made down to 10 μm or even less. One more important advantage of this technique as compared, e.g., to diffusion-limited growth in inorganic glass, is the low defect concentration due to a low temperature of synthesis (200–300 °C) and the well-defined and controllable surface structure.

Quantum dot heterostructures

Semiconductor heterostructures are crystalline systems with a continuous crystal lattice structure but discontinuous stoichiometry. They are organized by means of substitution of atoms in the crystal lattice of the parent crystal by other atoms with the same number of valent electrons. To keep a continuous crystal lattice it is necessary that the lattice symmetry and lattice constants of both components coincide. Heterostructures of group IV semiconductors (Si–Ge), III–V (e.g. GaAs–InAs) and II–VI (e.g. ZnS–ZnSe) compounds can be developed by means of molecular beam epitaxy or metal–organic chemical vapor deposition. First proposed several decades ago by Zh. I. Alferov and co-workers, heterostructures have found numerous applications in optoelectronics including light-emitting diodes and lasers [42]. The self-organized growth of quantum dot heterostructures on a crystal substrate due to strain-induced phenomena has been discovered recently [9, 43]. It provides broad opportunities for developing high-quality nanocrystals of the most important industrial semiconductors: III–V compounds, silicon, and germanium. This growth mode takes place in the case of a sub-monolayer heteroepitaxy with a noticeable lattice misfit of a monocrystal substrate with the growing layer. In this case the growing monolayer exhibits coherent growth, i.e. the structure of the layer reproduces the structure of the substrate. This means the monolayer experiences a strong pressure due to the lattice misfit. If this strain is compressive with respect to the monolayer, the latter becomes unstable and tends to a strain–relaxed arrangement. Remarkably, the strain relaxation occurs via the $2d \to 3d$ transition after which the strained monolayer divides into a number of hut-like crystallites. These crystallites have a well-defined pyramidal shape with hexagonal bases (Fig. 5.13). This effect has been found both for metal–organic chemical vapor deposition and for molecular beam epitaxy. The pyramid base typically measures $10 \times 10\,\mathrm{nm}^2$ or more, the height being in the range of 5 to 10 nm. Smaller dots would be hard to develop by this method. The characteristic size of nanocrystals was found to correlate with the misfit value: the larger the misfit the smaller the crystallites. For example, InAs pyramids on GaAs (7% misfit) have typical sizes of 12–20 nm base diameter and heights of 3–6 nm, whereas InP nanocrystals on GaInP (4% misfit) have a base size about 50 nm and heights of the order of 10–20 nm. Under certain conditions self-organization results in a regular 2-dimensional quantum dot array on a crystal surface [44]. Quantum dot heterostructures exhibit high luminescence quantum yield and therefore are promising for applications in light emitting diodes and lasers. Also, Si/Ge nanostructures can be used

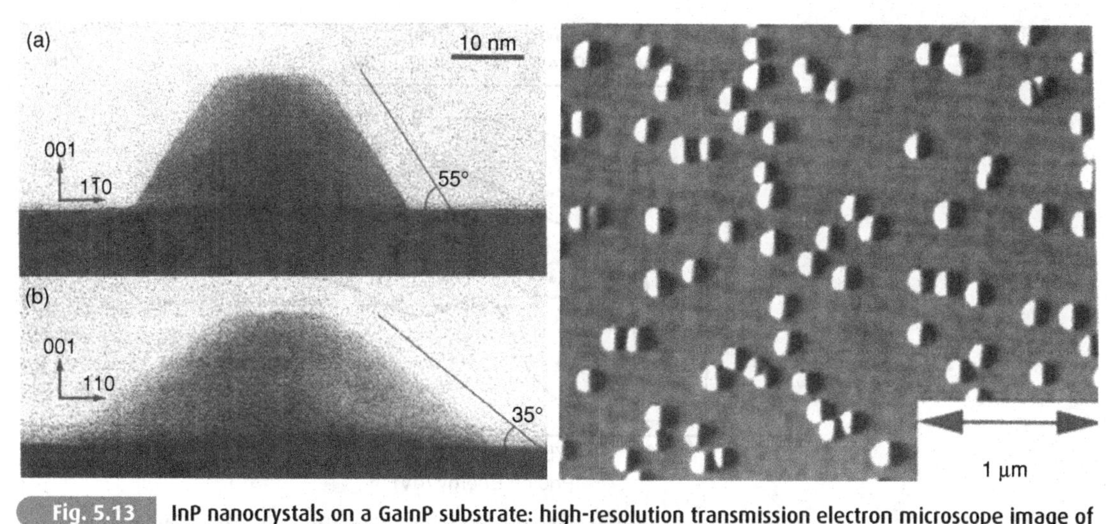

Fig. 5.13 InP nanocrystals on a GaInP substrate: high-resolution transmission electron microscope image of a single nanocrystal for two cross-sections (left) [46], atomic force microscope image on a larger scale (right) [47]. Copyright 1995 AIP[46]. Copyright 1995 Elsevier Science B.V. [47].

to get self-organized Ge quantum dots on Si substrates including formation of regular arrays [45].

5.5 Absorption spectra, electron–hole pair states and many-body effects

Size-dependent optical absorption spectra are well documented for many semiconductor materials in different matrices and on various substrates. A few examples related to a weak confinement regime in I–VII nanocrystals (Fig. 5.3) and to a strong confinement regime in II–VI nanocrystals (Fig. 5.10) in glasses have been shown in previous sections. The II–VI nanocrystals offer an opportunity to trace modification of optical absorption spectra in a wide range of sizes covering smooth evolution from bulk-like to cluster-like behavior. Such an example for CdS nanocrystals is shown in Figure 5.14. Larger nanocrystals (radius about 10 nm) feature pronounced narrow excitonic absorption bands followed by continuous interband absorption (compare to Fig. 4.9). With decreasing size, absorption onset shifts to the short-wavelength side, the absorption spectrum consisting of a number of wide bands resulting from optical transition involving discrete electron and hole energy states. These bands are very wide even at very low temperature, which is just the case in Figure 5.14 where data for $T = 2\,\mathrm{K}$ are shown. A large absorption bandwidth results from inevitable inhomogeneous broadening. It arises, first of all, from the finite size distribution. Further physical mechanisms resulting in inhomogeneous broadening are variations in stoichiometry, shape, surface structure, defect concentration, charge, local environment and others. To evaluate contributions of individual crystallites to the absorption spectrum of an ensemble, the spectrally selective techniques were applied.

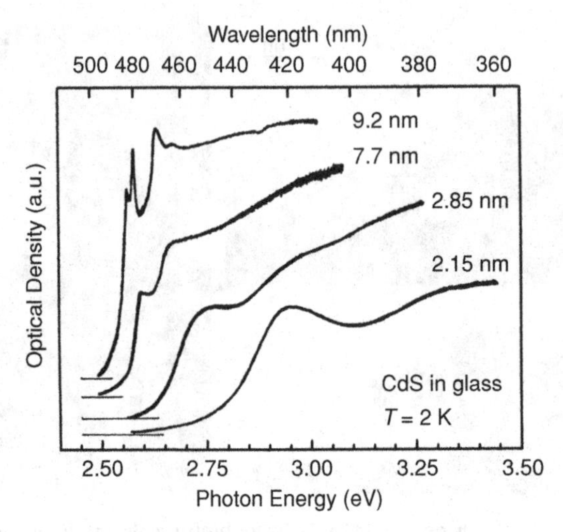

Fig. 5.14 **Absorption spectra of CdS nanocrystals in a glass matrix for four different mean radii** (9.2, 7.7, 2.85, 2.15 nm) **measured at** 2 K **[8, 34].**

In Figures 5.15 and 5.16 the internal structures of a wide absorption band of CdSe nanocrystals in a glass matrix revealed using nonlinear pump-and-probe spectroscopy are shown. In this technique, a strong, narrow-band pump radiation from a tuneable solid state or dye laser selectively saturates the optical transition which is resonant to the pump wavelength, whereas another weak, broad-band radiation is used to probe the optical transmission spectrum of the exited sample in a broader wavelength range. At the limit of lower pump power, the spectral hole "burned" in the absorption spectrum is two times wider (full width at half maximum) than the intrinsic homogeneous width of the resonantly saturated optical absorption band. In this case one bandwidth comes from the intrinsic absorption bandwidth whereas one more bandwidth is added in the course of optical readout of the population of states under investigation. One can learn only from the accuracy of the homogeneous width whether the state probed is populated or not. This technique is discussed in detail, e.g. in [8].

Upon steady growth of pump power, the width and depth of the burned hole exhibits steady growth since absorption of light by nanoparticles which are exactly resonant to pump wavelength appear to be saturated, and further excitation leads to saturation of more and more nanoparticles whose absorption spectrum overlaps with the excitation radiation. Excitation at the long-wave side of the inhomogeneous spectrum (so called "red-edge excitation") typically results in a single hole being burned. However, when the excitation wavelength tunes towards higher photon energy, higher energy electron-hole states can be probed (Fig. 5.16). In these experiments a complex internal structure of the wide absorption band in semiconductor quantum dots was found with a homogeneous spectrum of individual nanocrystals in the form of doublets. These doublets come from the complex valence band structure of II–VI compounds resulting in coupling of every electron state with a pair of hole states.

An example of assignment of optical transitions masked by inhomogeneous absorption is presented in Figure 5.17 where the absorption spectrum of semiconductor doped glass

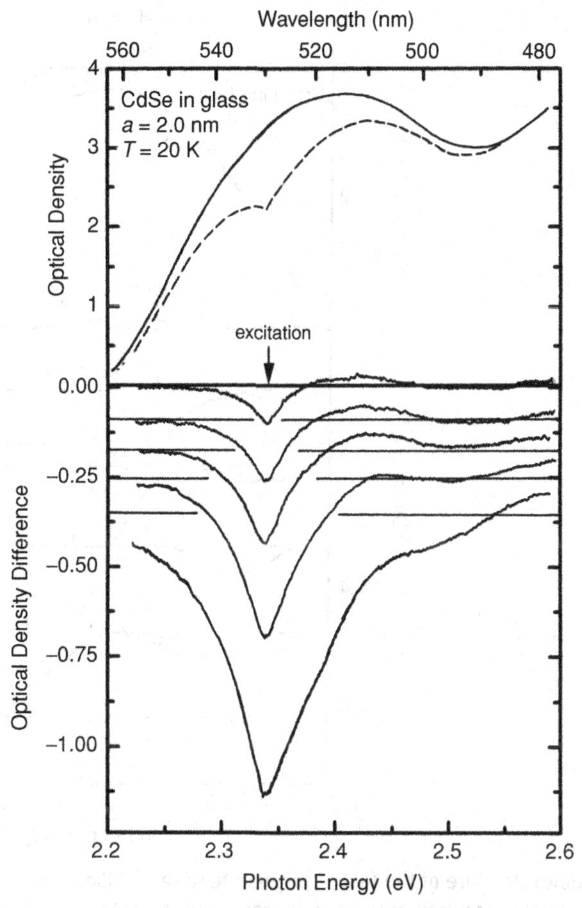

Fig. 5.15 **Spectral hole-burning due to selective absorption saturation in an ensemble of CdSe nanocrystals in a glass matrix [8, 48]. Mean nanocrystal radius is 2.0 nm, temperature is 20 K. The upper panel shows the absorption spectrum of the unexcited sample (solid line) and the spectrum recorded under excitation by a laser pulse (photon energy 2.34 eV, duration is 5 ns, intensity is 7 MW/cm^2, dashed line). The lower panel shows differential absorption spectra for successively growing excitation intensity from 90 kW/cm^2 to 7 MW/cm^2. Negative change in optical density corresponds to bleaching.**

is shown along with spectral position and relative weight of calculated optical transitions based on a "particle-in-a-box" model with the valence band structure of II–VI compounds thoroughly accounted for.

At the initial growth stages of colloidal nanocrystals in solution a peculiar phenomenon has been observed [51]. In the course of CdSe crystallite growth, in time new absorption peaks develop with the previous peaks being stable in their original spectral positions (Fig. 5.18). This observation can be interpreted as a manifestation of certain "magic sizes" inherent in crystallites in the size range close to the cluster regime. These magic sizes can be reasonably assigned to the shell-like atomic structure of smaller nanoparticles. Because

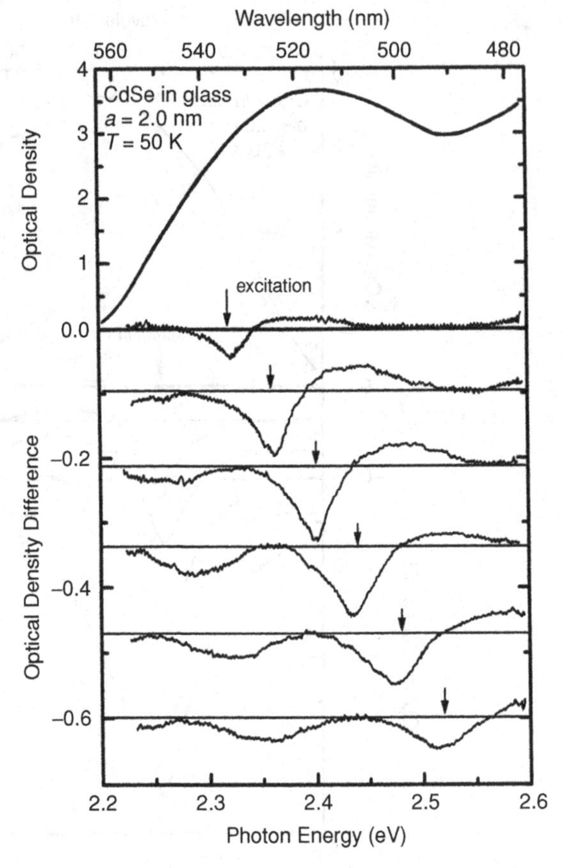

Fig. 5.16 **Doublet structure of the first absorption feature of CdSe nanocrystals in glass elucidated by means of a nonlinear pump-and-probe technique [8, 49]. Mean crystallite radius is 2.0 nm, temperature is 50 K. The upper panel presents the linear absorption spectrum. The lower panel presents differential absorption spectra recorded under monochromatic excitation by a tuneable laser (pulse duration is 5 ns, intensity is $100\,\mathrm{kW/cm^2}$) with simultaneous read-out by a weak broad-band probe beam. Negative differential optical density corresponds to bleaching.**

of these distinct atomic shells certain types of crystallites appear to be much more stable as compared to crystallites with an intermediate number of atoms and atomic shells. Then every given spectrum in Figure 5.18 represents a superposition of clusters with different sizes among the set of "magic" ones. Using Figure 5.7 for quick reference to estimate the size of those particles, one can see that the radius value is 1.5 nm to 1.7 nm. This value is compared with the parent bulk crystal lattice $a_L = 0.6$ nm to conclude that the magic size approach is indeed reasonable.

In bulk semiconductor crystals, hard optical excitation results in the formation of an electron–hole plasma which in turn, modifies the absorption spectrum drastically by means of band gap shrinkage, electron–hole states population and exciton screening [8, 52]. Therefore, the theory of nonlinear optical properties of resonantly excited bulk crystals to a large extent is the theory of dense electron–hole plasmas. In a nanocrystal, only a discrete number

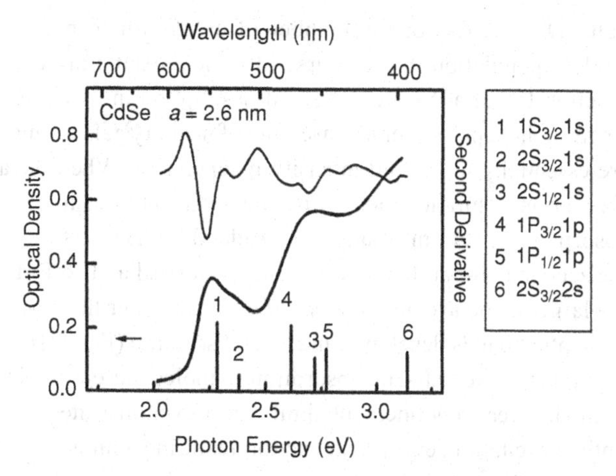

Fig. 5.17 Optical absorption spectrum of CdSe nanocrystals in a silicate glass matrix ($T = 300\,\mathrm{K}$), its second derivative (upper curve) and calculated energy positions and relative weights of the transitions shown by vertical lines. The assignment of the transitions is shown where 1S, 2S, and 1P labels stand for mixed hole states and 1s, 2s, and 1p labels stand for electron states in a spherical potential well [8, 50].

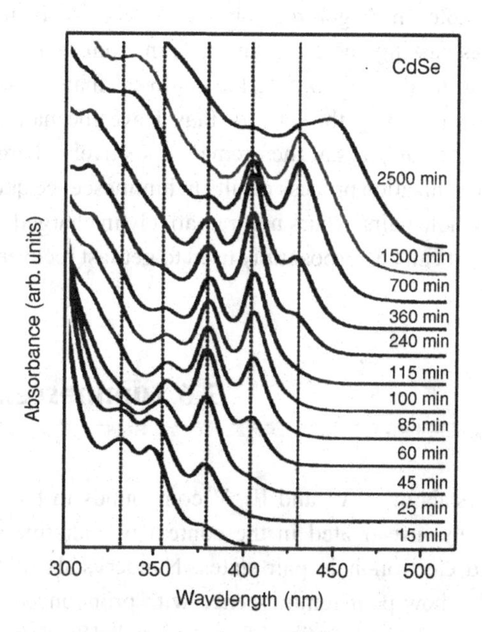

Fig. 5.18 Sequential (from bottom to top) absorption spectra of CdSe colloidal nanocrystallites observed in the course of their growth in a solution. Numbers indicate growth time in minutes for every spectrum. Reproduced with permission from [51]. Copyright Wiley-VCH.

of electron-hole pairs can be generated. Then, the standard theory of electron–hole states in quantum dots provides an energy spectrum of the first electron–hole pair. It is the energy spectrum of the first electron–hole pair that, along with the selection rules, determines the photon energies of resonant optical transition bands of nanocrystals for low-intensity light. Discreteness of the number of electron-hole pairs in every nanocrystal and their random population in the course of optical excitation lead to the Poissonian distribution function for an adequate statistical description in this case. When the average number of electron–hole pairs over an ensemble of nanocrystals is much less than 1, optical properties are essentially linear and intensity independent. When the average number approaches 1, absorption saturation occurs of the resonant transition relevant to the first pair. The absorption spectrum changes considerably. The first e–h-pair band bleaches whereas the second e–h-pair band becomes pronounced and so forth, i.e. the harder the excitation, the larger is the average number of e–h-pairs over the ensemble under consideration. New absorption bands develop as induced absorption (Fig. 5.16). To a large extent, this scenario is similar to excited-state absorption in molecular ensembles.

In the weak confinement limit multi-exciton states are created by means of intense optical excitation, every next exciton featuring a higher resonant energy of the lowest state because of interaction with the existing excitons. This results in a continuous blue shift of the exciton resonance absorption band with excitation intensity. It is well established for copper chloride nanocrystals in glasses and for nanocrystals of III–V compounds grown on a crystalline surface (see, e.g., [8] for detail).

In the strong confinement limit, creation of two electron–hole pairs in the same dot can promote an Auger recombination process. In this case one electron–hole pair recombines passing energy to an electron from another electron–hole pair. If the potential barrier at the dot–matrix interface is lower than the electron–hole transition energy, then an electron gaining this energy may leave the nanocrystal and enter into matrix states nearby. The nanocrystal then remains positively charged and no longer luminesces. Thus the photoionization process results in luminescence quenching. Recombination of newborn electron–hole pairs occurs non-radiatively in charged nanocrystals at very high rates about $10^{11}\ s^{-1}$. It can be purposefully used to get fast recovery of absorption in non-linear optical devices.

5.6 Luminescence

Nanocrystals of II–VI and III–V compounds in the strong confinement limit have been extensively investigated in the context of radiative and non-radiative recombination of confined electron–hole pair states. Nanocrystals of II–VI compounds in glass matrices typically show poor luminescence with pronounced photodegradation upon illumination because of photoionization of nanocrystals [8]. Even at low temperatures intrinsic band-edge emission is observed along with a wide emission band with large Stokes shift. This band is typically attributed to the recombination via the surface or interface state (Fig. 5.19, left panel). Colloidal nanoparticles of II–VI compounds in solutions or in a polymer also feature the presence of an additional wide band similar to glasses (Fig. 5.19, right panel). The core–shell design of II–VI quantum dots enables us to get high-quality nanocrystals with a quantum yield of intrinsic luminescence approaching 100%, by means of a potential barrier and chemical interface developed with a wide-gap shell with respect to a narrow-

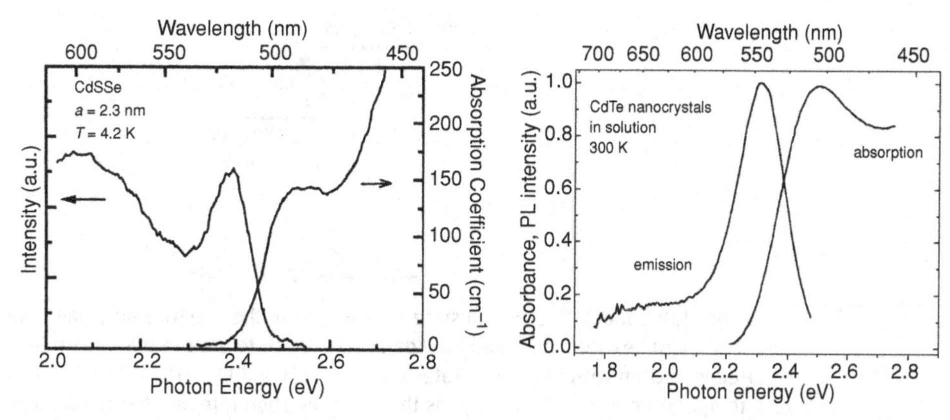

Fig. 5.19 **Absorption and emission spectra of II–VI nanocrystals in the strong confinement regime. Left panel: CdSSe nanocrystals with mean radius 2.3 nm in a Schott glass filter (adapted from [53]). Right panel: CdTe nanocrystals with mean radius 2.4 nm in aqueous solution (adapted from [54]).**

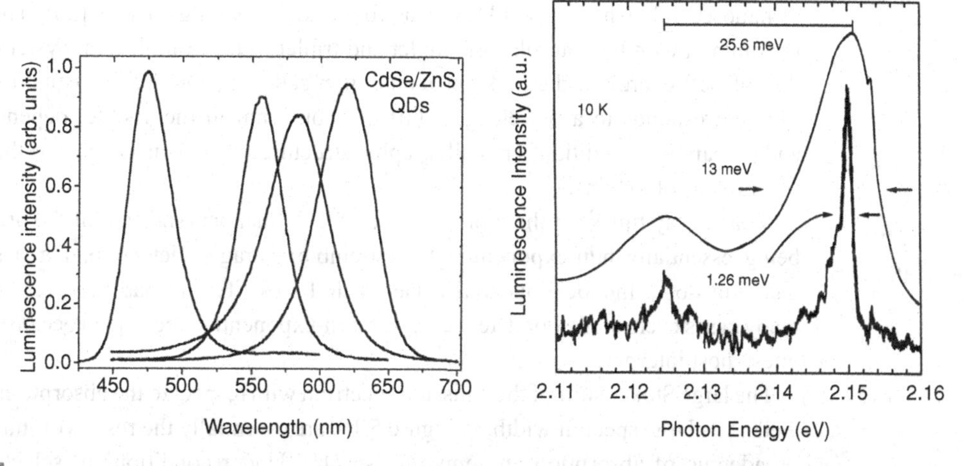

Fig. 5.20 **Photoluminescence of CdSe colloidal quantum dots in the strong confinement regime. Left panel: a set of emission spectra for a quantum dot ensemble with different mean sizes at room temperature under excitation by near UV light [55]. Right panel: Low-temperature luminescence data collected under conditions of ensemble-averaged red-edge excitation (upper curve) and single dot excitation (lower curve) [59]. The bar of 25.6 meV indicates the longitudinal optical phonon energy.**

band core crystallite. Remarkably, in accordance with the theory (Eq. (5.11) and Fig. 5.7) variation in size in fact provides wide tunability options. For example based on CdSe/ZnS core–shell nanocrystals, a spectral band from blue-green to red is feasible (Fig. 5.20).

In nanocrystals of II–VI (e.g. CdSe) and III–V (e.g. InP) compounds and in silicon nanocrystals, complex valence band structure results in a forbidden lowest energy transition. The relevant electron–hole state is referred to as "dark exciton". The temperature-dependent balance of population of this state and the next electron–hole pair state (Fig. 5.21) whose recombination represents an allowed transition results in peculiar temperature

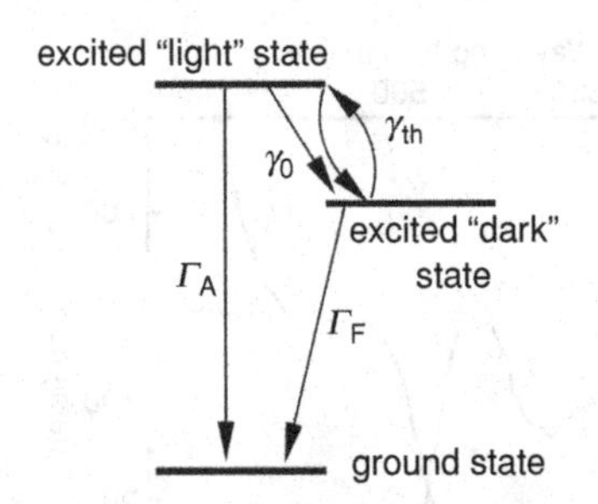

Fig. 5.21 **The three-state model. The ground state corresponds to the electron–hole pair number equal to zero. The "light" excited state and the "dark" excited state are the two lowest states of the band-edge exciton with respective rates $\Gamma_A(\sim 10^8\ s^{-1})$ and $\Gamma_F(\sim 10^6\ s^{-1})$. $\gamma_0 \approx 10^8\ s^{-1}$ is the zero-temperature relaxation rate. γ_{th} is the thermalization rate due to the interaction with acoustic phonons.**

dependence of the luminescence kinetics. At lower temperatures decay becomes noticeably slower. For example, for InP nanocrystals at room temperature the mean decay time is tens of nanoseconds whereas at 13 K it was found to be as long as 1 μs [56]. This has much in common with the interplay of singlet and triplet states in molecular systems. For CdSe the split-off energy of the dark versus the light state was found to be equal to a few meV (this corresponds to a wavelength shift of about 1nm in the visible) depending on size and preparation condition (crystallographic structure) [12]. This is equal to the kT value at several tens of kelvins.

Mean decay times for the luminescence of II–VI nanocrystals are of the order of 10^{-8} s being essentially non-exponential for ensemble averaged detection. For a single CdSe quantum dot a monoexponential decay with 19 ns lifetime has been reported [57]. In many cases even single dot detection gives non-exponential decay [58] because of complex guest–host interplay.

The large Stokes shift of the emission spectrum with respect to the absorption spectrum as well as its large spectral width, in Figure 5.19, are essentially the results of inhomogeneous broadening of absorption and emission spectra. Under conditions of selective red-edge excitation where only a small portion of crystallites are excited, narrower emission bands are observed (Fig. 5.20, right panel). Spectral width under these conditions to a large extent agrees with the spectral width evaluated by means of the pump-and-probe experiments (see Fig. 5.15). The Stokes shift in red-edge excited luminescence measures in tens of meV (a few nanometers on the wavelength scale).

In single dot photoluminescence experiments [59], emission lines are narrower than hole-burning data and data based on red-edge excitation of luminescence (Fig. 5.20, lower curve in the right panel). It is probable that, under spectral selection of a sub-ensemble in hole-burning and red-edge excitation experiments, this sub-ensemble still keeps inhomogeneity and dispersion in shape, surface structure, defect number and potential barrier, because such a sub-ensemble is still statistically big in spite of spectral selection. Ten-fold narrowing of emission bands is observed in single dot luminescence detection as compared to red-edge excitation and hole-burning. The spectral width and position of emission bands exhibit wandering with time because of multiple guest–host effects and reversible photoionization events. Therefore the spectrum widens with data acquisition time. Probably, the same reasons are responsible for the non-exponential decay of single quantum dot emission.

Silicon nanocrystals, buried in silica through oxidation of nanoporous silicon were found to exhibit noticeable visible emission with decay times in the microsecond range [60–62]. Their luminescent properties are described in terms of the quantum confinement effect on electron–hole spectra, increase of radiative transition probability due to confinement, along with decrease of non-radiative recombination rate because of statistical lack of defects in nanocrystals. Interestingly, silicon nanocrystals in space are considered as a possible source of the red emission detected in astronomical observations [63].

5.7 Probing the zero-dimensional density of states

The local density of electron states can be mapped over the surface by means of *scanning tunneling microscopy*. Known since 1982 [64], this technique has become a routine and powerful technique in examination of conductive nanostructure surfaces [65, 66]. The basic principle of a tunneling microscope is mapping of the tunnel current using a precisely scanned metal tip over the surface under investigation. In Chapter 3 (Section 3.4), we saw that the tunneling probability for a quantum particle reduces exponentially with increasing barrier width and the square root of the barrier height (Eq. (3.70)). In the case of a conductive tip over a conductive surface in a vacuum, the barrier height is determined by the electron *work function*. Equation (3.70) holds for free space over the barrier, that is for the case of the fairly large number of final states available for a tunneling particle. Resonance tunneling (Section 3.5) gives a further idea that tunneling is essentially controlled by the states available behind the barrier. In more detail, the *differential conductance* dI/dV appears to be directly proportional to the local density of final states behind the barrier. Therefore, in the case of localized electron states on the surface, differential conductance mapping will give an image of the electron density of states, which, in turn correlates with electron wave function.

To probe the zero-dimensional local density of electron states, scanning tunneling microscopy with variable voltage has been applied for hut-like InAs quantum dots (Fig. 5.22) [67, 68]. This technique is referred to as *scanning tunneling spectroscopy*. Quantum dots were found to exhibit sharp tunneling current maxima relevant to discrete electron states (Fig. 5.22(b)). Mapping of the differential conductance near resonant enhancement of current showed reasonable distribution of the electron local density of states (Fig. 5.22(c)) in good agreement with the calculations (Fig. 5.22(d)).

5.8 Quantum dot matter

In previous sections, the intrinsic properties of nanocrystals were discussed implying absence of any cooperative effect on the properties of a given nanocrystal ensemble. In what

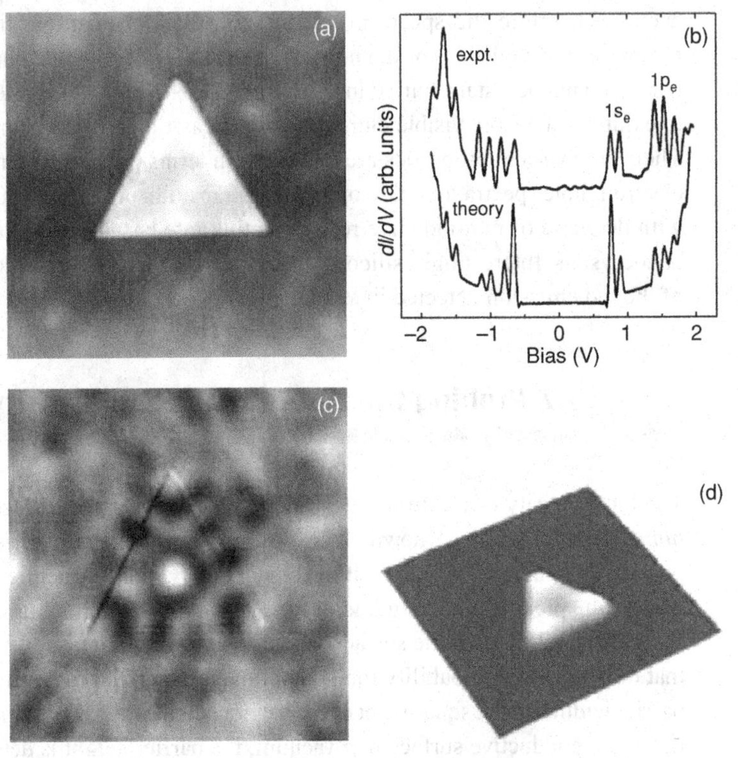

Fig. 5.22 **Imaging electron local density of states in an InAs quantum dot by means of scanning tunneling spectroscopy [67, 68]. (a) a scanning tunneling microscope image of a triangular InAs quantum dot, image size is** $93.8\,\mathrm{nm} \times 93.8\,\mathrm{nm}$; **(b) measured and calculated differential conductance** $\mathrm{d}I/\mathrm{d}V$; **(c)** $\mathrm{d}I/\mathrm{d}V$ **mapping for bias voltage** $V = 0.08\,\mathrm{V}$; **(d) calculated local density of electron state for the lowest state of an InAs quantum dot.**

follows the concept of quantum dot matter will be discussed, i.e. novel materials whose properties are controlled by both spatial confinement of electrons within a nanocrystal and spatial organization on a larger scale.

There are several ways to develop a nanocrystal superlattice, i.e. a structure consisting of identical nanocrystals with regular spatial arrangement. One method is to use zeolites which form a skeleton with regular displacement of extremely small cages, the size of a cage being typically about 1 nm. Using various zeolites as templates for semiconductor clusters, there is the possibility of studying regular three-dimensional cluster lattices with variable intercluster spacing. A pronounced shift to the red of the absorption spectrum with decreasing spacing has been observed for such structures [69]. The size range of embedded clusters is severely restricted because of the very small cages.

Several groups have reported on successful realization of periodic two- and even three-dimensional arrays of nanocrystals using the self-organization effect in strained heterostructures under condition of submonolayer epitaxy [45, 70]. In this case, however, the typical size of nanocrystals is equal to, or larger than 10 nm, the interdot spacing being of the order of 10 nm.

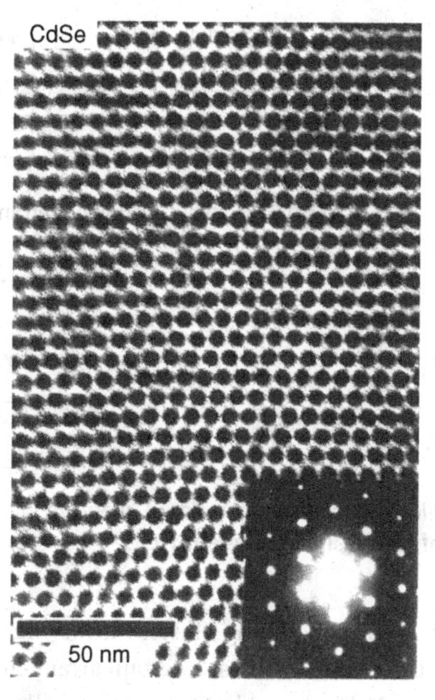

Fig. 5.23 A high-resolution transmission electron microscopy image and small-angle electron diffraction pattern (insert) of a three-dimensional quantum dot superlattice [73]. The superlattice consists of CdSe nanocrystals of 4.8 nm diameter each. Nanocrystals are assembled in a face-centered cubic lattice. Image corresponds to (101) crystallographic plane of the array.

Self-assembly of nanosize particles into a *colloidal crystal* promoted by van der Waals interaction in a monodisperse concentrated solution seems to be the most challenging route, resulting in close-packed three-dimensional structures. Since the first identification of natural colloidal crystals, namely a specific type of *virus* [71], colloidal crystal structure was found to be inherent in a number of natural and artificial objects [72]. For nanocrystals self-organized into a macroscopic colloidal crystal, a notation "*quantum dot solid*" has been introduced [73]. A face-centered cubic lattice of semiconductor quantum dots is shown in Figure 5.23. The optical properties were found to modify due to interdot interactions. Comparison of optical spectra for nanocrystals close packed in the solid with dots in a dilute matrix revealed that, although the absorption spectra are essentially identical, the emission line shape of the dots in the solid is modified and red-shifted indicating interdot coupling. Systematic studies of electronic energy transfer in CdSe quantum dot solids revealed long-range resonance transfer of electronic excitation from the more electronically confined states of the small dots to the higher excited states of the large dots. Foerster theory for long-range resonance transfer through dipole–dipole interdot interaction was used to explain electronic energy transfer in these close-packed nanocrystal structures. Energy transfer promotes luminescence quenching since it makes it possible for an electron to be trapped by a defect located in another crystallite.

In disordered solids, at a certain degree of disorder, a transition occurs from localized electron states to coexisting delocalized and localized ones separated by a border. With an increase in concentration, delocalization occurs with respect to nearest-neighbor wells

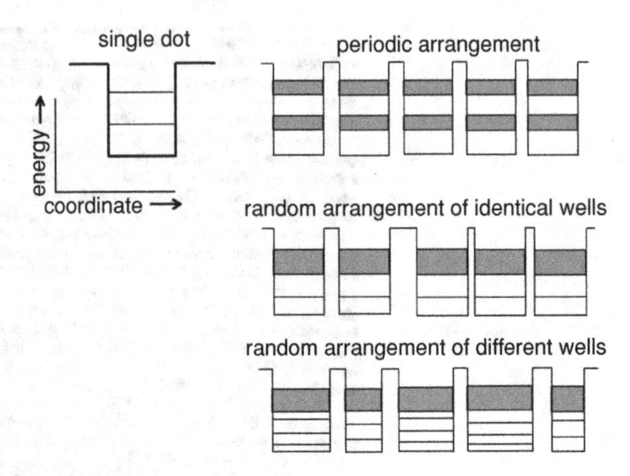

Fig. 5.24 **A sketch of electron levels in an isolated quantum well and in periodically and randomly arranged ensembles of identical and different wells.**

resulting in small conducting clusters involving several wells. At higher concentrations, the cluster size increases and at a certain arrangement electrical conductivity is established throughout a whole ensemble of potential wells. In the physics of disordered solids, this effect has been known since 1958 as the Anderson transition [74, 75]. It was discussed for disordered solids in Section 4.5. In the case of random ensembles there are two idealized models. The first one is the random arrangement of identical wells (the Lifshitz model [76]). The second one is the regular arrangement of different wells. By and large, quantum dot solids are expected to exhibit the principal effects known for normal solids, i.e. formation of energy bands in the case of periodic lattices and coexistence of localized and delocalized states in a dense random quantum dot ensemble (Fig. 5.24).

This consideration is supported by numerical modeling and experimental observations [77–79]. Numerical modeling was performed using displacement of 3375 (15^3) identical spherical potential wells randomly distributed within a fragment of a cubic lattice with period L (the Lifshitz model). The aim of the calculations was to estimate the number of delocalized states as a function of the concentration C of nanocrystals, their radius R and electron (hole) effective mass $m_e^*(m_h^*)$. A particle state is treated as delocalized when the energy overlap integral is larger than the difference of energy level shifts in the nearest-neighbor wells because of the influence of all other wells within the system under consideration. Random displacement of wells was described in terms of a deviation of their coordinates with respect to the nodes of a regular cubic lattice according to a Gaussian distribution. The maximal deviation value was chosen to be no larger than one half of the lattice period by means of the truncated distribution function. Statistical analysis has been performed over 500 configurations of a quantum dot ensemble for each of several sets of parameters $\{C, R, m^*\}$.

The results (Fig. 5.25) indicate the steady growth of delocalized electron states with increasing concentration of quantum dots. The fraction of delocalized states depends drastically on electron effective mass and dot size. It is not surprising since these two parameters determine the energy spectrum. In the case of larger dots and higher electron mass

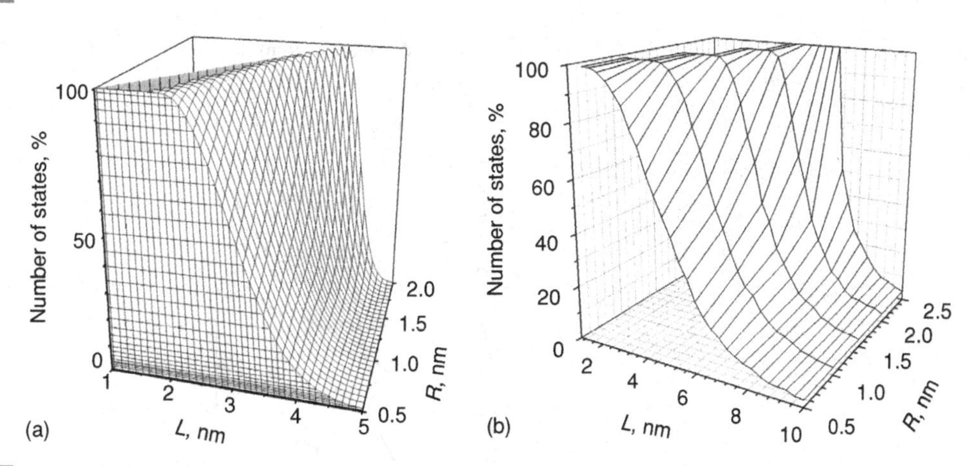

Fig. 5.25 Number of calculated delocalized electron states versus average interdot spacing L for several values of dot radius R. Electron effective mass is $0.13m_0$ (CdSe, left) and $0.065m_0$ (GaAs, right) [78].

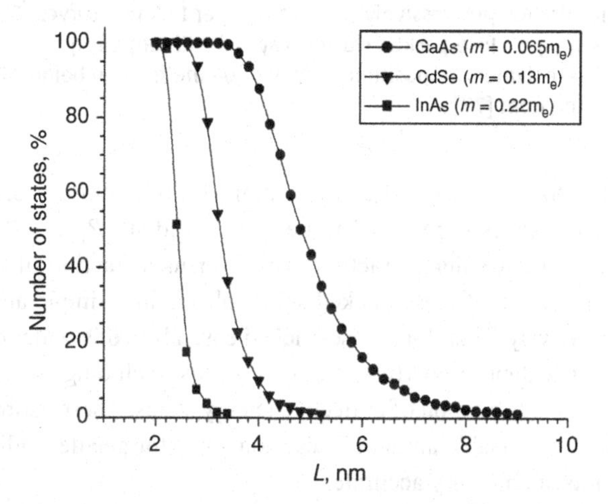

Fig. 5.26 Number of delocalized electron states versus average interdot spacing L in an ensemble of randomly arranged spherical potential wells with radius 1 nm. Electron effective mass of GaAs (circles), CdSe (triangles) and InAs (squares) was used in calculations. Courtesy of A. I. Bibik [78].

(e.g. InAs, ZnSe dots) even packing of particles up to volume fraction 0.5 (this corresponds to close packing into a simple cubic lattice, i.e. $L = 2R$) does not result in a noticeable fraction of delocalized states. This result provides an explanation as to why modification of absorption spectra with increasing concentration was not observed for larger CdSe nanocrystals [73]. Taking into account the presence of an organic or inorganic shell in all experiments with dense quantum dot ensembles (otherwise uncontrollable agglomeration and Oswald ripening will develop), even close-packed ensembles of larger crystallites may not satisfy the delocalization condition. Therefore, in agreement with our calculations, close-packed larger CdSe quantum dots reproduce properties of molecular solids with resonant long-range energy transfer. However, for smaller dots a noticeable

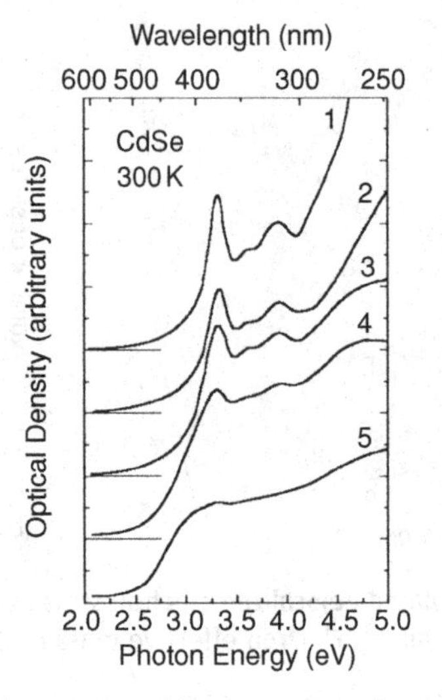

Fig. 5.27 **A set of absorption spectra of CdSe nanoparticles, average radius is 1.6 nm. Nanoparticle concentration progressively rises from upper to lower curves. Curve 5 corresponds to a CdSe film consisting of close-packed clusters capped with organic groups, curves 1–4 correspond to cluster/polymer composition, polymer volume fraction being 37%, 18%, 3%, 1%, respectively. Adapted from [77].**

delocalization occurs at concentrations far from close-packing, e.g. for $R \approx 1$ nm, 75% of electron states were found to be delocalized at $2R/L < 0.6$. This value corresponds to a quantum dot volume fraction of 0.1 versus a fraction of 0.52 and 0.74 relevant to maximal fraction of close-packed solid spheres in a simple and a face-centered cubic lattice, respectively. Therefore close-packed ensembles of smaller quantum dots will reproduce, to a large extent, properties of atomic solids, including not only energy and charge transfer but also a formation of collective energy states. These estimates should however be treated with a reasonable amount of caution since for smaller radii effective mass approximation results are not very accurate.

To verify the above predictions on the possible delocalization of electron states in a dense quantum dot ensemble, solid films from CdSe nanoparticles have been examined with polyethylene glycol acting as a spacer in order to control the distance between the particles [77, 80]. Absorption spectra of the original diluted solution feature very sharp, stable and reproducible resonances indicating cluster-like behavior or the magic-size regime discussed in Section 5.4 (compare Fig. 5.27 with Fig. 5.18). A systematic reversible modification of optical absorption spectra has been observed with increasing dot concentration from a set of discrete sub-bands inherent in isolated nanocrystals which gradually get broader and develop into a smooth band-edge absorption similar to that of bulk semiconductors. Note that absorption onset in this quantum dot solid is far to the blue as compared with the CdSe bulk crystal whose interband absorption starts at approximately 670 nm (see Table 4.1). It is noteworthy that the broad structureless absorption spectrum does not significantly change at lower temperature (Fig. 5.28). No distinct substructure has been resolved at 30 K. The results are interpreted in terms of an evolution from individual (localized) to collective

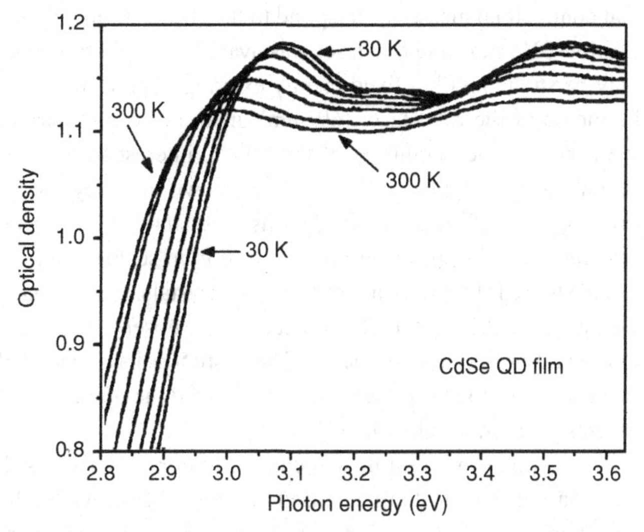

Fig. 5.28 **A portion of the absorption spectrum of close-packed CdSe quantum dots measured at different temperatures ranging from 300 K to 30 K. The insert shows spectra at 300 and 30 K in a wider range [79].**

electron states, delocalized within at least a finite number of nanocrystals indicating the possibility of an Anderson transition in a close-packed quantum dot ensemble. Other authors have also reported for smaller semiconductor quantum dots a systematic modification in the absorption spectra [81] and photoluminescence spectra a shift up to 30 nm to the long-wave side [82] in solid films.

These findings raise a number of further issues related to electron properties of quantum dot superlattices such as, e.g. electron band structure of a quantum dot super-crystal, renormalized electron (hole) effective masses within a superstructure, modified electron–hole interaction (super-exciton?) and others. First calculations of mini-band formation in three-dimensional quantum dot superlattices have been reported [83]. For a relatively large Ge dot size of 6.5 nm discrete electron energy levels evolve to minibands for an interdot distance of approximately 1 nm. For GaAs dots of similar size minibands are believed to occur for larger distances.

5.9 Applications: nonlinear optics

Semiconductor-doped glasses containing Cd and Pb chalcogenide nanocrystals show superior nonlinear absorption behavior due to saturation of the absorption relevant to creation of the first electron–hole pair. Absorption saturation in glasses doped with CdS_xSe_{1-x} semiconductor crystallites with $\overline{a} \approx a_B$ were studied intensively in the 1960s in connection with the use of commercial glasses as passive shutters in ruby lasers, at the initial stages of quantum electronics. At that time the sizes of crystallites were not measured and the fact that commercial glasses correspond to the case $\overline{a} \approx a_B$ only became known from later publications. No tuneable dye laser was available at that time and absorption saturation could only be studied with a fixed-energy exciting photon, which corresponded to the principal harmonic of the ruby laser (694 nm) or to the second harmonic of the neodymium laser (532 nm). At the beginning of the 1980s, interest in nonlinear absorption in selenocadmium glasses reappeared, since it was shown that they are a convenient model object for investigating different manifestations of quantum-size effects in quasi-zero-dimensional structures. Studies of absorption saturation in selenocadmium glasses using tuneable dye lasers showed that, in contrast to single crystals, glasses have an unusually wide spectral interval (0.2–0.3 eV) in which bleaching is observed in single-beam experiments and in pump-and-probe measurements. These studies have been followed by extensive experiments (see [8, 14] for review), which revealed the unique features of semiconductor-doped glasses as saturable absorbers.

Absorption saturation in semiconductor-doped glasses is due to population of the one-electron-hole-pair state in an ensemble of nanocrystals. Such glasses possess a number of advantageous features which make them superior Q-switchers and mode-locking elements in laser devices. The main advantages as compared to other saturable media are:

(i) the fast recovery time of bleaching in the nano- and picosecond range;
(ii) high ratio of the saturable absorption coefficient to the non-saturable background (10-fold and more absorption coefficient drop);
(iii) high photostability;
(iv) negligible spatial diffusion rate of excited species, which is important when the saturable absorber is used in optical processing;
(v) tuneable spectral range where nonlinear response is available;
(vi) broad-band simultaneous bleaching (typically tens of nanometers) under conditions of monochromatic pumping.

Fast recovery time, high saturable-to-non-saturable absorption contrast, good photostability and tuneability options are all necessary in all applications. Broad-band bleaching is crucial for mode-locking to get ultrashort light pulses. Notably, narrow-band spectral hole-burning, shown in Figures 5.15 and 5.16, is by no means typical for semiconductor-doped glasses. It was found to be inherent only in specially synthesized glasses where only nucleation and the normal growth stage occur. In typical commercial glasses the competitive growth stage is widely used to give wide homogeneous width of absorption band which is of the same order of magnitude as inhomogeneous width from size dispersion [35, 84] even at low temperature (Fig. 5.29).

Semiconductor-doped glasses exhibit genuine absorption saturation at room temperature in a wide spectral range with low non-saturable background and fast recovery time. The spectral range in which absorption saturation occurs may be as wide as 80 nm, the spectral range where bleaching occurs in pump-and-probe experiments being 20–30 nm, an increase in transmission can be as large as 10^4–10^6 times while the recovery time is in the subnanosecond range [85, 86]. A typical example is given in Figure 5.30, which represents the power-dependent optical density of commercial glass containing CdS_xSe_{1-x} crystallites.

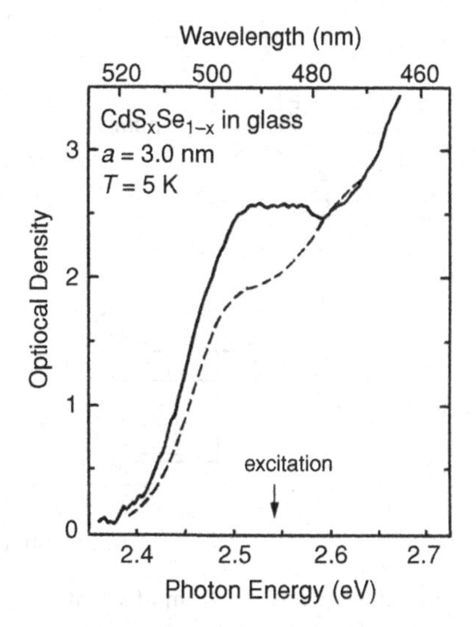

Fig. 5.29 **Modification of the absorption spectrum of a typical commercial-based semiconductor-doped glass under conditions of laser excitation [84]. Solid line – original spectrum, dashed line – spectrum of excited sample, arrow indicates position of the laser wavelength. Laser pulse duration is** 10 ns, **intensity is about** 1 MW/cm^2.

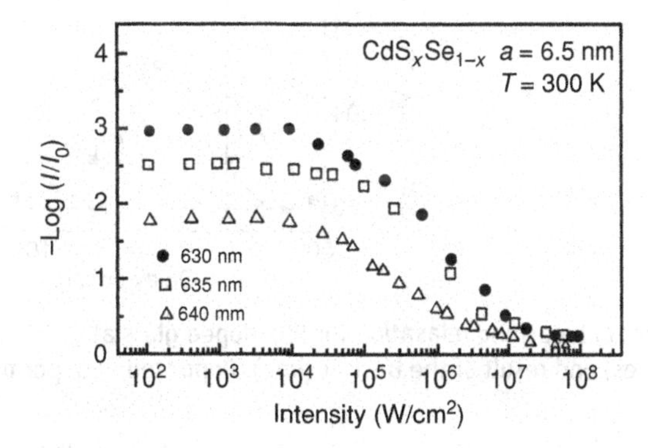

Fig. 5.30 **Absorption saturation in commercial glasses doped with** CdS_xSe_{1-x} **nanocrystals at the room temperature measured with nanosecond laser pulses [85].** I_0 **is the incident intensity,** I **is the transmitted power density.**

Lead sulfur (PbS) nanocrystals in glasses offer the unique possibility of tuning optical absorption within a 1–2 μm spectral range (Fig. 5.31). The parent bulk crystal is a narrow-band semiconductor with very large exciton Bohr radius $a_B^*(\text{PbS}) = 18$ nm. Therefore a genuine strong confinement regime ($a \ll a_B^*$) holds for $a \gg a_L$ when the crystal lattice exhibits bulk-like properties. The PbS-doped glasses were found to exhibit pronounced absorption saturation with recovery time in the picosecond range under conditions of laser excitation (Fig. 5.32).

These glasses were successfully applied to Q-switching and mode-locking for a number of solid-state lasers in the spectral range 1 to 2 μm. Q-switching provides laser pulses with

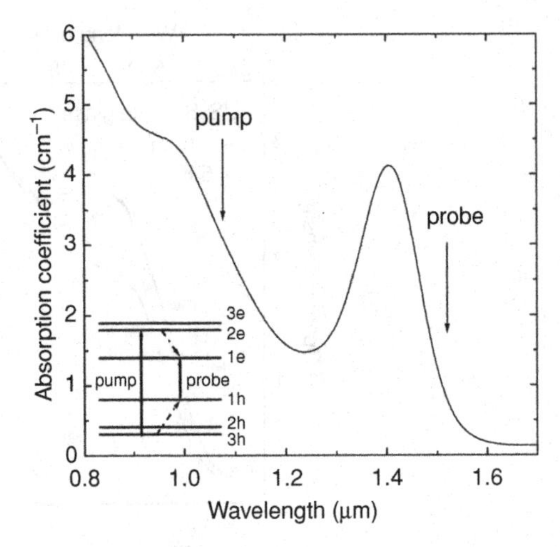

Fig. 5.31 Room-temperature absorption spectrum of silicate glass doped with PbS QDs. Inset is the energy-level diagram for PbS QDs. Reprinted with permission from [87]. Copyright Elsevier Ltd.

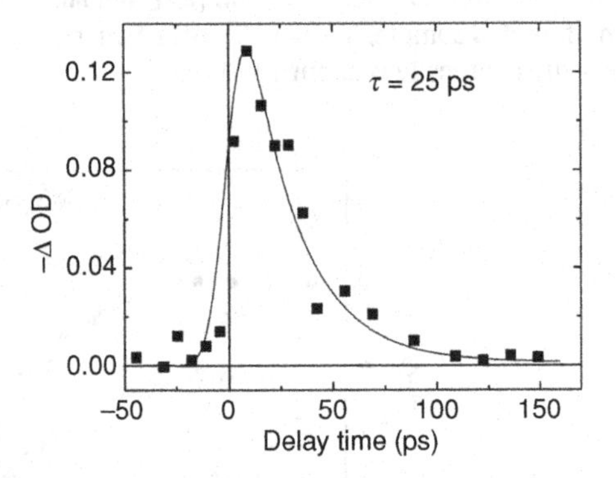

Fig. 5.32 Kinetics of bleaching relaxation for PbS-doped glass at 1.524 μm after the pump at 1.08 μm (squares) and result of the best fit (line). Reprinted with permission from [87]. Copyright Elsevier Ltd.

duration of the order of 10 ns, whereas mode-locking is used to get pulses as short as a few picoseconds. A representative example is presented in Figure 5.33 of a 10 ps pulse shape and its spectral width for a laser based on a Cr^{4+}:YAG crystal as active medium with a PbS-quantum-dot-doped glass filter as a mode-locking component [87]. Figure 5.34 presents a summary of the successful applications of PbS-doped glasses in solid-state lasers [87–92].

5.10 Applications: quantum dot lasers

At the beginning of the laser era in the 1960s, the first solid-state and gas lasers were rather large and inefficient. At that time there was no prospect in sight that sometime compact,

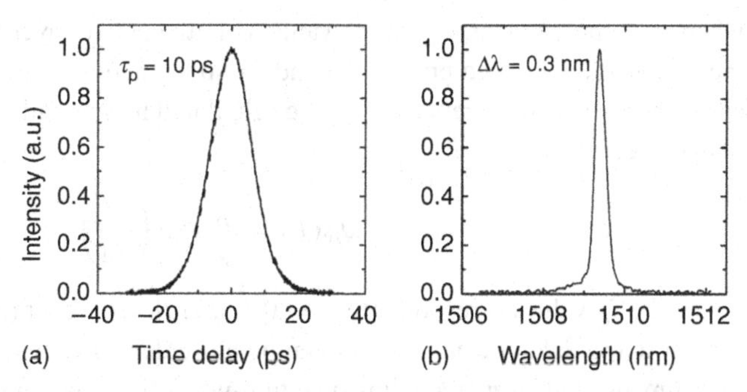

Fig. 5.33 Application of PbS-quantum-dot-doped glass for passive mode locking in a solid state Cr:YAG laser with optical pumping by semiconductor diodes. (a) Intensity autocorrelation, dotted curve is fit assuming an ideal sech^2 pulse shape, (b) spectrum of mode-locked pulses. Reprinted with permission from [87]. Copyright Elsevier Ltd.

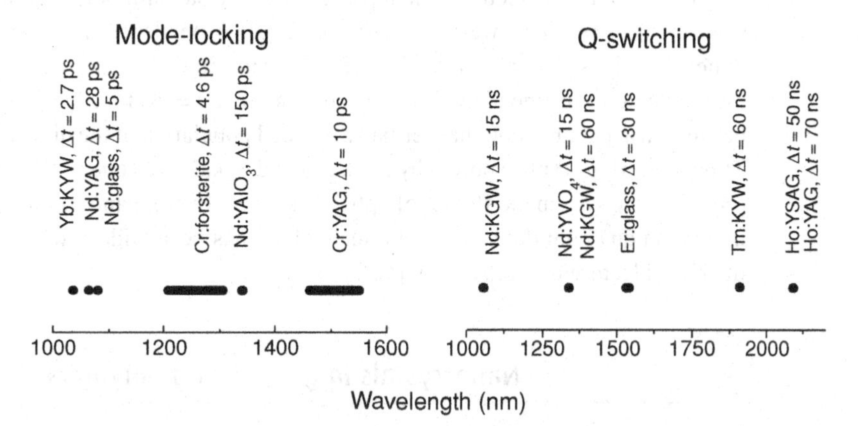

Fig. 5.34 A sketch with summary of mode-locking and Q-switching performed for a number of solid-state lasers by means of PbS quantum-dot-doped glasses (according to data reported in [87–92]).

hardly visible by the naked eye, semiconductor lasers could enter our daily lives. In fact, CD-players and laser printers, which are massively produced and consumed nowadays, contain small semiconductor lasers and laser arrays. Ten or even more per cent efficiency "from a wall plug" is a typical commercial demand.

Optical properties of quantum dots such as the discrete density of states and tailored energies of optical transitions are definitely advantageous for laser applications. Additionally, there is one more peculiar advantage that is inherent in quasi-zero-dimensional structures only, and is not the case for other nanostructures and bulk materials. This is the temperature-independent threshold current.

The threshold current is the main technical parameter of an injection laser. The rapid increase of threshold current with temperature is inherent in all types of semiconductor lasers preventing development of highly efficient semiconductor lasers and arrays operating at room temperature. The value of the threshold current can be expressed as,

$$J_{\mathrm{th}} = \frac{ed}{\eta} \int r_{\mathrm{sp}}(E)\mathrm{d}E, \tag{5.12}$$

where e is the elementary charge value, d is the active layer thickness, η is the quantum yield, and $r_{sp}(E)$ is the energy-dependent spontaneous emission rate. In the majority of cases the threshold current can only be calculated numerically. Its temperature dependence obeys a relation,

$$J_{th}(T) = J_{th}^{0} \exp\left(\frac{T}{T_0}\right), \tag{5.13}$$

where the J_{th}^{0} value corresponds to $T = 0°C$ and T_0 is the characteristic temperature. In 1982 Arakawa and Sakaki examined the temperature dependence of the J_{th} value for models of low-dimensional lasers including quantum-well (2D), quantum-wire (1D) and quantum-dot (0D) structures [93]. The temperature dependence of J_{th} arises from the thermally induced spread of electrons and holes over a wider energy interval. This energy spread results in exhaustion of electron and hole populations near the band extrema (i.e. the bottom of the conduction band for electrons and top of the valence band for holes). Lower dimensionality was found to result in weakening of the $J_{th}(T)$ dependence, i.e. T_0 is larger for lower dimensionalities. It is a remarkable fact that a zero-dimensional laser was found to possess the temperature-independent threshold current, i.e. $T_0 = \infty$ for a quantum-dot laser. This is because in a quantum dot, the thermally induced population of the higher states is inhibited. This pioneering result reported by Arakawa and Sakaki was one of the cornerstones which promoted the systematic studies of optics of semiconductor nanocrystals afterwards, and its impact on quantum dot science and technology has been widely acknowledged including the Nobel lecture by Zh. Alferov [42].

Nanocrystals in glasses and polymers

Optical gain and lasing in semiconductor nanocrystals were observed for the first time in 1991 by Vandyshev *et al.* [94] using CdSe nanocrystals in a glass matrix of size $a \approx a_B$. Lasing was observed at $T = 80\,K$ at a wavelength of 640 nm under pumping with second-harmonic radiation of a picosecond YAG:Nd laser (532 nm). This report was followed by further experimental studies relevant to CdSe nanocrystals in glasses [95, 96]. The qualitative features of the expected gain spectrum are as follows. The gain can occur in a rather broad spectral range including photon energies well below the absorption onset. The low-energy edge of the red-shifted emission is determined by a transition from the lower two-electron-hole–pair state (biexciton) $|1s1s1S_{3/2}1S_{3/2}\rangle$ to the higher exciton state $|1p1P_{3/2}\rangle$ which becomes allowable due to Coulombic interaction. The build-up of the gain spectrum under successively growing pump intensity observed in the experiments is qualitatively in good agreement with the computational result (Fig. 5.35). Thus the phenomenon of optical gain in nanocrystals can actually be attributed to the biexciton to exciton recombination. In Figure 5.36 the photoluminescence dynamics is presented of colloidal CdSe nanocrystals forming a solid film. Unlike glass matrices, such films offer higher dot concentration and therefore higher gain.

Nanocrystals of II–VI compounds offer optical gain in the visible range because of the large energy band gap of the parent crystals. Many potential applications are considered

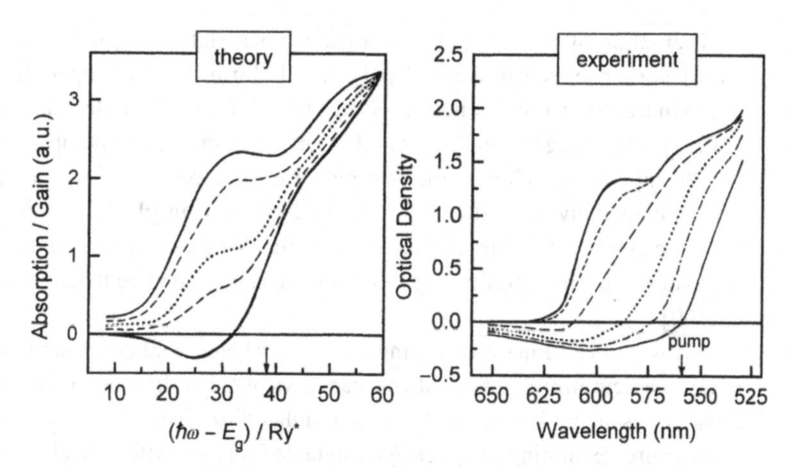

Fig. 5.35 Calculated (left) and observed (right) absorption spectra of CdSe nanocrystals [95]. Upper curves in each panel are the linear absorption spectra. The other curves correspond to successively growing (theory) population and (experiment) excitation intensity, from top to bottom. The mean nanocrystal radius is $\overline{a} = 2.5\,\mathrm{nm}$, the Gaussian size distribution in calculation was given to be $\Delta a = 0.1\overline{a}$.

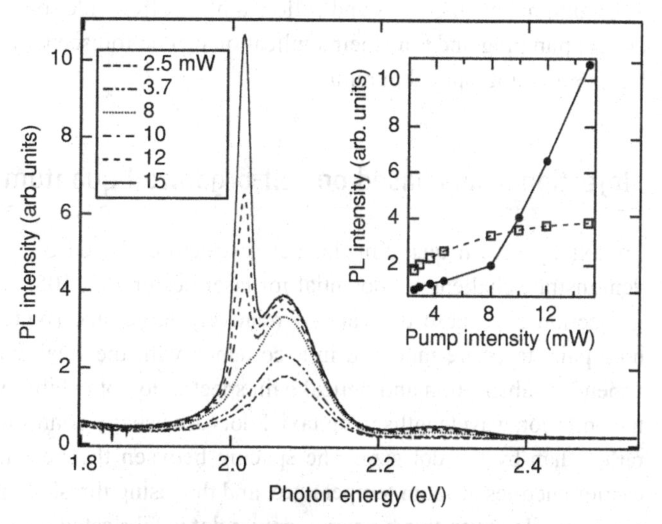

Fig. 5.36 Development of a sharp stimulated emission band as a function of pump intensity in photoluminescence spectra of films ($T = 80\,\mathrm{K}$) fabricated from TOPO-capped CdSe nanocrystals with radius $2.1\,\mathrm{nm}$ [97]. The inset shows superlinear intensity dependence of the stimulated emission (circles) with a clear threshold compared to the sublinear dependence of the PL intensity outside the sharp stimulated emission peak (squares). Reprinted with permission from AAAS.

for optical communication spectral ranges near 1.5 and 1.3 μm. Among II–VI compounds, crystals of mercury chalcogenides posess a narrow band gap suitable for infrared applications. However their synthesis is not as advanced as Cd and Zn chalcogenides. For these wavelengths nanocrystals of narrow-band semiconductor materials like PbS, PbSe, InAs, InSb in the strong confinement regime are appropriate. Chen *et al.* [98] reported on the observation of optical gain from InAs nanocrystal quantum dots which emit at 1.55 μm and are embedded in a novel polymer platform. Room-temperature optical gain at the ground exciton transition of PbS quantum-dot-doped glasses has been obtained [99] in the spectral range from 1317 to 1352 nm by changing the pump wavelength from 900 to 980 nm, corresponding to the next higher exciton resonance. The optical gain is 80 cm^{-1}. It is essentially limited by low dot concentration in glasses (typically 0.1–1%). Higher gain can be obtained by means of a different synthesis approach. Sol–gel synthesis offers nanocrystal concentrations up to 10% and has been applied to get higher gain for PbSe dots [100].

It would be rather challenging to try to develop a silicon laser based on the promising luminescent properties of silicon nanocrystals. The first report on optical gain in silicon nanocrystals by Pavesi *et al.* [101] stimulated extensive research in the search for silicon structures exhibiting an optical gain suitable for lasers with optical pumping [102]. However, to date, experimental performance of a silicon laser with silicon as gain medium has not been reported. Problems have been highlighted in elucidation of the optical gain value in a thin waveguiding layer containing silicon nanocrystals [103]. Meanwhile, silicon has been introduced into laser devices but as a Raman-active medium to shift the original laser wavelength rather than as an optical gain material [104].

Quantum dots in glasses and colloidal films offer wide spectral tuneability but they need optical pumping and thus their application can be foreseen only in combination with other lasers providing optical pumping.

Injection lasers based on self-organized quantum dot heterostructures

Epitaxially grown quantum dot heterostructures based on group III–V semiconductors demonstrate challenging potential for laser design [42, 105, 106]. As has been mentioned in Section 5.3, these dots are of relatively large size (typically 10 nm) and electron–hole pair states are modified in accordance with the weak confinement model, i.e. size-dependent absorption and optical gain spectra do not exhibit wide tuneability. As a result, the emission wavelengths in epitaxial dots are usually controlled by chemical composition rather than by the dot size. The spacing between their electronic states is smaller than carrier energies at room temperature, and the lasing threshold is therefore still temperature sensitive. However, this type of quantum dot is fabricated by means of versatile technology suitable for mass industrial production. Unlike colloidal or glass structures, epitaxial growth offers the possibility of developing injection lasers readily. Desirable wavelengths in the near infrared range, including optical communication wavelengths of 1.3 and 1.5 μm, can be performed by means of feasible band-gap tuning in ternary compounds. For example, InGaAs/InAs quantum dots on a GaAs substrate emit in the 1.0–1.3 μm wavelength range, which could be extended to 1.55 μm. When grown on an InP substrate, InGaAs/InAs quantum dot emission covers the 1.4–1.9 μm wavelength range.

Quantum dot injection lasers with multiple vertically coupled dots to get higher gain have been experimentally developed [42, 105–107] (Fig. 5.37). These lasers can operate in a continuous wave (cw) regime at room temperature with moderate threshold currents.

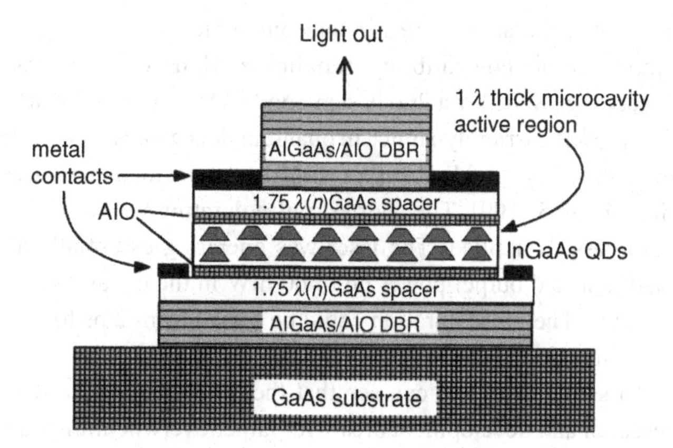

Fig. 5.37 Design of a vertically emitting microcavity laser with vertically coupled InGaAs quantum dots reported by Ledentsov *et al.* [105]. DBR stands for Distributed Bragg Reflector.

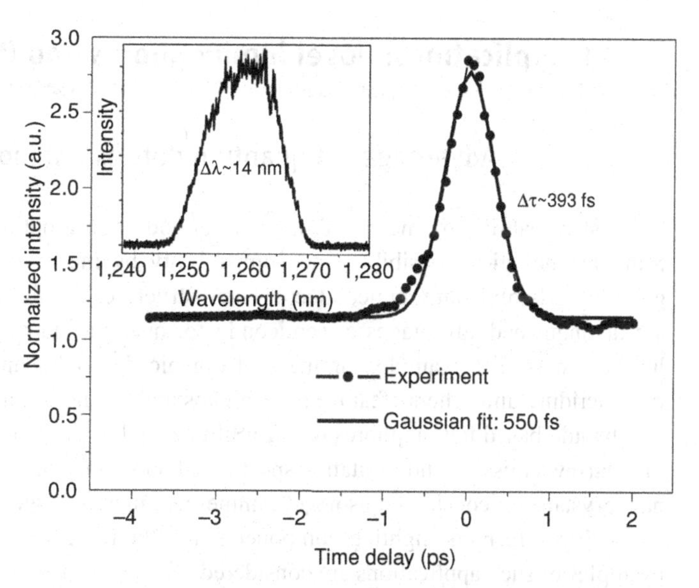

Fig. 5.38 The measured intensity autocorrelation trace and the optical spectrum (insert) of a passively mode-locked quantum dot injection laser using a quantum dot saturable absorber for mode-locking. Reprinted with permission from [109]. Copyright 2005, AIP.

As compared to quantum well lasers, quantum dot injection lasers were actually found to feature relaxed temperature dependence of the threshold current. For a 1.3-μm laser Lui *et al.* reported a threshold current of 17 A/cm^2 with cw output power 100 mW! [107].

Quantum dot materials and structures show a rather wide optical gain spectrum, mainly because of inhomogeneous broadening. Wide gain bandwidth makes these materials particularly promising for the amplification of femtosecond pulses. They have the potential for generating sub-100 fs optical pulses provided the whole bandwidth is coherently engaged and dispersion effects are minimized. Compact semiconductor optical amplifiers are of considerable interest in telecommunication circuitry because of their high gain and optoelectronic compatibility. Rafailov *et al.* have demonstrated a high-gain amplification exceeding 18 dB for a 200 fs semiconductor quantum dot amplifier [108].

Ultrafast carrier dynamics in quantum dots makes it possible to integrate a mode-locking quantum dot saturable absorber with a quantum dot injection laser. This was done for the first time in 2005 [109]. The shortest pulse duration was equal to 390 fs with no special pulse compression applied. The laser was operating essentially above the threshold providing distinguished output power up to 60 mW in the cw and up to 45 mW in the mode-locked regimes. The pulse duration could be varied from 2 ps to as short as 400 fs at the 21 GHz pulse repetition rate.

To summarise, one can see that the field of quantum dot lasers has become a mature research and development area with rather overwhelming output and challenging promise. It represents an impressive example of strong practical impact from what was originally basic and academic research activity.

5.11 Applications: novel luminophores and fluorescent labels

Advantages of quantum dots as luminophores

Spectral tuneability by means of size control and doping options, wide excitation spectrum, technological feasibility and compatibility by means of solid polymer and sol–gel films, all-solid-state semiconductor heterostructures feasibility and high photostability are distinguished advantages of semiconductor quantum dots as compared to traditional luminophores. For example, luminescent organic dyes (rhodamine, fluorescein, coumarine, acridine and others) feature poor photostability and a narrow excitation spectrum. Lanthanide-based luminophores (e.g. ZnS:Eu, ZnS:Tb, Eu-doped glasses and others) feature narrow emission and excitation spectra and poor tuneability. Therefore semiconductor nanocrystals are considered as novel luminescent materials with potential applications as spectral transformers, lighting components and fluorescent labels. In this section a few examples of such applications are considered.

Spectral converters

Spectral converters can be developed based on semiconductor quantum dots to enhance the sensitivity of silicon photodetector and photovoltaic cells. Typically, the spectral sensitivity of a silicon photosensitive device is governed by interband absorption onset from the long-wave side (approximately 1 μm) and by a rapid surface recombination rate at the short-wave edge (<500 nm) because very high interband absorption well above the band-gap energy results in a very thin absorption layer where surface defects promote efficient recombination paths. Typical spectral response is shown in Figure 5.39 by a dashed line. A film cover absorbing light at wavelengths $\lambda < 500$ nm and emitting light in the range of 700–800 nm

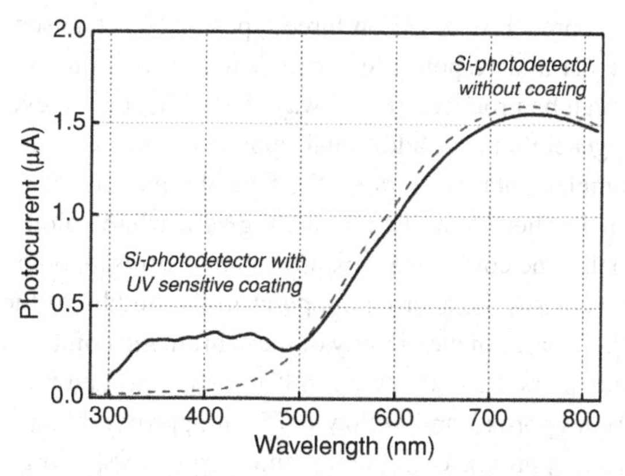

Fig. 5.39 **The spectral sensitivity of a silicon commercial photodiode as made (dashed line) and enhanced by a CdS:Mn nanocrystalline spectral transformer (solid line).**

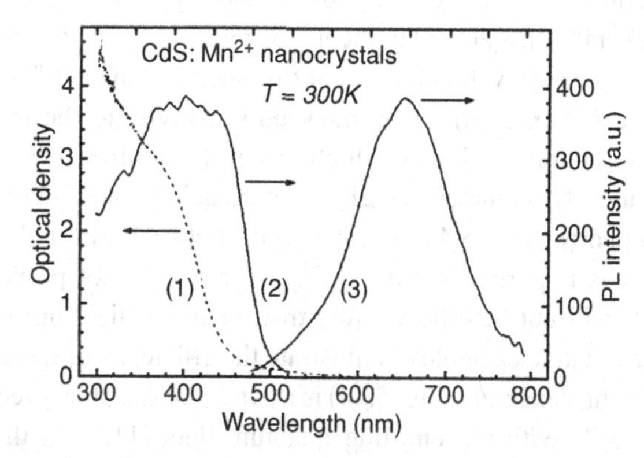

Fig. 5.40 **Optical properties of $CdS : Mn^{2+}$ nanocrystals in a polymer film [110]: (1) absorption, (2) excitation (emission at 650 nm) and (3) emission spectra (excitation at 415 nm).**

will therefore enhance sensitivity to short-wave radiation. The proper candidate for such an application is a doped semiconductor quantum dot material, CdS:Mn. These quantum dots feature a luminescence excitation spectrum controlled by size-dependent absorption (Fig. 5.40, curves 1 and 2), whereas the emission spectrum is determined by the properties of Mn^{2+} ions (Fig. 5.40, curve 3).

White light sources

All solid-state lighting to replace currently used incandescent and gas discharge bulb lamps is considered to be a feasible goal for the near future on a global scale. In this context, semiconductor quantum dot luminophores combined with commercial monochrome light emitting diodes (LEDs) are competitive candidates for all solid-state white light sources.

White light can be generated by a mixture of red (R), green (G), and blue (B) light sources whose intensity is properly balanced to account for human eye sensitivity. The

RBG-approach is based on three types of eye cone sensors occupying the central part of the retina and responsible for high-resolution color vision in daylight. Any identifiable color can be produced in this way as is done, e.g. in every color display screen. For white light generation, an additional approach can be used. It is based on the use of a pair of complementary colors on the CIE chromaticity diagram.[1] Two complementary colors together, when properly weighted, give a white color. In the chromaticity diagram (Fig. 5.41) the line connecting a pair of points corresponding to wavelengths of complementary colors crosses the achromatic point in the middle of the diagram with coordinates (0.33; 0.33). The area in the vicinity of the achromatic point corresponds to white color. Examples of such pairs are blue–green (485 nm approx.) and reddish orange (590 nm approx), blue (470 nm approx.) and yellow (575 nm approx.), blueish green (495 nm approx.) and red (630 nm approx.). Using complementary colors, two types of luminescent sources rather than three RBG colors can give a white light source. Moreover, human eye sensitivity substantially differs for daylight and night light conditions. In the night, RBG-sensors are not efficient because of their low sensitivity. Instead, rod-like sensors occupying the wide peripheral retina area play the major role. Rods do not exhibit high resolution but provide higher sensitivity. It is important that spectral curves of cone and rod sensitivity are rather different (Fig. 5.42). Rods show no sensitivity to the red (for wavelengths greater than 600 nm) but instead feature highest sensitivity around 500 nm [111].

Currently, commercial LEDs use blueish electroluminescent crystal and cerium luminophores (Fig. 5.43) to give white light. However the spectral range around 507 nm which is important for dark-adapted vision is not properly included. This means such LEDs will not be efficient for street or traffic lighting. Figures 5.44 and 5.45 show two representative examples to illustrate the efficiency of quantum dots in white light sources.

The first example (Fig. 5.44) is a combination, blue–green emitting semiconductor quantum wells with red emitting quantum dots [112]. In this hybrid device, cyan-emitting InGaN/GaN quantum wells (photoluminescence peak at 490 nm) pump red-emitting CdSe/ZnS core/shell quantum dots (photoluminescence peak at 650 nm) via both radiative and non-radiative energy transfer. Both types of emitters together form a complementary pair producing white light. It is important that the darkness-efficient spectral range around 500 nm is purposely included in the spectrum.

The second example shows a possibility of performing white light emission by means of only one type of visibly emitting luminophore (Fig. 5.45). Typically, if no special measure (e.g., the core/shell design) has been applied, semiconductor quantum dots in the strong confinement regime feature a broad luminescence band with large Stokes shift along with narrow-band intrinsic emission with small Stokes shift. The broad emission band is reproducible in many cases and is attributed to the surface trap states. Ozel *et al.* [113] have recently proposed the use of two-band emission for white light generation. Using metal nanoparticles selective enhancement/inhibition of intrinsic and trap luminescence have been demonstrated to optimize color balance. The role of metal nanoparticles is the subject of extensive discussion in Chapter 16.

[1] CIE stands for Commission Internationale de l'Eclairage.

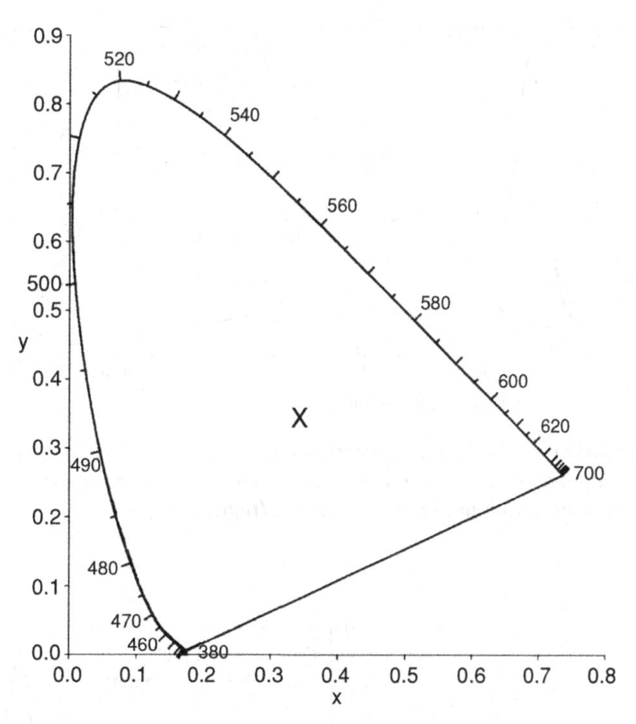

Fig. 5.41 The CIE 1931 chromaticity diagram. Numbers on the perimeter are wavelengths in nanometers. A cross in the middle shows the achromatic point.

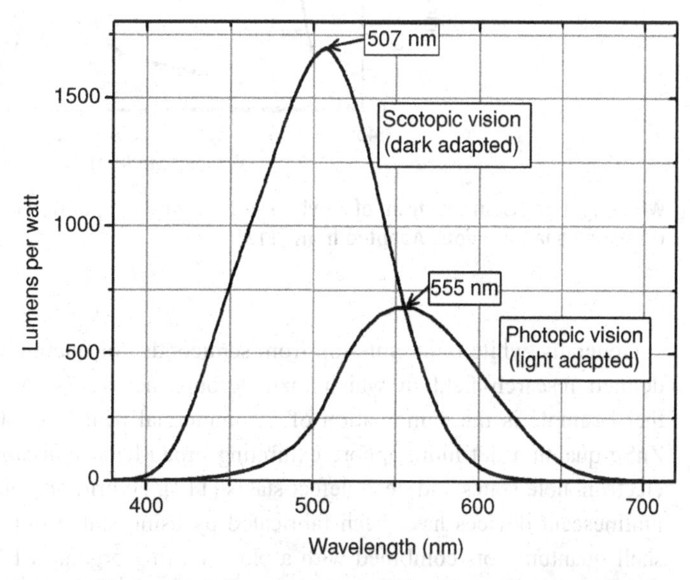

Fig. 5.42 Scotopic (dark-adapted) and photopic (light-adapted) vision efficiency.

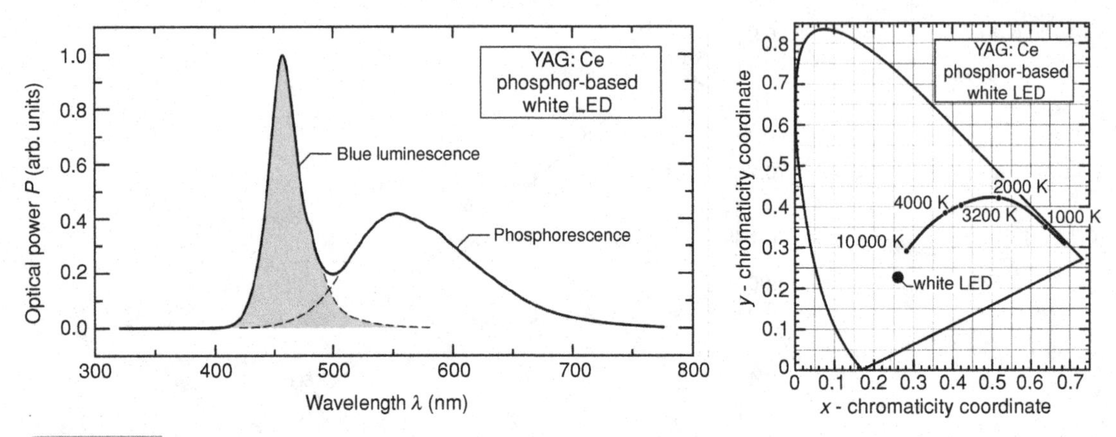

Fig. 5.43 **Emission spectrum of a commercial white diode manufactured by Nichia corporation (left) and position of its spectrum on the chromaticity diagram along with the line indicating the black body spectrum with indicated temperatures (right) [111].**

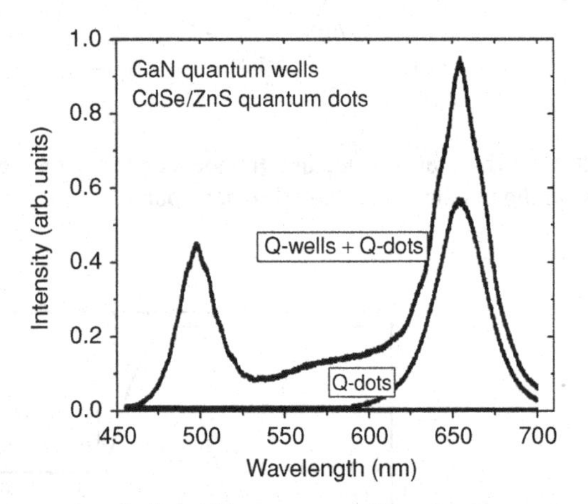

Fig. 5.44 **White light emission spectrum of a hybrid system consisting of GaN quantum well + CdSe/ZnS core-shell quantum dots. Adapted from [112].**

Generally, white light emission from semiconductor quantum dots has become a well-defined research field in which many groups across the world are involved. A further example is the combination of a commercial near-UV (400 nm) InGaN LED with ZnSe quantum dot luminophore exhibiting broad-band emission from confined intrinsic electron–hole states and deep defect states [114]. Hybrid organic/inorganic white electroluminescent devices have been fabricated by using stable red-emitting CdSe/ZnS core–shell quantum dots combined with a blue emitting organic LED consisting of the PFH-MEH polymer, poly(9,9-diocty(fluorene-2,7-diyl)-co-(2-methoxy-5-(2′-ethy(hexoxy)-1,4-phenylenevinylene) and Alq_3 chelate complex [115]. However the overall efficiency of the device measures 0.24% to date. Further issues on this topic are discussed in [116].

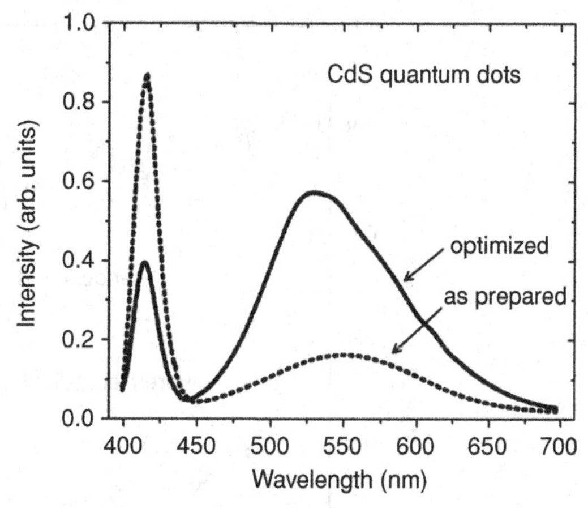

Fig. 5.45 White light emission spectrum from CdS quantum dots. The dashed line shows the original spectrum, the solid line shows the optimized spectrum by means of plasmon-assisted enhancement of the long-wave band using silver nanoparticles. Adapted from [113].

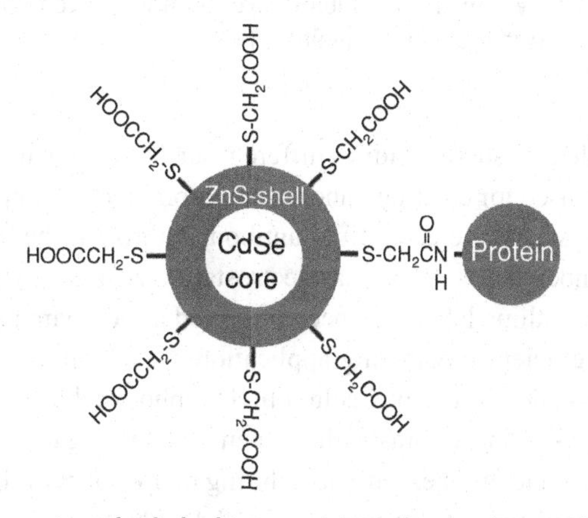

Fig. 5.46 A CdSe/ZnS core–shell quantum dot linked to a protein molecule via organic groups as proposed by Chan and Nie [118].

Fluorescent labels

Colloidal nanocrystals are proposed as efficient fluorescent labels in the form of bioconjugates (Fig. 5.46) for biosensing, fluorescent probes for biological molecules, fluorescent labels in biology and medicine and high-sensitivity biological imaging [117, 118]. Quantum dots were found to possess a narrower emission spectrum, wider excitation spectrum and superior photostability as compared with conventional probes like rhodamine and fluorescein molecules. A wider excitation spectrum offers the possibility of using different colors of labeling with imaging by means of a single laser. A case is reported [119] that four different sequences of DNA have been linked to four nanocrystal samples having different colors of emission in the range 530 to 640 nm (2.339 eV to 1.937 eV). Water-soluble nanocrystals

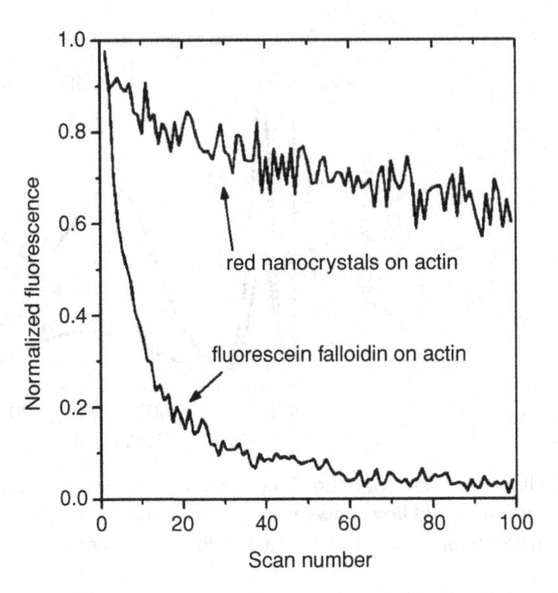

Fig. 5.47 **Photostability comparison of luminescence of fluorescein-falloidin-labeled actin fibers compared with nanocrystal-labeled actin fibers [117].**

with siloxane shells to adjust different surface charge in the outer coat of the CdSe nanocrystals are used for creating nanocrystal–biomolecule conjugates [120]. CdSe–ZnS core–shell nanocrystals were coupled to antibodies through the use of an avidin bridge adsorbed to the nanocrystal surface via electrostatic self-assembly [121]. Superior photostability of nanocrystalline labels has been observed as compared to fluorescein (Fig. 5.47).

For efficient biolabeling applications, semiconductor core/shell colloidal nanocrystals are desirable in aqueous solution. The photostability of the luminescence for such solutions was found to drastically depend on the organic acids used [122]. The proper acids permit stable luminescent yield during many hours of illumination by a typical commercial continuous wave laser source (Fig. 5.48). Note a small increase at the very beginning of illumination.

Electroluminescent structures

Several groups have reported on electroluminescence of thin films with a high concentration of colloidal quantum dots. Reviews on II–VI electroluminescent materials can be found in [8, 12]. Electroluminescent films containing silicon nanocrystals have also been developed and investigated [123]. However, overall efficacy of such structures is too low to promise competitive commercial devices. Quantum dot hetrostructures of larger III–V nanocrystals which were developed as injection laser active components can be treated also as electroluminescent devices if the current is below the threshold. However these structures only emit infrared radiation. Application of epitaxial quantum dot heterostructures in

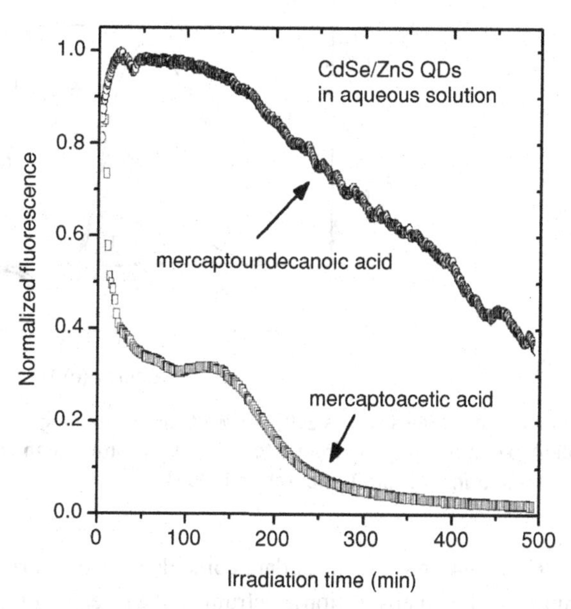

Fig. 5.48 **Photoluminescence intensity of CdSe/ZnS core–shell quantum dots in aqueous solution with mercaptoundecanoic and mercaptoacetic acids for prolonged excitation by a cw Nd:YAG laser ($5\,\mathrm{mW/mm^2}$ at $530\,\mathrm{nm}$) [122].**

electroluminescent devices can be foreseen only if the wider band-gap materials like GaN can be used.

5.12 Applications: electro-optical properties

Electroabsorption

The application of strong (10^4–10^6 Vcm^{-1}) external electrical fields to an ensemble of isolated semiconductor nanocrystals results in broadening and red shift of the absorption bands (the quantum-confined Stark effect). The optical transitions broaden and shift to the red due to gradual field-induced ionization of excitons. Changes in optical transmission are much higher for quantum dots than for bulk crystals and can be purposefully used in electro-optical devices. Figure 5.49 shows the change in optical density of isolated CdSe nanocrystals for various values of applied external bias. Nanocrystals of 1.8 nm diameter have been embedded in a poly(methylmethacrylate) 250 nm thick film. A semitransparent contact of a SnO_2:Sb layer and an aluminum layer were used for voltage application across the sample. Optical density has been measured with and without an external field in the reflection geometry through a semitransparent layer with light double passing through the sample. The measured data for change in optical density ΔD corresponds to a broadening and red shift of the first absorption peak, as expected for isolated nanocrystals. A careful analysis of the data allows us to resolve the negative signals from the second and third optical transitions

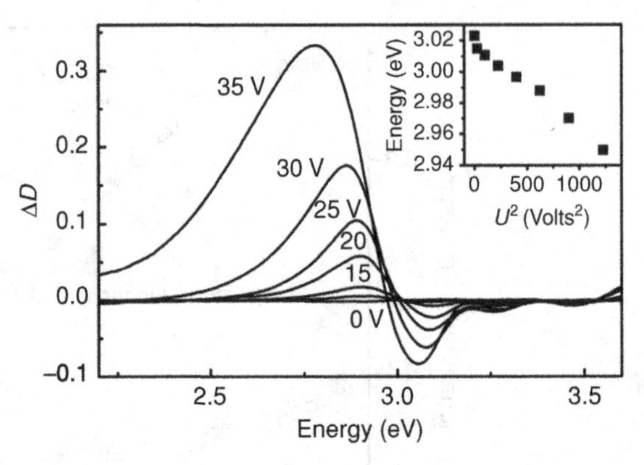

Fig. 5.49 **Change in optical density of a 200 nm PMMA film containing 1.8 nm diameter CdSe nanocrystals at applied external voltage U from 0 to 35 V [79]. Positive change in optical density corresponds to decrease in transmission. Temperature is 300 K.**

at 3.26 and 3.47 eV. The peak data coincide with the corresponding energies of the room-temperature linear absorption spectrum. In the inset of Figure 5.49 the peak photon energy of the first confined optical transition is plotted versus the squared external bias U. The nearly linear dependence $E \sim U^2$ is characteristic for the quantum-confined Stark effect [8, 12].

External electric field effect on luminescence

An external electric field modifies substantially the photoluminescence of quantum dots resulting in luminescence red shift and quenching (Fig. 5.50). Unlike Figure 5.49, the differential absorption spectrum features a rather symmetrical shape which correlates with change in nanocrystal size. The photoluminescence intensity drop is more than one order of magnitude. A possible mechanism of photoluminescence quenching is the spatial separation of electron and hole wave functions in the presence of the electric field. Relaxed overlap of electron and hole wave functions makes the probability of their radiative recombination lower. Then the radiative recombination rate decreases and in the presence of a parallel non-radiative recombination path the luminescence quantum yield drops. If such a mechanism is the case, then a systematic steady modification of luminescence decay rate should be observed. However, time-resolved photoluminescence studies have not revealed a detectable modification of the decay law in the presence of an external electric field. This means that the above model is not plausible. Instead, the following mechanism can be considered. The electric field may cause, or promote, photostimulated "darkening" of a portion of the nanocrystals thus removing this portion from contribution to photoluminescence yield. The remaining "light" nanocrystals continue to emit photons with the same kinetics as if the electric field were absent. The higher the field, the higher is the portion of "dark" nanocrystals. Thus a drop in quantum yield upon application of the external electric field is not from interplay of competitive radiative and non-radiative paths, but rather from the ratio of luminescent and non-luminescent nanocrystals. A photo-induced "darkening" of

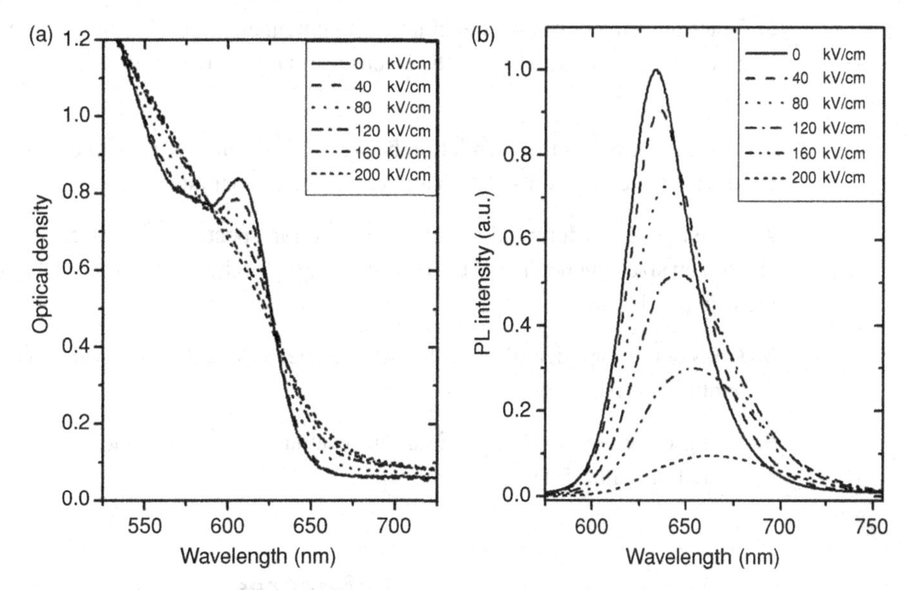

Fig. 5.50 **Electric field effect on (a) optical absorption and (b) emission spectra of CdSe nanocrystals in a polymer film [124]. The mean radius of nanocrystals is 2 nm, temperature is 300 K.**

nanocrystals may occur, e.g. as a result of electron (or a hole) capture by a surface state. This capture is expected to be electric-field promoted since the field separates electron and hole wave functions. Capturing of an electron is more probable since the heavier hole is typically located in the center of a spherically symmetrical dot.

To summarize the material discussed in this chapter, the physics, chemistry and technology of semiconductor nanocrystals (quantum dots) has become a mature field of research and development with promising applications in novel lasers, luminophores, electroluminescent devices and optical and electro-optical modulators.

Problems

1. Compare formulas (5.1) and (5.6) with Eq. (2.98) discussed in Chapter 2 for a particle in a spherical potential barrier with infinite wall and explain the similarities and differences.

2. Try to estimate the absorption blue shift in a quantum dot due to electron confinement from Heisenberg's uncertainty relation, taking the electron effective mass equal to $0.1\, m_0$ and the dot radius equal to 1.5 nm.

3. Consider the applicability criteria for dielectric permittivity in Eq. (5.9) for an interacting electron and a hole inside a quantum dot.

4. Explain why the Coulomb correction in Eq. (5.9) for an electron–hole pair in a spherical dot is always greater than the exciton Rydberg energy for the bulk parent crystal.

5. Trace the tendency in the quantum confinement effect on absorption blue shift of various semiconductor materials versus the electron and hole effective masses and the exciton Bohr radius.

6. Trace and explain the correlation between the original band-gap energy of a bulk semiconductor material and its shift with decreasing nanocrystal size.

7. Based on the differential nonlinear absorption spectra in Figure 5.16 determine a conclusion on size dependence of the energy splitting of the highest hole states $1S_{3/2}$ and $2S_{3/2}$ identified in Figure 5.17.

8. Consider the applicability of the Ioffe–Regel localization criterion (see Section 4.5) to a quantum dot ensemble.

9. Using effective mass approximation, evaluate the mean nanocrystal size for the data presented in Figure 5.18.

References

[1] A. I. Ekimov, A. A. Onushchenko and V. A. Tsekhomskii. Exciton light absorption by CuCl microcrystals in glass matrix. *Sov. Glass Physics and Chemistry*, **6** (1980), 511–512.

[2] A. I. Ekimov and A. A. Onushchenko. Quantum-size effect in optical spectra of semiconductor microcrystals. *Semiconductors*, **16** (1982), 1215–1219.

[3] Al. L. Efros and A. L. Efros. Interband light absorption in a semiconductor sphere. *Semiconductors*, **16** (1982), 1209–1214.

[4] L. E. Brus. A simple model for the ionization potential, electron affinity, and aqueous redox potentials of small semiconductor crystallites. *J. Chem. Phys.*, **79** (1983), 5566–5571.

[5] R. Rossetti, S. Nakahara and L. E. Brus. Quantum size effects in the redox potentials, resonance Raman spectra, and electronic spectra of CdS crystallites in aqueous solution. *J. Chem. Phys.*, **79** (1983), 1086–1088.

[6] L. Banyai and S. W. Koch. *Semiconductor Quantum Dots* (Singapore: World Scientific, 1993).

[7] U. Woggon. *Optical Properties of Semiconductor Quantum Dots* (Berlin: Springer, 1996).

[8] S. V. Gaponenko. *Optical Properties of Semiconductor Nanocrystals* (Cambridge: Cambridge University Press, 1998).

[9] D. Bimberg, N. N. Ledentsov and M. Grundmann. *Quantum Dot Heterostructures* (Chichester: John Wiley and Sons, 1999).

[10] L. Jacak, P. Hawrylak and A. Wojs. *Quantum Dots* (Berlin: Springer-Verlag, 1998).

[11] V. I. Klimov (Ed.). *Semiconductor and Metal Nanocrystals: Synthesis and Electronic and Optical Properties*, vol. 1 (New York: Marcel Dekker, 2004).

[12] S. V. Gaponenko, H. Kalt and U. Woggon. *Semiconductor Quantum Structures. Part 2. Optical Properties.* (Ed. C. Klingshirn) (Berlin: Springer Verlag, 2004).
[13] U. Woggon and S. V. Gaponenko. Excitons in quantum dots. *Phys. Stat. Sol. (b)*, **189** (1995), 286–343.
[14] S. V. Gaponenko. Optical processes in semiconductor nanocrystallites. *Semiconductors*, **30** (1996), 577–619.
[15] A. P. Alivisatos. Semiconductor clusters, nanocrystals, and quantum dots. *Science*, **271** (1996), 933–937.
[16] M. Nirmal and L. Brus. Luminescence photophysics in semiconductor nanocrystals. *Accounts for Chem. Phys.*, **32** (1999), 407–414.
[17] S. A. Empedocles, R. Neuhauser, K. Shimizu and M. G. Bawendi. Photoluminescence from single semiconductor nanostructures. *Adv. Mater.*, **11** (1999), 1243–1256.
[18] A. Eychmüller. Structure and photophysics of semiconductor nanocrystals. *J. Phys. Chem.*, **104** (2000), 6514–6528.
[19] V. Klimov. *Handbook of Nanostructured Materials and Nanotechnology* (Ed. H. S. Nalwa) (Orlando: Academic Press, 2000), p. 451.
[20] A. D. Yoffe. Semiconductor quantum dots and related systems: electronic, optical, luminescence and related properties of low-dimensional systems. *Adv. Phys.*, **50** (2001), 1–208.
[21] S. V. Gaponenko, A. M. Kapitonov, V. N. Bogomolov, A. P. Prokifiev, A. Eychmueller and A. Rogach. Electrons and photons in mesoscopic structures: Quantum dots in a photonic crystal. *JETP Letters*, **68** (1998), 142–147.
[22] A. I. Ekimov, Al. L. Efros and A. A. Onushchenko. Quantum size effects in semiconductor microcrystals. *Solid State Comm.*, **56** (1985), 921–924.
[23] L. E. Brus. Electron-electron and electron-hole interactions in small semiconductor crystallites, the size dependence of the lowest excited electronic state. *J. Chem. Phys.*, **80** (1984), 4403–4409.
[24] L. E. Brus. Electronic wave functions in semiconductor cluster, experiment and theory. *J. Phys. Chem.*, **90** (1986), 2555–2560.
[25] Y. Kayanuma. Wannier exciton in microcrystals. *Solid State Comm.*, **59** (1986), 405–408.
[26] H. M. Schmidt and H. Weller. Quantum size effects in semiconductor crystallites, calculation of the energy spectrum for the confined exciton. *Chem. Phys. Lett.*, **129** (1986), 615–618.
[27] Y. Kayanuma. Quantum-size effects of interacting electrons and holes in semiconductor microcrystals with spherical shape. *Phys. Rev. B*, **38** (1988), 9797–9805.
[28] V. A. Kuzmitskii, V. I. Gael and I. V. Filatov. Quantum-chemical calculation of electron structure and excited states of ZnS and CdS clusters. *J. Appl. Spectr.*, **63** (1996), 714–723.
[29] L. M. Ramaniah and S. V. Nair. Optical absorption in semiconductor quantum dots, a tight-binding approach. *Phys. Rev. B*, **47** (1993), 7132–7139.
[30] M. V. Rama Krishna and R. A. Friesner. Quantum confinement effects in semiconductor clusters. *J. Chem. Phys.*, **95** (1991), 8309–8321.

[31] T. Vossmeyer, L. Katsikas, M. Giersing, I. G. Popovic, K. Diesner, A. Chemseddine, A. Eychmüller and H. Weller. *J. Phys. Chem.*, **98** (1994), 7665–7670.

[32] A. I. Ekimov. Optical properties of semiconductor quantum dots in glass matrix. *Physica Scripta*, **39** (1991), 217–222.

[33] B. G. Potter and J. H. Simmons. Electronic states of semiconductor clusters: homogeneous and inhomogeneous broadening of the optical spectrum. *Phys. Rev. B*, **37** (1988), 10838–10845.

[34] H. Mathieu, T. Richard, J. Allegre, P. Lefebvre, G. Arnaud, W. Granier, L. Boudes, J. L. Marc, A. Pradel and M. Ribes. Quantum confinement effects of CdS nanocrystals in a sodium borosilicate glass prepared by the sol–gel process. *J. Appl. Phys.*, **77** (1995), 287–293.

[35] S. Gaponenko, U. Woggon, M. Saleh, W. Langbein, A. Uhrig, M. Muller and C. Klingshirn. Nonlinear-optical properties of semiconductor quantum dots and their correlation with the precipitation stage. *J. Opt. Soc. Amer. B*, **10** (1993), 1947–1955.

[36] C. B. Murray, D. J. Norris and M. G. Bawendi. Synthesis and characterization of nearly monodisperse CdE (E = S, Se, Te) semiconductor nanocrystallites. *J. Am. Chem. Soc.*, **115** (1993), 8706–8715.

[37] A. Mews, A. V. Kadavanich, U. Banin and A. P. Alivisatos. Structural and spectroscopic investigations of CdS/HgS/CdS quantum-dot quantum wells. *Phys. Rev. B*, **53** (1996), R13242–R13245.

[38] J. Schrier and L. Wang. Electronic structure of nanocrystal quantum-dot quantum wells. *Phys. Rev. Lett. B*, **73** (2006), 245332.

[39] S. Nizamoglu and H. V. Demir. Onion-like (CdSe)ZnS/CdSe/ZnS quantum-dot quantum-well heteronanocrystals for multi-color emission. *Optics Express*, **16** (2008), 3515–3526.

[40] V. S. Burakov, E. A. Nevar, M. I. Nedel'ko and N. V. Tarasenko. Formation of zinc oxide nanoparticles during electric discharge in water. *Techn. Phys. Lett.*, **34** (2008), 679–681.

[41] W. Faschinger. Fundamental doping limits in wide-band II–VI compounds. *J. Cryst. Growth*, **159** (1996), 221–228.

[42] Zh. I. Alferov. The double heterostructure: Concept and its applications in physics, electronics and technology. Nobel Lecture. 2000.

[43] R. Nötzel, N. N. Ledentsov, L. Däweritz, M. Hohenstein and K. Ploog. Direct synthesis of corrugated superlattices. *Phys. Rev. Lett.*, **67** (1991), 3812–3815.

[44] J. Stangl, V. Holý and G. Bauer. Structural properties of self-organized semiconductor nanostructures. *Rev. Mod. Phys.*, **76** (2004), 725–783.

[45] D. Grützmacher, T. Fromherz, G. Dais, J. Stangl, E. Müller, Y. Ekinci, H. H. Solak, H. Sigg, R. T. Lechner, E. Winterberger, S. Birner, V. Holy and G. Bauer. Three-dimensional Si/Ge quantum dot crystals. *Nano Letters*, **7** (2007), 3822–3826.

[46] K. Georgsson, N. Carlsson, L. Samuelson, W. Seifert and L. R. Wallenberg. Transmission electron microscopy investigation of the morphology in InP Stranski-Krastanow islands grown by metalorganic chemical vapor deposition. *Appl.Phys. Lett.*, **67** (1995), 2981–2982.

[47] N. Carlsson, K. Georgsson, L. Montelius, L. Samuelson, W. Seifert and L. R. Wallenberg. Improved size homogeneity of InP-on GaInP Stranski-Krastanow islands by growth on a thin GaP interface layer. *J. Cryst. Growth*, **156** (1995), 23–29.

[48] S. Gaponenko, U. Woggon, A. Uhrig, W. Langbein and C. Klingshirn. Narrow-band spectral hole burning in quantum dots. *J. Lumin.*, **60** (1994), 302–307.

[49] U. Woggon, S. Gaponenko, W. Langbein, A. Uhrig and C. Klingshirn. Homogeneous linewidth of electron-hole-pair states in II-VI quantum dots. *Phys. Rev. B*, **47** (1993), 3684–3689.

[50] A. I. Ekimov, F. Hache, M. C. Schanne-Klein, D. Ricard, C. Flytzanis, I. A. Kudryavtsev, T. V. Yazeva, A. V. Rodina and Al. L. Efros. Absorption and intensity-dependent photoluminescence measurement on CdSe quantum dots: assignment of the first electronic transitions. *J. Opt. Soc. Amer. B*, **10** (1993), 100–110.

[51] S. Kudera, M. Zanella, C. Giannini, A. Rizzo, Y. Li, G. Gigli, R. Cingolani, G. Ciccarella, W. Spahl, W. J. Parak and L. Manna. Sequential growth of magic-size CdSe nanocrystals. *Adv. Mater.*, **19** (2007), 548–552.

[52] C. Klingshirn. *Semiconductor Optics* (Berlin: Springer Verlag, 1995).

[53] A. Uhrig, L. Banyai, S. Gaponenko, A. Worner, N. Neuroth and C. Klingshirn. Linear and nonlinear optical studies of CdSSe quantum dots. *Z. Physik D*, **20** (1991), 345–348.

[54] A. M. Kapitonov, A. P. Stupak, S. V. Gaponenko, E. P. Petrov, A. L. Rogach and A. A. Eychmueller. Photoluminescence of CdTe nanocrystals in aqueous solution. *J. Phys. Chem.*, **103** (1999), 10109–10113.

[55] M. V. Artemyev, L. I. Gurinovich and S. V. Gaponenko. (unpublished data) 2004.

[56] O. I. Micic, J. Sprague, Z. Lu and A. J. Nozik. Highly efficient band-edge emission from InP quantum dots. *Appl. Phys. Lett.*, **68** (1996), 3150–3152.

[57] O. Labeau, P. Tamarat and B. Lounis. Temperature dependence of the luminescence lifetime of single CdSe/ZnS quantum dots. *Phys. Rev. Lett.*, **90** (2003), 257404.

[58] G. Schlegel, J. Bohnenberger, I. Potapova and A. Mews. Fluorescence decay time of single semiconductor nanocrystals. *Phys. Rev. Lett.*, **88** (2002), 137401.

[59] S. A. Empedocles, D. J. Norris and M. G. Bawendi. Photoluminescence spectroscopy of single CdSe quantum dots. *Phys. Rev. Lett.*, **77** (1996), 3873–3876.

[60] A. G. Cullis, L. T. Canham and P. D. J. Calcott. The structural and luminescence properties of porous silicon. *J. Appl. Phys.*, **82** (1997), 909–965.

[61] S. V. Gaponenko, E. P. Petrov, U. Woggon, O. Wind, C. Klingshirn, Y. H. Xie, I. N. Germanenko and A. P. Stupak. Steady-state and time-resolved spectroscopy of porous silicon. *J. Luminescence*, **70** (1996), 364–376.

[62] S. V. Gaponenko, I. N. Germanenko, E. P. Petrov, A. P. Stupak, V. P. Bondarenko and A. M. Dorofeev. Time-resolved spectroscopy of visibly emitting porous silicon. *Appl. Phys. Lett.*, **64** (1994), 85–87.

[63] A. Witt, K. Gordon and D. Furton. Silicon Nanoparticles: Source of Extended Red Emission? *Astrophys. J.*, **501** (1998), L111–L115.

[64] G. Binning, H. Rohrer, Ch. Gerber and E. Weibel. Surface studies by scanning tunneling microscopy. *Phys. Rev. Lett.*, **57** (1982), 57–60.

[65] R. Wiesendanger. *Scanning Probe Microscopy and Spectroscopy* (Cambridge: Cambridge University Press, 1994).

[66] E. L. Wolf. *Principles of Electron Tunneling Spectroscopy* (Oxford: Oxford University Press, 1989).

[67] O. Millo, D. Katz, Y. W. Cao and U. Banin. Scanning tunneling spectroscopy of InAs nanocrystal quantum dots. *Phys. Rev. B*, **61** (2000), 16773–16776.

[68] K. Kanisawa, M. J. Butcher, Y. Tokura, H. Yamaguchi and Y. Hirayama. Local density of states in zero-dimensional semiconductor structures *Phys. Rev. Lett.*, **87** (2001), 196804.

[69] G. D. Stucky and J. E. MacDougall. Quantum confinement and host/guest chemistry: probing a New Dimension. *Science*, **247** (1990), 669–678.

[70] J. Tersoff, C. Teichert and M. G. Lagally. Self-organization in growth of quantum dot superlattices. *Phys. Rev. Lett.*, **76** (1996), 1675–1678.

[71] R. C. Williams and K. Smith. A crystallizable insect virus. *Nature*, **45** (1957), 119–121.

[72] P. Pieranski. Colloidal crystals. *Contemp. Physics*, **24** (1983), 25–73.

[73] C. B. Murray, C. R. Kagan and M. G. Bawendi. Self-organization of CdSe nanocrystallites into three-dimensional quantum dot superlattices. *Science*, **270** (1995), 1335–1338.

[74] P. W. Anderson. Absence of electron diffusion in certain random lattices. *Phys. Rev.*, **109** (1958), 1492–1508.

[75] B. I. Shklovskii and A. L. Efros. *Electron Properties of Doped Semiconductors* (Moscow: Nauka; Berlin: Springer, 1979).

[76] I. M. Lifshitz. On the structure of energy spectrum and quantum states in disordered condensed systems. *Sov. Phys. Uspekhi*, **83** (1964), 617–640.

[77] M. V. Artemyev, A. I. Bibik, L. I. Gurinovich, S. V. Gaponenko and U. Woggon. Evolution from individual to collective electron states in a dense quantum dot ensemble. *Phys. Rev. B*, **60** (1999), 1504–1507.

[78] A. I. Bibik. Influence of quantum size effects on electronic and optical spectra of thin film heterostructures and nanocrystal assemblies. Ph.D. thesis, Minsk (2007).

[79] M. V. Artemyev, U. Woggon, H. Jaschinski, L. I. Gurinovich and S. V. Gaponenko. Spectroscopic study of electronic states in an ensemble of close-packed CdSe nanocrystals. *J. Phys. Chem.*, **104** (2000), 11617–11621.

[80] M. V. Artemyev, A. I. Bibik, L. I. Gurinovich, S. V. Gaponenko, H. Jaschinski and U. Woggon. Optical properties of dense and diluted ensembles of semiconductor quantum dots. *Physica Status Solidi (b)*, **224** (2001), 393–396.

[81] A. Eychmueller. Private communication 2000.

[82] A. Rizzo, Y. Li, S. Kudera, F. D. Sala, M. Zanella, W. J. Parak, R. Cingolani, L. Manna and G. Gigli. Blue light emitting diodes based on fluorescent CdSe/ZnS nanocrystals. *Appl. Phys. Lett.*, **90** (2007), 051106.

[83] O. L. Lazarenkova and A. A. Balandin. Mini band formation in a quantum dot crystal. *J.Appl.Phys.*, **89** (2001), 5509–5515.

[84] A. Uhrig, L. Banyai, S. Gaponenko, A. Worner, N. Neuroth and C. Klingshirn. Linear and nonlinear optical studies of CdSSe quantum dots. *Z. Physik D*, **20** (1991), 345–348.

[85] L. G. Zimin, S. V. Gaponenko and V. Yu. Lebed. Room-temperature optical nonlinearity in semiconductor-doped glasses. *Phys. Stat. Sol. (b)*, **150** (1988), 653–656.

[86] A. L. Rogach (Ed.). *Semiconductor Nanocrystals Quantum Dots* (New York: Springer, 2008).

[87] A. A. Lagatsky, C. G. Leburn, C. T. A. Brown, W. Sibbett, A. M. Malyarevich, V. G. Savitski, K. V. Yumashev, E. L. Raaben and A. A. Zhilin. Passive mode-locking of a Cr^{4+}: YAG laser by PbS quantum-dot-doped glass saturable absorber. *Optics Commun.*, **241** (2004), 449–454.

[88] M. S. Gaponenko, I. A. Denisov, V. E. Kisel, A. M. Malyarevich, A. A. Zhilin, A. A. Onushchenko, N. V. Kuleshov and K. V. Yumashev. Diode-pumped $Tm:KY(WO_4)_2$ laser passively Q-switched with PbS-doped glass. *Appl. Phys. B*, **93** (2008), 787–791.

[89] V. G. Savitsky, A. M. Malyarevich, K. V. Yumashev, V. L. Kalashnikov, B. D. Sinclair, H. Raaben and A. A. Zhilin. Experiment and modeling of a diode-pumped 1.3-μm $Nd:YVO_4$ laser passively Q-switched with PbS-doped glass. *Appl. Phys. B*, **79** (2004), 315–320.

[90] A. A. Lagatsky, A. M. Malyarevich, V. G. Savitski, M. S. Gaponenko, K. V. Yumashev, A. A. Zhilin, C. T. A. Brown and W. Sibbett. Passive mode locking of $Yb^{3+}:KY(WO_4)_2$ laser with PbS quantum-dot-doped glass. *IEEE Photonics Technol. Lett.*, **18** (2006), 259–261.

[91] A. A. Lagatsky, A. M. Malyarevich, V. G. Savitski, M. S. Gaponenko, K. V. Yumashev, A. A. Zhilin, C. T. A. Brown and W. Sibbett. Passive mode locking of $Yb^{3+}:KY(WO_4)_2$ laser with PbS quantum-dot-doped glass. *IEEE Photonics Technol. Lett.*, **18** (2006), 259–261.

[92] A. M. Malyarevich, K. V. Yumashev and A. A. Lipovskii. Semiconductor-Doped Glass Saturable Absorbers for Near-Infrared Solid-State Lasers (review). *J. Appl. Phys.*, **103** (2008), 081301.

[93] Y. Arakawa and H. Sakaki. Multidimensional quantum well laser and temperature dependence of its threshold current. *Appl. Phys. Lett.*, **40** (1982), 939–941.

[94] Y. V. Vandyshev, V. S. Dneprovskii, V. I. Klimov and D. K. Okorokov. Laser generation in semiconductor quasi-zero-dimensional structure on a transition between size quantization levels. *JETP Lett.*, **54** (1991), 441–444.

[95] Y. Z. Hu, S. W. Koch and N. Peyghambarian. Strongly confined semiconductor quantum dots: Pair excitations and optical properties. *J. Luminescence*, **70** (1996), 185–202.

[96] U. Woggon, O. Wind, F. Gindele, E. Tsitsishvili and M. Mueller. Optical transmission in CdSe quantum dots: From discrete levels to broad gain spectra. *J. Luminescence*, **70** (1996), 269–280.

[97] V. I. Klimov, A. A. Mikhailovsky, S. Xu, A. Malko, J. A. Hollingsworth, C. A. Leatherdale, H. J. Eisler and M. G. Bawendi. Optical gain and stimulated emission in nanocrystal quantum dots. *Science*, **290** (2000), 314–317.

[98] G. Chen, R. Rapaport, D. T. Fuchs, L. Lucas, A. J. Lovinger, S. Vilan, A. Aharoni and U. Banin. Optical gain from InAs nanocrystal quantum dots in a polymer matrix. *Appl. Phys. Lett.*, **87** (2005), 251108.

[99] K. Wundke, J. Auxier, A. Schulzgen, N. Peyghambarian and N. F. Borrelli. Room-temperature gain at 1.3 μm in PbS-doped glasses. *Appl. Phys. Lett.*, **75** (1999), 3060–3063.

[100] R. D. Schaller, M. A. Petruska and V. I. Klimov. Tunable near-infrared optical gain and amplified spontaneous emission using PbSe Nanocrystals. *J. Phys. Chem. B*, **107** (2003), 13765–13768.

[101] L. Pavesi, L. D. Negro, C. Mazzoleni, G. Franzo and F. Priolo. Optical gain in silicon nanocrystals. *Nature*, **408** (2000), 440–444.

[102] L. Pavesi, S. Gaponenko and L. Dal Negro (Eds.). *Towards the First Silicon Laser. NATO Science Series* (Dordrecht: Kluwer, 2003).

[103] J. Valenta, I. Pelant and J. Linnros. Waveguiding effects in the measurement of optical gain in a layer of Si nanocrystals. *Appl. Phys. Lett.*, **81** (2002), 1396–1398.

[104] H. S. Rong, A. Liu, R. Jones, O. Cohen, D. Hak, R. Nicolaescu, A. Fang and M. Paniccia. An all-silicon Raman laser. *Nature*, **433** (2005), 292–294.

[105] N. N. Ledentsov, V. M. Ustinov, V. A. Shchukin, P. S. Kop'ev, Zh. I. Alferov and D. Bimberg. Quantum dot heterostructures: Fabrication, properties, lasers (review). *Semiconductors*, **32** (1998), 343–365.

[106] V. M. Ustinov, A. E. Zhukov, A. Y. Egorov and N. A. Maleev. *Quantum Dot Lasers* (New York: Oxford Univ. Press, 2003).

[107] H. Y. Liu, D. T. Childs, T. J. Badcock, K. M. Groom, I. R. Sellers, M. Hopkinson, R. A. Hogg, D. J. Robbins, D. J. Mowbray and M. S. Skolnick. High-performance three-layer 1.3-μm InAs-GaAs quantum-dot lasers with very low continuous-wave room-temperature threshold currents. *IEEE Photon. Technol. Lett.*, **17** (2005), 1139–1141.

[108] E. U. Rafailov, M. A. Cataluna and W. Sibbett. Mode-locked quantum-dot lasers. *Nature Photonics*, **1** (2007), 395– 401.

[109] E. U. Rafailov, M. A. Cataluna, W. Sibbett, N. D. Il'inskaya, Yu. M. Zadiranov, A. E. Zhukov, V. M. Ustinov, D. A. Livshits, A. R. Kovsh and N. N. Ledentsov. High-power picosecond and femtosecond pulse generation from a two-section mode-locked quantum-dot laser. *Appl. Phys. Lett.*, **87** (2005), 081107.

[110] L. I. Gurinovich, M. V. Artemyev, A. P. Stupak and S. V. Gaponenko. Luminescence of CdS nanoparticles doped with Mn. *Physica Status Solidi (b)*, **224** (2001), 191–194.

[111] E. F. Schubert. *Light Emitting Diodes* (Cambridge: Cambridge University Press, 2006).

[112] S. Nizamoglu, E. Sari, J.-H. Baek, I. H. Lee and H. V. Demir. White light generating resonant nonradiative energy transfer from epitaxial InGaN/GaN quantum wells to colloidal CdSe/ZnS core/shell quantum dots. *New J. Physics*, **10** (2008), 123001.

[113] T. Ozel, I. M. Soganci, S. Nizamoglu, I. O. Huyal, E. Mutlugun, S. Sapra, N. Gaponik, A. Eychmuller and H. V. Demir. Giant enhancement of surface-state emission in white luminophor CdS nanocrystals using localized plasmon coupling. *New J. Physics*, **10** (2008), 083035.

[114] H. S. Chen, S. J. Wang, C. J. Lo and J. Y. Chi. White-light emission from organics-capped ZnSe quantum dots and application in white-light-emitting diodes. *Appl. Phys. Lett.*, **86** (2005), 131905.

[115] Y. Li, A. Rizzo, M. Mazzeo, L. Carbone, L. Manna, R. Cingolani and G. Gigli. White organic light-emitting devices with CdSe/ZnS quantum dots as a red emitter. *Appl. Phys. Lett.*, **97** (2005), 113501.

[116] S. Sapra, S. Mayilo, T. A. Klar, A. L. Rogach and J. Feldmann. Bright white light emission from semiconductor nanocrystals: by chance and by design. *Adv. Mater.*, **19** (2007), 569–575.

[117] M. Jr. Bruchez, M. Moronne, P. Gin, S. Weiss and A. P. Alivisatos. Semiconductor nanocrystals as fluorescent biological labels. *Science*, **281** (1998), 2013–2016.

[118] W. C. W. Chan and S. Nie. Quantum dot bioconjugates for ultrasensitive nonisotopic detection. *Science*, **281** (1998), 2016–2018.

[119] D. Gerion, W. J. Parak, S. C. Williams, D. Zanchet, C. M. Micheel and A. P. Alivisatos. Sorting fluorescent nanocrystals with DNA. *J. Am. Chem. Soc.*, **124** (2002), 7070–7074.

[120] A. P. Alivisatos. The use of nanocrystals in biological detection. *Nature Biotechnology*, **22** (2004), 47–52.

[121] J. K. Jaiswal, H. Mattousi, J. M. Mauro and S. M. Simon. Long-term multiple color imaging of live cells using quantum dot bioconjugates. *Nature Biotechnology*, **21** (2003), 47–51.

[122] M. V. Artemyev, L. I. Gurinovich, A. A. Lutich and S. V. Gaponenko. Synthesis of photostable cadmium halcogenide nanocrystals and investigation of their stability under prolonged illumination by UV and visible light. *J. Appl. Spectr.*, **73** (2006), 506–509.

[123] P. M. Fauchet. Photoluminescence and electroluminescence from porous silicon. *J. Luminescence*, **70** (1996), 294–309.

[124] L. I. Gurinovich, A. A. Lutich, A. P. Stupak, S. Ya. Prislopsky, E. K. Rusakov, M. V. Artemyev, S. V. Gaponenko and H. V. Demir. Luminescence of cadmium selenide quantum dots and nanorods in the external electric field. *Semiconductors*, **43** (2009), 1045–1053.

6 Nanoplasmonics I: metal nanoparticles

This chapter provides a brief introduction to optical properties of metal nanoparticles in terms of plasma-based optical response, interband transitions and size-dependent properties. The chapter is important for understanding contemporary research in *nanoplasmonics* including surface-enhanced emission and scattering of light near metal surfaces and nanobodies which will be the subjects of Chapter 16. Amazingly, nanoplasmonics is actually an ancient field of science and technology in spite of the fact that the notation for this trend in science only emerged just a decade or so ago. The first systematic studies of brilliant colors of dispersed metal colloids date back to Michael Faraday (1857). Purposeful applications of optical properties of metal nanoparticles are well known for example, to get colors in stained glass, which dates back to ancient Roman times. Gold and copper nanoparticles have been used routinely for decades in the glass industry in red glass production.

Prior to going through this chapter, it is advisable to recall the description of the dielectric function of a gas of non-interacting charged particles (Section 3.3) and the introduction to the electron theory of solids given in Sections 4.1–4.3. For a comprehensive description of optical properties of metal nanoparticles the books by Kreibig and Vollmer [1] and Maier [2] are recommended.

6.1 Optical response of metals

. . . if the electrons are treated as free . . . , we can not only account qualitatively for the phenomenon observed by Wood, but also can predict correctly the approximate values of the critical wave-length.

Clarence Zener, 1933

The optical properties of metals are determined mainly by the response of free electrons occupying states in the conduction band. This response can in turn be reasonably described in terms of the dielectric function of a free electron system and, additionally, by absorption of light owing to possible electron transitions to upper states in the higher bands with respect to the conduction band of a given metal crystal. The role of the crystal lattice in such considerations is reduced to modification of the electron mass to give the effective mass m_e^* instead of the free electron mass m_0, and to the development of upper states for optical transitions. These characteristics are discovered in the electron theory of metals and then used in the theory for the optical properties of metals.

Consider first the dielectric response of an electron gas with electron concentration N. As in Section 3.3, it can be described in terms of the dielectric function $\varepsilon(\omega)$ defined from the relation,

$$\mathbf{D} = \varepsilon_0 \mathbf{E} + \mathbf{P} = \varepsilon_0 \varepsilon(\omega) \mathbf{E}, \tag{6.1}$$

where $\mathbf{E}$ is the electric field which gives rise to polarization described by the $\mathbf{P}$ vector to give the electric displacement vector $\mathbf{D}$ in the medium under consideration (see Eq. (3.48) and the material equation (2.30)).

Similar to Chapter 3, we depart from the single particle equation of motion which, unlike Eq. (3.49) should contain a damping factor to account for collisions described by the damping constant, $\Gamma = \tau_{\text{coll}}^{-1}$, with τ_{coll}^{-1} being the average time between successive collisions. The equation of motion for an oscillating field,

$$\mathbf{E}(t) = \mathbf{E}_0 \exp(-i\omega t), \tag{6.2}$$

reads,

$$m \frac{\mathrm{d}^2 x}{\mathrm{d}t^2} + m\Gamma \frac{\mathrm{d}x}{\mathrm{d}t} = -e\mathbf{E}_0 \exp(-i\omega t). \tag{6.3}$$

Its solution has the form,

$$\mathbf{x}(t) = \frac{e}{m(\omega^2 + i\Gamma\omega)} \mathbf{E}_0 \exp(-i\omega t), \tag{6.4}$$

with the complex amplitude accounting for possible phase shift between the driving field and the medium response. Further, evaluating the polarization,

$$\mathbf{P} = -Ne\mathbf{x} = -\frac{Ne^2}{m(\omega^2 + i\Gamma\omega)} \mathbf{E}_0 \exp(-i\omega t), \tag{6.5}$$

and the $\mathbf{D}(\mathbf{E})$ function,

$$\mathbf{D}(\mathbf{E}) = \varepsilon_0 \left(1 - \frac{Ne^2}{\varepsilon_0 m(\omega^2 + i\Gamma\omega)} \right) \mathbf{E}, \tag{6.6}$$

one arrives at the expression for $\varepsilon(\omega)$,

$$\varepsilon(\omega) = 1 - \frac{Ne^2}{\varepsilon_0 m(\omega^2 + i\Gamma\omega)} = 1 - \frac{\omega_{\mathrm{p}}^2}{\omega^2 + i\Gamma\omega}, \tag{6.7}$$

with the familiar notation of plasma frequency ω_{p} (see Eq. 3.54),

$$\omega_{\mathrm{p}}^2 = \frac{Ne^2}{\varepsilon_0 m}. \tag{6.8}$$

Comparing expression (6.7) with Eq. (3.54) one can see there is the additional term $i\Gamma\omega$ in the denominator describing damping. This expression can be written in the form dividing the real and the imaginary part of the dielectric function,

$$\varepsilon(\omega) = 1 - \frac{\omega_{\mathrm{p}}^2}{\omega^2 + i\Gamma\omega} = 1 - \frac{\omega_{\mathrm{p}}^2}{\omega^2 + \Gamma^2} + i \frac{\omega_{\mathrm{p}}^2 \Gamma}{\omega(\omega^2 + \Gamma^2)} \equiv \varepsilon_1(\omega) + i\varepsilon_2(\omega). \tag{6.9}$$

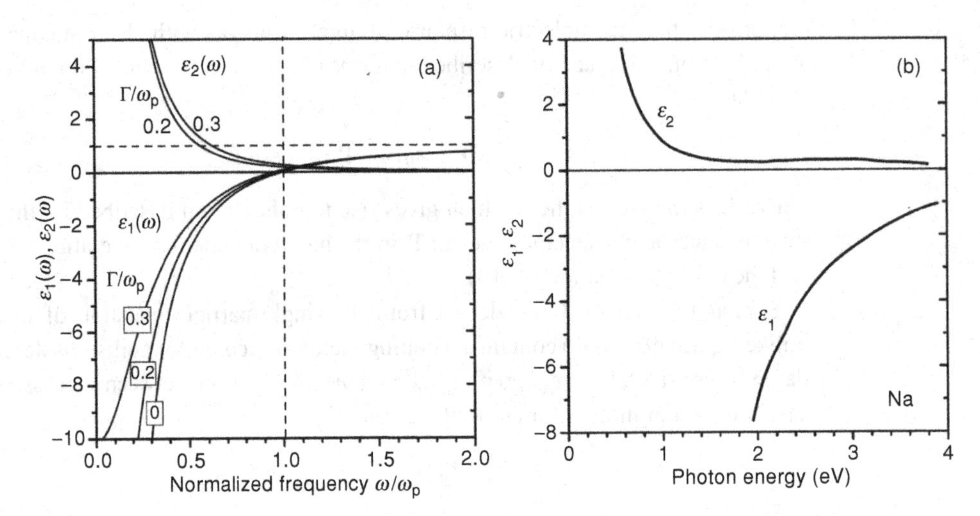

Fig. 6.1 **(a) Real $\varepsilon_1(\omega)$ and imaginary $\varepsilon_2(\omega)$ parts of the dielectric function of an electron gas with different values of damping Γ. (b) Experimentally found $\varepsilon_1(\omega)$ and $\varepsilon_2(\omega)$ functions for solid bulk sodium metal according to [1].**

The real $\varepsilon_1(\omega)$ and imaginary $\varepsilon_2(\omega)$ parts of the dielectric function (6.9) of the electron gas with different values of damping Γ are plotted in Figure 6.1a. It is reasonable to analyze the dielectric function (6.9) for different frequencies with respect to ω_p and Γ. Consider frequencies $\omega < \omega_p$, where metals retain their metallic features and the case $\Gamma \ll \omega_p$ which holds in typical experimental conditions.

For large frequencies close to ω_p, the inequality $\omega \gg \Gamma$ is plausible. In this case $\varepsilon(\omega)$ is predominantly real, and,

$$\varepsilon(\omega) \approx 1 - \frac{\omega_p^2}{\omega^2} \tag{6.10}$$

holds. The nearly real value of the dielectric function means negligible absorption. However, in reality, at high frequencies in common metals interband transitions do contribute to the imaginary part of $\varepsilon(\omega)$ resulting in absorptive losses. Sodium metal represents a rare example for which real and imaginary parts of the dielectric function (Figure 6.1b) look actually similar to those given by Eq. (6.9).

For other frequencies, the dielectric function is essentially complex. It is related to the complex refractive index $n = n_1 + in_2$ as $[n(\omega)]^2 = \varepsilon(\omega)$ whence,

$$n_1(\omega) = \sqrt{\frac{\sqrt{\varepsilon_1^2 + \varepsilon_2^2}}{2} + \frac{\varepsilon_1}{2}}, \quad n_2(\omega) = \sqrt{\frac{\sqrt{\varepsilon_1^2 + \varepsilon_2^2}}{2} - \frac{\varepsilon_1}{2}}. \tag{6.11}$$

The imaginary part $n_2(\omega)$ of the complex refractive index defines the value of κ introduced in Chapter 3 (Eq. 3.65 and Table 3.5) which determines electric field amplitude $E(L)$ evanescence $E(L) \propto \exp(-\kappa L)$ and the skin layer thickness as κ^{-1}, namely,

$$\kappa = \frac{\omega}{c} n_2(\omega). \tag{6.12}$$

Table 6.1. Parameters of free electrons in selected metals. Source (if not specified): the skin depth, the mean free path, and the threshold energy for interband absorption [1], free electron concentration, plasmon energy $\hbar\omega_p$, Fermi energy and Fermi velocity [3], refractive index [4]. Temperature is 273 K.

Element	Na	Al	Cu	Ag	Au
Skin depth (2 eV) [nm]	38	13	30	24	31
Skin depth (3 eV) [nm]	42	13	30	29	37
Skin depth (4 eV) [nm]	48	13	29	82	27
Mean free path ℓ [nm]	34	16	42	52	42
Free electron concentration, N [cm^{-3}]	$2.65 \cdot 10^{22}$	$18.0 \cdot 10^{22}$	$8.45 \cdot 10^{22}$	$5.85 \cdot 10^{22}$	$5.9 \cdot 10^{22}$
Plasmon energy $\hbar\omega_p$	5.71	15.3	9.3 [8]	9.0 [5]	8.55 [6]
Fermi velocity, v_F [cm/s]	$1.07 \cdot 10^8$	$2.02 \cdot 10^8$	$1.57 \cdot 10^8$	$1.39 \cdot 10^8$	$1.39 \cdot 10^8$
Fermi energy, E_F [eV]	3.23	11.63	7.00	5.48	5.51
Threshold energy for interband absorption [eV]	2.1	1.5	2.1	3.9	2.4
Refractive index (560 nm)	–	1.02-i6.85	0.83-i2.60	0.12-i3.45	0.31-i2.88

Then the absorption coefficient α entering into the Bouguer law for light intensity I evanescence at a distance L,

$$I(L) = I(0)\exp(-\alpha L), \tag{6.13}$$

reads $\alpha = 2\kappa = 2\omega n_2/c$ since intensity is proportional to E^2. The real and imaginary parts of the complex refractive index for a number of metals are listed in Table 6.1.

In the regime of very low frequencies, where $\omega \ll \Gamma \ll \omega_p$, the real and the imaginary parts have comparable magnitudes resulting in close values of the real n_1 and imaginary n_2 parts of the complex refractive index with,

$$n_1 \approx n_2 = \sqrt{\frac{\varepsilon_2}{2}} = \sqrt{\frac{\omega_p^2}{2\omega\Gamma}}. \tag{6.14}$$

In this region, metals are mainly absorbing, with the absorption coefficient,

$$\alpha = \frac{\omega_p}{c}\sqrt{\frac{2\omega}{\Gamma}}. \tag{6.15}$$

Let us now introduce the notion of the electron mean free path ℓ. This is the path length of an electron with mean velocity $\bar{v}$ during the time between two successive scattering events $\tau = \Gamma^{-1}$. Electrons obey the *Fermi–Dirac distribution function*,

$$f(E) = \overline{N}(E) = \frac{1}{\exp\dfrac{E - E_F}{k_B T} + 1}, \tag{6.16}$$

and only those electrons whose kinetic energy E lies in the vicinity of the Fermi energy E_F can actually experience scattering. Neither electron with much lower energy can change its

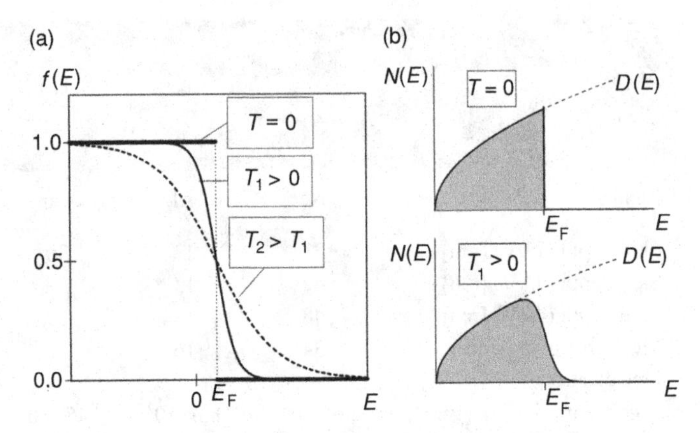

Fig. 6.2 **(a) Fermi–Dirac distribution function $f(E)$ at different temperatures. (b) Electron density of states $D(E)$ (dashes) and the product $f(E)D(E)$ (solid line). The shaded area under the curve $f(E)D(E)$ equals the total number of electrons in accordance with Eq. (6.18).**

energy and/or momentum since there are no appropriate free states. Therefore, the Fermi velocity v_F determined by the electron Fermi energy $E_\mathrm{F} = m_\mathrm{e}^* v_\mathrm{F}^2/2$ is a good evaluation for $\bar{v}$ to give,

$$\ell = v_\mathrm{F}\Gamma^{-1} = \frac{1}{\Gamma}\sqrt{\frac{2E_\mathrm{F}}{m_\mathrm{e}^*}}, \tag{6.17}$$

where m_e^* is the electron effective mass.

The Fermi energy for a given electron density N can be found by means of statistical physics based on the statement that the given total density of electrons within the conduction band should be equal to the integral,

$$N = \int_0^{E^*} \bar{N}(E,T)D(E)\mathrm{d}E, \tag{6.18}$$

where $D(E)$ is the electron density of states defined by Eq. (2.26). The integration range here starts at the bottom of the conduction band where the kinetic energy equals zero and extends high enough to account for all occupied states. At $T = 0$ the Fermi energy is the highest energy an electron can possess, then the E_F value can be readily elucidated being the upper point of the integration range. It reads,

$$E_\mathrm{F} = \frac{\hbar^2}{2m_\mathrm{e}^*}(3\pi^2 N)^{2/3}, \tag{6.19}$$

whence,

$$\ell = \frac{\hbar}{\Gamma m_\mathrm{e}^*}\pi^{2/3}(3N)^{1/3}. \tag{6.20}$$

In Table 6.1 the mean free path value for selected bulk metals is presented along with the skin depths κ^{-1} (see also Table 3.5 in Chapter 3 for κ values for other metals), Fermi velocity and Fermi energy.

The above consideration of the optical properties of a free electron plasma is often referred to as the *Drude model* to emphasize that it is essentially based on the same assumptions as the classical Paul Drude model of metal conductivity [7]. Conductivity $\sigma(\omega)$ couples electric field and current density via Ohm's law,

$$\mathbf{J} = \sigma \mathbf{E}, \tag{6.21}$$

and is determined by electron charge, concentration, mass and momentum as,

$$\sigma = \frac{Nep(\omega)}{m}, \tag{6.22}$$

where $p(\omega) = mv(\omega)$ is an electron momentum. Based on Eq. (6.3) rewritten in the form,

$$\frac{\mathrm{d}p}{\mathrm{d}t} + \Gamma p = -eE_0 \exp(-i\omega t), \tag{6.23}$$

the conductivity reads,

$$\sigma(\omega) = \sigma_0 \frac{\Gamma}{\Gamma - i\omega}. \tag{6.24}$$

Now we can establish the linkage between the dielectric function and the conductivity [2]. The polarization vector $\mathbf{P}$ describes the electric dipole moment per unit volume inside the material, caused by the alignment of microscopic dipoles with the electric field. It is related to the internal charge density via $\nabla \cdot \mathbf{P} = -\rho$. Charge conservation ($\nabla \cdot \mathbf{J} = -\partial\rho/\partial t$) requires that the internal charge and current densities are linked via,

$$\mathbf{J} = \frac{\partial \mathbf{P}}{\partial t}. \tag{6.25}$$

Noting that $\partial/\partial t \to -i\omega$ and using relations (6.1), (6.21) and (6.25) we arrive at,

$$\varepsilon(\omega) = 1 + \frac{i\sigma(\omega)}{\varepsilon_0 \omega}. \tag{6.26}$$

In metals, the optical response does not reduce to free electron response in the conduction band. It is essentially influenced by a possibility of optical transitions by electrons in deeper, i.e. core levels. These *interband transitions* alter the dielectric function considerably. This contribution can be described in terms of *dielectric susceptibility* χ defined via a relation $\mathbf{P} = \chi \varepsilon_0 \mathbf{E}$, whence a simple relation between dielectric permittivity and dielectric susceptibility reads,

$$\varepsilon(\omega) = 1 + \chi(\omega). \tag{6.27}$$

For electrons in a metal, the dielectric function can be written as,

$$\varepsilon(\omega) = 1 + \chi^{\mathrm{free}}(\omega) + \chi^{\mathrm{IB}}(\omega), \tag{6.28}$$

where the free electrons contribution $1 + \chi^{\mathrm{free}}(\omega)$ is given by Eq. (6.7) and the interband contribution $\chi^{\mathrm{IB}} = \chi_1^{\mathrm{IB}} + i\chi_2^{\mathrm{IB}}$ is generally a complex function. Its imaginary part χ_2^{IB} describing the direct energy dissipation (absorption) becomes large only for frequencies where interband transitions occur, whereas the real part χ_1^{IB} is noticeable also for lower

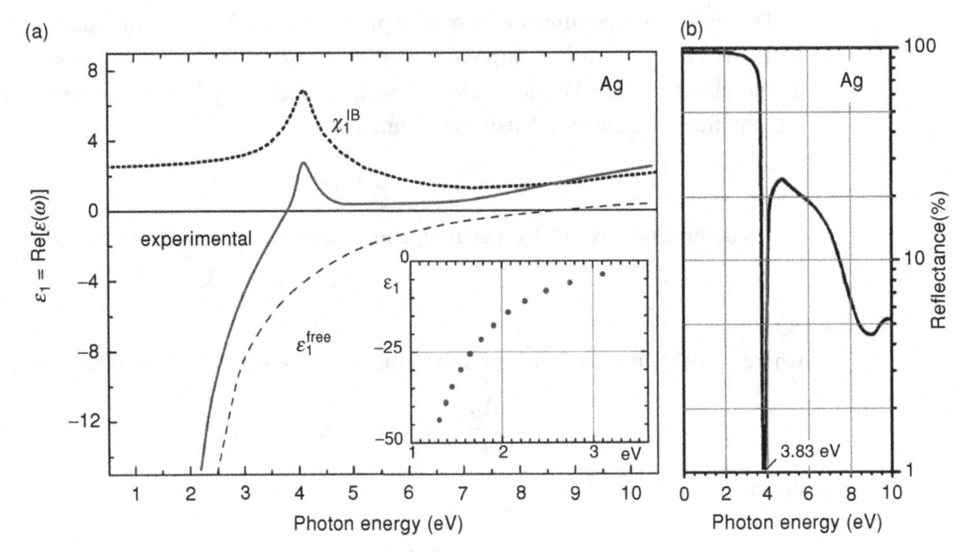

Fig. 6.3 Optical properties of silver (adapted from [8]). (a) Experimentally observed real part ε_1 of the dielectric function of bulk solid silver (solid line) and its decomposition into free electron gas contribution $\varepsilon_1^{\mathrm{free}}$ (long dashes) and interband transitions contribution χ_1^{IB} (short dashes). The insert shows experimental data in a wider range. (b) Experimentally observed reflectance spectrum.

frequencies. The theory for interband transitions is based on the electron band theory of metals and can be found in detail in the book by Bassani and Paraviccini [9].

Figure 6.3 presents the experimentally observed real part ε_1 of the dielectric function of bulk solid silver and its decomposition into free electron gas contribution $\varepsilon_1^{\mathrm{free}}$ and interband transitions contribution χ_1^{IB} (short dashes) [8]. One can see that interband transitions drastically alter the dielectric function. Due to the χ_1^{IB} contribution the $\mathrm{Re}[\varepsilon_1(\omega)] = 0$ point is redshifted by about 5 eV from approximately 9 eV ($\lambda \approx 140$ nm) to 3.7 eV ($\lambda \approx 330$ nm). Note, the low-frequency part of the dispersion curve for $\hbar\omega < 3$ eV can be satisfactorily fitted by the free electron gas formula $\varepsilon = 1 - \omega_{\mathrm{p}}^2/\omega^2$ with $\omega_{\mathrm{p}} \approx 4.5$ eV [5].

For the noble metals (e.g. Au, Ag, Cu), further extension of the free electron gas model is necessary for higher frequencies $\omega > \omega_{\mathrm{p}}$. Though the response is essentially determined by free s-electrons, the filled d-band close to the Fermi surface causes a highly polarized environment. This polarization background of the ion cores can be described by adding the term,

$$\mathbf{P}_\infty = \varepsilon_0(\varepsilon_\infty - 1)\mathbf{E}, \tag{6.29}$$

to the right part of Eq. (6.1). The contribution of the core electrons is therefore described by a dielectric constant ε_∞ (typically $1 \leq \varepsilon_\infty \leq 10$), and we can write,

$$\varepsilon(\omega) = \varepsilon_\infty - \frac{\omega_{\mathrm{p}}^2}{\omega^2 + i\omega\Gamma}. \tag{6.30}$$

For silver, the parameters in this formula are $\varepsilon_\infty = 5,\ \hbar\omega_{\mathrm{p}} \approx 9$ eV, $\hbar\Gamma = 0.02$ eV [5].

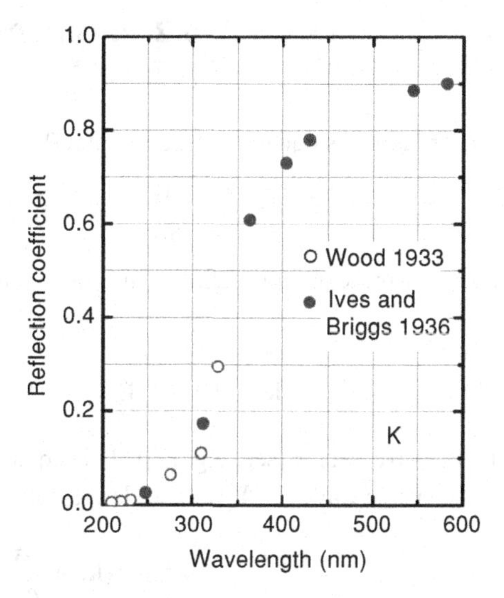

Fig. 6.4 Early data on metal potassium reflection reported by Wood [10] and Ives and Briggs [13].

Among solid metals, the alkali group of metals (Li, Na, K, Cs, Rb) exhibits the best resemblance with the free electron gas model, with no considerable contribution from the core electrons. It is apparent from comparison of Figures 6.1 and 6.3 where the dielectric functions of bulk Na and Ag are presented. Notably, in accordance with the free electron gas model which predicts propagation of electromagnetic waves for high frequencies $\omega > \omega_p$, alkali metals do exhibit remarkable transparency and minor reflectance in the short-wave range. Furthermore, since for alkali metals the plasma frequency values are relatively low ($\hbar\omega_p = 3.8$ eV for K, 5.7 eV for Na, 7.12 eV for Li [3]), their transparency observation is feasible in common experiments since it occurs in the affordable near-UV spectral range. The very first report on this amazing property of alkali metals dates back to the pioneering paper by Robert Wood [10] (Fig. 6.4) which was followed promptly by the elegant explanation by Clarence Zener [11]. Although interband transitions are present in the optical response of alkali metals, their contribution is minor. For example in Na, interband transitions starting at 2.1 eV upward (Table 6.1) just slightly disturb the dielectric functions (Fig. 6.1b). For alkali metals the low-energy threshold for interband transitions $\hbar\omega_{IB}$ is simply related to the electron Fermi energy E_F as $\hbar\omega_{IB} = 0.64E_F$ [12]. One can check the applicability of this relation for Na using the data in Table 6.1.

Noble metals like Cu, Ag, Au, which will often be referred to hereafter because of their use in nanoplasmonics, consist of atoms with completely filled 3d-, 4d-, and 5d-shells and just a single electron in the 4s-, 5s-, and 6s-bands, respectively. In Group 1B metals the contribution from interband transition is noticeable. In accordance with the interband absorption contribution to the real part of the Ag dielectric function (Fig. 6.3) a dip in reflectance occurs, apparent also from Table 3.4 in Chapter 3. More detail on band structure of various metals can be found in the book [12].

6.2 Plasmons

From the curl Maxwell's equations (see Chapter 2, Eqs. (2.32), (2.33)),

$$\nabla \times \mathbf{E} = -\frac{\partial \mathbf{B}}{\partial t}, \quad \nabla \times \mathbf{H} = -\frac{\partial \mathbf{D}}{\partial t} + \mathbf{J}, \tag{6.31}$$

one can arrive at the equation defining the propagation of electromagnetic waves in the form,

$$\mathbf{k}(\mathbf{k} \cdot \mathbf{E}) - k^2\mathbf{E} = -\varepsilon(\mathbf{k}, \omega)\frac{\omega^2}{c^2}\mathbf{E}, \tag{6.32}$$

where $\mathbf{k}$ is the wave vector (see, e.g. [2]). This equation allows the two types of waves to occur. For transverse waves $\mathbf{k} \cdot \mathbf{E} = 0$ holds yielding the generic dispersion relation,

$$k^2 = \varepsilon(\mathbf{k}, \omega)\frac{\omega^2}{c^2}. \tag{6.33}$$

In Chapter 3 we saw that inserting the free electron gas dielectric function $\varepsilon = 1 - \omega_p^2/\omega^2$ in Eq. (6.33) gives rise to the dispersion law (see Eq. (3.56) and Fig. 3.9b),

$$\omega^2 = c^2k^2 + \omega_p^2, \tag{6.34}$$

implying propagation of transverse electromagnetic waves with frequency $\omega > \omega_p$. Now we look at yet another possibility for electromagnetic waves to occur in plasma. Equation (6.32) allows for longitudinal waves to occur when $\varepsilon(\mathbf{k}, \omega) = 0$ holds. For the case of an ideal electron gas (or another type of ideal plasma) this happens at the plasma frequency. For a non-ideal plasma with damping, the dielectric function tends to zero at the plasma frequency in the small damping limit. For the longitudinal mode, from Eq. (6.1),

$$\mathbf{D} = \varepsilon_0\varepsilon(\omega)\mathbf{E} = 0 = \varepsilon_0\mathbf{E} + \mathbf{P}, \tag{6.35}$$

one has $\mathbf{P}/\varepsilon_0 = -\mathbf{E}$ which means that a longitudinal wave occurs by means of a pure depolarization field in the plasma. In this case the plasma oscillates as a single entity with plasma frequency ω_p, all electrons moving in phase. The quantum of these oscillations is called the *volume plasmon.* In metals, longitudinal oscillations occur not only at the plasma frequency ω_p but in every case when the dielectric function equals zero, which becomes possible because of the complicated character of the $\varepsilon(\omega)$ function in real solids. Because of the longitudinal character of these oscillations, volume plasmons can not be directly excited by the standard optical excitation since a longitudinal wave does not couple with a transverse one. In the experiments, plasmon excitation is observed, using reflectance or transmittance of electron beams, as oscillations in the electron energy loss spectra with a period equal to $\hbar\omega_p$.

There is another type of wave which is important in metal optics. It is referred to as *surface plasmon polariton mode* and occurs at the interface between a dielectric and a metal. For such a mode to occur, the negative dielectric function inherent in a metal is mandatory. We start consideration with the familiar Helmholtz equation (see Eq. (2.47) in

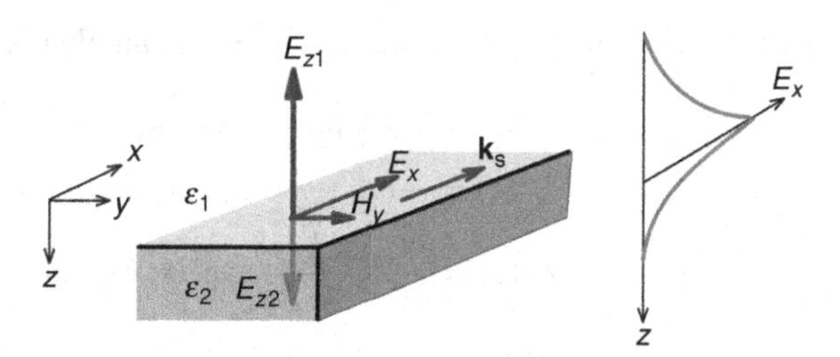

Fig. 6.5 The interface of two media with dielectric functions ε_1 and ε_2 and the x-component of electric field vector E_x evanescing exponentially in both media along the z-axis from the point $z = 0$. $\mathbf{k}_s$ stands for the wave vector of the surface plasmon polariton.

Chapter 2),

$$\nabla^2\mathbf{E} + k_0^2\varepsilon(z)\mathbf{E} = 0, \tag{6.36}$$

implying that the dielectric function changes in a stepwise manner from ε_1 to ε_2 at the point $z = 0$ (Fig. 6.5). Here $k_0 = \omega/c$ is the wave number value inherent in a vacuum. The wave we are looking for is characterized by the electric field $\mathbf{E}(x, y, z) = \mathbf{E}(z)\exp(ik_x x)$. The k_x is treated as a complex value. The Helmholtz equation then takes the form,

$$\frac{\partial^2\mathbf{E}(z)}{\partial z^2} + \left[k_0^2\varepsilon(z) - k_x^2\right]\mathbf{E} = 0. \tag{6.37}$$

This is the basic equation for waveguide theory. A thorough analysis can be found in the books [2, 14]. Omitting cumbersome derivations, we restrict ourselves to the final results.

Analysis of Eq. (6.36) together with Eqs. (6.31) leads to two sets of solutions with different polarization properties of the propagating waves. The first set is the *transverse magnetic modes* (TM-modes, or p-modes). For these modes only the field components E_x, E_z, and H_y are nonzero. The second set is the *transverse electric modes* (TE-modes, or s-modes). For these modes only H_x, H_z and E_y are nonzero. Transverse magnetic modes are described by the equation for H_y,

$$\frac{\partial^2 H_y}{\partial z^2} + \left[\frac{\omega^2}{c^2}\varepsilon(z) - k_x^2\right] H_y = 0, \tag{6.38}$$

and TE-modes are described by the similar one for E_y,

$$\frac{\partial^2 E_y}{\partial z^2} + \left[\frac{\omega^2}{c^2}\varepsilon(z) - k_x^2\right] E_y = 0. \tag{6.39}$$

For the interface geometry sketched in Fig. 6.5 there are two sets of equations describing TM- and TE-modes. For the TM-modes these are,

$$\begin{aligned}
H_y(z) &= A_1 \exp(ik_x x)\exp(-k_{z1}z),\\
E_x(z) &= iA_1\frac{k_{z1}}{\omega\varepsilon_0\varepsilon_1}\exp(ik_x x)\exp(k_{z1}z),\\
E_z(z) &= -A_1\frac{k_x}{\omega\varepsilon_0\varepsilon_1}\exp(ik_x x)\exp(k_{z1}z),
\end{aligned} \tag{6.40}$$

in medium 1 ($z < 0$, dielectric constant $\varepsilon_1 > 0$ is purely real and positive), and,

$$\left.\begin{aligned} H_y(z) &= A_2 \exp(ik_x x)\exp(k_{z2}z), \\ E_x(z) &= -iA_2\frac{1}{\omega\varepsilon_0\varepsilon_2}k_{z2}\exp(ik_x x)\exp(k_{z2}z), \\ E_z(z) &= -A_2\frac{k_x}{\omega\varepsilon_0\varepsilon_2}\exp(ik_x x)\exp(-k_{z2}z), \end{aligned}\right\} \tag{6.41}$$

in medium 2 ($z > 0$, dielectric function $\varepsilon_2 < 0$ is purely real but negative) [2]. Here k_{z1} and k_{z2} stand for the z-component of the wave vector in medium 1 and 2, respectively. The condition of continuity of H_y and $\varepsilon_i E_z$ at the interface requires that $A_1 = A_2$ and,

$$\frac{k_{z1}}{k_{z2}} = -\frac{\varepsilon_1}{\varepsilon_2}. \tag{6.42}$$

The values k_{z1} and k_{z2} can both be positive if the dielectric permittivities of the two media have opposite signs, i.e. one of these media should necessarily be a dielectric whereas another one should necessarily be a metal. One more condition arises from Eq. (6.38) which H_y must satisfy. This yields a pair of equations,

$$k_{z1}^2 = k_x^2 - \frac{\omega^2}{c^2}\varepsilon_1, \quad k_{z2}^2 = k_x^2 - \frac{\omega^2}{c^2}\varepsilon_2, \tag{6.43}$$

which, combined with Eq. (6.42) give three equations for the three unknown variables, k_x, k_{z1} and k_{z2}. After simple arithmetic we find that,

$$k_1 = \frac{\omega}{c}\varepsilon_1\sqrt{\frac{1}{\varepsilon_1+\varepsilon_2}}, \quad k_2 = -\frac{\omega}{c}\varepsilon_2\sqrt{\frac{1}{\varepsilon_1+\varepsilon_2}}, \tag{6.44}$$

and

$$k_x = \frac{\omega}{c}\sqrt{\frac{\varepsilon_1\varepsilon_2(\omega)}{\varepsilon_1+\varepsilon_2(\omega)}}. \tag{6.45}$$

Equation (6.44) defines the evanescence factors for E_z and E_x (shown in the right-hand part of Fig. 6.5). Equation (6.45) represents the dispersion relation of the *surface plasmon polariton mode* (SPP-mode). Remarkably this expression is valid also if the metal dielectric function is complex. If medium 1 is a vacuum, i.e. $\varepsilon_1 = 1$ then Eq. (6.45) reads,

$$k_x = \frac{\omega}{c}\sqrt{\frac{\varepsilon_2(\omega)}{1+\varepsilon_2(\omega)}}. \tag{6.46}$$

Notably, TE-modes can meet the continuity requirement only if $A_1 = A_2 = 0$, which means there is no TE-mode propagating along the interface. Thus *surface plasmon polaritons exist in TM-polarization only*.

For surface plasmon polariton mode wave number k_x to be real and positive, not only the condition $\varepsilon_2 < 0$ should hold but also the more rigid condition,

$$\varepsilon_2 < -\varepsilon_1$$

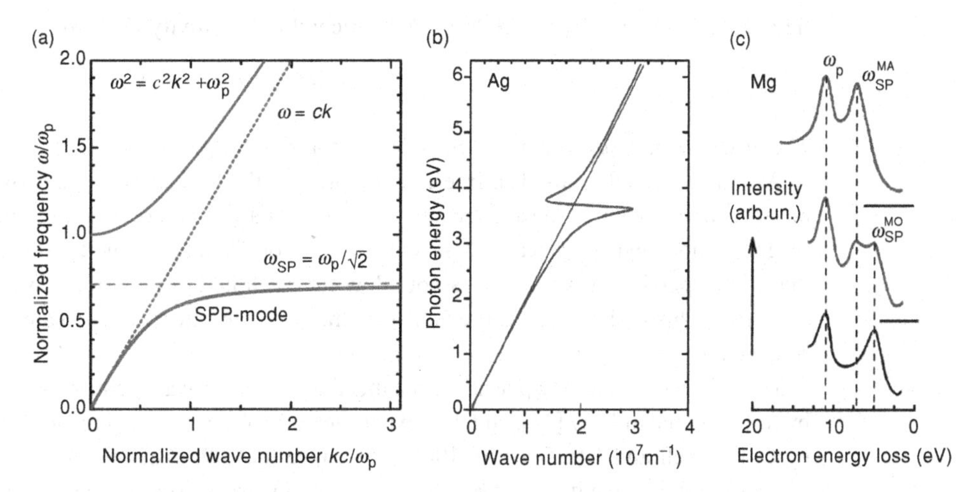

Fig. 6.6 **The surface plasmon polariton (SPP) mode at the vacuum (air) interface with (a) an ideal free electron plasma, (b) a silver surface and (c) a magnesium surface. Panel (a) shows three types of dispersion laws $\omega(k)$. Straight dotted line represents the dispersion law of electromagnetic waves in a vacuum $\omega = ck$ ("the light cone"). The SPP dispersion curve emerges from the light cone and then substantially deviates from it with asymptotic limit of the surface plasmon frequency $\omega_{SP} = \omega_p/\sqrt{2}$. For $\omega > \omega_p$ propagation of electromagnetic waves is allowed with the dispersion law $\omega^2 = c^2k^2 + \omega_p^2$. In the range $\omega_p/\sqrt{2} < \omega < \omega_p$ no propagating mode exists. Panel (b) shows calculations by Maier [2] made for the Ag/air interface. The straight line corresponds to the vacuum dispersion law $\omega = ck$. The dispersion curve for the Ag/air interface deviates considerably from that for an ideal plasma because of the contribution from interband transitions in the range 3.5–4 eV. Panel (c) shows electron energy loss spectra recorded for a Mg thin film [15]. Along with the volume plasmons (ω_p) there is a surface plasmon at the metal/air interface (ω_{SP}^{MA}). Upon the observation time from the top to the bottom this band is replaced by the lower one (ω_{SP}^{MO}) relevant to the metal/oxide Mg/MgO interface because of film oxidation.**

is mandatory. For a metal/air or metal/vacuum interface this means $\varepsilon_2 < -1$. For the ideal plasma of free electrons from Eq. (6.10) one can see this condition is fulfilled for frequencies $\omega < \omega_p/\sqrt{2}$. The frequency $\omega_{SP} = \omega_p/\sqrt{2}$ is referred to as the *surface plasmon frequency.*

For an air/metal interface with the assumption of an ideal free electron plasma, insertion of the dielectric function in the form of Eq. (6.10) into Eq. (6.46) gives,

$$k = \frac{\omega}{c}\sqrt{\frac{1 - \omega_p^2/\omega^2}{2 - \omega_p^2/\omega^2}}. \tag{6.47}$$

This dependence is presented in Figure 6.6a using dimensionless frequency and wave number. Note, the SPP-mode lies entirely outside the light cone. The dispersion law is also presented described by Eq. (6.34) for the familiar electromagnetic mode $\omega^2 = c^2k^2 + \omega_p^2$ existing for $\omega > \omega_p$. The surface plasmon polariton (SPP) dispersion curve for smaller wave numbers tends to the light line $\omega = ck$ and therefore gains the features of an ordinary electromagnetic wave coming from a dielectric at grazing incidence. In the opposite limit of infinitely large wavenumbers, the SPP dispersion curve approaches the surface plasmon frequency $\omega_{\rm SP}$ which for an arbitrary dielectric with permittivity ε_1 reads,

$$\omega_{\rm SP} = \frac{\omega_{\rm p}}{\sqrt{1+\varepsilon_1}}. \tag{6.48}$$

The group velocity $v_{\rm g} = {\rm d}\omega/{\rm d}k$ tends to zero for $\omega \to \omega_{\rm SP}$.

Within the model of an ideal free electron plasma, the frequency range between surface plasmon frequency and plasma frequency represents a gap where no propagating electromagnetic mode exists. In real metals, the contribution from interband transitions modifies the SPP dispersion curve and does not allow for unlimited growth of the wavenumber at $\omega \to \omega_{\rm SP}$. This is shown in Figure 6.6(b) for the silver/air interface calculated by S. Maier based on the realistic silver dielectric function.

The existence of surface plasmons at a metal/air interface was predicted for the first time by R. H. Ritchie in 1957 [16], and then experimentally observed in energy loss spectroscopy by C. J. Powell and J. B. Swan in 1960 [15]. Since the SPP dispersion curve lies entirely outside the light cone it cannot be excited directly by an external incident electromagnetic wave. The direct excitation of surface plasmons by means of electron impact is the most efficient experimental technique. It was used in the pioneering experiments shown in Figure 6.6c, where resonant electron energy losses were observed at three frequencies indicating the intrinsic metal volume plasmon, the surface plasmon at the metal/air interface and the surface plasmon at the metal/oxide interface, with the corresponding development of the third band instead of the second one in the course of the experiment because of the inevitable oxidation of the metal surface in air.

In the context of photonics, optical rather than electron beam techniques of plasmon excitation are desirable. For optical SPP excitation, various techniques are used based on conversion of an evanescent electromagnetic field which can be generated under conditions of light tunneling. The problem of light tunneling and development of evanescent waves will be the subject of Chapter 10. Here we only mention the typical experimental solutions used without going into detail. These include prism coupling based on frustrated total reflection, near-field excitation using leakage mode, as in near-field optical microscopy and excitation using a diffraction grating grooved directly on top of a metal surface.

The prism coupling scheme to generate a surface plasmon polariton is shown in Figure 6.7. A SPP-mode can be generated only for frequencies corresponding to the portion of the SSP-curve confined between the air and the glass curves. Further restriction comes from the phase matching condition which reads [2],

$$k_{\rm SPP} = \frac{\omega}{c} n_{\rm glass} \sin\theta, \tag{6.49}$$

where θ is the angle of incidence defined by the total internal reflection condition $\sin\theta > 1/n_{\rm glass}$, which means that the incident wave vector projection of the electromagnetic wave in glass on the interface plane should correspond to the wave number in the SPP dispersion curve. Combining the two conditions and accounting for Eq. (6.47) gives the SPP frequency accessible in this scheme as,

$$\omega_{\rm SPP} = \omega_{\rm p} \sqrt{\frac{n_{\rm glass}^2 \sin^2\theta - 1}{2n_{\rm glass}^2 \sin^2\theta - 1}}. \tag{6.50}$$

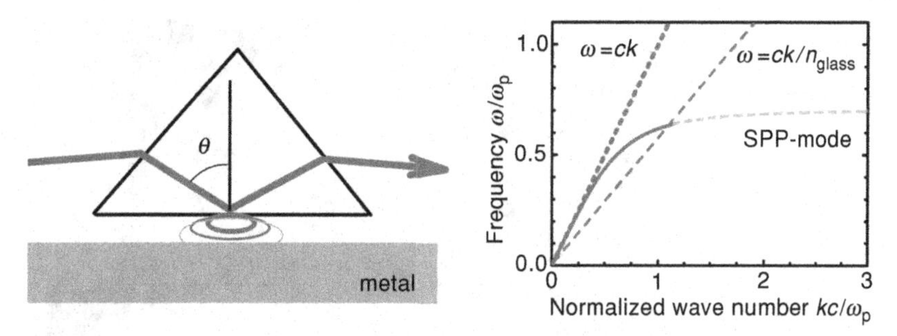

Fig. 6.7 **A prism coupling scheme used to excite a surface plasmon polariton at the air/metal interface. An evanescent field under conditions of total internal reflection generates the SPP-mode at the interface. It is possible only for frequencies corresponding to the SSP-curve confined between the air and the glass curves. The part of the SPP curve inaccessible by external excitation is shown with dashes.**

Thus the generated SPP mode is the leakage mode for total internal reflection and its excitation does manifest itself as a decrease in the total reflection coefficient when the above excitation conditions are met. More detail on SPP-mode excitation by means of frustrated total reflection can be found in the original work by E. Kretschmann [17] as well as in the review by Zayats and Smolyaninov [18] and in the book by Raether [19].

6.3 Optical properties of metal nanoparticles

Nanoparticles of semiconductor crystalline materials (quantum dots) were the subject of close consideration in Chapter 5. Their pronounced size-dependent intrinsic optical properties resulting from the spatial confinement of electron and hole motion were highlighted. In this section, their metallic counterparts will be discussed. For metal nanoparticles extrinsic effects rather than intrinsic ones are responsible for various size-dependent optical phenomena, although intrinsic size-dependent properties are also detectable. By *intrinsic* effects here we imply size-dependent electron energies, transition probabilities etc. Their study originated from the pioneering papers by H. Frölich [20] and R. Kubo [21], where non-trivial properties of electron heat capacity were predicted based on discreteness of the electron energy spectrum in smaller nanoparticles. Interestingly, not only in *technology*, where metal nanoparticles had become important much earlier than their semiconductor counterparts (note the stained glasses since Roman times mentioned in the introduction to this chapter), but also in *science*, these became the subject of close consideration much earlier in comparison with semiconductors.

Extrinsic effects imply the crucial role of an ambient dielectric medium hosting nanoparticles. For metal nanoparticles extrinsic effects play the major role in their optical response, whereas the intrinsic size-dependent properties are of secondary importance. In this context, metal nanoparticles differ considerably from semiconductor nanoparticles discussed in Chapter 5.

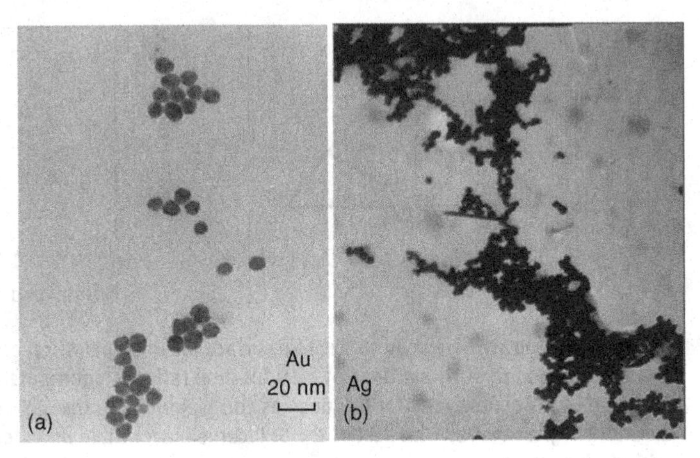

Fig. 6.8 **Images of (a) gold and (b) silver nanoparticles obtained by means of transmission electron microscopy.**

In this section optical properties of metal nanoparticles will be briefly discussed. A couple of typical examples of nanometer-sized silver and gold particles are shown in Figure 6.9.

Surface plasmons in a nanoparticle

Consider a metal nanoparticle whose size is much smaller than the electromagnetic wavelength. When such a particle is embedded in an oscillating electromagnetic field it experiences a nearly space-independent external field effect on its electron subsystem. The time dependence is ignored at the moment. Under the impact of the external field, the internal field inside the particle reads (see, e.g. [22]),

$$E_{\text{int}} = E_0 \frac{3\varepsilon_1}{\varepsilon + 2\varepsilon_1}, \tag{6.51}$$

where ε is the particle dielectric permittivity and ε_1 is the host medium dielectric permittivity. This relation is known from electrostatics as a direct consequence of the *Laplace equation*, $\nabla^2\Phi = 0$ for the potential Φ and further evaluation of $\mathbf{E} = -\nabla\Phi$. It holds in optics in the long-wave approximation, i.e. a particle size is implied to be small as compared to the electromagnetic wavelength so that the electric field amplitude is seen as a constant value within a particle. Its application implies also that the role of the magnetic component of the electromagnetic wave may not be considered [1]. If an alternating field is considered, the electron subsystem oscillates in the field resulting in emerging surface charges and relevant polarization along the field.

Within the same electrostatic problem, a particle *polarizability*, α, is defined as,

$$\mathbf{p} = \varepsilon_0\varepsilon_1\alpha\mathbf{E}_0, \tag{6.52}$$

where $\mathbf{p}$ is a particle dipole moment, and reads [1, 2],

$$\alpha = 4\pi R^3 \frac{\varepsilon - \varepsilon_1}{\varepsilon + 2\varepsilon_1}, \tag{6.53}$$

where R is a particle radius.[1] One can see that the πR^3 term is defined by a particle volume.

Now consider that the ambient medium has a real positive dielectric permittivity. Then if a particle material has a dielectric permittivity which can take negative values and is frequency dependent, then the internal field will exhibit a resonance. This will happen whenever,

$$|\varepsilon(\omega) + 2\varepsilon_1| = \text{minimum} \tag{6.54}$$

occurs. Considering particle material permittivity as a frequency-dependent complex function,

$$\varepsilon(\omega) = \varepsilon'(\omega) + i\varepsilon''(\omega), \tag{6.55}$$

the resonance condition reads,

$$[\varepsilon'(\omega) + 2\varepsilon_m]^2 + [\varepsilon''(\omega)]^2 = \text{minimum}. \tag{6.56}$$

If $\varepsilon''(\omega) \ll 1$, or if its frequency dependence is weak $(\partial\varepsilon''(\omega)/\partial\omega \ll \partial\varepsilon'(\omega)/\partial\omega)$, then the minimum in Eq. (6.54) occurs at

$$\varepsilon_1(\omega) = -2\varepsilon, \tag{6.57}$$

which is known as the *Fröhlich condition*.[2] The relevant localized oscillating electromagnetic mode is referred to as the *dipole surface plasmon.* For the dielectric permittivity of a metal in the simplest form of Eq. (6.10) the resonance condition reads,

$$\omega_{\mathrm{SP}} = \frac{\omega_{\mathrm{p}}}{\sqrt{1+2\varepsilon}}, \tag{6.58}$$

which in the case of a vacuum, or air ambient medium with $\varepsilon = 1$, gives the surface plasmon frequency,

$$\omega_{\mathrm{SP}} = \frac{\omega_{\mathrm{p}}}{\sqrt{3}} = 0.577\ldots\omega_{\mathrm{p}}, \tag{6.59}$$

where the subscript "SP" stands for "surface plasmon". Along with the dipole surface plasmons, higher-order resonances are inherent in a metal nanoparticle forming the following sets of resonant frequencies and dielectric permittivities [1, 23]:

$$\begin{aligned} \varepsilon_{(N)} &= -\frac{N+1}{N}\varepsilon_1, \quad N = 1, 2, 3, \ldots \left(\varepsilon_{(1)} = -2\varepsilon_1,\ \varepsilon_{(2)} = -\frac{3}{2}\varepsilon_1,\ \varepsilon_{(3)} = -\frac{4}{3}\varepsilon_1, \ldots\right) \\ \omega_N &= \frac{\omega_{\mathrm{p}}}{\sqrt{1-\varepsilon_{(N)}}} = \omega_{\mathrm{p}}\sqrt{\frac{N}{2N+1}}\varepsilon, \left(\omega_1 = \frac{1}{\sqrt{1+2\varepsilon_1}}\omega_{\mathrm{p}},\ \omega_2 = \frac{\sqrt{2}}{\sqrt{2+3\varepsilon_1}}\omega_{\mathrm{p}}, \ldots\right) \end{aligned} \tag{6.60}$$

[1] In the previous chapter, for spherical semiconductor particles the notation a was used for a particle radius. Here R is applied to emphasize that it is rather a geometrical value but not the quantum well size as was the case for a spherical quantum dot in Chapter 5 (starting from notations for quantum well sizes in Chapters 3 and 4).

[2] The condition (6.57) was derived by H. Fröhlich in his book *Theory of dielectrics* (Oxford University, London 1949) for the frequency of polarization oscillations due to lattice vibrations in small dielectric crystals.

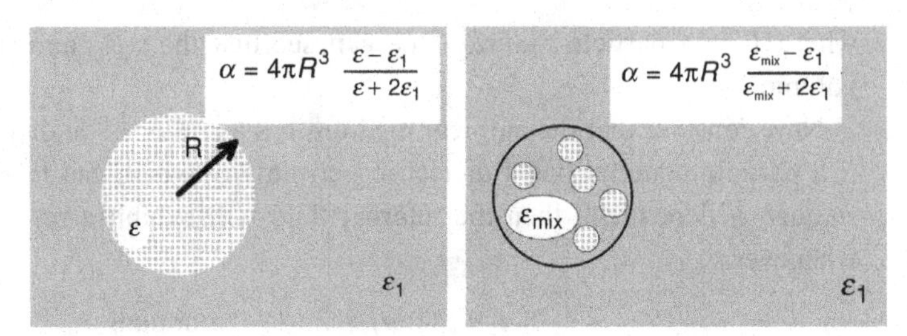

Fig. 6.9 **Application of Eq. (6.53) to (left) a homogeneous and (right) a composite particle in a host medium.**

Notably, the dielectric function of a metal particle features an extremum defined by the property of a particle material (plasma frequency) and the property of the host medium. Therefore, monitoring these resonances offers the possibility of evaluating changes in the environment and may be exploited in optical sensors.

Now we are in a position to discuss the dielectric function of a medium containing dispersed small particles. With the above mentioned supposition that particles are small enough ($R/\lambda \ll 1$), the dielectric function of the composite mixture $\varepsilon_{\text{mix}}(\omega)$ is related to the dielectric permittivity of the host medium ε_1 and the dielectric function of inclusions $\varepsilon(\omega)$ in a rather elegant form,

$$\frac{\varepsilon_{\text{mix}}(\omega) - \varepsilon_1}{\varepsilon_{\text{mix}}(\omega) + 2\varepsilon_1} = f\frac{\varepsilon(\omega) - \varepsilon_1}{\varepsilon(\omega) + 2\varepsilon_1}. \tag{6.61}$$

Here f is the total volume fraction of the dispersed inclusions which for identical spherical particles reads, $f = \frac{4}{3}\pi R^3 N$ with N being the particle concentration. Equation (6.61) was derived in 1904 by J. C. Maxwell-Garnett [24] and is therefore referred to as the *Maxwell-Garnett formula*. Notably, the structure of left- and right-hand parts in this formula resembles the term in Eq. (6.53) coupling a particle's polarizability with its volume. Formally, Eq. (6.61) states that it is this term that changes in proportion to f in a composite medium. It has rather apparent physical content (Fig. 6.9). Consider a solid particle of the ε_1-material buried in ε-medium (Fig. 6.9(a)) versus a composite particle consisting of a number of spherical ε_1-material sub-particles buried in ε-medium with fraction f. A composite particle is further embedded in ε-medium (Fig. 6.9(b)). Then looking at Eq. (6.53) (it is definitely applicable not only to metal particles but to any particle of a polarizable material), one arrives at the conclusion that the polarizability of that composite particle will be f times the polarizability of the solid one.

The dielectric function of the mixture (it is complex in the general case) can be extracted from Eq. (6.61) in the form [25],

$$\varepsilon_{\text{mix}}(\omega) = \varepsilon_1 \frac{1 + 2F(\omega, f)}{1 - F(\omega, f)}, \tag{6.62}$$

where,

$$F(\omega, f) = f\frac{\varepsilon(\omega) - \varepsilon_1}{\varepsilon(\omega) + 2\varepsilon_1}. \tag{6.63}$$

In the case of a composite medium with more than a single material dispersed therein in the form of small particles, the right-hand part of Eq. (6.61) modifies accordingly to account for the additive contribution of every component with partial fraction f_j to give,

$$\frac{\varepsilon_{\text{mix}}(\omega) - \varepsilon_1}{\varepsilon_{\text{mix}}(\omega) + 2\varepsilon_1} = \sum_{j=1}^{M} (f_j \frac{\varepsilon_j(\omega) - \varepsilon_1}{\varepsilon_j(\omega) + 2\varepsilon_1}). \tag{6.64}$$

The Maxwell-Garnett formula had two important precursors. The first one is the *Clausius–Mossotti formula* derived in 1850 by R. Clausius and in 1879 by O. Mossotti. It states that the value of $(\varepsilon - 1)/(\varepsilon + 2)$ where ε is the dielectric permittivity of the matter is directly proportional to the density of that matter. Therefore the Maxwell-Garnett formula is often considered as an extension of the Clausius–Mossotti formula. The latter actually transforms into Eq. (6.61) if the dielectric permittivity ε is replaced by the normalized value $\varepsilon/\varepsilon_1$. Another precursor is the Lorentz–Lorenz formula (1880)[3] with n^2 instead of ε which is a version of the Clausius–Mossotti formula for the optical range where, typically, $\mu = 1$ holds and the dielectric function is then $\varepsilon(\omega) = [n(\omega)]^2$ where n is the refractive index.

Consider the application range of Eqs. (6.51), (6.53), (6.61) and (6.62). First, the relatively low external fields imply that the polarization of the matter should be linear with the field **E**. Second, any type of spatial retardation effect over particle volume should be absent. These equations are equally valid for dielectric particles as well as for metal particles.

Consider the application of the Maxwell–Garnett formula to metal particles. With no interband transition to be accounted for, the complex dielectric function $\varepsilon(\omega)$ of a metal can be taken in the simpler form of Eq. (6.30). Insertion of Eq. (6.30) into Eq. (6.62) allows us to elucidate the complex dielectric function of the composite medium,

$$\varepsilon_{\text{mix}}(\omega) = \varepsilon'_{\text{mix}}(\omega) + i\varepsilon''_{\text{mix}}(\omega). \tag{6.65}$$

The real and the imaginary parts of this function are plotted in Figure 6.10 for the parameters of silver $\varepsilon_\infty = 5$, $\hbar\omega_p \approx 9$ eV, $\hbar\Gamma = 0.02$ eV. One can see that the imaginary part of the dielectric function basically resembles the typical behavior inherent in elementary absorption bands; for example for a gas of atoms. The imaginary part will accordingly give rise to the absorptive band whereas the real part will define the refraction with abnormal dispersion inside that band. The spectral position of this absorptive-like behavior is determined by the metal parameters and by its volume fraction. For the specific case of Figure 6.10 it corresponds to approximately 440 nm wavelength. Similar to atomic absorption lines, the real part of the dielectric function at first grows with growing frequency (normal dispersion), then reaches its maximal value near a frequency corresponding to the half-maximum of the imaginary part, reduces (anomalous dispersion), crosses the zero axis near the frequency ω_{01} corresponding to the maximum in the imaginary part, and acquires the negative sign thereafter.[4] For the case under consideration, the zero value of the real

[3] Hendrick Lorentz (1853–1928) was a Dutch physicist. Ludwig Lorenz (1829–1891) was a Danish physicist.

[4] For every system interacting with an electromagnetic field the real and imaginary parts of the dielectric function are coupled via the *Kramers–Kronig relations* originating from the causality principle. The case under consideration also obeys these relations.

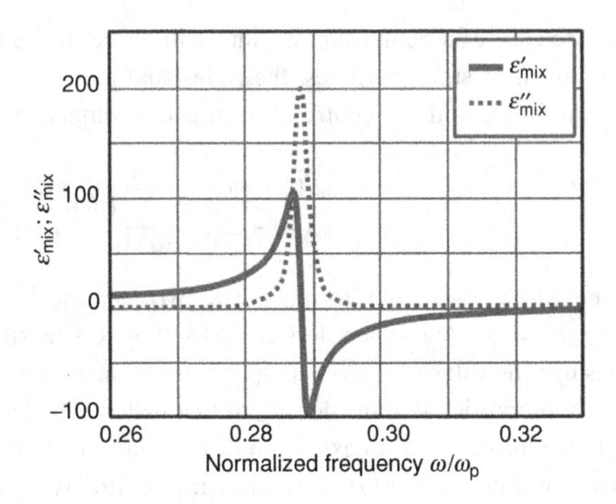

Fig. 6.10 **The real (solid line) and the imaginary (dots) parts of the complex dielectric function of a composite medium consisting of Ag nanoparticles dispersed in a medium with $\varepsilon = 2.56$ with volume fraction $f = 0.2$ [26]. The silver dielectric function obeys Eq. (6.30) with parameters $\varepsilon_\infty = 5$, $\hbar\omega_p \approx 9$ eV, $\hbar\Gamma = 0.02$ eV.**

part of the dielectric function occurs at the two frequencies [26]

$$\omega_{01} = \omega_p \sqrt{\frac{1-f}{\varepsilon_\infty + 2\varepsilon_1 - f(\varepsilon_\infty - \varepsilon_1)}},$$
$$\omega_{02} = \omega_p \sqrt{1 + \frac{\varepsilon_\infty(\varepsilon_\infty + 2\varepsilon_1 - f(\varepsilon_\infty - \varepsilon_1))}{\varepsilon_1(\varepsilon_\infty + 2\varepsilon_1 - f(\varepsilon_\infty - \varepsilon_1))}}, \tag{6.66}$$

determined by the metal parameters ω_p, ε_∞, the ambient medium parameter ε_1, and the composite medium under consideration parameter f. In derivation of Eqs. (6.66) the small terms proportional to Γ^2 were neglected. In the case under consideration $\omega_{01} = 0.288\omega_p$, $\omega_{02} = 0.355\omega_p$.

In Figure 6.10 interband transitions are not accounted for thoroughly. Their contribution is partially accounted for by introducing $\varepsilon_\infty > 1$ which shifts the zero-point frequency of $\varepsilon(\omega)$ towards more realistic lower values as compared with the plasma frequency. For real metals, interband transition gives rise to an increase in absorption for higher frequencies. As a representative example, the optical density spectrum of a sol of gold particles is presented exhibiting a pronounced maximum followed by considerable absorption for shorter waves (higher frequencies).[5] Interband absorption (Table 6.1) becomes noticeable from 500 nm onwards to shorter wavelength as shown in the experimental data in Figure 6.11(a). The

[5] The transmission coefficient T for light intensity and the optical density D are actually the values that are measured in a typical experiment. Optical density reads $D = -\log T$ (sometimes $D = -\ln T$ is used). Therefore D is directly proportional to the absorption coefficient α in the case of a homogeneous medium where the relation $T = (1-R)^2 \exp(-aL)$ holds, R being the intensity reflection coefficient at the boundaries. For a heterogeneous medium elucidation of α as a medium parameter from transmission measurements is not straightforward because of coexisting elastic and inelastic scattering. One can see that assignment of the absorption coefficient to a metal particle is not possible at all.

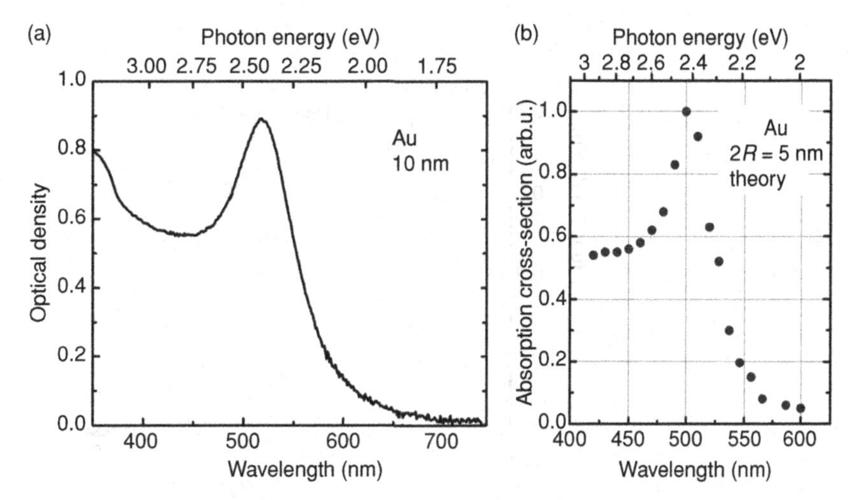

Fig. 6.11 **(a) Measured optical density spectrum of sol of Au nanoparticles imaged in Figure 6.8 in a liquid solution and (b) the absorption cross-section calculated by Kreibig and Vollmer [1].**

relevant calculations reported by Kreibig and Vollmer [1] are presented in Figure 6.11b. A concordance of observed and calculated spectra is evident. Because of the optical density spectra similar to those shown in Figures 6.10 and 6.11, glasses containing gold or copper nanoparticles exhibit a red color, are extensively used in glass production and are referred to as the "golden ruby" and the "copper ruby" in the glass industry. The larger width of the band observed in the experiments as compared to the calculated one (Fig. 6.11) can be tentatively attributed to the size-dependent damping which is discussed in Section 6.4.

The Maxwell-Garnett theory implies that nanoparticles embedded in a homogeneous host medium occupy a very small volume fraction $f \ll 1$. The approach to a composite medium description without this restriction has been developed by Bruggeman [27]. For arbitrary values of $0 < f < 1$ he derived the relation,

$$f\frac{\varepsilon(\omega)-\varepsilon_{\mathrm{mix}}(\omega)}{\varepsilon(\omega)+2\varepsilon_{\mathrm{mix}}(\omega)}+(1-f)\frac{\varepsilon_1-\varepsilon_{\mathrm{mix}}(\omega)}{\varepsilon_1(\omega)+2\varepsilon_{\mathrm{mix}}(\omega)}=0. \tag{6.67}$$

It is noteworthy that the Maxwell-Garnett and Bruggeman theories are the so-called *effective medium mean field* theories. The "effective medium" notion means that the composite mixture under consideration can be ascribed with the dielectric function as a whole, i.e. the mixture is seen as a homogeneous medium for electromagnetic wave. This approximation is only valid provided the scattering cross-section by a single particle is negligibly small as compared to the wavelength scale. The "mean field" notion implies that variation in electric field across a particle is neglected which, again, implies a particle size much smaller than the wavelength, and, additionally the influence of neighboring particles on the polarizability of a given particle is neglected. Besides the Maxwell-Garnett and Bruggeman formulas, other modifications of the effective medium mean-field theory of composite materials have been proposed. These are discussed and compared with respect to metal nanoparticles in dielectrics in [6].

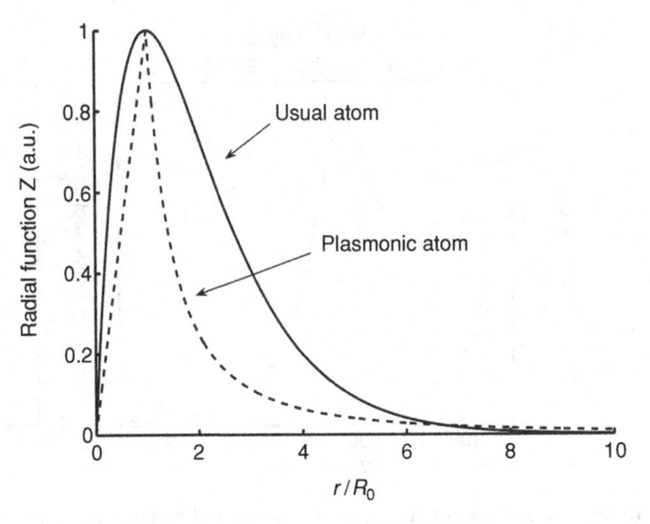

Fig. 6.12 Radial wave functions Z of a hydrogen ($2p$-state of an electron) and a plasmonic atom versus normalized distance from the centre. The parameter $R_0 = \sqrt{2}a_{\mathrm{B}}$ for a hydrogen atom (a_{B} is the Bohr radius). For a metal nanoparticle the parameter R_0 equals the particle radius. The solid line corresponds to a usual atom and the dashed line corresponds to the plasmonic atom. Reprinted with permission from [23]. Copyright 2007 Springer.

A metal nanoparticle viewed as a "plasmonic atom"

Klimov and Guzatov [23] proposed consideration of a metal nanoparticle as a "plasmonic atom". First, this notion emphasizes the peculiar properties of nanoparticles and suggests the possibility of treating them as peculiar elementary "bricks" in certain device or functional material applications. Second, definite analogies with the electric field spatial distribution for plasmon resonance and electron wave function in a hydrogen atom have been discovered. Looking at the radial part of the wave function relevant to an electron in a Coulomb potential and comparing it with the radial profile of an electric potential in a spherical metal particle, these authors found amazing similarities. Both functions exhibit steady growth at a distance from 0 to $\sqrt{2}a_{\mathrm{B}}$ for an atom (a_{B} is the Bohr radius, see Chapter 2) and from 0 to a particle radius R for a metal particle, with subsequent rapid evanescence outside.

These functions are presented in Figure 6.12. In more detail, the wave function of an electron reads,

$$\psi_{nlm}(r,\theta,\varphi) = Z_{nl}(r)Y_l^m(\theta,\varphi), \tag{6.68}$$

where Z_{nl} is the radial part and Y_l^m is the spherical harmonic (see also Section 2.6 in Chapter 2). For the $2p$ electron state as an example, the radial function reads,

$$Z_{21}(r) = N_{21}\left(\frac{r}{a_{\mathrm{B}}}\right)\exp\left(-\frac{r}{2a_{\mathrm{B}}}\right), \tag{6.69}$$

where N_{21} is the normalizing constant and a_{B} is the Bohr radius defined by Eq. (2.109). The energy spectrum obeys a discrete series (Eq. (2.108)),

$$E_n = -\frac{\mathrm{Ry}}{N^2}, \quad N = 1, 2, 3, \ldots, \tag{6.70}$$

and is determined by the ionization energy Ry (Eq. (2.109)) and the principal quantum number N.

For a plasmonic "atom", the simplest dispersion law in the form of Eq. (6.10) gives the spectrum of plasmonic oscillations (Eq. (6.60)),

$$E_n = \hbar\omega_{\mathrm{p}}\sqrt{\frac{N}{2N+1}}, \quad N = 1, 2, 3, \ldots, \tag{6.71}$$

in a vacuum, whereas the electric field potential will have the form, as in Eq. (6.68) with the radial function,

$$Z_N = A_N \begin{cases} (r/R)^N, & r \leq R \\ (R/r)^{N+1}, & r > R, \end{cases} \tag{6.72}$$

where A_N is some constant, and N is an integer number.

This similarity between an electron in an atom and the electric field of an electromagnetic wave in a spatially confined object further extends the list of analogies between electromagnetic waves and electron properties, discussed in Chapter 3.

6.4 Size-dependent absorption and scattering

To this point, no dependence of the optical properties on metal nanoparticle size has arisen. The only indication of size was the pre-requisite condition of $R \ll \lambda$ implied in the very prefix "nano". In Chapter 5 we saw for semiconductor nanocrystals the pronounced size dependence arise mainly from the size-dependent electron and hole energy spectra when the size approaches their de Broglie wavelengths. This is the case for a few nanometer crystallites. Discrete electron spectra in metal nanoparticles are a subject of interest since the pioneering papers of Fröhlich and Kubo [20, 21]. However, unlike semiconductor nanoparticles, in metals the discreteness of the spectrum is so negligible that it can be identified only at liquid helium temperatures in extremely small particles of the order of 1 nm [28]. This is because of the relatively large electron effective mass ($m_{\mathrm{e}}^* \geq m_0$) and shorter de Broglie wavelength (see Problem 7). For 10 nm particles energy spacing between neighboring electron levels falls within the sub-millielectronvolt range and therefore the quantum size effect is not pronounced *in optics*. Instead, the *electrical* manifestations of quantized electron spectra in metal nanoparticles are examined extensively [29].

There is another mechanism of size dependence of the optical response inherent in metal nanoparticles. It is the electron confinement effect on the mean free path value ℓ. It is defined as the path length between two successive scattering events. It is determined by the product of the Fermi velocity v_{F} and the inverse damping rate Γ^{-1}, as is seen in Eq. (6.17). When a particle size decreases, extra scattering events at the surface contribute to the damping constant. If the particle size becomes smaller than $\ell = v_{\mathrm{F}}\Gamma^{-1}$, the contribution from surface scattering can even dominate over the other scattering processes. Thus the size-dependent damping rate $\Gamma(R)$ enters into the dielectric function and therefore the optical response of metal particles dispersed in a dielectric ambient medium acquire dependence on the particle

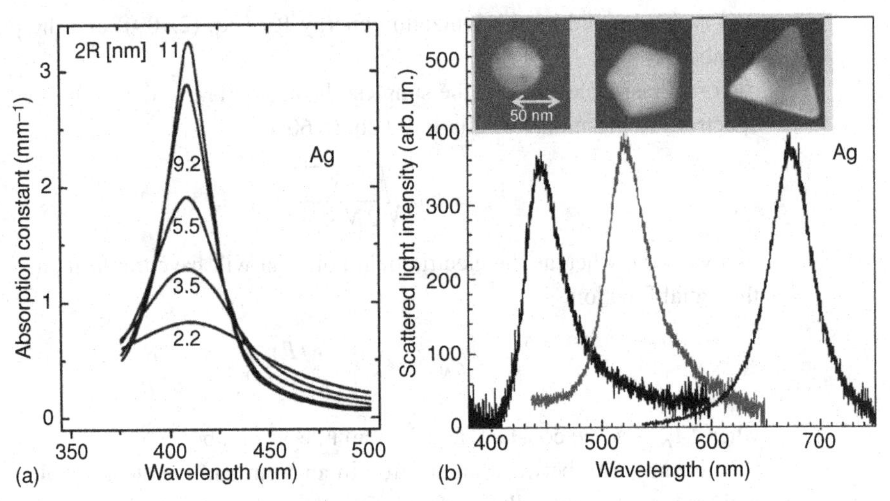

Fig. 6.13 **Size and shape effects on the optical properties of silver particles. (a) Absorption constant calculated for nanoparticles of various radii using dielectric function of silver in the form of Eq. (6. 67) with size-dependent damping (adapted from [1]). The particle radius is much smaller than wavelength. The size effect manifests itself via broadening because of the enhanced damping rate in smaller particles. (b) Observed scattering of light by a single particle for three different sizes/shapes (adapted from [30]). Particle size is no longer negligible at the wavelength scale. Size and shape effects manifest themselves as characteristic resonant scattering. Copyright 2002 AIP [30].**

size. Note that typical mean free path values are 40–50 nm (see data for Ag, Au and Cu in Table 6.1). These values define the scale where size dependence should occur.

The size-dependent damping constant can be written as,

$$\Gamma(R) = \Gamma_0 + A v_F / R, \tag{6.73}$$

where A is the phenomenological factor accounting for the specific scattering mechanism at the surface. This approach to the size–dependent optical response of metal nanoparticles is referred to as the *Kreibig model* and leads to size dependence of the particle dielectric function [1],

$$\varepsilon(\omega, R) = \varepsilon_{\text{bulk}}(\omega) + \omega_p^2\left(\frac{1}{\omega^2+\Gamma_0^2} - \frac{1}{\omega^2+\Gamma(R)^2}\right) + i\frac{\omega_p^2}{\omega}\left(\frac{\Gamma(R)}{\omega^2+\Gamma(R)} - \frac{\Gamma_0}{\omega^2+\Gamma_0^2}\right). \tag{6.74}$$

Optical absorption calculated for silver nanoparticles of various radii based on Eq. (6.73) is plotted in Figure 6.13(a). One can see that the size effect is essentially reduced to a broadening of the spectrum, which is not surprising since it is damping rate which determines the width of the resonant absorption in spectroscopy. Further details on size-dependent scattering are related to accounting for particle shape, the appearance of surface phonons and their contribution to the electron–phonon scattering rate, and for larger particles ($R > \ell$) the existence of loop-like scattering paths with possible phase recovery after multiple scattering events. A review of the relevant works can be found in [1].

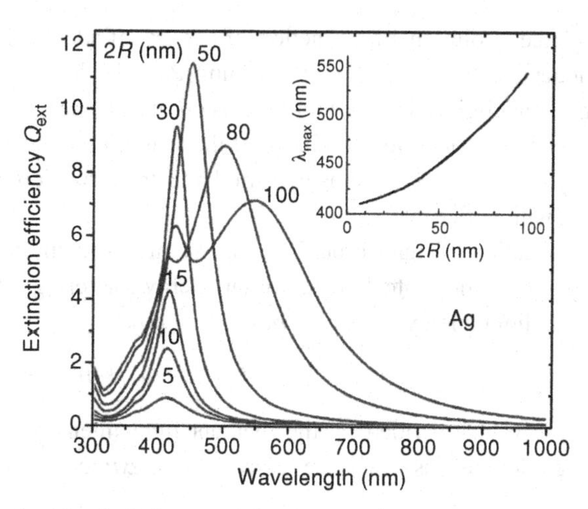

Fig. 6.14 **Extinction efficiency for identical silver particles dispersed in a medium with refractive index 1,5 for various diameters indicated near every curve (adapted from [36]). The insert shows absorption peak wavelength dependence on particle diameter.**

Size-dependent plasmonic modes develop in non-spherical particles such as, e.g. a three-axial ellipsoid. The relevant electrostatic problem dates back to the 1940s [31, 32]. The three characteristic length parameters give rise to three dipole modes, five quadrupole modes, seven octopole modes etc. The total number of modes of a given multiplicity is $2N + 1$ where 2^N equals the multiplicity of mode, i.e. $2^1 = 2$ for a dipole, $2^2 = 4$ for a quadrupole and so on. For dipole modes (denoted by the subscript "1" in ε_1) the three resonant values of dielectric function read [23],

$$\varepsilon_1^{(n)} = 1 - \left(\frac{a_1 a_2 a_3}{2} I_n\right)^{-1}, \quad n = 1, 2, 3, \tag{6.75}$$

where

$$I_n = \int_0^\infty \frac{1}{\left(u + a_n^2\right)} \frac{1}{\left(u + a_1^2\right)^{1/2} \left(u + a_2^2\right)^{1/2} \left(u + a_3^2\right)^{1/2}} du. \tag{6.76}$$

For cubic particles the spectrum is more complex. For example a few primary resonances are [33],

$$\omega/\omega_p = 0.46225, 0.54473, 0.58722, 0.66372, 0.74953, 0.83918.$$

For larger particles whose diameter is no longer negligibly small compared to optical wavelengths, the correct scattering theory should be thoroughly applied. It is related to the basic contribution by Gustav Mie dating back to 1908 [34]. Scattering of light by small particles is considered in detail in the book by Bohren and Huffman [35]. Application of the Mie theory to metal particles was discussed by Kreibig and Vollmer [1]. Here we restrict ourselves to just a couple of representative examples of how scattering can modify the optical properties of metal particles. Figure 6.13(b) presents the shape and size effects in the range 50–100 nm on scattering of light by a single metal particle. The effect of size/shape on optical properties in this size range is dominating. Figure 6.14 shows the calculated evolution of extinction efficiency of identical silver nanoparticles dispersed homogeneously in a dielectric medium. Particle diameters from 5 to 100 nm cover the range of negligibly small nanoparticles on the wavelength scale, as well as larger particles for which the continuous build-up of the scattering contribution is evident. Let us discuss briefly the important notions used in the characterization of optical properties of disperse media with noticeable scattering [35].

Propagation of light through a disperse medium can be described in terms of the light energy absorption rate W_{abs} and light energy scattering rate W_{scat}. The sum of these give the total light energy *extinction rate*,

$$W_{\text{ext}} = W_{\text{abs}} + W_{\text{scat}}. \tag{6.77}$$

These values divided by the incident light intensity I_{inc} give rise to the values with squared dimensions which are referred to as, *extinction cross-section*,

$$C_{\text{ext}} = W_{\text{ext}}/I_{\text{inc}}, \tag{6.78}$$

absorption cross-section,

$$C_{\text{abs}} = W_{\text{abs}}/I_{\text{inc}}, \tag{6.79}$$

and *scattering cross-section*,

$$C_{\text{scat}} = W_{\text{scat}}/I_{\text{inc}}. \tag{6.80}$$

These are coupled as Eq. (6.77) prescribes, i.e.

$$C_{\text{ext}} = C_{\text{abs}} + C_{\text{scat}}. \tag{6.81}$$

With respect to a single particle, these values can be compared to its geometrical square S ($S = \pi R^2$ for a spherical one) to give the respective *efficiency factors*,

$$Q_{\text{ext}} = \frac{C_{\text{ext}}}{S}, \quad Q_{\text{abs}} = \frac{C_{\text{abs}}}{S}, \quad Q_{\text{scat}} = \frac{C_{\text{scat}}}{S}. \tag{6.82}$$

If the concentration of particles per unit volume N is small enough to ensure the negligible contribution of multiple scattering then the Bouguer law holds,[6]

$$I(L) = I_{\text{inc}} \exp(-\alpha_{\text{ext}} L), \tag{6.83}$$

with,

$$\alpha_{\text{ext}} = NC_{\text{abs}} + NC_{\text{scat}} = NC_{\text{ext}}. \tag{6.84}$$

The exponential law (6.81) is a direct consequence of the assumed linear attenuation of light intensity within infinitesimal length $\mathrm{d}x$ along light propagation direction x,

$$\mathrm{d}I(x) = -\alpha_{\text{ext}} I(x) \mathrm{d}x. \tag{6.85}$$

The assumption (6.85) is justified if the contribution of scattering is small, $NC_{\text{scat}}L \ll 1$.

[6] Multiple scattering of light in dispersed dielectric is considered in detail in Chapter 8 devoted to properties of non-periodic media.

For small particles (not necessarily metallic ones) absorption and scattering cross-sections are directly defined by polarizability (6.53) [35],

$$\begin{aligned} C_{\text{scat}} &= \frac{k^4}{6\pi}|\alpha|^2 = \frac{8\pi}{3}k^4R^6\left|\frac{\varepsilon-\varepsilon_1}{\varepsilon+2\varepsilon_1}\right|^2, \\ C_{\text{abs}} &= k\,\text{Im}[\alpha] = 4\pi kR^3\,\text{Im}\left[\frac{\varepsilon-\varepsilon_1}{\varepsilon+2\varepsilon_1}\right]. \end{aligned} \tag{6.86}$$

Looking back at Figure 6.14 one can see that in the diameter range 5–30 nm ($R \ll \lambda$) the extinction efficiency factor Q_{ext} monotonically rises with only minor long-wave shift of the spectral shape with size. Larger particles exhibit higher Q_{ext}, the rise in $Q_{\text{ext}}(R)$ being superlinear which is indicative that the extinction cross-section C_{ext} grows more rapidly than a single particle volume. Note that, contrary to Figure 6.13a, size-dependent damping was not included in the calculations. Between the diameter values 30 and 50 nm an increase in Q_{ext} occurs along with the noticeable long-wave shift of the extinction band which is indicative of scattering contribution. For a 50 nm diameter, contributions to extinction of absorptive and scattering parts are nearly equal. Note that the spectral position of the calculated curve in Figure 6.14 for 50 nm concords with experimentally observed (Fig. 6.13(b), the left curve corresponding to a sphere-like particle). Further increase in particle size results in development of a strong and wide long-wave band due to scattering with the remaining original narrow short-wave peak from the purely absorptive contribution.

In a dense ensemble of such particles multiple scattering effects along with interference of scattered waves will additionally contribute to the transmission and reflectance of light. Further, particle polydispersity will additionally broaden the extinction spectrum. Application of scattering theory to dense metal–dielectric structures will be discussed later in Chapter 11, related to nanoplasmonics.

6.5 Coupled nanoparticles

Coupling of close metal nanoparticles offers further options toward engineering of plasmonic resonances. Klimov and Guzatov [23] proposed to consider coupled metal nanoparticles as a "plasmonic molecule", contrary to isolated nanoparticles viewed as "plasmonic atoms".

Examples of such plasmonic molecules are shown in Figure 6.15, where two conical and spherical metal nanoparticles are located at a distance comparable with their size (all characteristic sizes in the problem are much less than the light wavelength). The notion of a "plasmonic molecule" is justified only provided the new properties arise as it happens in an atomic molecule (see, e.g. splitting of energy levels in close quantum wells in Chapter 3). Klimov and Guzatov did show that this is the case using a pair of identical spherical and hyperbolic particles [23, 37, 38]. In Figure 6.15c resonant dielectric function values are plotted versus spacing between two identical nanoparticles. At the very right axis the arrows show the three modes inherent in asingle isolated nanoparticle, $\varepsilon_{\text{res}} = -2, -\frac{3}{2}, -\frac{4}{3}$. One

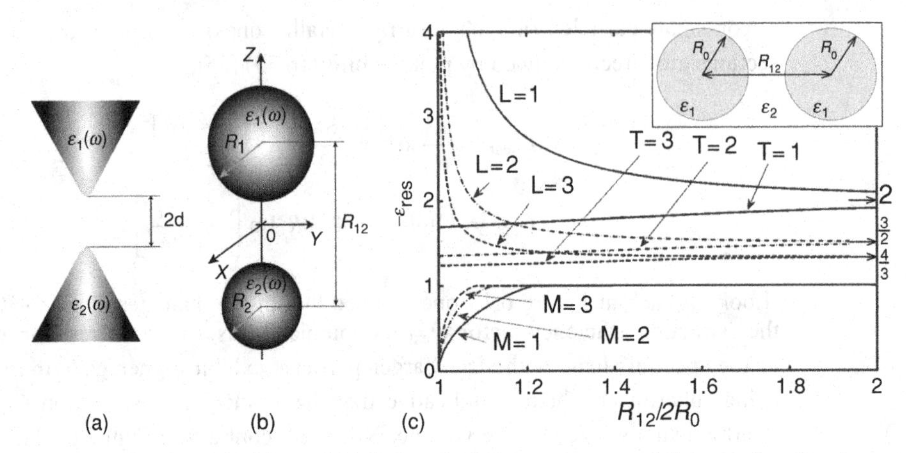

Fig. 6.15 **Plasmonic molecules. (a), (b) possible designs [23]; (c) resonances of dielectric function for a pair of two identical metal spherical particles (adapted from [37]).**

can see each of these resonant modes splits into two modes with dielectric function values rapidly diverging as the particles get closer. These three pairs of modes are labeled T_1,L_1, T_2,L_2 and T_3,L_3. Moreover, the additional type of resonance is revealed, labeled as M_1, M_2 and M_3 . These modes have no analog in a single particle. Thus the notions "plasmonic atoms" and "plasmonic molecules" acquire reasonable physical content and can be used in photonic engineering not only alone, but arranged in groups or coupled with other spatial structures as well. Kreibig and Vollmer introduced the notion "cluster matter" for a dense ensemble of nanometer sized metal particles and demonstrated numerous experimental examples of development of additional low-frequency extinction bands in addition to the single original band inherent in isolated nanoparticles [1]. This qualitatively agrees with calculations presented in Figure 6.15 in terms of the splitting of the principal resonant band corresponding to $\varepsilon(\omega) = -2$ with strong distant dependence of the L_1. It is this band which exhibits a pronounced low-frequency shift when coupled particles get closer. Note that higher absolute values of the negative dielectric function in Figure 6.15 do correspond to lower frequencies in $\varepsilon(\omega)$ (see Figs. 6.1 and 6.3).

6.6 Metal–dielectric core–shell nanoparticles

A dielectric nanoparticle covered with a metal shell or, vice versa, a metal particle covered with a dielectric shell represents yet another type of building unit for photonic nanoengineering with desirable optical resonances. Their optical properties depend on the intrinsic properties of a dielectric and a metal as well as a dielectric core size/shape and a metal shell thickness.

For a spherical coated particle, the polarizability derived from electrostatics reads [35],

$$\alpha = 4\pi R^3 \frac{(\varepsilon_{\text{shell}} - \varepsilon_1)(\varepsilon_{\text{core}} + 2\varepsilon_{\text{shell}}) + f(\varepsilon_{\text{core}} - \varepsilon_{\text{shell}})(\varepsilon_1 + 2\varepsilon_{\text{shell}})}{(\varepsilon_{\text{shell}} + 2\varepsilon_1)(\varepsilon_{\text{core}} + 2\varepsilon_{\text{shell}}) + f(2\varepsilon_{\text{shell}} - 2\varepsilon_1)(\varepsilon_{\text{core}} - \varepsilon_{\text{shell}})}, \tag{6.87}$$

where R is the outer radius, $f = (R_{\text{core}}/R)^3$ is the core volume fraction and $\varepsilon_{\text{core}}$, $\varepsilon_{\text{shell}}$, and ε_1 are the core, shell and ambient medium dielectric functions, respectively. Amazingly, a particle becomes invisible ($\alpha = 0$) if the numerator in Eq. (6.87) equals zero, i.e.

$$\frac{(\varepsilon_1 - \varepsilon_{\text{shell}})}{(\varepsilon_1 + 2\varepsilon_{\text{shell}})} = f\frac{(\varepsilon_{\text{core}} - \varepsilon_{\text{shell}})}{(\varepsilon_{\text{core}} + 2\varepsilon_{\text{shell}})}. \tag{6.88}$$

Equation (6.87) has been derived for arbitrary values of dielectric permittivities without any special assumptions on their metallic or dielectric features. Let us look at the condition of resonant increase in polarizability which may occur when the denominator in the right-hand part of Eq. (6.87) takes the minimal value. Let us check the extreme possibility of a zero value of the denominator, i.e.

$$(\varepsilon_{\text{shell}} + 2\varepsilon_1)(\varepsilon_{\text{core}} + 2\varepsilon_{\text{shell}}) + f(2\varepsilon_{\text{shell}} - 2\varepsilon_1)(\varepsilon_{\text{core}} - \varepsilon_{\text{shell}}) = 0. \tag{6.89}$$

Equation (6.89) is a linear equation with respect to $\varepsilon_{\text{core}}$ but a quadratic one with respect to $\varepsilon_{\text{shell}}$. It means that, unlike the core, the shell will always possess two surface modes – one for the inner interface and another one for the outer interface. If the core is a metal, and the shell and the ambient medium are both dielectric, then the conditions of surface plasmon oscillations on the metal surface are [36],

$$\varepsilon'_{\text{core}} = \frac{2\varepsilon'_{\text{shell}}\left[\varepsilon'_{\text{shell}}(f-1) - \varepsilon_1(2+f)\right]}{\varepsilon'_{\text{shell}}(1+2f) + 2\varepsilon_1(1-f)}, \quad \varepsilon''_{\text{core}} = 0, \tag{6.90}$$

where primes on ε' and ε'' denote the real and the imaginary part of the proper dielectric function, respectively. Note, the ambient medium is implied to have a purely real dielectric function. If the core size is small as compared to the shell thickness (i.e. $f \to 0$) then Eq. (6.90) reduces to the familiar condition of surface plasmon resonance at the surface of a metal particle inside the medium with dielectric permittivity $\varepsilon'_{\text{shell}}$,

$$\varepsilon'_{\text{core}} = -2\varepsilon'_{\text{shell}}. \tag{6.91}$$

If the shell thickness is negligibly small as compared with the core radius (i.e. $f \to 1$), then Eq. (6.90) gives a similar condition with respect to a metal nanoparticle inside the medium with dielectric permittivity ε_1,

$$\varepsilon'_{\text{core}} = -2\varepsilon_1. \tag{6.92}$$

In the intermediate case of finite f, the effect of a dielectric coating is pronounced as the spectral shift of surface plasmon resonance, the value of this shift being dependent on coating and ambient medium permittivities (the reader may care to solve Problem 10).

Consider now the dielectric core with metal shell. The quadratic equation (6.89) with respect to $\varepsilon_{\text{shell}}$ has two solutions (simultaneously $\varepsilon''_{\text{shell}} = 0$ is met) [36],

$$\varepsilon'_{\text{shell}(\pm)} = \frac{1}{4(1-f)}\Big\{-2f(\varepsilon'_{\text{core}} + \varepsilon_1) - (\varepsilon'_{\text{core}} + 4\varepsilon_1) \pm \sqrt{\left[2f(\varepsilon'_{\text{core}} + \varepsilon_1) + (\varepsilon'_{\text{core}} + 4\varepsilon_1)\right]^2 - 16(1-f)^2\varepsilon'_{\text{core}}\varepsilon_1}\Big\}. \tag{6.93}$$

The two solutions correspond to surface plasmon oscillations at $\varepsilon_{\text{core}}/\varepsilon_{\text{shell}}$ and $\varepsilon_{\text{shell}}/\varepsilon_1$ interfaces. Interestingly, the relative position on the frequency scale of metal/core and metal/medium surface plasmons is defined by the inequalities,

$$1 < \varepsilon_{\text{core}} < 4\varepsilon_1, \tag{6.94}$$

$$\varepsilon_{\text{core}} > 4\varepsilon_1 > 4. \tag{6.95}$$

Here we consider the core material permittivity as purely real and omit the prime label. In terms of refractive indices, Eqs. (6.94) and (6.95) give $n_{\text{core}} < 2n_1$ and $n_{\text{core}} > 2n_1$, respectively. In the first case, the solution $\varepsilon'_{\text{shell}(+)}$ corresponds to plasmons on the outer shell/medium interface whereas $\varepsilon'_{\text{shell}(-)}$ corresponds to plasmons at the inner shell/core interface. The outer plasmon resonance occurs at $\varepsilon'_{\text{shell}(-)} \approx -2\varepsilon_1$. The inner plasmon resonance occurs at $\varepsilon'_{\text{shell}(+)} \approx -\frac{1}{2}\varepsilon_{\text{core}}$. Therefore, in terms of absolute values, $1 < |\varepsilon'_{\text{shell}(+)}| < |\varepsilon'_{\text{shell}(-)}| = 2\varepsilon_1$ holds and using the simplified purely real metal dielectric function in the form $\varepsilon = 1 - \omega_{\text{p}}^2/\omega^2$, one can see the inner plasmons have higher frequency compared with the outer plasmons (Fig. 6.16a). In the second case, the outer plasmon resonance occurs at $\varepsilon'_{\text{shell}(-)} \approx -\frac{1}{2}\varepsilon_{\text{core}}$, whereas the inner plasmon resonance occurs at $\varepsilon'_{\text{shell}(+)} \approx -2\varepsilon_1$. Now, in terms of absolute values, $2 < |\varepsilon'_{\text{shell}(+)}| < |\varepsilon'_{\text{shell}(-)}| = \frac{1}{2}\varepsilon_{\text{core}}$ holds and the outer plasmon acquires the higher frequency as compared with the inner one (Fig. 6.16(b)). Figure 6.16(c) shows the calculated spectra for these two cases for a 7.9 nm silver shell over a 20 nm core. The reader is asked to identify the inner and the outer resonances themselves (Problem 12).

It is interesting to trace the size dependence of extinction spectra for core–shell particles. An increase of the dielectric core size when the metal shell thickness is constant results in rapid long-wave shift of the resonant bands [36]. Such behavior qualitatively resembles the scattering characteristics of solid metal particles (see Fig. 6.13(b)). Amazingly, if the dielectric core size is kept constant but the shell thickness increases, the resonant extinction bands move in the opposite direction. Although the overall size of a particle increases, the resonant wavelengths decrease. This is shown in Figure 6.16d. Such behavior is non-trivial and can not be foreseen from the theory of light scattering by solid particles. The behavior predicted in the theory has been observed in experiments by Oldenburg *et al.* [39]. The reason for this unusual behavior may, at least partially, result from the size-dependent scattering of electrons in thinner shells.

Notably, the extinction band for nanoshell structures is rather wide as compared to the solid metal nanoparticles (see Figs. 6.11, 6.12). The reason for the extra broadening is definitely efficient damping by the outer as well as the inner nanoshell/dielectric interface. Kachan and Ponyavina [40] derived the modified electron mean free path ℓ which determines the damping rate Γ via the relation $\Gamma = v_{\text{F}}/\ell$. These authors considered the equal probability of scattering of an electron from any given point inside the shell by both interfaces and then found the average value of the mean free path in the form,

$$\ell(R, h) = R\left[\frac{1}{1+h^2} - \frac{h}{2} - \frac{1}{4}\frac{(1-h^2)}{(1+h^2)} \cdot (1-h)\ln\frac{(1-h)}{(1+h)}\right], \tag{6.96}$$

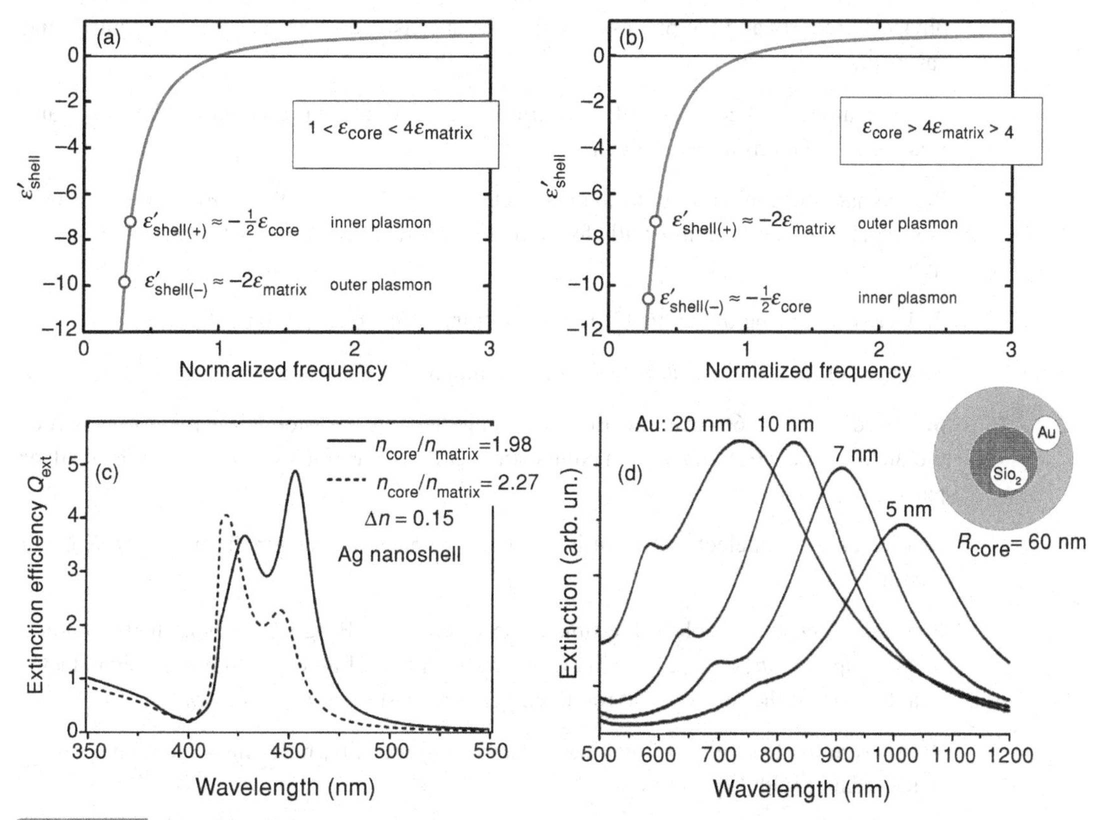

Fig. 6.16 **Optical properties of spherical nanoparticles with a dielectric core and a metal shell. (a), (b) The real part of the dielectric function of an ideal metal and the relative positions of the inner and outer surface plasmon resonances, according to the solutions of Eq. (6.93), the normalized frequency is ω/ω_p. (c) Extinction spectra calculated for the two combinations of refractive indices of core and matrix corresponding to the different mutual positions of the inner and outer resonances for silver nanoshells [36]. Core radius is 10 nm, silver shell thickness is 7.9 nm. (d) Calculated extinction spectra of a number of core–shell silica/gold particles in air with core radius 60 nm and varying shell thickness between 5 to 20 nm (indicated near every curve). Adapted from. Reprinted with permission from [39]. Copyright 1998, Elsevier B.V.**

where R is the outer radius of the structure and h is the ratio of the core and outer radii $h = R_{core}/R$. When the core radius tends to zero the mean free path given by Eq. (6.96) tends to R.

Further primary coupled structures by analogy to plasmonic molecules consisting of solid metal particles could be coupled metal nanoshells with dielectric cores.

Problems

1. At first glance, a different plasma frequency in different metals could be treated as the reason for their colors. However this is not the case since plasma frequency actually lies beyond the visible. Looking at the metal parameters listed in Table 6.1 elucidate the true physical property and the process defining the sheens known as "coppery", "auburn" and "argentine" ("silvery").

2. Compare the reflectance and dielectric function of silver presented in Figure 6.3 and find the correlations between them.

3. Estimate the applicability of the parameters $\varepsilon_\infty = 5$, $\hbar\omega_p \approx 9$ eV, $\hbar\Gamma = 0.02$ eV proposed in [5] for the experimentally found real part of the dielectric function for silver shown in Figure 6.3.

4. Derive an analog of Eq. (6.47) for an arbitrary dielectric medium with $\varepsilon_1 \neq 1$.

5. Based on Eq. (6.47) elucidate the inverse function $\omega(k)$ which is plotted in Figure 6.6a.

6. Based on Eq. (6.66) examine the variation in the peak position of the imaginary part of the dielectric function inherent in composite media with various volume fractions of silver particles.

7. Elucidate the dielectric permittivity of the composite medium from the Bruggeman formula (6.67).

8. Using data from Table 6.1 estimate the electron de Broglie wavelength in common metals implying $m_e^* = m_0$, $E = E_F$. Compare with that of an electron in a semiconductor quantum dot of the same size at room temperature, assuming $m_e^* = 0.1m_0$.

9. Analyze the feasibility of making dielectric and metal particles invisible with coatings in accordance with Eq. (6.88).

10. Using Eq. (6.90) consider a spherical metal nanoparticle with dielectric coating layer embedded in a host dielectric medium and evaluate the effect of the coating on surface plasmon resonance for different ratios of coating and host medium dielectric permittivity. Consider purely real and frequency-independent permittivities of dielectrics for simplicity. Using various tables in this book find out numerical estimates for Al particles covered with an oxide shell embedded in air, water, silica and titania.

11. Using Eq. (6.93) evaluate inner and outer plasmon resonances for core–shell–matrix parameters: silica–gold–air, GaAs–silver–silica, ZnSe–copper–titania.

12. Identify the inner and outer surface plasmon resonances for the spectra presented in Figure 6.16c.

References

[1] U. Kreibig and M. Vollmer. *Optical Properties of Metal Clusters* (Berlin: Springer, 1995).

[2] S. A. Maier. *Plasmonics: Fundamentals and Applications* (Berlin: Springer Verlag, 2007).

[3] C. Kittel. *Introduction to Solid State Physics* (New York: Wiley and Sons, 1978).

[4] J. A. Dobrowolski. Optical properties of thin films and coatings. In: *Handbook of Optics I*, Ed. M. Bass (New York: McGraw-Hill, 1995).
[5] B. Johnson and R. W. Christy. Optical constants of the noble metals. *Phys. Rev. B*, **6** (1972), 4370–4379.
[6] C. G. Granqvist and O. Hunderi. Optical properties of ultrafine gold particles. *Phys. Rev.*, **16** (1977), 3513–3533.
[7] P. Drude. Zur Elektronentheorie der Metalle. *Ann. Phys.*, **1** (1900), 566–613.
[8] H. Ehrenreich and H. R. Philipp. Optical properties of Ag and Cu. *Phys. Rev.*, **28** (1962), 1622–1629.
[9] F. Bassani and G. P. Parravicini. *Electronic States and Optical Transitions in Solids* (London: Pergamon, 1975).
[10] R. W. Wood. Remarkable optical properties of the alkali metals. *Phys. Rev.*, **44** (1933), 353–360.
[11] C. Zener. Remarkable optical properties of the alkali metals. *Nature*, **132** (1933), 968–968.
[12] N. W. Ashcroft and N. D. Mermin. *Solid State Physics* (Orlando: Saunders College Publishing, 1976).
[13] H. E. Ives and H. B. Briggs. The optical constants of potassium. *J. Opt. Soc. Amer.*, **26** (1936), 238–246.
[14] A. Yariv. *Optical Electronics in Modern Communications* (Oxford: Oxford Univeristy Press, 1997).
[15] C. J. Powell and J. B. Swan. Effect of oxidation on the characteristic loss spectra of aluminum and magnesium. *Phys. Rev.*, **118** (1960), 640–643.
[16] R. H. Ritchie, Plasma losses by fast electrons in thin films. *Phys. Rev.*, **106** (1957), 874–881.
[17] E. Kretschmann. Die Bestimmung optischer Konstanten von Metallen durch Anregung von Oberflächenplasmaschwingungen. *Z. Physik*, **241** (1971), 313–324.
[18] A. V. Zayats and I. I. Smolyaninov. Near-field photonics: surface plasmon polaritons and localized surface plasmons. *J. Opt. A: Pure Appl. Opt.*, **5** (2003), 816–850.
[19] H. Raether. *Surface Plasmons* (Berlin: Springer-Verlag, 1988).
[20] H. Frölich. Die spezifische warme der kleiner metallteilchen. *Physica*, **4** (1937), 406–410.
[21] R. Kubo. Electronic properties of metallic fine particles. *J. Phys. Soc. Jap.*, **17** (1962), 975–980.
[22] J. D. Jackson. *Classical Electrodynamics* (New York: Wiley, 1975).
[23] V. V. Klimov and D. M. Guzatov. Plasmonic atoms and plasmonic molecules. *Appl. Phys. A*, **89** (2007), 305–314.
[24] J. C. Maxwell-Garnett. Colours in Metal Glasses and in Metallic Films. *Philos. Transact. Royal Soc. London*, **203** (1904), 385–420.
[25] A. N. Oraevsky and I. E. Protsenko. Optical properties of heterogeneous media. *Quantum Electronics*, **31** (2001), 252–256.
[26] P. N. Dyachenko and Yu. V. Miklyaev. One-dimensional photonic crystal based on a nanocomposite "metal nanoparticles – dielectric". *Kompyuternaya Optika*, **31** (2007), 31–34.

[27] D. A. G. Bruggeman. Berechnung verschiedener physikalischer Konstanten von heterogenen Substanzen. I. Dielektrizitetskonstanten und Leitfehigkeiten der Mischkorper aus isotropen Substanzen. *Ann. Phys.*, **416** (1935), 636–664.

[28] W. P. Halperin. Quantum size effects in metal particles. *Rev. Mod. Phys.*, **58** (1986), 533–606.

[29] J. von Delft and D. C. Ralph. Spectroscopy of discrete energy levels in ultrasmall metallic grains. *Physics Reports*, **345** (2001), 61–173.

[30] J. J. Mock, M. Barbic, D. R. Smith, D. A. Schultz and S. Schultz. Shape effects in plasmon resonance of individual colloidal silver nanoparticles. *J. Chem. Phys.*, **116** (2002), 6755–6759.

[31] J. A. Stratton. *Electromagnetic Theory* (New York, McGraw-Hill, 1941).

[32] E. W. Hobson. *The theory of Spherical and Ellipsoidal Harmonics* (Cambridge: Cambridge University Press, 1931).

[33] R. Fuchs. Theory of the optical properties of ionic crystal cubes. *Phys. Rev. B*, **11** (1975), 1732–1740.

[34] G. Mie. Beiträge zur Optik trüber Medien, speziell kolloidaler Metallösungen. *Ann. Phys.*, **25** (1908) 377–445.

[35] C. F. Bohren and D. R. Huffman. *Absorption and Scattering of Light by Small Particles* (New York: John Wiley & Sons, 1985).

[36] S. M. Kachan. *Effect of Coherent Resonant Interactions on Optical Properties of Metal-Dielectric Nanostructures.* Ph.D. Thesis. Stepanov Institute of Physics, Minsk, 2007.

[37] D. V. Guzatov. Spontaneous Radiation of Atoms and Molecules Near Metal Nanobodies with Complex Shape. Ph.D. Thesis, Russian Academy of Sciences P. I. Lebedev Physical Institute, 2007.

[38] V. V. Klimov and D. V. Guzatov. Strongly localized plasmon oscillations in a cluster of two metallic nanospheres and their influence on spontaneous emission of an atom. *Phys. Rev. B*, **75** (2007), 024303.

[39] S. J. Oldenburg, R. D. Averitt, S. L. Westcott and N. J. Halas. Nanoengineering of optical resonances. *Chem. Phys. Lett.*, **288** (1998), 243–247.

[40] S. M. Kachan and A. N. Ponyavina. Resonance absorption spectra of composites containing metal-coated nanoparticles. *J. Mol. Struct.*, **563–564** (2001), 267–272.

7 Light in periodic structures: photonic crystals

> *"I have discussed in a recent paper [1] the propagation of waves in an infinite laminated medium . . . , and have shown that, however slight the variation, reflexion is ultimately total, provided the agreement be sufficiently close between the wavelength of the structure and the half-wavelength of the vibration."*
>
> *Lord Rayleigh, 1888* [2]

7.1 The photonic crystal concept

Since the time when de Broglie published his hypothesis on the wave properties of matter particles in 1923, the wave mechanics of matter has become a well-developed field of science. It has provided an explanation for the properties of atoms, molecules and solids. Furthermore, it predicted novel properties of artificial solids like quantum wells and quantum wires. As we have seen in Chapters 2 and 3, there are many common features and phenomena in wave mechanics and wave optics. At the very dawn of wave mechanics, it essentially borrowed much from wave optics.

Nowadays, the reverse process manifests itself in science. Results of quantum mechanics which are direct consequences of the wave properties of electrons and other quantum particles are transferred to classical electromagnetism, and to wave optics. These are results that are not related directly to spin and charge. Such transfers have formed a new emerging field in modern optics of inhomogeneous media with the concept of a photonic crystal at the heart of the field. The very concept of photonic crystals has been developed using an analogy between propagation of an electron in a periodic potential and an electromagnetic wave in a medium with periodic alteration of refractive index in space. In natural crystals, the periodic displacement of ions in the crystal lattice forms a periodic potential for an electrons. The results of electron motion in such a potential are propagation of Bloch waves rather than plane waves, development of energy bands separated by band gaps, modification of electron mass, formation of Brilloiun zones, conservation of quasi-momentum rather than momentum. All these issues are results for a single-particle quantum theory of solids. Many of them have been discussed in Chapter 4. The concept of the photonic crystal implies the systematic consideration of electromagnetic waves in structures where the refractive index features periodicity in one, two or three directions. This consideration in many instances replicates the one-particle quantum theory of solids. Such consideration constitutes the rather wide field of photonics with multiple applications in optical communication, laser technologies, optoelectronics etc.

To give a definition, *a photonic crystal is a medium with periodic change of dielectric function in one, two or three dimensions.* It creates band structure and band gaps for electromagnetic waves as a periodic Coulomb potential gives rise to bands for an electron in a crystal lattice. The physical reason for these phenomena is the interference of waves experiencing multiple scattering from periodically arranged scatterers. Band gaps arise because of the formation of standing waves. Band structure and band gaps are essentially the classical electromagnetic phenomena that can be consistently discussed and understood in terms of wave optics. In a sense, the notation "photonic" when speaking about photonic crystals is somewhat misleading since light quanta are not involved in the very basic features of photonic crystals. It should be taken as indication of the field of *photonics* rather than of photons.

There are two essential precursors of the photonic crystal concept. The first one is formation of a reflection band in a periodically layered medium, which has been known for more than a century and dates back to Rayleigh's research [1, 2]. In a periodically laminated thick (as compared to wavelength) slab, a reflection band always develops independently of the amplitude of refractive index change. Smaller alteration simply needs a thicker slab to obtain high reflectance. Another precursor is dynamic X-ray diffraction with the particular features of specular reflection for selected directions, originating from standing wave formation. This was identified for the first time by C. Darwin [3].[1] The new realm that the photonic crystal concept has introduced is the properties of light waves and light–matter interactions in the case of two- and three-dimensional (i.e. beyond Rayleigh's theory) periodic lattices with high refractive index alteration (100% and more for light waves versus 0.01% in X-ray physics). The essential hint on the remarkable properties of three-dimensional lattices for light propagation was revealed by K. Ohtaka in 1979 [4]. The notation "photonic crystal" was introduced by E. Yablonovitch in 1989 [5]. He was inspired by the ingenuous idea of V. P. Bykov on the inhibition of spontaneous emission of light in periodic media [6]. The latter will be a subject for close consideration in Chapter 14.

In this chapter, an introduction to the wave optics of periodic media will be provided, with a number of characteristic examples based on mainly numerical calculations. For the comprehensive theory of photonic crystals including calculation techniques the reader is referred to books and reviews [7–12].

7.2 Bloch waves and band structure in one-dimensionally periodic structures

We start with the wave equation for an electromagnetic wave and consider the properties of electromagnetic waves in a medium with periodic step-wise dielectric function $\varepsilon(x + a) = \varepsilon(x)$ (Fig. 7.1), i.e. a multilayer infinite stack consisting of the two types of materials with

[1] Charles Galton Darwin is a grandson of Charles Darwin, the outstanding biologist.

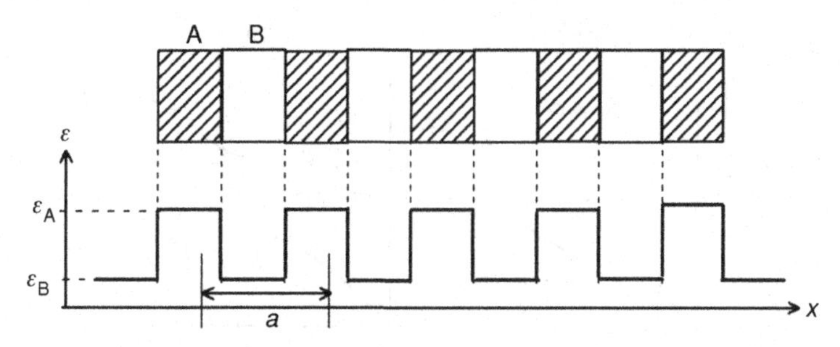

Fig. 7.1 **An optical analog to the quantum-mechanical Kronig–Penney model.**

different dielectric permittivities. The wave equation for an electric field, $E(x,t)$ reads,

$$\frac{\mathrm{d}^2}{\mathrm{d}x^2}E(x,t)+\frac{1}{c^2}\varepsilon(x)\frac{\partial^2}{\partial t^2}E(x,t)=0. \tag{7.1}$$

Substituting $E(x,t)$ in the form,

$$E(x,t)=E(x)\mathrm{e}^{i\omega t}, \tag{7.2}$$

we arrive at the familiar Helmholtz equation (see Eq. 2.47),

$$\frac{\mathrm{d}^2}{\mathrm{d}x^2}E(x)+\varepsilon(x)\frac{\omega^2}{c^2}E(x)=0. \tag{7.3}$$

Now we have to reproduce the steps performed in Chapter 3 in terms of parallel consideration of electrons and electromagnetic waves using the analogies of a step-wise potential barrier and a step-wise change in refraction index n. In an ideal case of purely refractive material without dissipative losses and with magnetic permittivity $\mu=1$, the relation $n=\sqrt{\varepsilon}$ holds. Since we have already considered the Kronig–Penney model for electrons in Chapter 4 (Section 4.1) we are in a position to write down the solution of Eq. (7.3) in the form of electromagnetic Bloch waves,

$$E(x)=E_k(x)\mathrm{e}^{ikx}, \tag{7.4}$$

where k is the *Bloch wave number* and $E_k(x)=E_k(x+a)$ is periodic with the same period as $\varepsilon(x)$. Subscript k in $E_k(x)$ means this function depends on k. Similar to the case of electrons, the states differing in wave number by the value $k_N=k\pm\frac{2\pi}{a}N$ are equivalent.

As in the case of the single-particle Schrödinger problem, the periodicity of the medium gives rise to breaks in the dispersion curve $\omega(k)$ and to formation of intervals on the k-axis where a solution of Eq. (7.3) in the form of plane waves Eq. (7.4) does not exist. The break points of the $E(x)$ function correspond to standing waves with electromagnetic energy concentration, either in the sublattice of the high-refractive, or in the sublattice of the low-refractive material. These properties are displayed in Figures 7.2–7.4.

Figure 7.2(a) shows the dispersion curve for an electromagnetic wave in a vacuum,

$$\omega=ck, \tag{7.5}$$

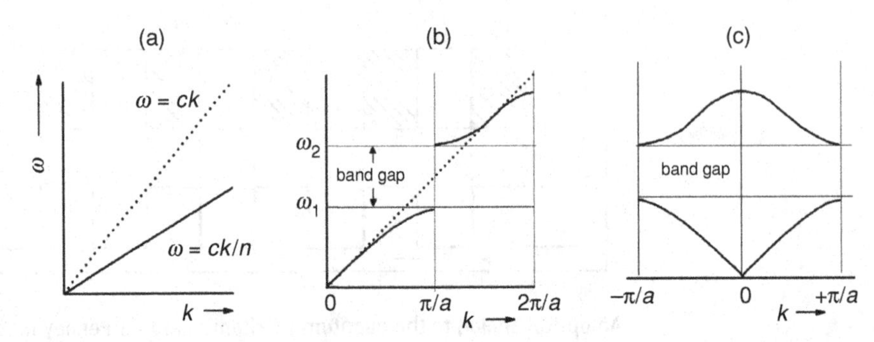

Fig. 7.2 Dispersion curves for electromagnetic waves (a) in a homogeneous and (b), (c) in a periodic medium. Only the parts of the curves relevant to positive k values are displayed in (a) and (b). The diagram displayed in the (c) panel is obtained from the diagram shown in (b) by means of translation of the two uppermost curves (for positive k and for negative k) to the left and to the right by $\pm 2\pi/a$ values. It is instructive to compare this figure with the electron analog in Figure 4.2.

and in a continuous medium with refraction index $n > 1$,

$$\omega = ck/n. \tag{7.6}$$

In a medium with periodic $n(x)$, breaks appear in the dispersion curve for every value of the wave number satisfying the condition,

$$k_N = N\pi/a, \tag{7.7}$$

where N is a positive or negative integer number. Figure 7.2 shows the range $[0, 2\pi/a]$ with the break at $k = \pi/a$. All wave numbers differing by the value $2N\pi/a$ appear to be equivalent. As was explained in Chapter 4 for the case of electrons, this equivalence leads to the concept of the Brilloiun zone, i.e. an interval of width $2\pi/a$ which contains all non-equivalent k values. It is convenient, as it was for electrons, to choose Brillouin zones symmetrically with respect to the $k = 0$ point, i.e. for the Nth zone,

$$(N-1)\frac{\pi}{a} < |k| < N\frac{\pi}{a}. \tag{7.8}$$

The first Brillouin zone is the interval $[-\pi/a, +\pi/a]$. The second zone consists of the two symmetrical intervals $[-2\pi/a, -\pi/a]$ and $[+\pi/a, +2\pi/a]$. The third one also consists of the two symmetrical intervals $[-3\pi/a, -2\pi/a]$, $[+2\pi/a, +3\pi/a]$, and so on. As in the case of an electron, it is convenient to move all parts of the dispersion curve into the first Brillouin zone. Then we arrive at the so-called reduced dispersion curve presented in Figure 7.2(c). This presentation is referred to as *photonic band structure*.

For small k values ($k \ll \pi/a$) the dispersion curve reproduces that of a homogeneous medium. Then with growing k the wave velocity $d\omega/dk$ reduces and at $k \to \pi/a$ the value of $d\omega/dk$ tends to zero. At $k = \pi/a$ a standing wave develops which may take two types. The first type corresponds to the concentration of the field in the sublattice with high refractive index, whereas the second type corresponds to the concentration of the field in the sublattice with low refraction index. The first type occurs near the bottom of the band gap (ω_1) whereas

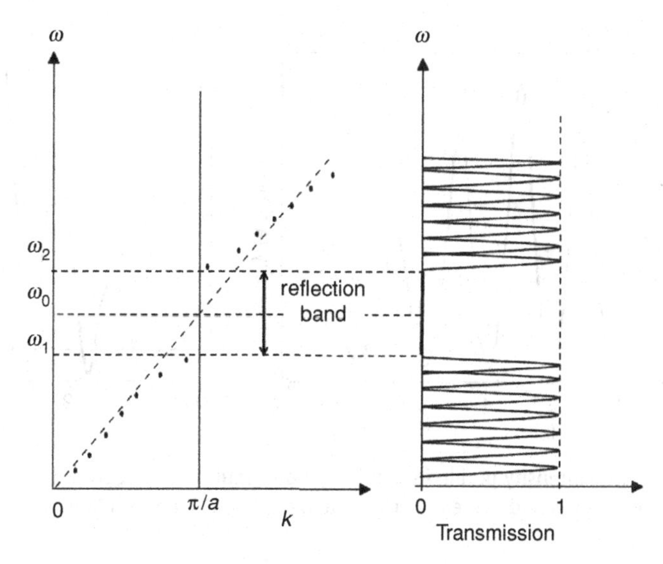

Fig. 7.3 **Properties of a finite-length stack of two materials with different refractive indexes. (a) Frequency versus wave number dependence and (b) optical transmission spectrum. The number of transmission peaks within one band equals the number of layers in the stack.**

the second type occurs at the top of the band gap (ω_2). Therefore the lower band is sometimes referred to as the *dielectric band*, whereas the higher band is referred to as the *air band* implying a dielectric/air lattice. By and large, the $\omega(k)$ function is close to the linear function for k, far from the break points and tends to constant value at the break points.

Therefore the continuous linear dispersion curve inherent in a homogeneous medium breaks into nonlinear portions in a periodic medium. For continuous portions of the dispersion curve, Bloch waves represent the solutions whereas within the forbidden gaps the wave solutions do not exist. Instead an evanescent wave develops. It is rather instructive to consider the properties of a finite multilayer periodic structure. For a finite slab with periodic dielectric function, the dispersion curve reduces to a discrete set of points (Fig. 7.3, left panel). The number of such points equals the number of periods in the slab. These discrete points generate the corresponding set of narrow transmission sub-bands with transmission $T = 1$ in the middle of every sub-band (Fig. 7.3, right panel). Within the band the gap transmission coefficient equals zero and, accordingly, the reflection coefficient R equals 1. Generally, reflection and transmission spectra are complementary in an ideal purely refractive medium without dissipation losses, i.e. $R + T = 1$ holds everywhere.

Figures 7.4 and 7.5 show spatial intensity distributions computed for a finite periodic slab for various frequencies. In Figure 7.4 the concentration of the field is evident either in high- or in low-refractive layers for the maxima of the two transmission bands adjacent to the band gap. In Figure 7.5 the evanescent field is presented for the frequency inside the gap. The evanescence of the field inside the structure coexists with high reflection of a wave coming from outside.

The complete reflection band is sometime referred to as *stop-band*. Noteworthy, for a one-dimensional periodic slab the stop-band with high reflection always develops independently

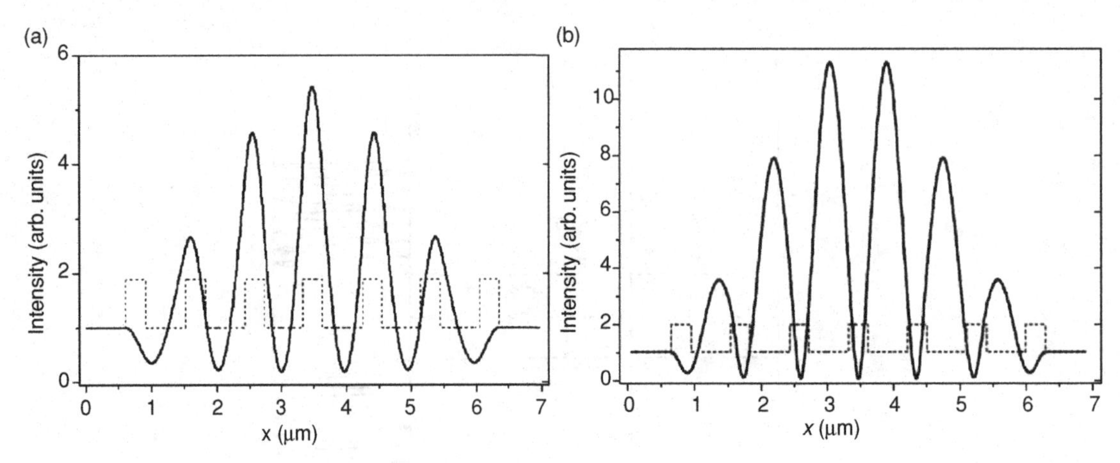

Fig. 7.4 Spatial intensity profile in a finite periodic multilayer structure for transmission bands in the close vicinity to band gap edges (a) ω_1 and (b) ω_2 [13]. Dashed lines show the refractive index profiles.

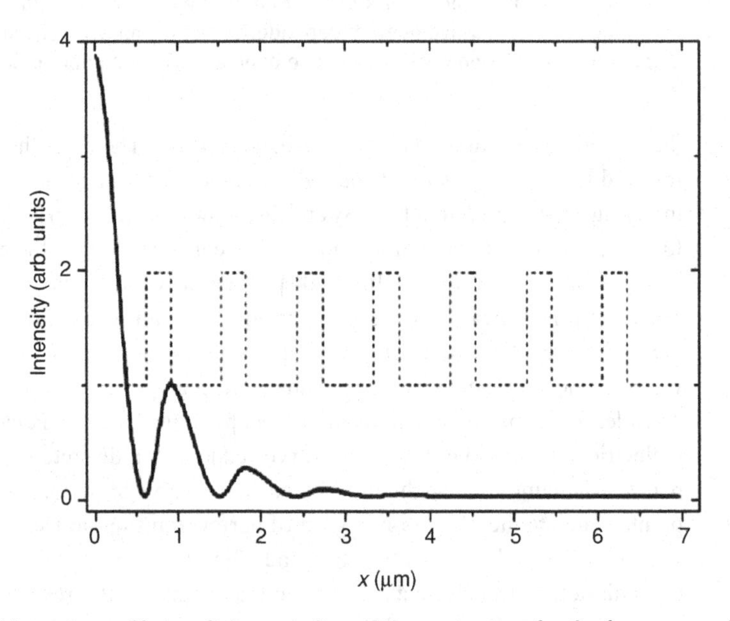

Fig. 7.5 The spatial intensity profile in a finite periodic multilayer structure for the frequency value in the center of the band gap ω_0 [13]. The dashed line shows refractive index profile.

of the size of the refractive index difference between the layers. The longer the periodic sequence of layers the higher is the reflection in the stop-band. The higher the refractive index difference, the wider is the stop-band. Remarkably, these phenomena had been predicted and explained by Rayleigh in 1887 [1]. He derived the analytic expression for the stop-band width for a finite "linearly laminated" structure, emphasizing that total reflection of waves occurs not only for a wavelength equal to double the period of the structure, but

Partly with a view to this question, I have discussed in a recent paper* the propagation of waves in an infinite laminated medium (where, however, the properties are supposed to vary continuously according to the harmonic law), and have shown that, however slight the variation, reflexion is ultimately total, provided the agreement be sufficiently close between the wave-length of the structure and the half wave-length of the vibration. The number of alternations of structure necessary in order to secure a practically perfect reflexion will evidently depend upon the other circumstances of the case. If the variation be slight, so that a single reflexion is but feeble, a large number of alternations are necessary for the full effect, and a correspondingly accurate adjustment of wave-lengths is then required. If the variation be greater, or act to better advantage, so that a single reflexion is more powerful, there is no need to multiply so greatly the number of alternations; and at the same time the demand for precision of adjustment becomes less exacting. The application of this principle to the case of an actual crystal, supposed to include a given number of alternations, presents no difficulty. At perpendicular incidence symmetry requires (and observation verifies) that the reflexion vanish; but, as the angle of incidence increases, a transition from one twin to the other becomes more and more capable of causing reflexion. Hence if the number of alternations be large, the spectrum of the reflected light is at first limited to a narrow band (whose width determines in fact the number of alternations). As the angle of incidence increases, the reflexion at the centre soon becomes sensibly total, and at the same time

* "On the Maintenance of Vibrations by Forces of Double Frequency, and on the Propagation of Waves through a Medium endowed with a Periodic Structure," Phil. Mag. Aug. 1887.

Phil. Mag. S. 5. Vol. 26. No. 160. *Sept.* 1888. S

Fig. 7.6 A fragment of Rayleigh's paper "On the remarkable phenomenon of crystalline reflexion described by Professor Stokes", published in Philosophical Magazine in 1888 [2].

for all values of wavelength λ in a medium with density ρ_0 within the interval,

$$\frac{(\lambda/2)^2}{a^2} = 1 \pm \frac{1}{2}\frac{\rho_1}{\rho_0}, \tag{7.9}$$

where ρ_1 is the density of another medium in the laminated structure. Rayleigh considered mechanical vibration but formulated the final results in terms of optical properties. A copy of Rayleigh's paper with the discussion of this issue is presented in Figure 7.6. More detail on the optical properties of one-dimensional periodic structures can be found in [14, 15]. Multilayer thin film coatings on dielectric substrates have been known as dielectric optical mirrors and interference narrow-band filters since the 1940s.

There is one particular, practically important case of multilayer periodic structures, namely, alternating layers of the two materials with different refractive indexes n_1, n_2 and thicknesses d_1, d_2, but with the same optical density for each layer, i.e.

$$n_1 d_1 = n_2 d_2 \equiv nd. \tag{7.10}$$

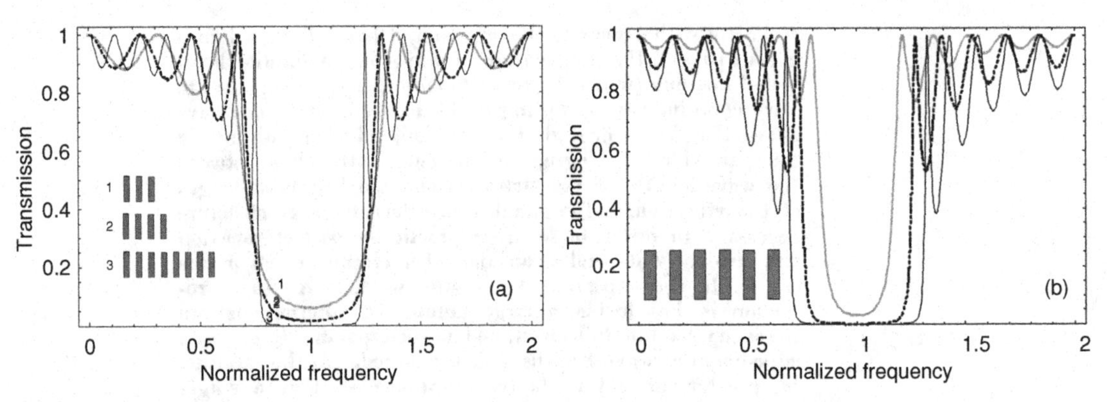

Fig. 7.7 **Transmission spectra of finite quarter-wave multilayer structures, consisting of alternating layers with refractive indexes n_2 and n_1 embedded in an ambient medium with refractive index $n_2 < n_1$ [13]. Every structure begins and ends with a highly refractive layer whose refractive index is n_1. (a) $n_1/n_2 = 2$, number of layers is 5 (gray line), 7 (dashed line), and 15 (solid line); (b) 11 layers with different $n_1/n_2 = 1.5$ (gray line), 2 (dashed line), and 2.5 (solid line). The structures are shown in the insets.**

In this case the central (midgap) frequency ω_0 in the reflection band is determined from the so-called quarter-wave condition,

$$\lambda_0/4 = nd, \tag{7.11}$$

and reads,

$$\omega_0 = 2\pi c/\lambda_0 = \pi c/2nd. \tag{7.12}$$

For quarter-wave periodic structures the transmission spectrum is a periodic function of frequency with period $2\omega_0$. Moreover, for *finite* periodic quarter-wave structures the analytical expression for the optical transmission coefficient $T_{QW}(\omega)$ has been derived [14]:

$$T_{QW}(\omega) = \frac{1 - 2R_{12} + \cos\pi\tilde{\omega}}{1 - 2R_{12} + \cos\pi\tilde{\omega} + 2R_{12}\sin^2\left[N \arccos\left(\dfrac{\cos\pi\tilde{\omega} - R_{12}}{1 - R_{12}}\right)\right]}, \tag{7.13}$$

where $\tilde{\omega} = \omega/\omega_0$ is the midgap-normalized dimensionless frequency, and R_{12} is the reflection coefficient at the $n_1 \leftrightarrow n_2$ refractive step (see Eq. 3.22),

$$R_{12} = \left(\frac{n_1 - n_2}{n_1 + n_2}\right)^2 = \left(\frac{1 - n_1/n_2}{1 + n_1/n_2}\right)^2. \tag{7.14}$$

Representative spectra for different numbers of periods and different n_1/n_2 values are given in Figure 7.7. One period is shown at the frequency scale normalized with respect to ω_0. Within every period on the frequency scale, the transmission spectrum of a quarter-wave structure is symmetrical with respect to points ω_0, $3\omega_0$, $5\omega_0$ etc. Spectral symmetry can be written in a general form as,

$$T(\omega + 2N\omega_0) = T(\omega), \quad N = 1, 2, 3, \ldots \tag{7.15}$$

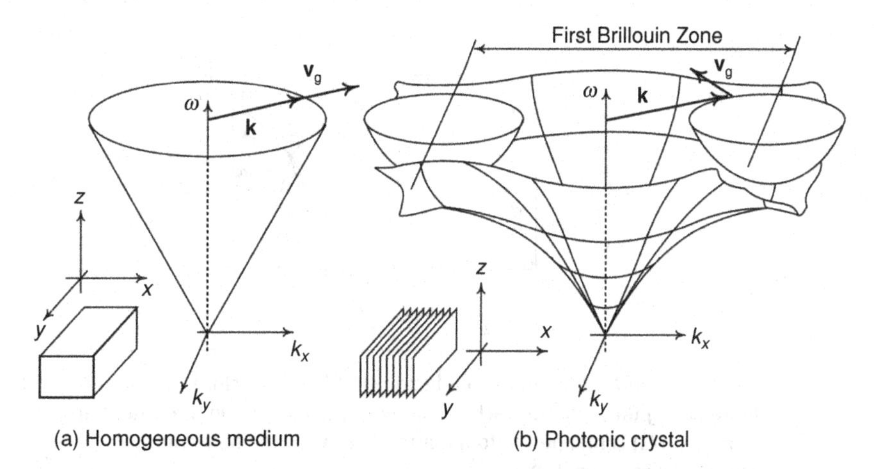

Fig. 7.8 A 3D representation of the photonic band structures of (a) an isotropic homogeneous non-dispersive medium and (b) a 1D photonic crystal. Only 2D slices of the wave vector space are depicted. Insets show the orientation of the media. A photonic crystal band structure (b) is presented only for one basic polarization. Adapted from [16].

The relative width of the reflection band for a quarter-wave periodic structure reads [7],

$$\frac{\Delta\omega_{\text{gap}}}{\omega} = \frac{4}{\pi}\frac{|n_1 - n_2|}{n_1 + n_2} = \frac{4}{\pi}\sqrt{R_{12}}, \tag{7.16}$$

i.e. it is unambiguously defined by the n_2/n_1 value, which is often referred to as the *refractive index contrast.*

7.3 Multilayer slabs in three dimensions: band structure and omnidirectional reflection

In Section 7.2 a one-dimensionally periodic structure was considered in one-dimensional space only. However, practical structures like laminated multilayer stacks are actually three-dimensional (3D) objects. In this section we consider properties of multilayer stacks (so-called 1D photonic crystals slabs) at oblique and grazing incidence of light.

A 3D representation of the dispersion law of a sample isotropic homogeneous medium with frequency-independent refractive index and band structure of an infinite multilayer periodic stack is presented in Figure 7.8 [16]. For a homogeneous medium, the dispersion law forms a cone. Every cross-section in the plane containing the frequency axis reproduces the curve shown in Figure 7.2(a). For a periodic medium, this occurs within the $\omega - k_y$ and $\omega - k_z$ cross-sections, whereas the $\omega - k_x$ cross-section reproduces the band structure shown in Figure 7.2(c).

For a homogenous medium in the 3D consideration, the dispersion law reads,

$$\mathbf{k}^2 = \frac{\omega^2}{c^2}n^2, \tag{7.17}$$

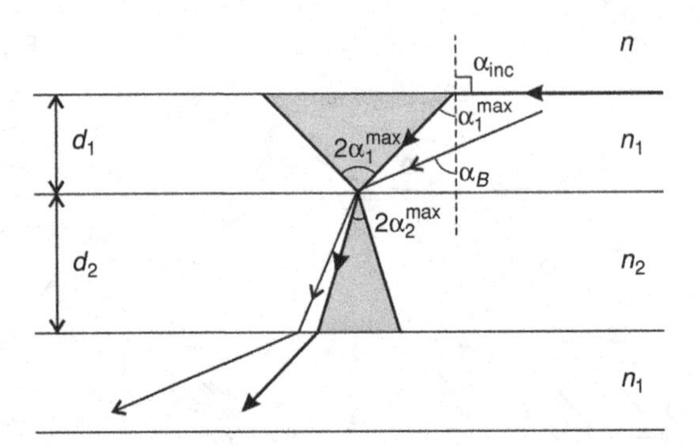

Fig. 7.9 **Schematic representation of a dielectric multilayer structure. The light rays refracting and propagating through the stack are shown. The full domain of incident angles in α_{inc} in the range $[-\pi/2, +\pi/2]$ is mapped onto the internal cone of the half-angle $\alpha_1^{max} = \arcsin(n/n_1)$ (light gray area). Adapted from [19].**

which agrees reasonably with the purely one-dimensional expression (7.6). The group velocity $\mathbf{v}_g$ is always parallel to $\mathbf{k}$,

$$\mathbf{v}_g = \nabla_k \omega(\mathbf{k}). \tag{7.18}$$

In a periodic medium band gaps develop for light with a finite k_x value. The band structure is essentially different for different polarizations. In the long-wavelength limit, the band structure asymptotically tends to the light cone inherent in a homogeneous anisotropic uniaxial crystal. So-called *form birefringence* develops [17, 18]. For higher frequencies, new bands, which are separated by frequency and angular gaps, appear in the band structure (Fig. 7.8(b)).

Notably, the energy flow direction is no longer parallel to the wave vector (Fig. 7.8(b)). In a periodic medium, the velocity of the energy flow integrated over a unit cell is still identical to the group velocity [21]. However, due to the strongly non-circular shape of the constant-frequency surfaces in the **k**-space, the group velocity Eq. (7.18) is no longer parallel to the wave vector.

Although multilayer thin film coatings have been used routinely in optical technologies and devices for decades, the photonic crystal concept brought new attention to this field. In 1998 groups from the USA [20], Belarus [21] and Great Britain [22] independently considered the possibility of obtaining 100% omnidirectional reflection from a semi-infinite multilayer structure. They found it was possible provided that refractive indexes n_1, n_2 are large enough and differ considerably from the refractive index, n, of the ambient medium (Fig. 7.9). This means that under certain conditions, a dielectric multilayer structure exhibits angular-independent reflection, as does a metallic mirror. The latter, however, features unavoidable dissipative losses. Such losses are absent in dielectric structures. Therefore, an omnidirectional reflector based on dielectric materials can in many cases substitute metal analogs to positive effect.

The properties of Bloch waves inside a periodic dielectric mirror can be discussed in terms of separate consideration of the TE (transverse electric) modes and TM (transverse magnetic) modes. For TE-modes the electric field vector is parallel to the layer interfaces,

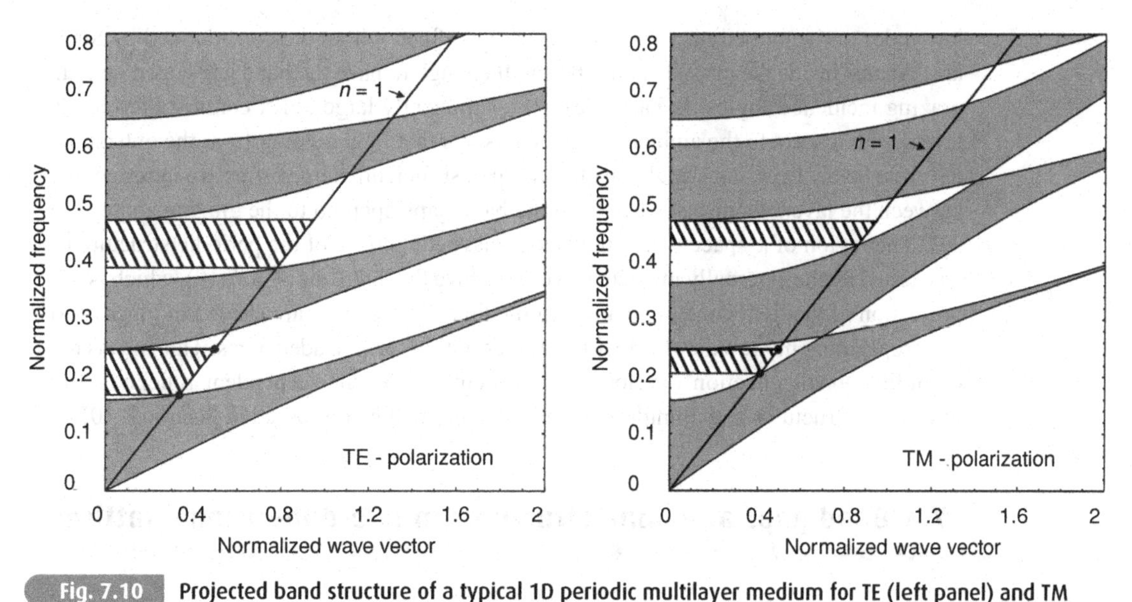

Fig. 7.10 Projected band structure of a typical 1D periodic multilayer medium for TE (left panel) and TM (right panel) polarizations. The parameters of the medium are $n_1 = 1.4$, $n_2 = 3.4$, filling factor $d_1/(d_1 + d_2) = 0.5$. The frequency is normalized as $\omega(d_1 + d_2)/2\pi c \equiv (d_1 + d_2)/\lambda_0$, and the tangential component of the wave vector $\mathbf{k}_\perp$ is normalized as $|\mathbf{k}_\perp|(d_1 + d_2)/2\pi$. The gray areas correspond to the propagating states, whereas the white areas contain the evanescent states only. The shaded areas correspond to omnidirectional reflection bands. The solid lines are the ambient-medium light-lines. Adapted from [19].

i.e. to the yz-plane in Figure 7.8. For TM-modes the magnetic field vector is parallel to the yz-plane. A projected band structure of an infinite periodic system of layers is given in Figure 7.10. This is a projection of a 3D band structure (Fig. 7.8(b)) onto the $\omega - k_y$ plane, where k_y is the tangential component of the wave vector $\mathbf{k}$, assuming the plane of incidence is x–y. The refractive indices correspond to the SiO_2/Si structure in the near IR region. The thicknesses of the layers are chosen as $d_1 = d_2$. An infinite periodic structure can support both propagating and evanescent Bloch waves. Gray areas correspond to the propagating states, whereas white areas are band gaps where only evanescent states exist.

When ω and $\mathbf{k}$ of a wave entering the periodic structure at an angle α_{inc} from an ambient homogeneous medium with refractive index n are within the band gaps, an incident wave undergoes strong reflection. Reflectivity depends on frequency and angle of incidence. This is seen in Figure 7.10. Photonic band gaps rapidly move to higher frequencies with increasing incident angle, denoted by the increase of the tangential component of the wave vector. Notably, the TM band gaps vanish when approaching the Brewster light-line $\omega = c\,|\mathbf{k}_\perp|\,/n_1 \sin\alpha_B$, where $\alpha_B = \arctan(n_1/n_2)$ is the Brewster angle. The TM polarized wave propagates without any reflection from the n_1 to the n_2 layer, and from the n_2 to the n_1 layer at the Brewster angle α_B. These properties of band structure restrict the angular aperture of a polarization-insensitive range of high reflectance. Therefore, omnidirectional reflectance can be achieved due to the restricted number of modes that can be excited by externally incoming waves inside a semi-infinite periodic structure. Waves coming from the low-index ambient medium ($n < n_1; n_2$) are confined within the cone defined by Snell's law. Angles inside the crystal should be small enough to have the band gaps open up to the grazing incidence angles. In particular, (i) a sufficiently large index contrast between the layers with respect to the ambient medium ensures that light coming from the outside will never go below Brewster's angle inside, and (ii) a sufficiently large refractive index contrast between the layers themselves can keep the band gaps open up to the grazing angles [19–22]. The region of k-space where the electromagnetic modes of the periodic structure can be excited by the externally incident wave lies above the light-line (Fig. 7.10), which is a 2D projection of the light cone of an ambient medium. These areas are shaded in Figure 7.10. No propagating mode exists for incident waves within these shaded areas. This means total omnidirectional reflection develops. Experiments on angular-dependent reflection from multilayer structures and omnidirectional reflection will be discussed in Section 7.10.

7.4 Band gaps and band structures in two-dimensional lattices

In two- and three-dimensional lattices with periodic dielectric function, the relation,

$$\varepsilon(\mathbf{r} + \mathbf{T}) = \varepsilon(\mathbf{r}) \tag{7.19}$$

holds. Here $\mathbf{T}$ is the translation vector (see Chapter 4, Eq. (4.6)). For a non-magnetic medium wave equations for the electric and magnetic fields take the form,

$$\nabla \times \left(\nabla \times \frac{1}{\varepsilon(\mathbf{r})}\mathbf{D}(\mathbf{r})\right) - \frac{\omega^2}{c^2}\mathbf{D}(\mathbf{r}) = 0, \tag{7.20}$$

$$\nabla \times \left(\nabla \times \frac{1}{\varepsilon(\mathbf{r})}\mathbf{H}(\mathbf{r})\right) - \frac{\omega^2}{c^2}\mathbf{H}(\mathbf{r}) = 0, \tag{7.21}$$

where $\mathbf{D}$ is the electrical displacement vector, and $\mathbf{H}$ is the magnetic field. When equations (7.20) and (7.21) are written in the form of eigenvalue problems, the corresponding operator in the equation for the magnetic field is Hermitian whereas the operator for the electric field is not. Therefore, calculations of photonic band structure for two- and three-dimensionally periodic materials are performed using solutions of Eq. (7.21) rather than Eq. (7.20). Then the electric field is recovered by means of the relation,

$$\mathbf{E}(\mathbf{r}) = \left(\frac{-ic}{\omega\varepsilon(\mathbf{r})}\right) \nabla \times \mathbf{H}(\mathbf{r}). \tag{7.22}$$

The calculation techniques are described in detail in [8, 9]. Different 2- and 3-dimensional lattices are the subject of numerical modeling based on dielectric constants (refractive indexes) of real materials, typically ranging from 1 (vacuum) to 13 (silicon, gallium arsenide) (see Table 3.1 in Chapter 3). Optimization of the geometrical structure of the lattices is performed to search for wider gaps within a bigger solid angle, ideally to search for complete (i.e. onmidirectional) gaps. This task is not easy. A one-dimensionally periodic structure, as was noted in the previous section, has been known since 1887 to always exhibit the gap. In 2- and 3-dimensional periodic lattices, special topology and the highest possible refractive

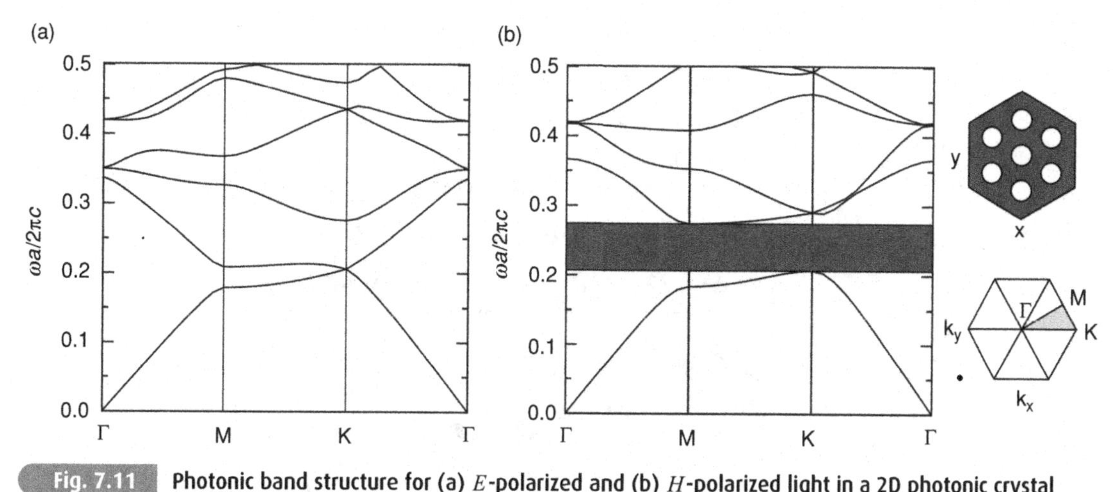

Fig. 7.11 Photonic band structure for (a) E-polarized and (b) H-polarized light in a 2D photonic crystal created by a triangular lattice of air holes ($\varepsilon = 1$) with the radius $r = 0.3a$ (where a is the lattice period) in a dielectric with permittivity $\varepsilon = 12$; the band gap is darkened. The top right inset shows a cross-sectional view of a fragment of the 2D photonic crystal. The bottom right inset shows the corresponding Brillouin zone, with the irreducible zone shaded. Reprinted with permission from [12]. Copyright 2007, Elsevier B.V.

index ratio of the constituent materials are necessary. Combinations of Si/air and GaAs/air offer refractive index ratios of 3.45 and 3.6, respectively. However these materials are not transparent in the visible range (see Table 4.1 for electronic band gaps). GaAs can be used for wavelengths longer than 820 nm and Si for wavelengths longer than 1000 nm.

When comparing properties of electromagnetic waves in periodic lattices with those of electrons, one essential difference should be emphasized. The electric charge inherent in electrons gives rise to a large scattering cross-section at charged ions in a crystal lattice. Therefore, the periodic potential gains high amplitude and an omnidirectional electronic band gap readily develops. It manifests itself straightforwardly as optical transparency of the large number of natural and artificial crystalline dielectric and semiconductor materials. Unlike electrons, electromagnetic waves scatter on inhomogeneities in the dielectric permittivity. This inhomogeneity in the visible range is approximately three times at its maximum (see Table 3.1). There is a pronounced tendency for decrease in dielectric permittivity with decreasing wavelength. For X-rays and gamma-rays the situation is even worse: 10^{-4} for X-rays and even less for gamma rays. On the other hand, for radiofrequencies dielectric permittivity up to 10^2 is feasible.

Consider a few selected examples of model two-dimensional structures. Figure 7.11 presents the band structure for a two-dimensional triangular lattice of air holes in a dielectric with a relatively large dielectric permittivity, $\varepsilon = 12$, relevant to silicon. The results are presented in terms of dimensionless frequency $\omega a/2\pi c \equiv a/\lambda_0$, where λ_0 is the light wavelength in a vacuum. Light propagates in the plane normal to the holes. In this case all Bloch modes can be separated into two sub-groups: the Bloch modes for which the electric field is parallel to the hole axis (referred to as E-polarized) and the Bloch modes for which the magnetic field is parallel to the hole axis (H-polarized modes). In terms of traditional

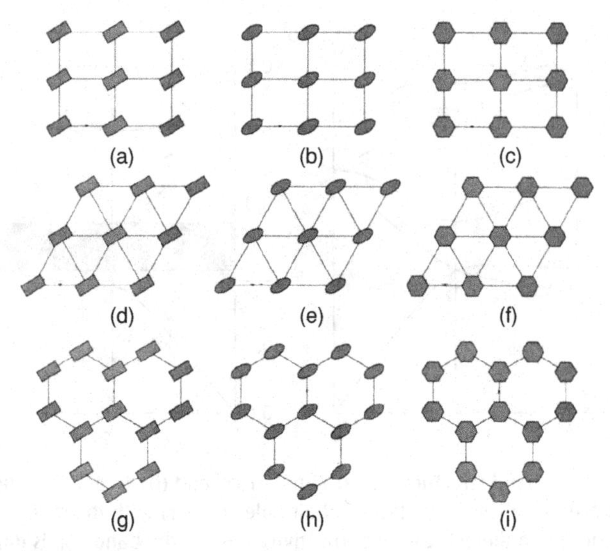

Fig. 7.12 **Model 2-dimensional lattices of rods examined in [23]: (a), (b), (c) rectangular; (d), (e), (f) triangular; (d), (h), (i) honeycomb.**

TE and TM notations, H-polarized light corresponds to the TE-mode (Transverse Electric) and E-polarized light corresponds to the TM-mode (Transverse Magnetic).

The band structure diagrams in Figure 7.11 for two-dimensional lattices represent the dispersion law $\omega(k)$ for the first Brillouin zone, as in the case of a one-dimensional periodic structure. However two-dimensional geometry needs three-dimensional (multi-value) functions $\omega(k_x, k_y)$ to be plotted, which is rather cumbersome. Therefore the following presentation style is generally adopted for two-dimensional periodic structures. The origin of coordinate system ω, k corresponds to $\omega = 0,\ k = 0$ and is indicated as the Γ-point in the Brillouin zone. Then the $\omega(k)$ function is presented with k varying along the $\Gamma \to M$ direction. Afterwards the **k** vector changes in value and direction between the M and K points at the edge of the first Brillouin zone. Note that at the very beginning, for small frequency and wave number, the dispersion curve is nearly linear as it was in the one-dimensional case (compare with Fig. 7.2(c)). This means that propagation of light can be interpreted in terms of an effective medium with constant refractive index. This is the familiar case of the long-wave approximation of complex media in optics where the wavelength is much larger than inhomogeneity sizes. The omnidirectional forbidden gap on the frequency axis in Figure 7.11 means an absence of $\omega(k)$ points within a finite range of frequencies. It is shown in the panel (b) as a gray band. Presentation of the band structure in terms of dimensionless frequency offers direct translation of results along the electromagnetic wave scale.

Note, in Figure 7.11 an omnidirectional (two-dimensional) band gap develops for H-polarized light only. Joannopoulos *et al.* [8] have shown the gap for E-polarized modes develops in the structure under consideration for $r/a = 0.5$. Formation of complete omnidirectional gaps for both types of modes in two dimensions is actually feasible with high-refractive materials, depending on the crystal geometry, shape of an individual scatterer and volume filling factor of the scatterers. Figure 7.12 shows a variety of two-dimensional

Table 7.1. Parameters of selected 2-dimensional photonic crystals consisting of rods with $\varepsilon = 12.96$ in air [23]

Notation in Figure 7.12	Lattice type	Scatterer type	Band gap $\Delta\omega/\omega$	Surface filling factor f
a	Rectangular	Rectangular	0.15	0.67
b	Rectangular	Circular	0.04	0.71
c	Rectangular	Hexagonal	0.025	0.71
d	Triangular	Rectangular	0.09	0.68
e	Triangular	Circular	0.2	0.85
f	Triangular	Hexagonal	0.23	0.70
g	Honeycomb	Rectangular	0.06	0.43
h	Honeycomb	Circular	0.11	0.2
i	Honeycomb	Hexagonal	0.11	0.2

lattices of rods that were the subject of extensive theoretical analysis [23]. Table 7.1 presents the widths of complete band gaps in the dimensionless units $\Delta\omega/\omega$ evaluated for GaAs ($\varepsilon = 12.96$) rods in air along with the surface filling factor f (the portion of the square filled with rods). A triangular lattice was found to offer wider gaps as compared with rectangular and honeycomb ones.

7.5 Band gaps and band structure in three-dimensional lattices

For three-dimensional dielectric lattices a number of theoretical investigations have been reported for close-packed dielectric spheres, as well as their three-dimensional replicas (inverted lattices) [8, 9, 24–28]. Analysis of various topologies for three-dimensional lattices has revealed the following features of band-gap formation. For the same lattice structure (simple cubic, face-centered cubic, and body-centered cubic lattices were examined), the structures comprising a small volume fraction ($f = 0.2$–0.3) of a dielectric in air appear to be more advantageous. For similar f values, those structure are advantageous that feature continuous topology of high-refractive fraction. For example, the cubic lattice of smaller (as compared to the lattice period) dielectric spherical particles feature a wider band gap when the particles are connected by cylindrical rods.

Among the variety of cubic lattices, the diamond lattice features a wider band gap. The diamond structure is a kind of face-centered cubic lattice. For the diamond lattice consisting of close-packed dielectric spherical particles, the band gap was found to open for all directions when the refractive index contrast $n_{\text{diel}}/n_{\text{air}} > 2$ [25, 26]. The maximal value of band gap width depends strongly on volume filling factor f. For example, for a diamond lattice built of dielectric balls with $n_{\text{diel}}/n_{\text{air}} = 3.6$ (GaAs), the band gap opens at $f = 0.2$, rises to $\Delta\omega/\omega = 0.15$ at $f = 0.35$, and then falls to zero at $f = 0.7$. For a diamond lattice built of spherical air cavities in a dielectric with $n_{\text{diel}}/n_{\text{air}} = 3.6$, the

Table 7.2. Values of the full gap width ($\Delta\omega/\omega$) and midgap position (a/λ, where a is the lattice parameter) for different combinations of materials and volume filling factors f

Sphere	Background	$\varepsilon_1 : \varepsilon_2$	f(%)	$\Delta\omega/\omega$ (%)	a/λ
Si	Air	12:1	43	13	0.45
Si	Silica	12:2.1	42	5	0.41
Air	Si	1:12	50	12	0.40
Silica	Si	2.1:12	50	4	0.38

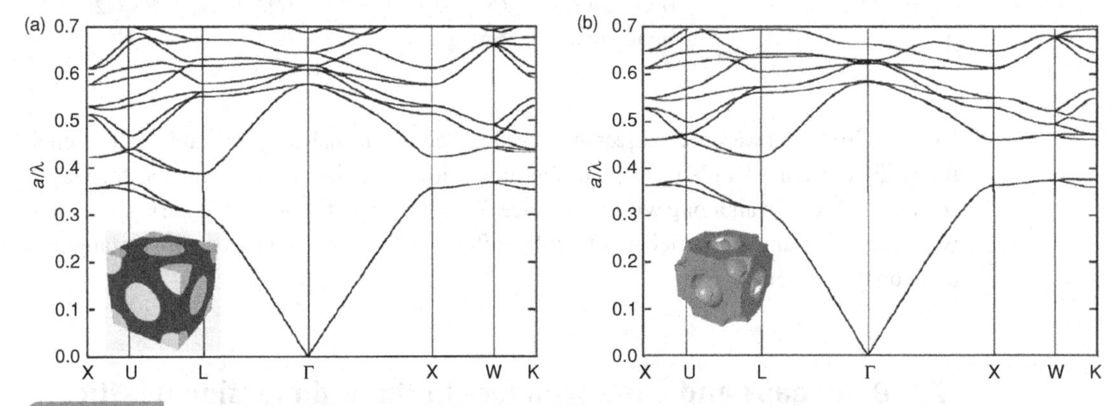

Fig. 7.13 **Photonic band diagrams of (a) a silicon/silica composite diamond structure and (b) structure made of air spheres in silicon resulting from the removal of the silica spheres from the former. The filling fraction for silicon is 50%. The insets show the corresponding real space structures. Reprinted with permission from [28]. Copyright 2001 AIP.**

band gap opens at $f = 0.35$, then widens up to $\Delta\omega/\omega = 0.28$ at $f = 0.8$ and then falls with growing filling factor. In the context of filling factors it is instructive to remind ourselves of the f values for various types of close-packed balls: $f = 0.52$ for a primitive cubic lattice, 0.68 for the body-centered cubic lattice and 0.73 for the face-centered cubic lattice [27].

Examples of computed band structures for diamond-like lattices exhibiting an omnidirectional band gap are shown in Figure 7.13 [28]. Silicon/silica and air/silicon structures were examined. As in Figure 7.11, the dimensionless frequency $\omega a/2\pi c \equiv a/\lambda$ is used, where λ is the wavelength value in a vacuum. Note that as in the case of 1D- and 2D-periodic structures, for low frequencies (long wavelengths) the dispersion is nearly linear, i.e. an effective medium approach can be applied to describe the propagation of electromagnetic waves in periodic structures. Remarkably, it works until $\lambda/a > 3$ holds, where λ is the wavelength in a vacuum.

7.6 Multiple scattering theory of periodic structures

The traditional theory of electromagnetic waves in periodic structures considers properties of waves in infinite space with translational symmetry. However, there is another approach based on the multiple scattering theory in which, instead of translational symmetry, a finite number of spatially arranged scatterers are considered. In this approach, instead of the band structure, the reflected and transmitted wave amplitudes and the relevant intensities are calculated. This approach is the same in essence as that considered for a finite number of layers to understand the build-up of transmission and reflections bands and the development of band structures for a one-dimensional periodic space. In the two- and three-dimensional cases, calculations are rather sophisticated and can only be performed numerically. The approach is described in [29], its application to photonic crystals is considered in [30, 31].

To compute the spectral features of the transmission and reflection of waves of a scattering medium consisting of a system of correlated scatterers, one should take into account interference cooperative effects, namely, coherent rescattering by particles and interference of the scattered waves. In this way, transmitted and reflected wave amplitudes can be derived for a single layer of close-packed identical dielectric spherical or cylindrical particles. Thus the characteristic features of two-dimensional periodic structures can be revealed. To go further towards understanding spectral features of three-dimensional periodic structures, the so-called quasicrystalline approximation is reasonable. It implies the structure consists of sequential layers, each layer containing close-packed identical spherical particles. In a multilayer system, not only the incident radiation field and the fields scattered by other particles of the same monolayer contribute to the effective field for a given particle, but fields from particles of other layers do also. This is the main difference of the multilayer system as compared to a single monolayer system. Based on the assumption of statistical independence of the individual monolayers, it is possible to first find the scattering amplitude of a single monolayer by taking into account multiple rescattering on particles within the layer, and then to account for the re-irradiation of different monolayers of the sample under consideration.

An example of this approach to close-packed dielectric spherical particles is shown in Figure 7.14. Refractive index contrast corresponds to silica balls and air voids. It corresponds to artificial *opals* that will be considered in more detail in Section 7.10. These results are rather instructive. In Figure 7.14 build-up of selective transmission and reflection with the growing number of layers is evident (compare with one-dimensional analog in Fig. 7.7). In the long-wave limit, the transmission is nearly 1 and reflection is negligible (note, there is no traditional Fresnel reflection here). A first-order stop-band develops for $d/\lambda \approx 0.45$. Second-order selective transmission/reflection bands are seen at half the wavelength, i.e. at $d/\lambda \approx 0.9$. For longer wavelengths ($d/\lambda < 0.5$) transmission and reflection are complementary, i.e. $T + R = 1$ holds with good accuracy. For shorter wavelengths ($d/\lambda \approx 1$), intense incoherent scattering develops and transmission selectivity smears. This scattering does not contribute to coherent reflection, therefore the absolute values of the reflection coefficient for short-wave stop-bands are rather low. The large contribution of incoherent

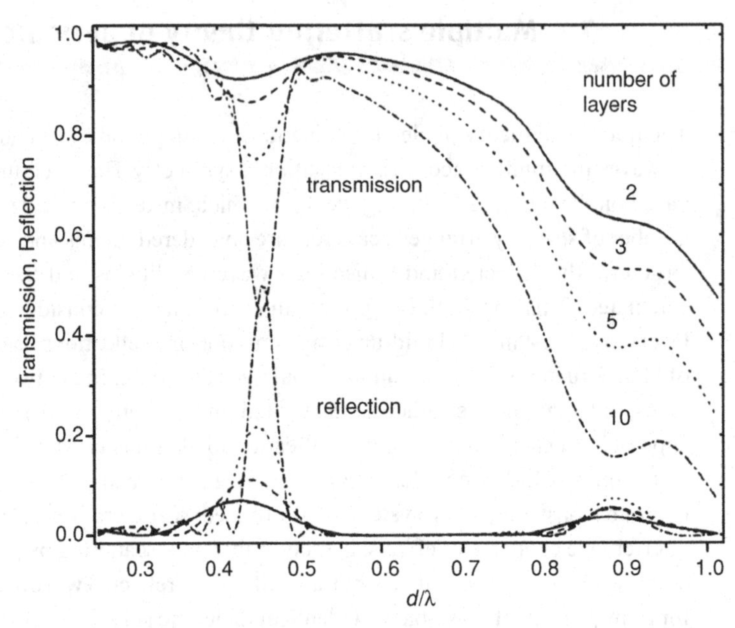

Fig. 7.14 Optical transmission and reflection spectra for a set of sequential layers consisting of close-packed spherical particles with cubic symmetry as a function of a dimensionless parameter d/λ, where d is the sphere diameter and λ is the wavelength in a vacuum. Numbers on the curves indicate the number of layers. Refractive index of spherical particles versus voids is 1.26. Adapted from [30].

scattering for shorter wavelengths lifts the complementarity of reflected and transmitted light and $T + R < 1$ holds.

7.7 Translation to other electromagnetic waves

Notably, all graphs in the previous section are plotted for a dimensionless normalized frequency and wavelength of electromagnetic waves. Only the dielectric permittivity and refractive index values were used inherent in materials within the visible or near IR range. This means, the results, conclusions and consideration in Sections 7.1–7.6 are equally valid for electromagnetic waves beyond the optical range. Periodic structures are considered for soft X-rays (wavelength of the order of 10 nm) in connection with high-resolution imaging in nanoelectronics. Activity for shorter wavelengths is not very extensive because of the negligible refractive index values for materials in the shorter wavelength range.

In the radiofrequency range the situation is quite the opposite for two reasons. First, higher dielectric constants of the order of 10^2 are feasible. Second, millimeter and centimeter wavelength scales readily enable performance of sophisticated topological configurations, in contrast to the optical range where nanotechnology is mandatory. In fact, experimental realization of full three-dimensional band gaps for electromagnetic waves were performed

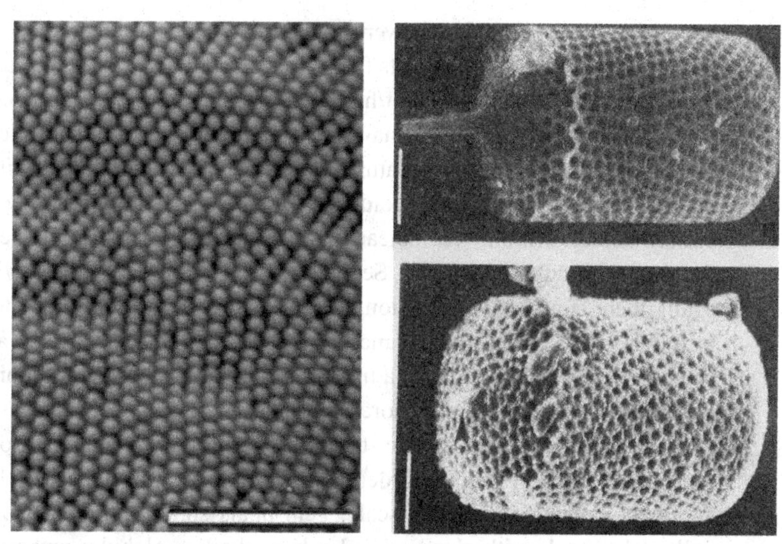

Fig. 7.15 **Periodic optical wavelength-scale structures in Nature. Left: An antireflecting corneal surface of a butterfly (*Vanessa kershawi*) eye. The scale bar represents 2 μm [36]. Right: Diatom water-plant fossils. The scale bars represent 10 μm [39].**

for the first time exactly in the millimeter range [32]. For advances and challenges in radiofrequency electromagnetic crystal structures [33–35] are recommended.

7.8 Periodic structures in Nature

Periodic structures on the scale of optical wavelengths with pronounced interference-based colors can be found in nature [36, 37]. For example, one-dimensional periodicity of coatings over wings can be found in several types of butterfly, in the feathers of peacocks and on the shells of many beetles. Isaac Newton outlined the role of interference in a peacock's shining colors in 1730. Structures with two-dimensional periodicity are present in the eyes of a number of moths and other insects and even more complex periodic structures are inherent in the cornea of human beings and mammalia [38]. Interestingly, two-dimensional periodicity is inherent in very ancient living objects, for example, diatom water plants that appeared about 500 million years ago. Two-dimensional periodicity can be found in pearls that consist of a layered package of cylinders.

The functionality of creatures' structures has been optimized over hundreds of millions of years of evolution in Nature. The presence of optical wavelength-scale periodicity is believed to be the result of such optimization. For example, the regular porous structure of insect eyes (Fig. 7.15(a)) and mammalian corneas form an antireflective interface and, probably, cut-off filter for ultraviolet radiation. At the same time, porous topology allows physico-chemical exchange processes to be feasible. Diatomic aquatic-plant fossils (Fig. 7.15(b)) feature pronounced micrometer-scale periodicity [39]. These are believed to gain a nanoporous photonic crystal fiber-like structure for the sake of more efficient light harvesting [40]. These fossils were found to reproduce photonic crystal fibers applicable for a variety of laser experiments [41]. An alternative reason, however, may just be related to an optimal mechanical weight/hardness combination. The rationale for interference colors is not straightforward. We may speculate, interferential (absorptionless) coloration, as compared to absorptive, in creatures is characterized by the following advantages. First, photochemical processes are avoided since no absorption of light occurs. This makes coloration durable without photobleaching, and therefore without the necessity for renewing the source of color frequently. Second, interference offers colors without heating since, again, there is no light absorption. This is again helpful for durable coloration as well as (possibly) for optimal heat balance in a hot climate. There is also a hypothesis that the sometime periodic structure of a light-emitting butterfly provides more efficient control of the emitted light cone of the fluorescent species [42].

Three-dimensionally periodic structures are present in nature as colloidal crystals. Biological colloidal crystals were identified for the first time as viruses [43]. When speaking about the optical wavelength scale, gem opals should be emphasized. These are silica colloidal crystals with close-packed spherical silica globules and voids filled by another inorganic compound [44]. The iridescent color of gem opals is caused by light-wave interference in a three-dimensional lattice.

7.9 Experimental methods of fabrication

One-dimensional periodic structures

Multilayer films featuring one-dimensional periodicity are routinely produced for commercial applications in optical devices such as mirrors and band-pass and narrow-band filters. The most typical and cheap way is vacuum deposition of polycrystalline materials (Fig. 7.16a). Monocrystalline periodic structures are classical semiconductor heterostructures developed by means of epitaxial growth of semiconductor single crystals of thickness from a few nanometers to microns, using lattice-matched pairs of materials. Epitaxial growth can be performed by means of molecular beam epitaxy (MBE) or metal–organic vapor phase epitaxy (MOVPE), sometimes also referred to as metal–organic chemical vapor deposition (MOCVD). Multilayer periodic mirrors (often referred to as *Bragg mirrors*) were mentioned in Chapter 5 (Fig. 5.37) in the context of quantum dot laser design. Another approach is based on preliminary templating of a semiconductor surface with subsequent electrochemical or chemical etching through the template. A template is usually produced by means of lithography (Fig. 7.16(b)).

Synthesis of mesoporous materials with 2-dimensional periodicity

The *templated approach* offers the possibility of writing on a solid surface a regular or any pre-defined irregular pattern by means of electron-beam lithography, with subsequent

(a)

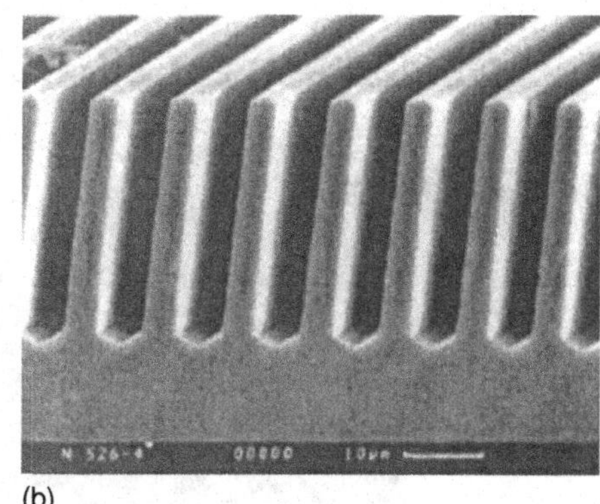
(b)

Fig. 7.16 Periodic structures with one-dimensional periodicity. (a) A multilayer dielectric mirror, period is about 500 nm [45]. (b) A silicon comb developed by means of templated etching. The period is about 10 μm [46].

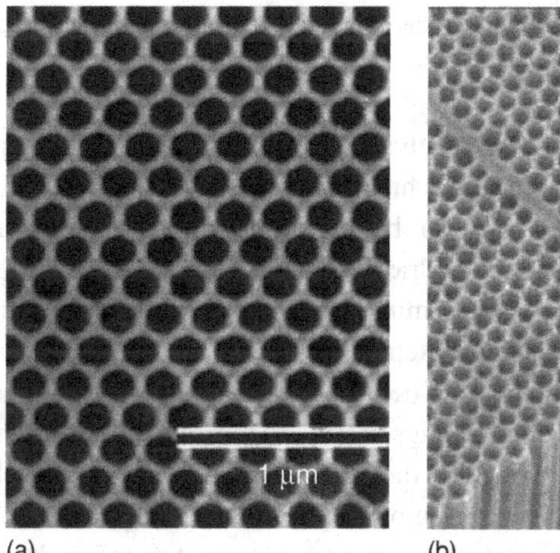

(a)

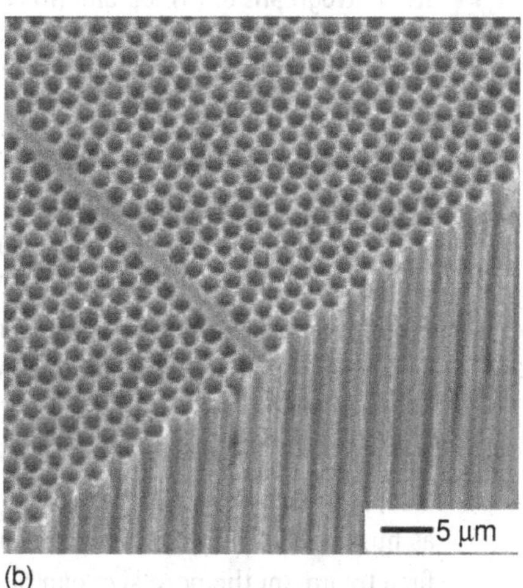

(b)

Fig. 7.17 Two-dimensional periodic structures developed by means of lithographical prepatterning and anodization. (a) SEM micrograph of regular pore configuration in an alumina layer with intervals of 200 nm, the film thickness is approximately 3 μm [47]. (b) Electron micrograph of a two-dimensional hexagonal porous silicon structure with a lattice constant $a = 1.5$ μm and a pore radius $r = 0.68$ μm, fabricated by anodization of p-type silicon [48]. Copyright 1997 AIP [47]. Copyright 2007 Elsevier B.V. [48].

electrochemical processing of the surface to get nanopore displacement defined by the original pattern. In spite of the exciting patterning opportunities offered by e-beam lithographic drawing there are at least two drawbacks in this approach which should be highlighted. First, it is expensive and multistage processing is required because of e-beam lithography patterning. Second, the thickness of a porous film is severely limited (typically to a few micrometers only) because the deeper part of the film under electrochemical reaction is less sensitive to the original pattern written on the surface. A number of representative examples of two-dimensional periodic structures developed by means of e-beam lithography templating are presented for nanoporous Al_2O_3 (alumina, Fig. 7.17(a)), and for

Fig. 7.18 **SEM micrographs of (a) top and (b) cross-sectional views for the representative sample of nanoporous alumina developed by means of self-organisation without prepatterning [51]. The pore diameter is approximately** 30 nm.

silicon (Fig. 7.17(b)). An evident advantage of the template-based technique is its applicability to any type of materials for which chemical processing results in pore development.

A *template-free technique* can be performed for the fabrication of regular two-dimensional arrays of deep cylindrical pores in alumina slabs. Anodizing of aluminum to get its porous oxide Al_2O_3 (alumina) has been shown to demonstrate self-organization in the course of electrochemical etching [49–51]. The regular structure with pronounced near-order nanopore arrangement develops for certain combinations of electric current, etching agent and temperature (Figs. 7.18). When self-organization is used, the porous sample thickness is determined by the processing time only. The thickness can be as large as hundreds of microns. Depending on the electrochemical treatment and preliminary surface treatment the pore size can be from a few nanometers to hundreds of nanometers. The surface domains visible in the top view in Fig. 7.18(a) are meant to reproduce the polycrystalline block structure of the original aluminum foil used in the fabrication process. Notably, nanoporous alumina is a mesoscopic material which was originally considered as the insulating component of semiconductor silicon microchips with metal aluminum conductors. No optical application was originally implied.

The fabrication process is performed using several stages. Stage 1 forms a regular preliminary pattern on the aluminum surface. The reverse side of the aluminum foil is protected against anodizing with a chemical-resistant varnish. The etching pits developed in stage 1 serve as the pore origination sites for further anodization stages. The duration of stage 1 is typically one hour. Stage 2 provides continuous development of an alumina film. The total etching process takes 10–20 hours to obtain pores of approximately 100 micron length. Stage 3 then removes the rest of the aluminum foil using an etchant to get a free-standing alumina film. Additionally, stage 4 can remove the alumina bottoms in the pores to produce hollow cylindrical pores throughout the sample. The adjustable anodisation parameters are: (i) electrolyte composition, (ii) voltage and (iii) temperature. An additional stage can also be used to widen the pores.

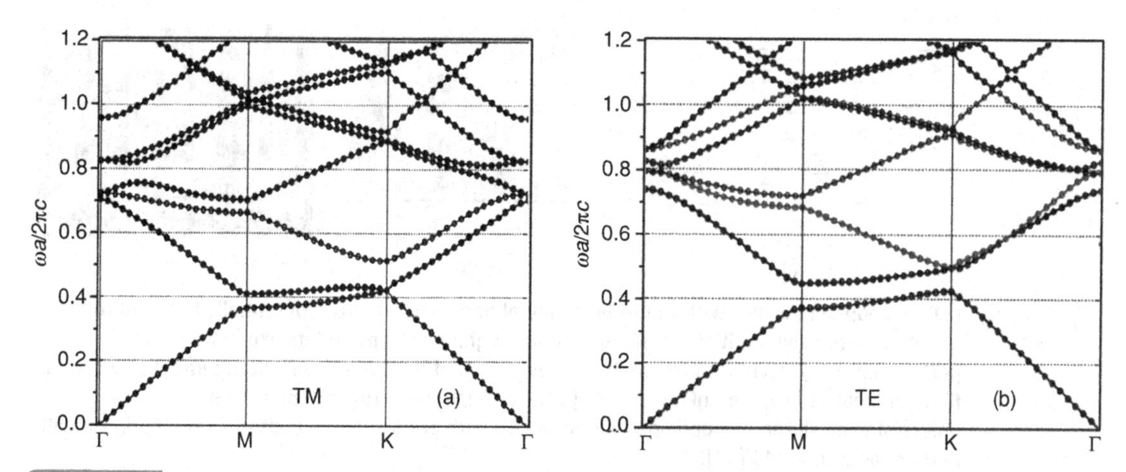

Fig. 7.19 **The photonic band structure of a two-dimensional triangular lattice of air cylinders with radius $R(\varepsilon = 1)$ in alumina ($\varepsilon = 2.72$) for $R/a = 0.3$ [52]. (a) TM-modes; (b) TE-modes.**

The formation mechanism for ordered arrays is supposed to be related to the mechanical stress model, i.e., the repulsive forces between neighboring pores at the metal/oxide interface promote the formation of hexagonally ordered pores during the oxidation process [49].

Calculations show that the refractive index contrast is not large enough for the alumina/air structure to give band gaps simultaneously for TE- and TM-modes [52]. An omnidirectional band gap develops for the TE-mode only (Fig. 7.19).

Direct laser writing by means of *multi-photon polymerization* can also give a two-dimensional periodic dielectric on a sub-micrometer scale. In this technique, a photoresist exhibiting an intensity threshold for exposure is illuminated by laser light whose photon energy is insufficient to expose the photoresist by a one-photon absorption process. If this laser light is tightly focused onto the resist, however, the light intensity inside a small volume element ("voxel") enclosing the focus may become sufficiently high to exceed the exposure threshold, by multi-photon processes. By scanning the focus onto and into the photoresist, in principle, any two- and three-dimensional connected structure consisting of these voxels may be written directly into the photoresist. In analogy to holographic lithography, only doubly connected structures can actually be fabricated. The size and shape of the exposed voxels depends on the isophotes of the microscope lens and the multiphoton exposure threshold of the photosensitive medium. The isophotes in the vicinity of the geometrical focus typically exhibit an ellipsoidal shape and voxels with a lateral diameter as small as 100 nm and an aspect ratio of about 2.7 can be realized [53].

This technique needs complex sequential micromechanical manipulation to get a three-dimensional structure by means of point-to-point photopolymerization. Another drawback is the limited number of materials suitable for photopolymerization.

Synthesis of materials with 3-dimensional periodicity

Although modern solid-state technologies presently existing in micro- and nanoelectronics offer several approaches resulting in high-quality one- and two-dimensional periodic

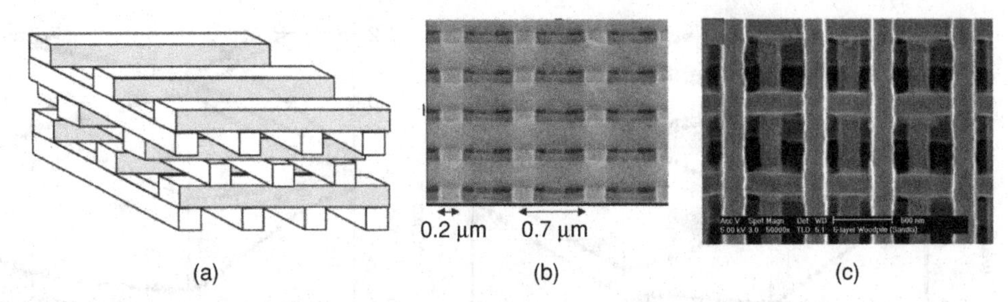

Fig. 7.20 **(a) A woodpile structure with three-dimensional periodicity that can potentially be performed by means of successive multiple masking–etching–alignment steps and its experimental performance: (b) Electron micrograph of a few layers of the "woodpile" arrangement feasible for III–V compounds like, e.g. InP, and GaAs [54]; (c) Electron micrograph of a five-layer polycrystalline silicon woodpile periodic structure with period** 450 nm **approx. [55]. Reprinted with permission from AAAS [54].**

dielectric lattices, fabrication of a three-dimensional periodic superstructure by means of these techniques still remains a challenging problem. A rather cumbersome multistep approach has been proposed to fabricate so-called "woodpile" structures (Fig. 7.20) of high-refractive semiconductor materials, gallium arsenide [54] and silicon [55] for the near infrared. Calculations have shown that woodpile structures of these semiconductors do exhibit a photonic band gap in all directions. For GaAs structures, it develops at $a/\lambda \approx 0.6$ and has a relative width $\Delta\omega/\omega \approx 0.1$.

In this approach, submicron lithography and etching are used to develop a single layer consisting of parallel grooves on a semiconductor substrate. A pair of such samples are then aligned to make a rectangular grid arrangement. Then the upper substrate is removed by etching. Afterwards the process is repeated until the desirable number of periods is fabricated. In this manner, only a few layers can be developed (typically no more than eight have been reported). The above layer-by-layer technique is rather complex and tricky because extremely careful and accurate alignment is crucial. Notably, in spite of the small number of periods, the structures shown in Figure 7.20 do exhibit pronounced angular-dependent and spectrally selective optical transmission/reflection bands owing to the extremely large refractive index ratio of the semiconductor–air interface exceeding 3. For comparison, in commercial multilayer mirrors used in laser optics this ratio is typically about 1.5.

One more approach is based on two-dimensional patterning on a crystal surface and subsequent electrochemical treatment with periodically alternating current [56, 57]. The pore diameter is typically dependent on current value. Therefore, periodic alteration of current gives rise to depth oscillation of the pore diameter. Thus three-dimensional periodicity can be achieved by means of two-dimensional periodicity coming from surface patterning and one-dimensional periodicity coming from periodic alteration of pore diameter. This approach has been demonstrated for silicon [58, 59] based on well-developed porous silicon electrochemistry (Fig. 7.21). However, such a three-dimensional periodic structure does not bear witness to a fully omnidirectional band gap even in spite of the high refractive index contrast inherent in the Si/air combination.

Non-template colloidal techniques are essentially based on synthesis of opal-like structures consisting of close-packed silica globules (colloidal crystals) [30, 60, 61], as well as

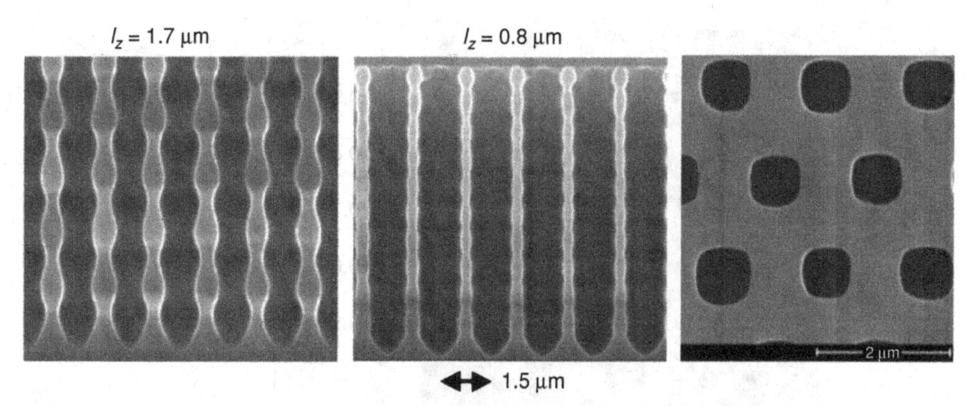

Fig. 7.21 Three-dimensional periodicity in electrochemically derived silicon structures using surface prepatterning and etching current alteration [59].

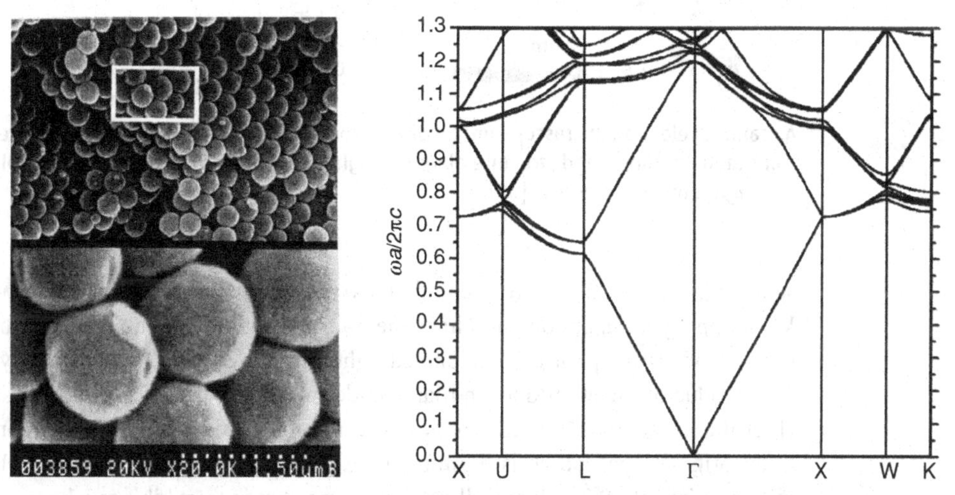

Fig. 7.22 Opal-based periodic structures. Left: Scanning electron microscopy image of a synthesized silica colloidal crystal. Lower panel shows the magnified image of a portion of the crystal inside a white frame [61]. Globule diameter is 250 nm. Right: Photonic band structure diagram according to calculations by Reynolds *et al.* [63].

using these artificial opals as templates for fabricating their three-dimensional replicas [62]. The latter features more optimal topology with respect to band-gap formation and offers flexibility of material choice in order to get higher refractive index contrast.

Natural opals consist of close-packed silica globules with interglobule space well filled by inorganic compounds. Unlike the natural counterpart, artificial opals (Fig. 7.22) possess air voids. Original close-packed spherical globules with refractive index about 1.4 do not exhibit an omnidirectional gap, although conditions for light propagation modify noticeably [63]. As voids form a continuous network, they can readily be impregnated with any liquid, as well as with polymers obtained after polymerization of liquid precursors and sol–gel oxide films developed from a liquid solution. Other options include development of a

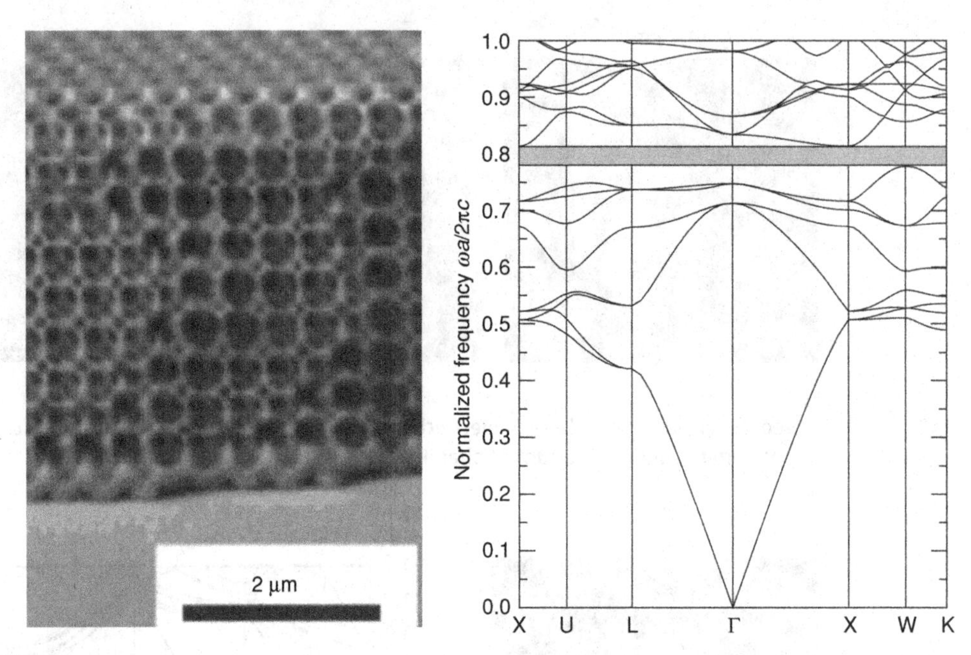

Fig. 7.23 A scanning electron microscopy image of a silicon replica of an artificial opal [62] (left) and the computed photonic band structure diagram (right) [64]. Copyright 2001 Nature Publ. Group [62]. Copyright 2007 Elsevier B.V. [64].

semiconductor (Si, CdSe, III–V compounds) polycrystalline phase inside the pores and also VO_2. Impregnation not only modifies the transmission/reflection features of original opals but also offers an opportunity to embed light-emitting species like organic dye molecules, semiconductor quantum dots and lanthanide ions using organic ligands or sol–gel films. Hydrofluoric acid can be used to remove the silica host if the photonic lattice complementary to an original opal lattice is desirable (Fig. 7.23). A three-dimensional dielectric lattice of approximately 0.3 volume filling factor was shown to exhibit much stronger photonic crystal properties compared with the close packing of dielectric balls (Section 7.5). A close-packed face-centered cubic lattice consisting of balls has an approximate 0.67 volume filling factor. Its replica thus fits the optimal filling factor quite well. Opal replicas can be readily impregnated by all the above mentioned species. However, even superior Si/air refractive index contrast can give rise to an omnidirectional band gap in second-order diffraction only ($a \approx \lambda$) as seen in the right panel of Figure 7.23 [64]. In the first order ($a \approx \lambda/2$) no omnidirectional gap develops.

A self-organization-based approach is advantageous as compared with template-alignment fabrication techniques in the sense that a three-dimensional lattice readily develops on a rather large scale in a few processing steps. These are (i) synthesis of monodisperse globules in a form of sol, (ii) sedimentation by means of gravity (sometimes by means of centrifugation), (iii) thermal treatment for hardening and (iv) drying. However, the approach still does not offer the optimal solution for photonic crystal engineering. First, crystallization occurs in the polycrystalline form, the single-crystalline domain typically

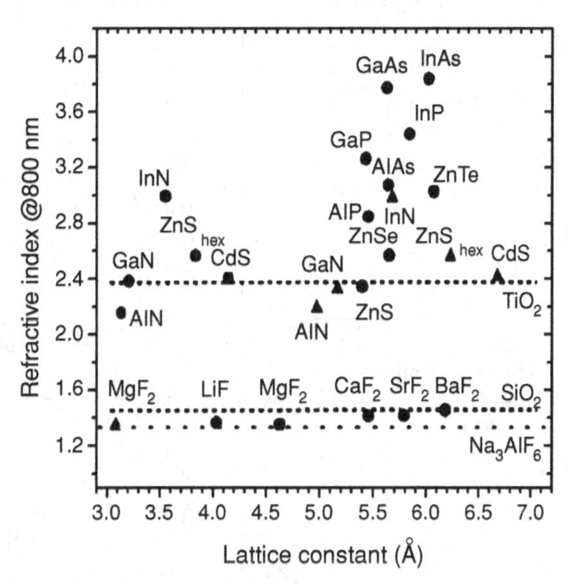

Fig. 7.24 **Refractive index versus lattice constant for a number of crystals (dots and triangles) along with refractive index for optical polycrystalline films TiO_2, SiO_2, and Na_3AlF_6 (dashed lines). All data are taken from [65] except Na_3AlF_6 [19]. Copyright 2000 AIP.**

not exceeding a few hundred micrometers. Second, face-centered cubic lattice and close-packing inherent in opals is not the optimal topology for band-gap opening. Even replicas of high-refractive compounds offer gaps only for higher diffraction orders (Fig. 7.22). However, at these high-order resonances the competitive non-resonant scattering arising from crystal imperfections are crucial and can smear coherent band-gap effects.

7.10 Properties of photonic crystal slabs

One-dimensional periodicity

Periodic multilayer dielectric mirrors have been used commercially in optical instruments and devices for many decades. Their properties with respect to normal or quasi-normal incidence are well understood and reproducible. Nevertheless, during the past decades even this well-defined field has experienced high research activity. In addition to traditional polycrystalline thin film vacuum deposition, fine epitaxial technology has come to the fore promising single crystalline periodic heterostructures for applications in compact solid-state lasers. An example of a compact semiconductor laser with single crystal periodic reflectors was given in Chapter 5 (Fig. 5.37). For generation of ultrashort laser pulses, broadband reflectors are desirable. Large reflection bandwidth, in turn, can be developed only by using a pair of materials with a large ratio of refractive indexes (see Eq. 7.16). Figure 7.24 presents available data for single crystal materials along with their refractive indexes and crystal lattice constants. Only these pairs of materials can be grown in the form

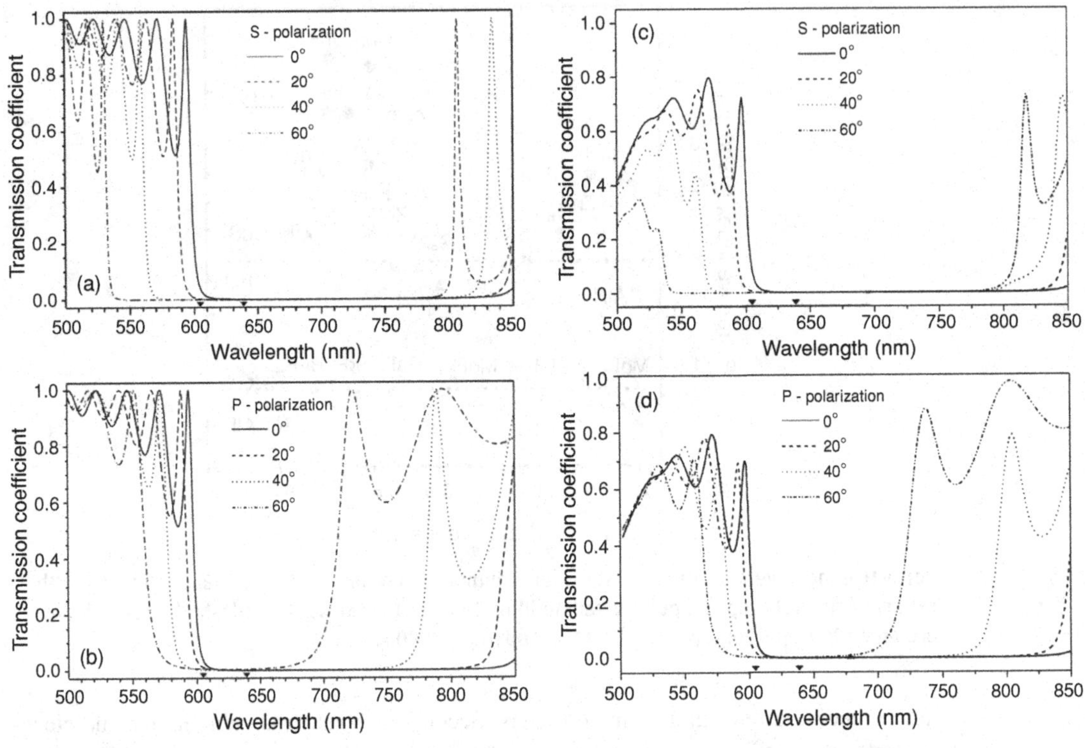

Fig. 7.25 Omnidirectional reflection in a periodic multilayer structure. (a), (b) Calculated and (c), (d) measured transmission spectra of a Na_3AlF_6/ZnSe 19-layer structure for (a), (c) *s*-polarized (TE-mode) and (b), (d) *p*-polarized (TM-mode) light at different angles of incidence (0°–solid line, 20°–dashed line, 40°–dotted line, 60°–dash-dotted line). The lower triangles indicate the edges of the calculated absolute omnidirectional band gap [19].

of a heterostructure, whose lattice constants differ negligibly to provide lattice matching and minimal mechanical strain. A greater than 400 nm wide reflection band has been reported for an AlGaAs/CaF_2 mirror with the reflection band centered at 850 nm [65].

A high refractive index ratio of a couple of optical materials is also desirable for omnidirectional reflection. Traditional vacuum deposition of dielectric multilayer structures meets the requirements for omnidirectional reflection for alternating layers of Na_3AlF_6 ($n = 1.34$) and ZnSe ($n = 2.5$) with air ($n = 1$) as the ambient medium [19]. A comparison of the calculated and measured transmission spectra (minimal transmission corresponds to maximal reflection) is presented in Figure 7.25 for various angles of incidence and polarization. Principally, the agreement of theory and experiment is satisfactory although the experimentally observed transmission values in its maxima are somewhat lower than in the theory because of certain deviations in layer thickness and structural imperfections, resulting in finite incoherent scattering that is not included in the calculation. Note, omnidirectional reflection occurs along with the rather wide reflection band for normal incidence. For the sample presented in Figure 7.25 it measures more than 250 nm in agreement with Eq. (7.16). This means that high reflection can extend over nearly the whole visible

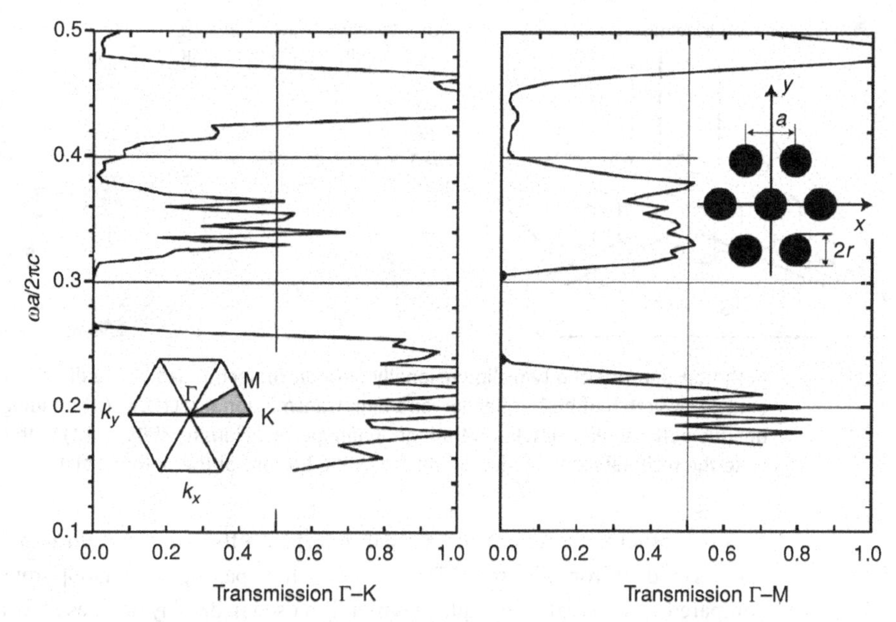

Fig. 7.26 **Transmission spectra of a 10-period slab of a two-dimensional photonic crystal with triangular lattice of silicon rods. Left-hand part corresponds to the Γ–K direction of the wave vector, right-hand part corresponds to the Γ–M direction. The inset on the left shows the first Brillouin zone in *k*-space with the irreducible zone shaded. The inset on the right shows a portion of the lattice in space. Adapted from [66].**

spectrum. To summarize, a dielectric multilayer stack can feature optical properties inherent in metal mirrors, i.e. high reflection coefficients in a wide spectrum and omnidirectional reflectance.

Two-dimensional periodicity: transmission bands

In a finite slab of a two-dimensional photonic crystal with periodicity in the *xy*-plane, transmission/reflection bands develop when light propagates within this plane. Figure 7.26 represents an example of the calculated transmission spectra for a triangle lattice of silicon rods ($n = 3.4$) for the two different directions of the wave vector, Γ–K and Γ–M calculated by Ochiai and Sakoda [66]. The number of periods in the slab equals 10 and the rod radius r measures a quarter of the period a. An omnidirectional reflection band (i.e. zero transmission) in the range $\omega a/2\pi c = a/\lambda = 0.26$–$0.30$ is seen, indicating formation of the two-dimensional electromagnetic band gap.

While in the *xy*-plane, two-dimensional photonic crystal slabs feature high reflectivity and in the *yz*- and *xz*-planes high transmission (i.e. low reflection bands) develops. This is shown schematically in Figure 7.27(a). For example, for a nanoporous alumina slab with thickness of the order of 100 μm, the transmission coefficient for the direction coinciding with a pore axis up to 98% has been reported for near field measurements [51]. This

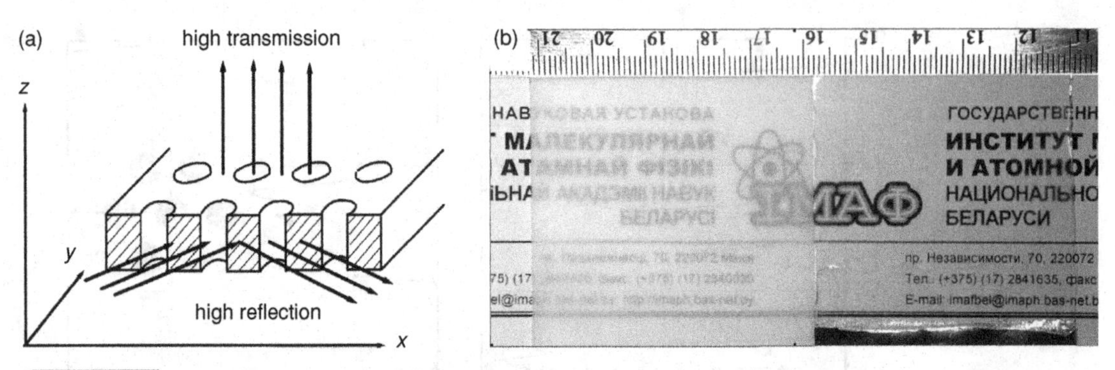

Fig. 7.27 **High transparency of a two-dimensionally periodic dielectric slab in the direction along the pore axis. (a) A sketch of high reflection and high transmission directions. (b) Imaging through an opaque polymer film (left) and through a nanoporous alumina slab (right) with size** 5 cm × 5 cm. **Note the residual piece of aluminium foil in the bottom of the alumina slab.**

means Fresnel reflection at the air/slab interface lifts on. The high transparency of such slabs is evident from Figure 7.27(b) where the transparency of a nanoporous alumina film is compared with that of an opaque polymer film used in display devices to form homogeneous backlight distribution.

Two-dimensional periodicity: antireflection effect

High transparency of 2d-periodic porous materials provides an instructive hint to the novel design of *antireflection coatings*. Traditional thin film antireflection designs fail to compensate reflection for high material/air refractive index ratio. High Fresnel reflection of more than 30% in the near infrared and in the visible for materials like silicon, germanium, gallium arsenide and gallium nitride is a serious problem for solar cells, light emitting diodes, semiconductor lasers, photodetectors, electro-optical modulators and other photonic components. In the 1990s an ingenious solution for efficient antireflection design was proposed, known as the "moth-eye design" [67]. It resembles the two-dimensional regular arrangement of pores or pillars inherent in moth eyes (Fig. 7.15). For a silicon surface patterned with 800 nm-period pores, reflectance is reduced to below 2% over the 1.5–5.5 μm band. Moth-eye surfaces are capable of maintaining low reflectance, even when the incident angle increases to 50°. For an antireflection effect in the visible, the submicrometer scale of the desirable surface lattice needs high-resolution lithography, typically based on the electron beam technique. This technique is expensive and cumbersome. Therefore alternative template-based techniques are sought. Regular nanoporous alumina films can be used as masks for development of antireflecting periodic structures on a silicon top [68]. Silica monodisperse colloids when arranged regularly on a silicon single crystal surface were also shown to be suitable as a template for an affordable and efficient antireflection solution [69]. Figure 7.28 shows the reflectivity of a silicon single crystal with and without coatings and the image of a moth-eye antireflection coating developed with a silica colloidal template.

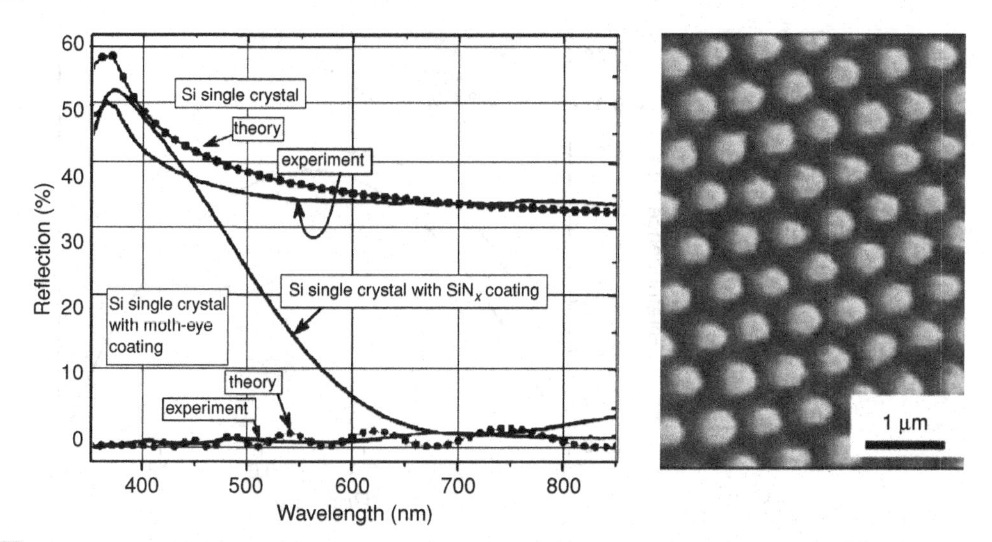

Fig. 7.28 **Effect of moth-eye surface relief on reflection at the Si/air interface. Left: reflection spectra of various structures. Right: Scanning electron microscope image of moth-eye silicon antireflection coating. Adapted with permission from [69]. Copyright 2008 AIP.**

Two-dimensional periodicity: form-anisotropy and birefringence

In Section 7.3 it was mentioned that in the long-wave regime a photonic crystal can be treated as a continuous medium with effective dielectric permittivity. In this size regime, so-called form birefringence arises. In this subsection we shall take a closer look at this phenomenon in two-dimensional periodic slabs e.g. nanoporous alumina.

For form-anisotropy birefringence to develop, not necessarily ideal periodic structures should be considered, but rather mesoporous materials with arranged elongated sub-wavelength pores with large aspect ratios are necessary. Form-anisotropy is the result of different boundary conditions on the elongated border of a pore for different polarizations in a set of aligned pores having the diameter much smaller and the length much larger than the wavelength of electromagnetic radiation. Such a behaviour of light in mesoporous structures results in different effective refractive indexes for different polarizations, which is usually described as birefringence. There are several approaches to birefringence modelling of mesoporous materials. The modified Bruggeman model [70] considers a set of rotational ellipsoids aligned in the same direction and parallel to the optical axis of an effectively anisotropic media. Notably, in this model a small refraction index contrast is implied. The Maxwell–Garnett (MG) theory [71] considers a binary material composed of two randomly distributed materials A and B with volume fractions p_a and p_b. It is assumed that the volume fraction of material A is small. The MG model may be generalized for the case of non-spherical pores by introducing the depolarization factor tensor. The boundary conditions [72] model considers a simplified structure consisting of air pores of square cross-section ordered in a square lattice. Such a structure offers the possibility of considering only two types of boundary condition: parallel or perpendicular to the walls. The value of the extraordinary refraction index n_e may be found for the

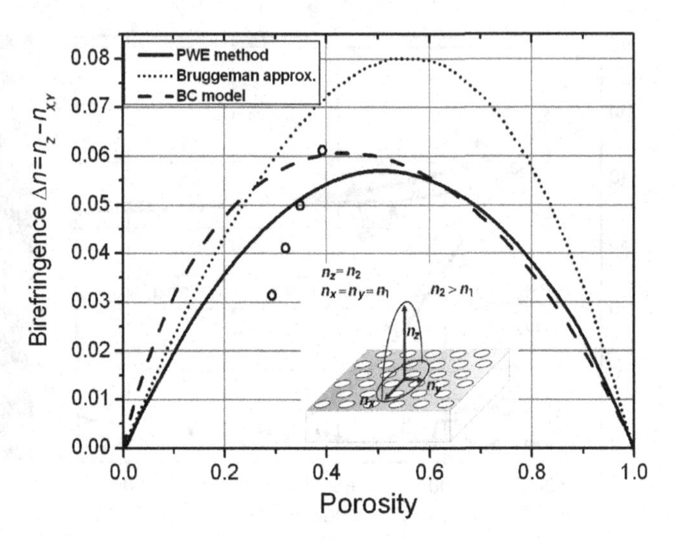

Fig. 7.29 **Modelling of birefringence in nanoporous alumina depending on the porosity (lines) and experimental values of the birefringence (circles) [52]. PWE stands for plane wave expansion, BC stands for boundary conditions. The inset shows an explanation of refractive indices used to define birefringence.**

case of a pure electric field direction parallel to the pore walls. For the ordinary refractive index n_o both parallel and perpendicular boundary conditions have to be taken into account.

The theory of photonic crystals elaborates another method based on the assumption of the ideal periodicity of the medium in two dimensions. This is the plane-wave expansion (PWE) method. This method is very efficient for simple 2D photonic crystal geometries (triangular, square, etc.) when $1/\varepsilon(\mathbf{r})$ may be found analytically or by uncomplicated Fourier transformation. In the long-wavelength limit, the PWE method may be used for the calculation of the effective refraction index of 2D porous nanostructures, taking into account the linear shape of the dispersion law at low frequencies. The slope of the dispersion curve may be used directly for the effective refractive index calculation.

Figure 7.29 shows the results of birefringence simulation in mesoporous alumina Al_2O_3 by different modelling approaches, along with the experimental data [52,73]. The form-birefringence reasonably reduces at zero and unit porosity which corresponds to a continuous medium. The maximum of birefringence appears at a value of porosity of 0.5 in the PWE approach and at slightly different porosities in other approaches. It is important to notice that only the refractive index contrast determines the maximum of the form-birefringence achievable by using a certain combination of materials. Birefringence monotonically grows with refractive index contrast, reaching a value of 0.35 when the difference in refractive indices of materials constituting the medium under consideration reaches 2.5 (GaAs–air). During the last few years birefringence caused by form-anisotropy has also been reported for porous silicon [74–76], and for porous GaP [77] in the visible and near infrared.

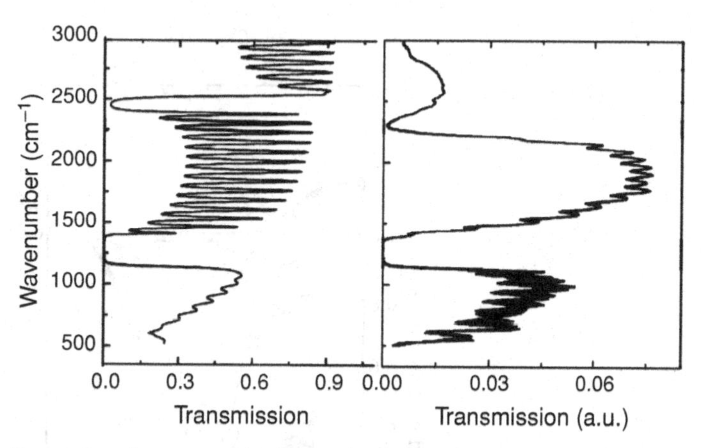

Fig. 7.30 Calculated (left panel) and measured (right panel) transmission spectra of a three-dimensional periodic silicon photonic crystal slab. Adapted from [59]. Transmission is calculated and measured along pore axis for a structure with $a = 1.5\ \mu m$, $l_z = 2\ \mu m$, whose image was presented in Figure 7.21.

Three-dimensional photonic crystal slabs

The calculated and measured transmission spectrum of a 3d-periodic structure developed by means of the modulated template-based etching of silicon is presented in Figure 7.30. The structure itself has been discussed in Section 7.9 (Fig. 7.21). Transmission along the pore axis has been examined. The two low-transmission bands are observed both in calculations and in experiments. Note, the absolute values of measured transmission for the propagating modes are still much lower than those obtained in calculations. This is probably because of crystalline imperfections in the course of fabrication. The specific geometry of the structure under consideration enables us to use a one-dimensional effective refractive index model. With the period along wave propagation equal to 2 μm, the characteristic selective transmission and reflection develops well in the infrared range for wavelengths of the order of 10 μm.

Another example of optical properties of a three-dimensional periodic structure is artificial opals, consisting of close-packed silica spherical globules forming a face-centered cubic lattice (Fig. 7.22). Opal samples feature an angular-dependent dip in the optical transmission spectrum (Fig. 7.31). This dip is accompanied by a reflection peak. The angular dependence conforms to the face-centered symmetry of the lattice. The dip in transmission spectrum gets deeper and wider when opals are inpregnated with various liquid or solid materials whose refractive index exceeds that of silica. Certain deviations from the theory have been found in these experiments, indicating that possibly the globules are not absolutely solid but feature a substructure consisting of smaller silica nanoparticles. Because of this internal substructure the globules themselves absorb impregnators and the whole opal + impregnator system features weaker modification of the transmission spectrum.

Note, every opal sample under investigation even in the best performance contains colloidal single crystal domains no bigger that a hundred micrometers. It is for this reason

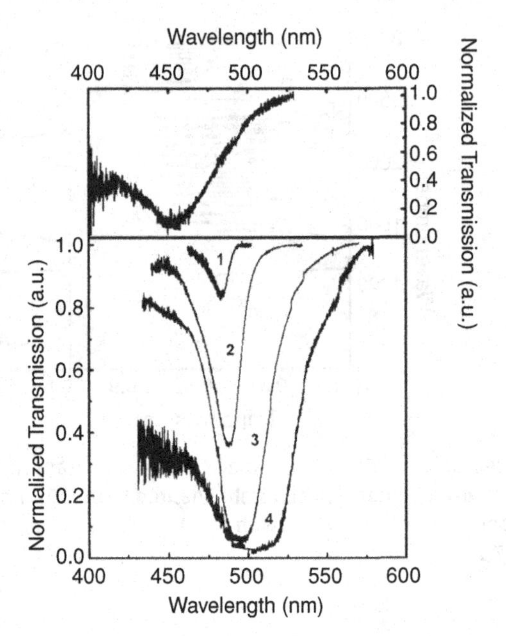

Fig. 7.31 **Optical transmission spectra of an artificial opal sample with (upper panel) empty voids and (lower panel) impregnated with various fillers: (1) methanol** ($n = 1.328$), **(2) ethanol** ($n = 1.361$), **(3) cyclohexane** ($n = 1.426$) **and (4) toluene** ($n = 1.497$) **[30].**

that pronounced sparkling iridescence occurs that is appreciated in jewelry, but not desired in optical device applications. Probably, the 80% reflection reported for SiO_2/TiO_2 in macroscopic averaged detection (about 1 mm^2) is the best value observed for surface-averaged opal properties. In microscopic detection, for a thin opal slab and smaller surface area inspected, an interference structure of transmission spectrum similar to that known for multilayer periodic stacks is observed.

7.11 The speed of light in photonic crystals

The concepts of phase velocity and group velocity as well as energy velocity in periodic layered media are subtle and require special examination.

A. Yariv and P. Yeh, 1984 [7]

What is the speed of light propagation in a photonic crystal? To consider possible answers to this question one should first define which speed we are searching for. Recalling Chapter 2 (Section 2.1) we can consider at least two types of velocities describing propagation of light. The first one is the phase velocity, which in a homogeneous medium reads,

$$\mathbf{v} = \frac{\omega}{k}\frac{\mathbf{k}}{k}, \quad k = |\mathbf{k}|. \tag{7.23}$$

It describes propagation of a wave front for a plane monochromatic wave. It has direction normal to the wave front and coinciding with the wave vector direction. The second one is the group velocity,

$$\mathbf{v}_g = \frac{d\omega}{d\mathbf{k}}. \tag{7.24}$$

This describes propagation of a finite group packet, which can be treated as a sum of plane monochromatic waves whose frequencies lie within a finite range $\Delta\omega$, and wave numbers lie within a finite range Δk.The length of such a packet ΔL is directly related to its spectrum via the relation,

$$\Delta L \Delta k \leq 1. \tag{7.25}$$

In a homogeneous isotropic medium without dispersion (dielectric function ε and refractive index $n = \sqrt{\varepsilon}$ are frequency- and angular-independent, the dispersion curve $\omega = \omega(k) = ck/n$ is straight), phase velocity and group velocity coincide. In a dispersive, anisotropic, or inhomogeneous medium these velocities diverge. In a dispersive medium with normal dispersion ($dn/d\omega > 0$) the relation $v_g < v$ holds. Accordingly, in the case of abnormal dispersion $dn/d\omega < 0$, the relation $v_g > v$ holds. In anisotropic media directions of phase velocity and group velocity do not coincide. Group and phase velocity can sometimes even have opposite directions. In homogeneous media group velocity describes energy transportation. In an inhomogeneous medium the relation between group velocity and energy velocity is to be carefully examined.

In a photonic crystal electromagnetic waves obey the form of Bloch waves rather than plane waves. For Bloch waves the equiphase surface cannot be defined rigorously, therefore the phase velocity concept is not rigorously applicable. We shall have a closer look at the group velocity and the energy velocity in a one-dimensional photonic crystal.

The energy flux density is given by the Poynting vector,

$$\mathbf{S} = \frac{c}{4\pi}[\mathbf{E} \times \mathbf{H}], \tag{7.26}$$

in the CGS-system, its direction is normal to both electric and magnetic fields vectors. Considering a one-dimensional periodic medium, the energy velocity $\mathbf{v}_E$ can be written in terms of the Poynting vector and the time-averaged electromagnetic energy density,

$$U = \tfrac{1}{4}(\mathbf{E} \cdot \varepsilon\mathbf{E}^* + \mathbf{H} \cdot \mu\mathbf{H}^*) \tag{7.27}$$

as,

$$\mathbf{v}_E = \frac{\frac{1}{a}\int_0^a \mathbf{S}dx}{\frac{1}{a}\int_0^a U dx} = \frac{\langle \mathbf{S} \rangle}{\langle U \rangle}, \tag{7.28}$$

where a is the period of the $\varepsilon(x)$ function of the medium and $\langle \ldots \rangle$ brackets denote an average over the unit cell. Yariv and Yeh [7] proved that for a one-dimensional periodic medium the equality

$$\mathbf{v}_E = \mathbf{v}_g, \tag{7.29}$$

holds, as in a homogeneous medium. Two remarks are noteworthy. First, electromagnetic Bloch waves are considered in periodic media and therefore Bloch wave vector and wave number are considered. Second, Eq. (7.29) holds rigorously in an infinite medium with ideal translational symmetry of its dielectric function.

The situation becomes rather different and much more complicated when a real *finite* periodic multilayer structure is considered, which is embedded in a continuous ambient medium. In this case the transmission spectrum contains discrete bands with finite widths, the total length of a slab is always finite and transmission is always accompanied by reflection (except at the transmission peak frequencies).

The peculiar circumstance of the problem is that a shorter (on a timescale) light pulse not only becomes wider on a frequency scale (to be compared with the transmission band width), but simultaneously becomes shorter on a length scale (to be compared with the total length of a photonic crystal slab). The light pulse length can become shorter as well as compared to the slab length. For example, a 150 fs pulse of a solid-state Ti:sapphire laser has geometrical extension about 20 μm (the full length at half maximum). It is equal to the length of a 100 period slab with a unit cell consisting, e.g., of a couple of materials with quarter-wave optical thickness corresponding to the midgap wavelength of 800 nm. This combination of photonic crystal and light pulse parameters is feasible in many experiments.

For propagation of light pulses through a finite slab of a one-dimensional periodic lossless medium D'Aguanno *et al.* [78] derived the following relation,

$$v_E(\omega) = v_g(\omega)\,|t(\omega)|^2\,, \tag{7.30}$$

where $|t(\omega)|^2 = T(\omega)$ is the intensity transmission coefficient (transmittance). This relation is valid within the so-called monochromatic regime, i.e. light pulse spectral width should not exceed the resonant transmission bandwidth. The pulse length is implied to be longer than the slab thickness. It holds both within transmission bands as well as within band gaps (minimal but finite transmittance occurs in the band gaps for a finite periodic slab).

The relation (7.30) has a very clear physical meaning. The group velocity describing pulse propagation, and the energy velocity describing energy transfer through a periodic slab are equal only at frequencies (wavelengths) strictly corresponding to the transmission peaks within pass bands. In every other case, energy velocity will always be smaller than the group velocity.

Let us have a closer look at the problem using numerical results for a specific slab. In Figure 7.32 energy velocity and group velocity are plotted along with the transmittance for a sample periodic structure on a normalized frequency axis ω/ω_0. The elementary cell is composed of a combination of half-wave–quarter-wave layers. The indexes of refraction are $n_1 = 1$ and $n_2 = 1.42857$, the respective thicknesses are $d_1 = \lambda_0/(4n_1)$ and $d_2 = \lambda_0/(2n_2)$ with $\omega_0 = 2\pi c/\lambda_0$. In the infinite structure the top of the pass band is at $\omega/\omega_0 = 0.6$. In this point the energy velocity equals zero (the dotted line in Fig. 7.32). The three other curves in this figure correspond to the finite structure with 20 periods. The transmittance spectrum consists of discrete bands with $|t(\omega)|^2 = 1$ in the peak of every band. This is similar to spectra shown in Figure 7.7. The group velocity for the finite slab is an oscillating function. Remarkably, oscillations of the group velocity anti-correlate with oscillations of

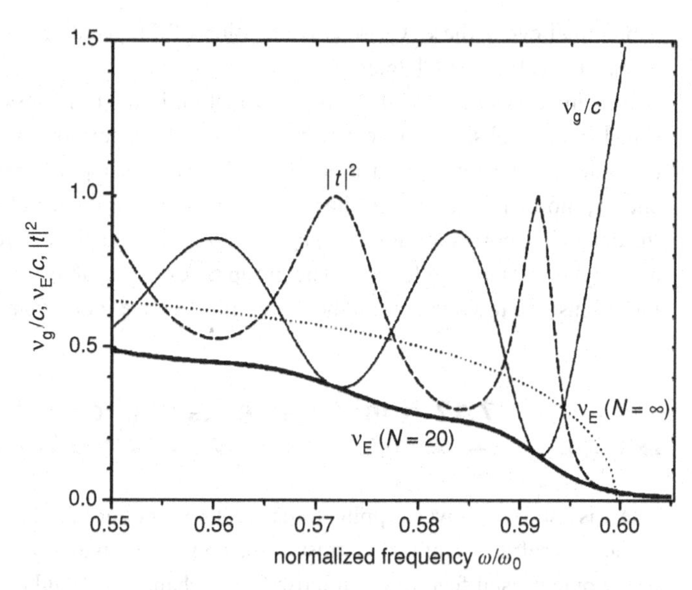

Fig. 7.32 **Normalized group velocity (v_g/c), energy velocity (v_E/c), and transmittance ($|t(\omega)|^2$) for a 20-period slab along with the energy velocity of the relevant infinite periodic medium [78]. See text for detail.**

transmittance. Maximal values of the group velocity occur at minimal transmittance and, vice versa, minimal values of the group velocity are inherent in the transmission maxima. This is the notable property of light propagation through a periodic slab. *Transmittance is equal to 1 for a light pulse propagating through a periodic slab at the expense of lower propagation velocity.* In other words, efficient transmission occurs because of the longer time the pulse spends inside the slab. It is rather reasonable and intuitive since the transmitted pulse forms as a result of multiple scattering and interference events. Constructive interference takes longer than simple propagation.

The product of the group velocity and transmittance gives the energy velocity which is shown by the thick solid line. The energy velocity equals the group velocity exactly in transmittance peaks only. Otherwise the energy velocity is lower than the group velocity everywhere. Notably, the energy velocity for a big but finite number of periods in a slab never merges with the energy velocity in the infinite structure (dotted line). This is because an infinite structure offers free propagation of electromagnetic Bloch waves, whereas every finite structure features a finite number of discrete transmission peaks corresponding to constructive interference of multiply scattered waves, including scattering events at the beginning and the end of the slab under consideration.

Close to the band edge, the group velocity exceeds the speed of light in a vacuum. This means the center of a probe light pulse passing through a periodic slab arrives at the detector earlier than its counterpart passing through a reference vacuum line of the same length. This paradox is rather peculiar but it is still in agreement with the causality principle. Superluminal propagation means pulse *reshaping* only, the energy velocity remains subluminal everywhere. Close consideration of the peculiar features of tunneling of light will be the subject of Chapter 10.

Slowing down of the light pulse propagation in the transmission maxima for a periodic slab has been observed in experiments [79]. A 30 period, GaAs/AlAs, one-dimensional periodic structure was used in studies of a 2 ps laser pulse time delay. An approximate energy, momentum and form invariance of the transmitted pulse has been observed with the delay time corresponding to $c/v_g = 13.5$. This value is referred to as the *group index* by analogy to refractive index c/v. The group index is tuneable and many orders of magnitude more sensitive to variations in the material refractive index than for bulk material.

7.12 Nonlinear optics of photonic crystals

There is a straightforward application in non-linear optics for the high sensitivity of photonic crystals to refractive index contrast in the constituent materials. If one or both materials in a photonic crystal feature a refractive index change with light intensity, then the photonic band structure becomes intensity-dependent with a pronounced modification in the transmission and reflection spectra. The concentration of light inside a photonic crystal and the corresponding slowing down of light propagation may further enhance nonlinear response. In the following pages we consider a few representative examples.

Epitaxial technology of thin film single crystal synthesis offers a possibility of developing lattice-matched periodic structures of semiconductor materials with minor misfit in the crystal lattice constants. A ZnS/ZnSe single crystal combination grown on a GaAs substrate is a representative example [80]. A periodic structure features typical reflection and transmission spectra discussed in Section 7.2 (Fig. 7.33, top panel). Zinc selenide has a narrower electronic band gap $E_g = 2.67$ eV with the interband absorption onset at room temperature at about 460 nm (arrow in the top panel). Zinc sulfide has a wider electronic band gap with no interband transitions within the visible range. Absorption by the ZnSe sublattice smears the interference pattern at the short-wave side of the reflection band peaking at approximately 500 nm. Such a structure ("Bragg reflector") offers the possibility to pump the ZnSe sublattice by laser radiation corresponding to interband optical transitions in ZnSe, whereas the ZnS sublattice remains unexcited. The smeared interference structure of the reflection spectrum favors efficient excitation because of the flat spatial distribution of excitation radiation inside the structure. Upon such excitation, an electron–hole plasma is created in every ZnSe layer resulting in modification of the absorption coefficient and refractive index. The latter modifies both because of absorption changes (via Kramers–Kronig relations) and because of the direct non-equilibrium electron–hole plasma dielectric function contribution. This contribution is in accordance with Figure 3.10 in Section 3.3, except that the plasma frequency at the relevant concentration of electron and holes (about 10^{19}–10^{20} cm^{-3}) in strongly excited semiconductors is far into the infrared range. Both contributions give rise to a negative addition to the original refractive index of ZnSe and the reflection spectrum shifts to shorter wavelengths (Fig. 7.33, bottom panel) indicating a large change in the refractive index $\Delta n = -0.05$.

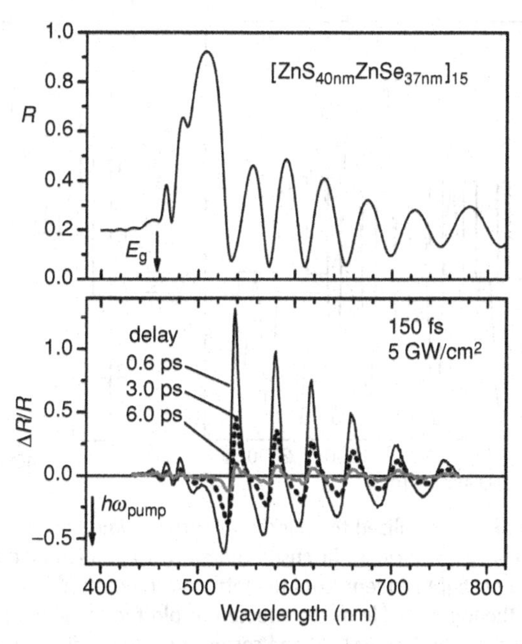

Fig. 7.33 **Optical properties of a 15-period ZnS/ZnSe multilayer stack. Top panel: linear optical reflection spectrum. Bottom panel: nonlinear differential reflection spectra recorded at different delays of probe versus pump pulse. A positive $\Delta R/R$ value corresponds to an increase in reflection [80].**

The GaAs substrate can be chemically removed, then nonlinear transmission measurements become feasible. For spectral range $\lambda > 500$ nm where absorption is negligible, the transmission spectrum modifies in accordance with the observed reflection modifications. In the spectral range $\lambda = 450-480$ nm, nonlinear transmission exhibits complex behavior related to electron–hole population dynamics and its influence on the absorption spectrum via electron and hole states filling (absorption saturation, i.e. bleaching) and band gap shrinkage (induced absorption near the band gap photon energy) [81]. The overall relaxation time of the nonlinear response is governed by the electron–hole plasma recombination rate. It was found to be approximately 3 ps and is attributed to many-body Auger recombination processes under conditions of electron and hole phase-space filling.

Tocci *et al.* proposed an asymmetric multilayer structure with ramp refractive index profile as a nonlinear optical diode whose transmission coefficient at higher intensities strongly depends on the light propagation direction (Fig. 7.34) [82]. In calculations, refractive indexes 2.7 and 1.6 at the left side of the structure monotonically grew to 2.8 and 1.7, respectively, at the right-hand end of the structure. Instead of steady growth of the refractive index along the structure, the growth of the geometrical layer thickness is possible with the same result. At least one of the two sublattices should exhibit third-order optical nonlinearity to obtain an increase in refractive index for higher intensities. More than one order of magnitude difference in transmittivity has been predicted (Fig. 7.33(b)) for the two different directions of light propagation.

Two- and three-dimensional nonlinear-optical photonic crystal structures can be developed by means of embedding polymeric, polycrystalline or other nonlinear materials in

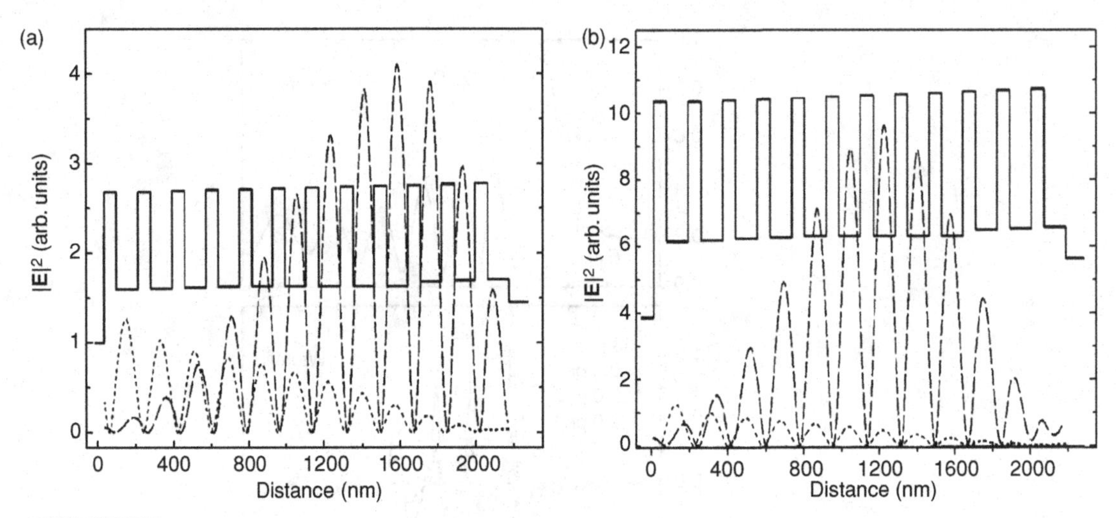

Fig. 7.34 $|\mathbf{E}|^2$ profile, normalized to an incident vacuum value of $|\mathbf{E}|^2 = 1$, inside the index ramp structure for (a) very low incident intensity (linear case) and (b) high incident intensity. The dotted line represents light incident from the left, and the dashed line represents light incident from the right. The index profile of the structure is plotted (solid line) for reference. The light wavelength is (a) 640 nm and (b) 641 nm and corresponds to the long-wave edge of the transmission band. Reprinted with permission from [82]. Copyright 1995 AIP.

porous two- and three-dimensionally periodic structures like nanoporous alumina, silicon or artificial opals. Artificial opals impregnated with Si and VO_2 were found to possess optical switching behavior in the picosecond time range [83, 84]. Switching in Si-impregnated opal results from electronic optical nonlinearity in the silicon. Switching in VO_2-impregnated opals results from structural phase transition of VO_2 occurring at temperatures around 70°C. Optical switching occurs via fast laser-induced heating of VO_2. Impregnation with liquid crystals [85, 86] offers tuneable optical properties of opals by means of an external electric field to get an electrically tuneable reflector or band-pass filter.

Second harmonic generation can be considerably enhanced in a periodic medium because of the modified phase matching condition. The standard phase matching condition that couples the wave vector of the principal harmonic $\mathbf{k}_\omega$ with that of the second harmonic radiation $\mathbf{k}_{2\omega}$ reads,

$$2\mathbf{k}_\omega = \mathbf{k}_{2\omega}. \tag{7.31}$$

In a photonic crystal this equality can be fulfilled with the accuracy of an arbitrary reciprocal lattice vector $\mathbf{G}$. This is because Bloch waves whose wave vectors differ by a reciprocal lattice vector are equivalent. Thus in a photonic crystal phase matching reads,

$$2\mathbf{k}_\omega = \mathbf{k}_{2\omega} + \mathbf{G}, \tag{7.32}$$

which offers much more flexibility in practical performance as compared to uniform media. The same relaxation of momentum conservation occurs for the sum frequency generation. For periodic structures based on nanoporous silicon with spatially modulated porosity, up to

10^2-fold enhancement of second harmonic generation has been reported [87, 88]. Though this approach has basically been known since 1970 [89] it has not achieved commercial-scale development. The total length of the structures is typically in the submillimeter range, whence the overall efficiency is not competitive with commercial frequency converters, even in the case of noticeable enhancement.

Yet another issue in nonlinear optical applications of photonic crystals has been revealed recently. It was found that in two-dimensional photonic crystals a subdiffraction propagation regime of laser pulses becomes feasible. In this regime efficient parametric amplification and second harmonic generation of narrow beams (of the width of a few wavelengths) occurs in nonlinear photonic crystals with $\chi^{(2)}$-nonlinearity [90, 91].

Problems

1. Compare the band structure diagrams for an electron (Chapter 4, Fig. 4.2) and an electromagnetic wave (Fig. 7.2) in a periodic structure and explain the difference.

2. Compare the evanescent waves in a metal (Chapter 3, Section 3.4) with those in a periodic dielectric structure. Reveal and compare the reasons responsible for development of evanescent waves.

3. Using the plots in Figure 7.10, evaluate the effective refractive index of a periodic structure under consideration in the long-wave limit.

4. Try to guess why periodic multilayer structures used as mirrors are often referred to as Bragg reflectors.

5. Recalling the traditional design of antireflection thin film coating based on a single quarter-wave layer of a material with refractive index $n = \sqrt{n_1 n_2}$, explain why it is not efficient for big n_1/n_2 values, like Si in air.

6. Apply standard mathematical tools in terms of the first and second derivatives to explore minima and maxima of the transmission spectrum represented by Eq. (7.13).

7. Compare the stop-band widths in Figures 7.7 and 7.25 with the values predicted by Eq. (7.16).

8. Compare the transmission spectra in Figure 7.25 with those in Figure 7.7. Observe and explain the difference.

9. Find out the limit to which the midgap transmission coefficient (Eq. 7.13) tends at a high number of periods in a quarter-wave stack by putting $\tilde{\omega} = 1, N \to \infty$. Prove that the relation holds [14]:

$$T(\omega)|_{\omega=\omega_0} \xrightarrow{N\to\infty} 4\left(\frac{n_i}{n_j}\right)^{2N}, \quad i, j = 1, 2, n_i < n_j.$$

10. Observe and explain why the multiple reflectance peaks in the ZnS/ZnSe periodic slab become closer for shorter wavelengths. Compare these with Figure 7.7.

References

[1] J. W. Strutt (Lord Rayleigh). On the maintenance of vibrations by forces of double frequency, and on the propagation of waves through a medium endowed with a periodic structure. *Phil. Mag., S.*, **24** (1887), 145–159.

[2] J. W. Strutt, (Lord Rayleigh). On the remarkable phenomenon of crystalline reflexion described by Professor Stokes. *Phil. Mag.*, S.5, **26** (1888), 256–265.

[3] C. G. Darwin. The theory of X-ray reflection, Part II. *Phil. Mag.*, **27** (1914), 675–690.

[4] K. Ohtaka. Energy band of photons and low-energy photon diffraction. *Phys. Rev. B*, **19** (1979), 5057–5067.

[5] E. Yablonovitch and T. J. Gmitter. Photonic band structure: The face-centered-cubic case. *Phys. Rev. Lett.*, **63** (1989), 1950–1953.

[6] V. P. Bykov. Spontaneous emission in a periodic structure. *Soviet Physics-JETP*, **35** (1972), 269–273.

[7] A. Yariv and P. Yeh. *Optical Waves in Crystals* (NewYork: Wiley, 1984).

[8] J. D. Joannopoulos, R. D. Meade and J. N. Winn. *Photonic Crystals: Molding the Flow of Light* (Princeton: Princeton University Press, 1995).

[9] K. Sakoda. *Optical Properties of Photonic Crystals* (Berlin: Springer, 2004).

[10] J.-M. Lourtioz, H. Benisty, V. Berger, J.-M. Gerard, D. Meystre and A. Tchelnokov. *Photonic Crystals* (Springer: Berlin, 2005).

[11] T. F. Krauss and R. M. De La Rue. Photonic crystals in the optical regime: past, present and future. *Progr. Quant. Electron.*, **23** (1999), 51–96.

[12] K. Busch, G. von Freymann, S. Linden, S. F. Mingaleev, L. Tkeshelashvili and M. Wegener. Periodic nanostructures for photonics. *Phys. Rep.*, **444** (2007), 101–202.

[13] S. V. Zhukovsky. *Propagation of Electromagnetic Waves in Fractal Multilayers*. Unpublished Ph. D. thesis, Institute of Molecular and Atomic Physics, Minsk (2004).

[14] J. M. Bendickson, J. P. Dowling and M. Scalora. Analytic expressions for the electromagnetic mode density in finite, one-dimensional, photonic band-gap structures. *Phys. Rev. E*, **53** (1996), 4107–4121.

[15] N. Stefanou, V. Yannopapas and A. Modinos. A new version of the program for transmission and band structure calculations of photonic crystals. *Comput. Phys. Commun.*, **132** (2000), 189–196.

[16] D. N. Chigrin and C. M. Sotomayor Torres. Periodic thin-film interference filters as one-dimensional photonic crystals. *Opt. and Spectroscopy*, **91** (2001), 484–489.

[17] P. Yeh. *Optical Waves in Layered Media* (NewYork: John Wiley and Sons, 1988).

[18] M. Born and E. Wolf. *Principles of Optics* (NewYork: Pergamon, 1980).

[19] D. N. Chigrin, A. V. Lavrinenko, D. A. Yarotsky and S. V. Gaponenko. All-dielectric one-dimensional periodic structures for total omnidirectional reflection and partial spontaneous emission control. *J. Lightwave Technol.*, **17** (1999), 2018–2024.

[20] Y. Fink, J. N. Winn, S. Fan, C. Chen, J. Michel, J. D. Joannopoulos and E. L. Thomas. A dielectric omnidirectional reflector. *Science*, **282** (1998), 1679–1684.

[21] D. N. Chigrin, A. V. Lavrinenko, D. A. Yarotsky and S. V. Gaponenko. Observation of total omnidirectional reflection from a one-dimensional dielectric lattice. *Appl. Phys. A*, **68** (1999), 25–28.

[22] P. S. J. Russell, S. Tredwell and P. J. Roberts. Full photonic bandgaps and spontaneous emission control in 1d-multilayer dielectric structures. *Opt. Commun.*, **160** (1999), 66–71.

[23] R. Wang, X.-H. Wang, B.-Y. Gu and G.-Z. Yang. Effects of shapes and orientations of scatterers and lattice symmetries on the photonic band gap in two-dimensional photonic crystals. *J. Appl. Phys.*, **90** (2001), 4307–4312.

[24] K. M. Ho, C. T. Chan and C. M. Soukoulis. Existence of a photonic gap in periodic dielectric structures. *Phys. Rev. Lett.*, **65** (1990), 3152–3155.

[25] E. N. Economou and M. M. Sigalas. Classical wave propagation in periodic structures: Cermet versus network topology. *Phys. Rev. B*, **48** (1993), 13434–13438.

[26] R. Biswas, M. M. Sigalas, G. Subramania and K. M. Ho. Photonic band gaps in colloidal systems. *Phys. Rev. B*, **57** (1998), 3701–3709.

[27] K. Busch and S. John. Photonic band gap formation in certain self-organizing systems. *Phys. Rev. E*, **58** (1998), 3896–4002.

[28] F. Garcia-Santamaria, C. Lopez, F. Meseguer, F. Lopez-Tejeira, J. Sanchez-Dehesa and H. T. Miyazaki. Opal-like photonic crystal with diamond lattice. *Appl. Phys. Lett.*, **79** (2001), 2310–2312.

[29] A. Ishimary. *Wave Propagation and Scattering in Random Media* (NewYork: Academic Press, 1978).

[30] V. N. Bogomolov, S. V. Gaponenko, I. N. Germanenko, A. M. Kapitonov, E. P. Petrov, N. V. Gaponenko, A. V. Prokofiev, A. N. Ponyavina, N. I. Silvanovich and S. M. Samoilovich. Photonic band gap phenomenon and optical properties of artificial opals. *Phys. Rev. E*, **55** (1997), 7619–7626.

[31] A. N. Ponyavina, S. M. Kachan and N. I. Silvanovich. Statistical theory of multiple scattering of waves applied to three-dimensional layered photonic crystals. *J. Opt. Soc. Amer. B*, **21** (2004), 1866–1875.

[32] E. Yablonovitch, T. J. Gmitter and K. M. Leung. Photonic band structure: The face-centered-cubic case employing non-spherical atoms. *Phys. Rev. Lett.*, **67** (1991), 2295–2298.

[33] *IEEE Transactions on Microwave Theory and Techniques*, Special Issue on: Electromagnetic Crystal Structures, Design, Synthesis, and Applications (Microwave), edited by A. Scherer, T. Doll, E. Yablonovitch, H. O. Everitt and J. A. Higgins, Vol. 47, issue 11, 1999.

[34] M. P. Kesler. Antenna design with the use of photonic band-gap materials as all-dielectric planar reflectors. *Microwave & Opt. Tech. Lett.*, **11** (1996), 169–176.

[35] I. Bulu, H. Caglayan and E. Ozbay. Designing materials with desired electromagnetic properties. *Microwave & Opt. Tech. Lett.*, **48** (2006), 2611–2615.

[36] A. R. Parker. 515 million years of structural color. *J. Opt. A*, **2** (2000), R15–R28.

[37] P. Vukusic and J. Sambles. Photonic structures in biology. *Nature*, **424** (2003), 852–855.

[38] D. B. Ameen, M. F. Bishop and T. McMullen. A lattice model for computing the transmittivity of the cornea and sclera. *Biophys. Journ.*, **75** (1998), 2520–2531.

[39] V. A. Nikolaev, D. M. Harwood and N. I. Samsonov. *Early Cretaceous Diatoms* (St-Petersburg: Nauka, 2001).

[40] T. Fuhrman, S. Lanwehr, M. El Rhabi-Kucki and M. Sumper. Diatoms as living photonic crystals. *Appl. Phys. B*, **78** (2004), 257–262.

[41] Yu. N. Kulchin, S. S. Voznesenskiy, O. A. Bukin, S. N. Bagaev and E. V. Pestriakov. Optical properties of natural biominerals-the spicules of the glass sponges. *Optical Memory and Neural Networks (Information Optics)*, **16** (2007), 189–197.

[42] P. Vukusic and I. Hooper. Directionally controlled fluorescence emission in butterflies. *Science*, **310** (2005), 1151.

[43] R. C. Williams and K. Smith. A crystallizable insect virus. *Nature*, **45** (1957), 119–121.

[44] P. Pieranski. Colloidal crystals. *Contemp. Physics*, **24** (1983), 25–73.

[45] A. P. Voitovich. Spectral properties of films. In: B. Di Bartolo and O. Forte (eds.), *Advances in Spectroscopy for Lasers and Sensing* (Dordrecht: Springer, 2006), 351–353.

[46] V. A. Tolmachev, L. S. Granitsyna, E. N. Vlasova, V. Z. Volchek, A. V. Naschekin, A. D. Remnyuk and E. V. Astrova. One-dimensional photonic crystal fabricated by means of vertical anisotropic etching of silicon. *Semiconductors*, **36** (2002), 998–1002.

[47] H. Masuda, H. Yamada, M. Satoh and H. Asoh. Highly ordered nanochannel-array architecture in anodic alumina. *Appl. Phys. Lett.*, **71** (1997), 2770–2772.

[48] A. Birner, R. B. Wehrspohn, U. M. Gösele and K. Busch. Silicon-based photonic crystals. *Advanced Materials*, **13** (2001), 377–382.

[49] A. P. Li, F. Mueller, A. Birner, K. Nielsch and U. Gösele. Hexagonal pore arrays with a 50–420 nm interpore distance formed by self-organization in anodic alumina. *J. Appl. Phys.*, **84** (1998), 6023–6028.

[50] N. V. Gaponenko. Sol-gel derived films in mesoporous matrices: porous silicon, anodic alumina and artificial opals. *Synthetic Metals*, **124** (2001), 125–130.

[51] A. A. Lutich, S. V. Gaponenko, N. V. Gaponenko, I. S. Molchan, V. A. Sokol and V. Parkhutik. Anisotropic light scattering in nanoporous materials: a photon density of states effect. *Nano Letters*, **4** (2004), 1755–1758.

[52] A. A. Lutich. *Optical Properties of Nanoporous Anodic Alumina*. Unpublished Ph.D. thesis, B. I. Stepanov Institute of Physics, Minsk (2008).

[53] M. Deubel, G. Von Freymann, M. Wegener, S. Pereira, K. Busch and C. M. Soukoulis. Direct laser writing of three-dimensional photonic crystal templates for telecommunications. *Nature Materials*, **3** (2004), 444–447.

[54] S. Noda, K. Tomoda, N. Yamamoto and A. Chutinan. Full three-dimensional photonic crystals at near-infrared wavelength. *Science*, **289** (2000), 604–606.

[55] M. J. A. de Dood, B. Gralak, A. Polman and J. G. Fleming. Superstructure and finite-size effects in a Si photonic woodpile crystal. *Phys. Rev. B*, **67** (2003), 035322.

[56] V. Parkhutik. Silicon anodic oxides grown in the oscillatory anodisation regime – kinetics of growth, composition and electrical properties. *Solid-State Electron.*, **45** (2001), 1451–1460.

[57] V. Pellegrini, A. Tredicucci, C. Mazzoleni and L. Pavesi. Enhanced optical properties in porous silicon microcavities. *Phys. Rev. B*, **52** (1995), R14328–14331.

[58] S. Matthias, F. Müller, C. Jamois, R. B. Wehrspohn and U. Gösele. Large-area three-dimensional structuring by electrochemical etching and lithography. *Adv. Mater.*, **16** (2004), 2166–2171.

[59] J. Schilling, F. Müller, R. B. Wehrspohn, U. Gösele and K. Busch. Dispersion relation of 3D photonic crystals based on macroporous silicon. *Mat. Res. Soc. Symp. Proc.*, **722** (2002), p. L.6.8.1.

[60] V. N. Bogomolov, S. V. Gaponenko, A. M. Kapitonov, A. V. Prokofiev, A. N. Ponyavina, N. I. Silvanovich and S. M. Samoilovich. Photonic band gap in the visible range in a three-dimensional solid state lattice. *Appl .Phys. A*, **63** (1996), 613–616.

[61] E. P. Petrov, V. N. Bogomolov, I. I. Kalosha and S. V. Gaponenko. Spontaneous emission of organic molecules in a photonic crystal. *Phys. Rev. Lett.*, **81** (1998), 77–80.

[62] Y. A. Vlasov, X. Z. Bo, J. C. Sturm and D. J. Norris. On-chip natural assembly of silicon photonic band gap crystals. *Nature*, **414** (2001), 289–293.

[63] A. Reynolds, F. Lopez-Tejeira, D. Cassagne, F. J. Garcia-Vidal, C. Jouanin and J. Sanchez-Dehesa. Spectral properties of opal-based photonic crystals having a SiO_2 matrix. *Phys. Rev. B*, **60** (1999), 11422–11426.

[64] O. Toader, S. John and K. Busch. Optical trapping, field enhancement and laser cooling in photonic crystals. *Opt. Expr.*, **8** (2001), 217–222.

[65] S. Schön, M. Haiml and U. Keller. Ultra-broadband AlGaAs/CaF_2 semiconductor saturable absorber mirrors. *Appl. Phys. Lett.*, **77** (2000), 782–784.

[66] T. Ochiai and K. Sakoda. Nearly free-photon approximation for two-dimensional photonic crystals slabs. *Phys. Rev. B*, **63** (2001), 125107.

[67] P. Lalanne and G. M. Morris. Antireflection behavior of silicon subwavelength periodic structures for visible light. *Nanotechnology*, **8** (1997), 53–60.

[68] Y. Kanamori, K. Hane, H. Sai and H.Yugami. 100 nm period silicon antireflection structures using a porous alumina membrane mask. *Appl. Phys. Lett.*, **78** (2001), 142–144.

[69] C. -H. Sun, P. Jiang and B. Jiang. Broadband moth-eye antireflection coating on silicon. *Appl. Phys. Lett.*, **92** (2008), 061112.

[70] D. A. G. Bruggeman. Berechnung verschiedener physikalischer Konstanten von heterogenen Substanzen. I. Dielektrizitetskonstanten und Leitfehigkeiten der Mischkorper aus isotropen Substanzen. *Annal. Physik*, **416** (1935), 636–664.

[71] J. C. M. Garnett. Colours in Metal Glasses and in Metallic Films. *Philos. Transact. Royal Soc. London*, **203** (1904), 385–420.

[72] O. Beom-hoan, C.-H. Choi, S.-B. Jo, M.-W. Lee, D.-G. Park and B.-G. Kang. Novel form birefringence modeling for an ultracompact sensor in porous silicon films using polarization interferometry. *IEEE Phot. Techn. Lett.*, **16** (2004), 1546–1548.

[73] A. A. Lutich, M. B. Danailov, S. Volchek, V. A. Yakovtseva, V. A. Sokol and S. V. Gaponenko. Birefringence of nanoporous alumina: Dependence on structure parameters. *Appl. Phys. B*, **84** (2006), 327–333.

[74] F. Genereux, S. W. Leonard, H. M. van Driel, A. Birner and U. Gösele. Large birefringence in two-dimensional silicon photonic crystals. *Phys. Rev. B*, **63** (2001), 161101.

[75] D. Kovalev, G. Polisski, J. Diener, H. Heckler, N. Kunzner, V. Y. Timoshenko and F. Koch. Strong in-plane birefringence of spatially nanostructured silicon. *Appl. Phys. Lett.*, **78** (2001), 916–918.

[76] L. A. Golovan, D. A. Ivanov, V. A. Melnikov, V. Y. Timoshenko, A. M. Zheltikov, P. K. Kashkarov, G. I. Petrov and V. V. Yakovlev. Form birefringence of oxidized porous silicon. *Appl. Phys. Lett.*, **88** (2006), 241113.

[77] V. V. Ursaki, N. N. Syrbu, S. Albu, V. V. Zalamai, I. M. Tiginyanu and W. B. Robert. Artificial birefringence introduced by porosity in GaP. *Semicond. Science Techn.*, **20** (2005), 745–748.

[78] G. D'Aguanno, M. Centini, M. Scalora, C. Sibilia, M. J. Bloemer, C. M. Bowden, J. W. Haus and M. Bertolotti. Group velocity, energy velocity, and superluminal propagation in finite photonic band-gap structures. *Phys. Rev. E*, **63** (2001), 036610.

[79] M. Scalora, R. J. Flynn, S. B. Reinhardt, R. L. Fork, M. J. Bloemer, M. D. Tocci, C. M. Bowden, H. S. Ledbetter, J. M. Bendickson, J. P. Dowling and R. P. Leavitt. Ultrashort pulse propagation at the photonic band edge: Large tunable group delay with minimal distortion and loss. *Phys. Rev. E*, **54** (2001), R1078–R1081.

[80] V. V. Stankevich, M. V. Ermolenko, O. V. Buganov, S. A. Tikhomirov, S. V. Gaponenko, P. I. Kuznetsov and G. G. Yakuscheva. Nonlinear Bragg structures based on ZnS/ZnSe superlattices. *Appl. Phys. B*, **81** (2005), 257–263.

[81] M. V. Ermolenko, S. A. Tikhomirov, V. V. Stankevich, O. V. Buganov, S. V. Gaponenko, P. I. Kuznetsov and G. G. Yakushcheva. Ultrafast nonlinear absorption and reflection of ZnS/ZnSe periodic nanostructures. *Photonics and Nanostructures*, **5** (2007), 101–105.

[82] M. D. Tocci, M. J. Bloemer, M. Scalora, J. P. Dowling and C. M. Bowden. Thin-film nonlinear optical diode. *Appl. Phys. Lett.*, **66** (1995), 2324–2326.

[83] V. G. Golubev, V. Yu. Davydov, N. F. Kartenko, D. A. Kurdyukov, A. V. Medvedev, A. B. Pevtsov, A.V. Scherbakov and E. B. Shadrin. Phase transition-governed opal–VO_2 photonic crystal. *Appl. Phys. Lett.*, **79** (2001), 2127–2129.

[84] D. A. Mazurenko, R. Kerst, J. I. Dijkhuis, A. V. Akimov, V. G. Golubev, D. A. Kurdyukov, A. B. Pevtsov and A. V. Sel'kin. Ultrafast optical switching in three-dimensional photonic crystals. *Phys. Rev. Lett.*, **91** (2003), 213903.

[85] S. Gottarda, D. S. Wiersma and W. L.Vos. Liquid crystal infiltration of complex dielectrics. *Physica B*, **338** (2003), 143–148.

[86] H.-S. Kitzerow, H. Matthias, S. L. Schweizer, H. M. van Driel and R. B. Wehrspohn. Tuning of the optical properties in photonic crystals made of macroporous silicon. *Adv. in Opt. Technol.*, **2008** (2008), Article ID 780784 (12 pages).

[87] L. A. Golovan', A. M. Zheltikov, P. K. Kashkarov, N. I. Koroteev, M. G. Lisachenko, A. N. Naumov, D. A. Sidorov-Biryukov, V. Yu. Timoshenko and A.B. Fedotov. Generation of the second optical harmonic in porous-silicon-based structures with a photonic band gap. *JETP Lett.*, **69** (1999), 300–305.

[88] T. V. Murzina, F. Yu. Sychev, E. M. Kim, E. I. Rau, S. S. Obydena, O. A. Aktsipetrov, M. A. Bader and G. Marowsky. One-dimensional photonic crystals based on porous n-type silicon. *J. Appl. Phys.*, **98** (2005), 123702.

[89] N. Bloembergen and A. J. Sievers. Nonlinear optical properties of periodic laminar structures. *Appl. Phys. Lett.*, **17** (1970), 483–486.

[90] K. Staliunas, Yu. Loiko, R. Herrero, C. Cojocaru and J. Trull. Efficient parametric amplification of narrow beams in photonic crystals. *Optics Letters*, **32** (2007), 1992–1994.

[91] C. Nistor, C. Cojocaru, Yu. Loiko, J. Trull, R. Herrero and K. Staliunas. Second-harmonic generation of narrow beams in subdiffractive photonic crystals. *Physical Review A*, **78** (2008), 053818.

8 Light in non-periodic structures

All periodic structures look alike. Every non-periodic object is essentially unique.
Unknown author.

Since every non-periodic medium is in a sense a unique structure, it seems at first glance that there is no regularity to be traced with respect to light propagation in such a medium. However this statement is not correct. *First*, in non-periodic media with absence of any regularity, i.e. in *random media*, the definite laws of light propagation resulting from random scattering can still be evaluated. There are analogies with the length-dependent resistance (and conductivity) of a conductor, coherent backscattering and Anderson localization of light. *Second*, there are well-identified classes of *aperiodic media* featuring definite geometrical regularities. These are *fractal media* with self-similar geometry and *quasi-periodic media*. Certain regularities of light propagation through aperiodic media with well-defined geometrical algorithms have been discovered to date and are still an issue of current research. *Finally*, it will be shown there are certain conservation relations that are valid for all structures independent of their spatial organization. These issues will all be the subject of consideration in this chapter.

8.1 The 1/L transmission law: an optical analog to Ohm's law

Consider the propagation of light waves in a medium which contains randomly distributed scatterers. It is refractive index inhomogeneity that makes the light wave scatter. Scatterers may differ in shape, size and refractive index of material. There is an important parameter of the scattering process, the *mean free path* ℓ, which denotes the average distance between two scattering events. If all scatterers are identical and can be characterized by the *scattering cross-section* σ, then the mean free path reads,

$$\ell = \frac{1}{N\sigma}, \tag{8.1}$$

where N is the scatterer's concentration. If the mean free path is large as compared to the light wavelength, one can imagine ray trajectories changing their direction after every scattering event. In this case, light propagation resembles to a large extent *diffusive transport* of mass-point particles. There are two different regimes of propagation depending on the mean free path versus sample thickness.

In the limit of low concentration of scatterers, when the mean free path is considerably larger than the thickness L of a sample under consideration and definitely much larger than the light wavelength,

$$\ell \gg \lambda, \quad \ell \gg L, \tag{8.2}$$

the single scattering regime of light propagation occurs. Every scattering event removes a portion of light from the flux and this portion will never reach a detector in the far field. The situation in general is similar to a beam of particles experiencing scattering. The single scattering notion means that a portion of particles is scattered once on average and this portion disappears from an observer. One can write,

$$\mathrm{d}I(x) = -\alpha I(x)\mathrm{d}x, \tag{8.3}$$

whence the transmission T reads,

$$T(L) = \frac{I(L)}{I(0)} = \exp(-\alpha L). \tag{8.4}$$

Here α is the constant coefficient with dimensions [cm^{-1}] characterizing the scattering rate. Equation (8.3) implies that scattering strictly along the original propagation direction has negligible probability. One can see Eq. (8.3) resembles the expression for light propagating in an absorbing medium and it is not surprising then that Eq. (8.4) coincides with the Bouger law. From the point of view of a distant observer, there is no difference between whether lightwaves partially change direction or a portion of light energy dissipates inside the medium. When speaking about particle scattering/absorption, then again, there is no difference for an observer whether particles experience elastic (changing the direction) or inelastic (absorbed) scattering.

The situation changes essentially when the *multiple scattering* regime occurs. That is the case of,

$$\lambda < \ell \ll L. \tag{8.5}$$

At first glance, multiple scattering should further decrease output light intensity as compared to the single scattering case. This would be correct if the analogy with absorption still worked. However it does not. If, in every scattering event, the probability of scattering is equal for all angles, then in the course of multiple scattering a portion of light originally scattered outside the propagation direction gains the finite probability of coming back into the originally propagating flux. When speaking in terms of particle scattering, multiple elastic scattering differs considerably from inelastic scattering, whereas in the single scattering mode elastic and inelastic scattering result in the same laws Eq. (8.3) and Eq. (8.4). Now, analysis of the problem leads to the universal relation,

$$T = \frac{I(L)}{I(0)} \propto \frac{\ell}{L}, \tag{8.6}$$

which means the light intensity is inversely proportional to the sample length L and directly proportional to the mean free path value ℓ [1, 2]. This relation is an optical analog to Ohm's law that can be formulated in terms of electrical conductivity per unit area g as,

$$g = \frac{1}{\rho L}, \tag{8.7}$$

where ρ is material resistivity per unit area and unit length. The similarity of expressions for light transmission and electrical current comes from the diffusive character of both phenomena. Electron and light transport in a medium with random scatterers can be described by the same equations developed in the diffusion theory. With respect to the light energy $W(\mathbf{r}, t)$, the diffusion equation reads [3],

$$\frac{\partial W(\mathbf{r}, t)}{\partial t} = D\nabla^2 W(\mathbf{r}, t) + S(\mathbf{r}, t), \tag{8.8}$$

where D is the diffusion coefficient and $S(\mathbf{r}, t)$ is the source function. The diffusion coefficient and the mean free path are related via transport velocity v_{tr} as,

$$D = \frac{1}{3} v_{\text{tr}} \ell. \tag{8.9}$$

When a finite slab of a turbid medium is considered in a homogeneous ambient environment, then at the medium/slab interfaces a portion R of incident light intensity reflects. To account for these reflections, Eq. (8.6) should be modified [3, 4] as,

$$T \propto \frac{\ell + z_{\text{e}}}{L + 2z_{\text{e}}} \tag{8.10}$$

where the parameter z_{e} is referred to as the *extrapolation length* and reads,

$$z_{\text{e}} = \frac{2\,(1 + R)}{3\,(1 - R)} \ell. \tag{8.11}$$

One can see reflectance modifies transmission for smaller L (but still $L > \ell$), whereas in the limit of $L \gg \ell$ Eq. (8.10) tends to Eq. (8.6).

Visual transmittivity of turbid media for various L/ℓ values is shown in Figure 8.1. Figure 8.2 shows a thorough examination of the diffusive light propagation theory experimentally performed for porous GaP samples of arbitrary length. GaP was chosen because of its relatively large refractive index in the visible, $n = 3.31$ (see Table 3.1 in Chapter 3), to enhance scattering cross-section. Perfect agreement of experimental data with the theory is evident. The transport mean free path value depends on the sample microstructure and was found to be smaller than the light wavelength (685 nm) in the experiment under consideration. This means that the diffusion theory describes perfectly the light transport in inhomogeneous media even when scattering is so strong that it is not possible to imagine mass-point-like trajectories of light propagation between scatterers. These trajectories are definitely helpful in the context of our daily experience since these are the light propagation regularities that give rise to Euclidian geometry based on our regularly used ray optic laws.

Diffuse propagation of light results in a time delay since every scattering event changes the propagation direction. This time delay can be evaluated based on Eq. (8.8) and Fick's law known in the general theory of diffusion [3],

$$I(L) = -D\nabla W(\mathbf{r}, t)|_{z=L}, \tag{8.12}$$

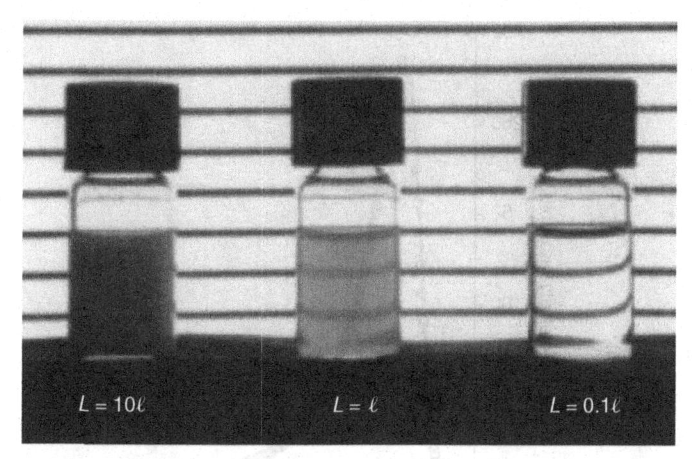

Fig. 8.1 Three vials filled with turbid liquid corresponding to different scattering regimes: left – opaqueness because of multiple scattering ($L/\ell = 10$); middle – intermediate regime ($L/\ell = 1$); right – weak scattering ($L/\ell = 0.1$). Illumination comes from the back. Adapted from [5].

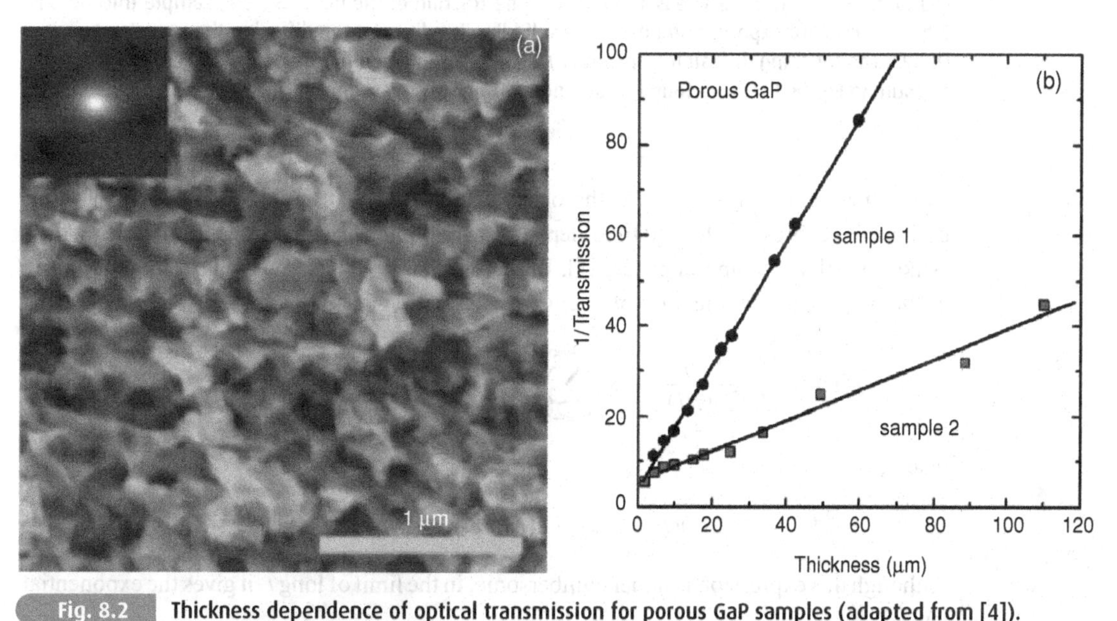

Fig. 8.2 Thickness dependence of optical transmission for porous GaP samples (adapted from [4]). (a) Scanning electron microscope image of a porous GaP (sample 1) and (inset) its autocorrelate. The lighter areas correspond to GaP and the dark parts to the pores. The typical size of the GaP entities is about 150 nm, as concluded from the autocorrelate. (b) The inverse total transmission as a function of the porous slab thickness, measured at 685 nm for the two samples with different porosities. The solid lines are calculated using diffusion theory with negligible absorption using Eq. (8.10). The transport mean free paths are found to be 0.17 ± 0.02 μm (sample 1) and 0.47 ± 0.05 μm (sample 2), respectively. Reprinted with permission from AAAS.

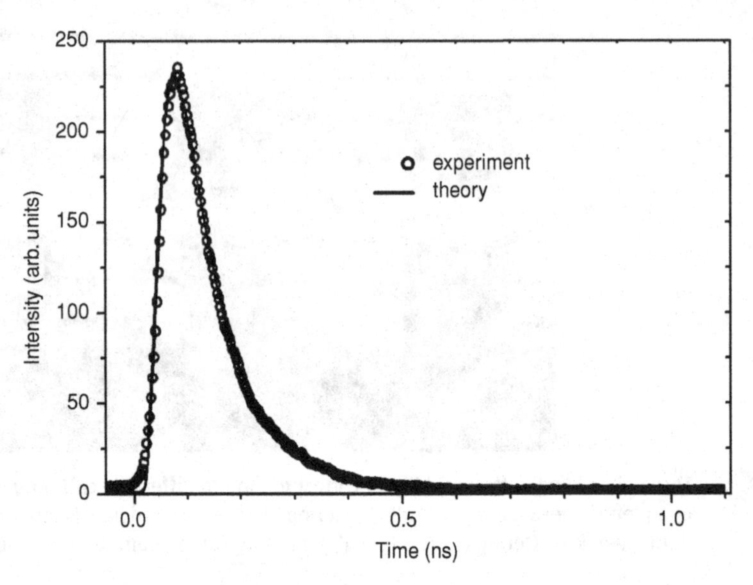

Fig. 8.3 **Time-resolved transmission through a slab of porous silica, infiltrated with the liquid crystal 8CB (adapted from [3]). Pore size is 100 nm, volume fraction of the pores is 38%, sample thickness is 2 mm. Circles are experimental data, the solid line is a fit from the diffusion theory using Eq. (8.12). The resulting diffusion constant is $D = 4.2 \cdot 10^3 \mathrm{m}^2/\mathrm{s}$ with 10% accuracy. Ten first terms in the sum in Eq. (8.13) were found to be enough to get convergence in the series.**

where the z-axis corresponds to the original light propagation direction. Wiersma *et al.* derived the expression for the time-dependent light intensity of a model light pulse with a δ-like spatial and temporal profile [3]. Based on Eqs. (8.8) and (8.11), the temporal profile of the output pulse in the limit of negligible dissipation losses reads,

$$I(L) = \frac{I(0)}{(4t)^{5/2}(\pi D)^{3/2}} \times \sum_{N=-\infty}^{+\infty} A\exp(-A^2/4Dt) - B\exp(-B^2/4Dt), \tag{8.13}$$

where,

$$A = (1-2N)(L+2z_e) - 2(\ell + z_e), \quad B = (2N+1)(L+2z_e). \tag{8.14}$$

Although this expression is rather cumbersome, in the limit of long t, it gives the exponential law,

$$I(L) \propto \exp\left(-\frac{\pi^2 D}{(L+2z_e)^2}t\right). \tag{8.15}$$

Experiments with 10 ps laser pulses passing through a porous dielectric slab have shown perfect agreement with the diffusion theory (Fig. 8.3). A 2-mm long porous slab exhibits a characteristic decay time of the order of 0.1 ns, whereas the same distance in a vacuum, light traverses only in 6.6 ps.

To account for possible absorptive losses ("inelastic scattering") described by the absorption length l_{abs}, the right-hand part of Eq. (8.12) is to be multiplied by the factor,

$$f_{abs} = \exp\left(-\frac{t}{\tau_{abs}}\right), \tag{8.16}$$

where the characteristic absorption time is $\tau_{abs} = l_{abs}^2/D$. It can also be represented as $\tau_{abs} = \frac{c}{n_{eff}}\frac{1}{l_{abs}}$, i.e. the time light needs to travel over the distance $l_{abs} \equiv \alpha^{-1}$ in a medium with effective refractive index n_{eff}. Here α is the absorption coefficient used in Eqs. (8.3) and (8.4) and c is the speed of light in a vacuum. Since we write about absorptive corrections to the multiple scattering propagation, a condition $l_{abs} \gg L \gg \ell$ is implied. Otherwise absorption will inhibit multiple scattering and diffusive transport of light will not be the case.

When light diffusively propagates through a turbid medium, the delay time for every portion of light is on average proportional to the scattering events number. Earlier portions correspond to the minimal scattering history. Therefore, vision through a turbid or even opaque medium is possible, provided short light pulses are used for illumination and synchronized time gates are arranged at the detector. This approach can be used in commercial devices to provide fog vision based on illumination of an object by short laser pulses, and time-gated image detection.

8.2 Coherent backscattering

When light falls from a continuous medium onto a scattering medium with pronounced multiple scattering, a peculiar phenomenon occurs, referred to as *coherent backscattering*. Coherent backscattering is an interference effect that manifests itself as doubling of the scattered intensity in the exact backscattering direction compared with all other directions. This enhancement occurs for any scattering material, but is difficult to observe in daily life because it occurs strictly in the direction towards the light source, fast decreasing as the angle of observation deviates from the angle of incidence. To understand the origin of coherent backscattering, consider a light wave that propagates along a certain path from a light source, through a random medium and then back towards the source (Fig. 8.4).

The physical picture of coherent backscattering is easily described in terms of wave vectors. We consider a plane wave with the wave vector $\mathbf{k}_0$ experiencing a sequence of elastic scattering events. These scatterings result in wave vector modifications $\mathbf{k}_0 \to \mathbf{k}_1 \to \mathbf{k}_2 \to \cdots \to \mathbf{k}_n$ where $\mathbf{k}_j$ is the wave vector emerging after the j-th scattering event. For every such scattering sequence, there is a reverse one determined by the set of wave vectors $\mathbf{k}_0 \to -\mathbf{k}_n \to -\mathbf{k}_{n-1} \to \cdots \to -\mathbf{k}_1 \to -\mathbf{k}_0$. In the case of backward scattering, the two waves arising from these two scattering sequences appear to be in phase at the input point A and can constructively interfere. This constructive interference occurs in a direction back to the light source, despite the fact that light has been randomly and multiply scattered, maybe thousands of times. The net effect of this constructive interference is an enhancement of the scattered intensity within a narrow cone around the exact backscattering direction.

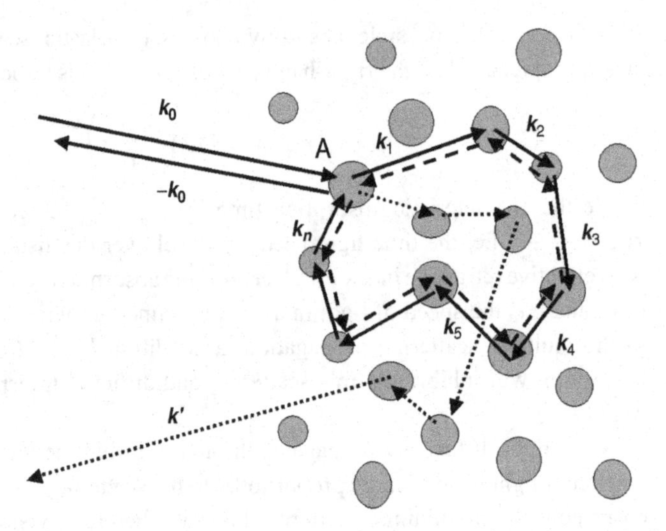

Fig. 8.4 **Multiple scattering in a turbid medium with random scatterers indicating constructive interference at point A resulting in enhanced backward scattering.**

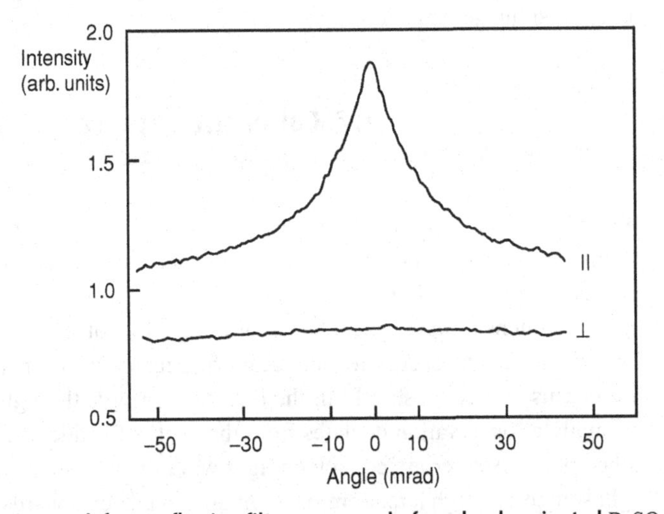

Fig. 8.5 **The backscattering peak for a reflective filter composed of randomly-oriented $BaSO_4$ microparticles typically used as a white reflectance etalon [8]. The analyzing polarizer is oriented either parallel (||) or perpendicular (⊥) to the incident laser radiation polarization.**

Coherent backscattering was observed for the first time in 1985, independently by two groups [6, 7]. Since then much experimental and theoretical work has been done resulting in a thorough understanding of the phenomenon [2, 8–10]. For various porous dielectric materials with negligible absorptive losses a pronounced sharp increase of scattering within the narrow cone around the incidence direction is readily observed (Figs. 8.5 and 8.6). Remarkably, unlike typical scattering behavior where scattered light is essentially non-polarized, the backscattering cone polarization of scattered light follows the original incident

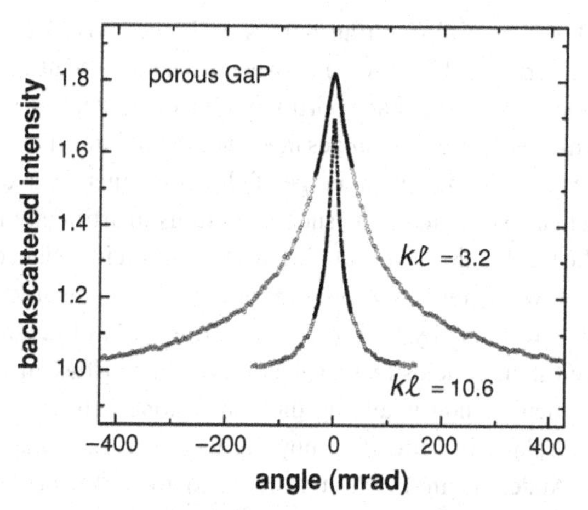

Fig. 8.6 The backscattered intensity normalized to the diffuse background as a function of angle for two samples of porous GaP [10]. Backscattered light is polarized parallel to the incident light. From the width of the cone, incorporating internal reflection corrections, the $k\ell$ values are inferred. The narrow cone: $k\ell = 10.6 \pm 0.9$, $R = 0.78 \pm 0.05$. The broad cone: $k\ell = 3.2 \pm 0.4$, $R = 0.67 \pm 0.06$.

light polarization (Fig. 8.5). This is a clear signature of the coherent optical effect. The backscattering enhancement found in experiments is typically slightly less than the ideal value of the enhancement factor equal to 2. That is probably because of small but finite absorptive losses ("inelastic scattering").

The width of the narrow backscattering cone is inversely proportional to the scattering mean free path ℓ,

$$W \approx \frac{0.7 n_e (1 - R)}{k\ell}, \tag{8.17}$$

where W is the cone full width at half-maximum height, n_e is the effective refractive index of the medium, R is the diffuse reflectance coefficient, and $k = 2\pi n_e/\lambda_0$ is the light wave number in the medium with λ_0 being the light wavelength in a vacuum [10]. This cone contains information about the deeper structure of the medium which cannot be accessed with normal optical techniques. These longer paths determine the sharp top of the cone.

8.3 Towards the Anderson localization of light

Diffusive transport of light energy still allows for light ray trajectories to be implied in the consideration of light propagation. What happens if light scatters so frequently that light rays have no chance of being plotted anywhere? The ultimate limit is known as the *Ioffe–Regel criterion* for mean free path versus wavenumber $k = 2\pi/\lambda$,

$$k\ell < 1, \tag{8.18}$$

which means diffusive transport is no longer possible since even a single oscillation can not be performed by a wave between successive scattering events. The wave appears to be confined within the space portion with a characteristic size of the order of the wavelength. This means the wave localizes in a random medium. This severe condition is hard to perform with respect to the optical range of electromagnetic waves in spite of many efforts that have been made so far to create such conditions in artificial media.

The localization criterion Eq. (8.18) was not introduced into electromagnetic theory until 1984. It was originally proposed for electrons in disordered solids within the framework of solid-state theory by A. F. Ioffe and A. R. Regel in 1927. Much later, in 1958, P. W. Anderson arrived at the conclusion that electrons cannot diffuse in a strongly disordered potential [11]. This phenomenon is among the cornerstone concepts of the theory of disordered solids and its place in solid-state physics has been recognized by the Nobel prize awarded to P. W. Anderson in 1977. It is referred to as the *Anderson localization*. With growing fluctuations, a transition occurs from the conducting to the insulating state, which is referred to as the *Anderson transition*. It was discussed in more detail earlier in Section 4.5. While being first introduced for electron conductivity processes, Anderson localization is possible for all types of waves provided the localization criterion is fulfilled in terms of sufficiently strong fluctuations of the physical parameters determining the speed of waves. For electrons, fluctuation of potential is the proper physical parameter, whereas for electromagnetic waves refractive index fluctuations are the relevant counterpart, as has been shown in Chapter 3.

Generally, localization occurs more readily in space with lower dimensionality. This is always the case for one-dimensional space provided a negligible disorder is present. This is because there are no independent scattering processes in a one-dimensional problem. In two dimensions, localization occurs under a certain degree of disorder, whereas in three dimensions the disorder should be even stronger for localization to occur.

S. John was the first to outline the possibility of Anderson localization of electromagnetic waves in 1984 [12]. This report was followed promptly by the elegant comment by P. W. Anderson [1] and since then localization of light has become a challenge for experimenters. However experimental observation of the Anderson localization of light is hard to perform and until the time of this book no straightforward report on the observation of light localization has been available. The principal obstacle is the relatively low refraction index of materials in the optical range. For the near infrared, the highest refractive index values are inherent in Ge ($n = 4$), Si ($n = 3.4$), and GaAs ($n = 3.37$); for the visible these are GaP ($n = 3.31$), ZnTe ($n = 2.98$), and TiO_2($n = 2.8$) (see Table 3.1). Not all of the above materials are readily available in the form of a sub-micrometer powder, not all of them are actually suitable for experiment. For example, Si and Ge smaller particles readily acquire an oxide shell in air.

Consider experiments towards the observation of Anderson light localization in detail [13–17]. D. Wiersma *et al.* [13] used GaAs powder with different average particle diameters (Fig. 8.7). A laser wavelength $\lambda = 1064$ nm was used, at which the absorption coefficient of pure GaAs is $\alpha < 1\ \mathrm{cm}^{-1}$ and the refractive index is 3.48. Upon reducing the average particle diameter these authors found three distinctive light propagation regimes. The first one is the known $T \propto \ell/L$ behavior inherent in typical diffusive light transport, as discussed in Section 8.1. This regime is clearly seen in Figure 8.8a. The mean free path evaluated

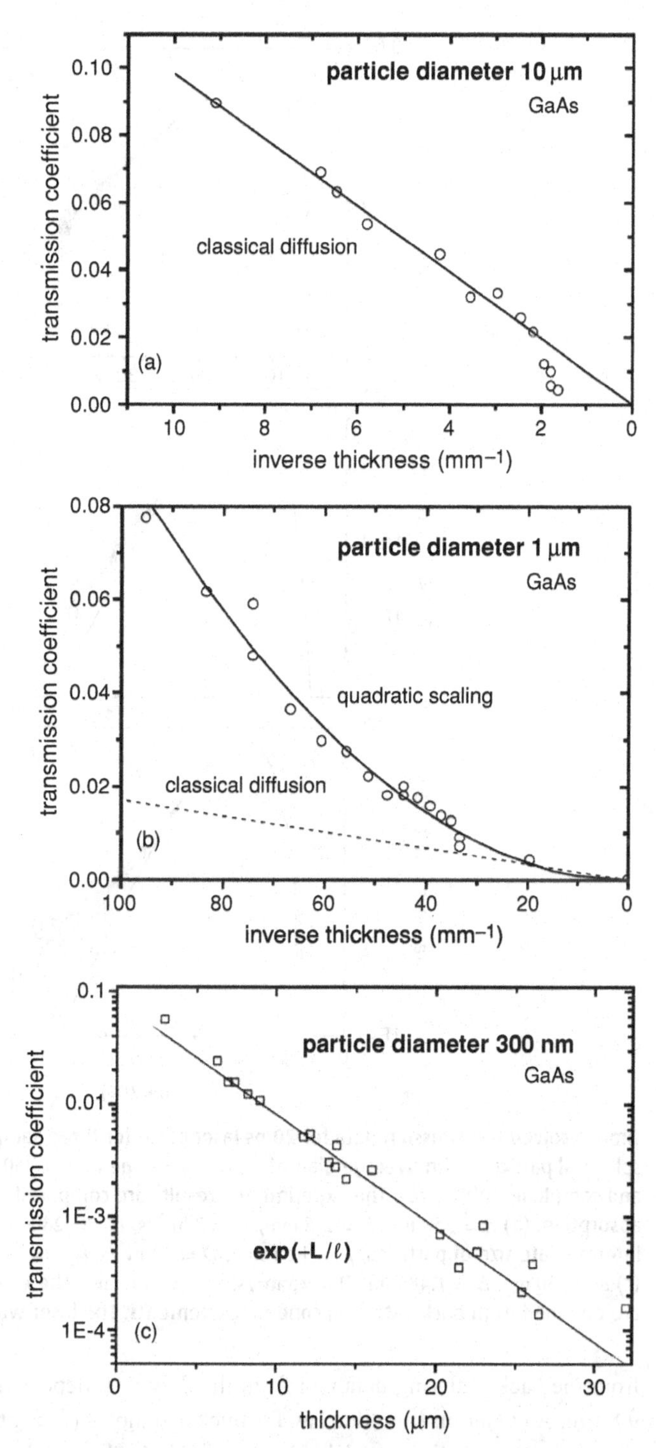

Fig. 8.7 Measured transmission coefficients for three sets of samples of GaAs powders with different average particle diameters (circles and squares) along with transmission versus thickness plotted for various models (lines). (a) particle diameter $d = 10\,\mu\text{m}$, (b) $d = 1\,\mu\text{m}$, (c) $d = 0.3\,\mu\text{m}$. The laser wavelength is 1064 nm. See text for more detail. Note change in scales in (c) as compared to (a) and (b). Reprinted with permission from [13]. Copyright 1997 *Nature* Publ. Group.

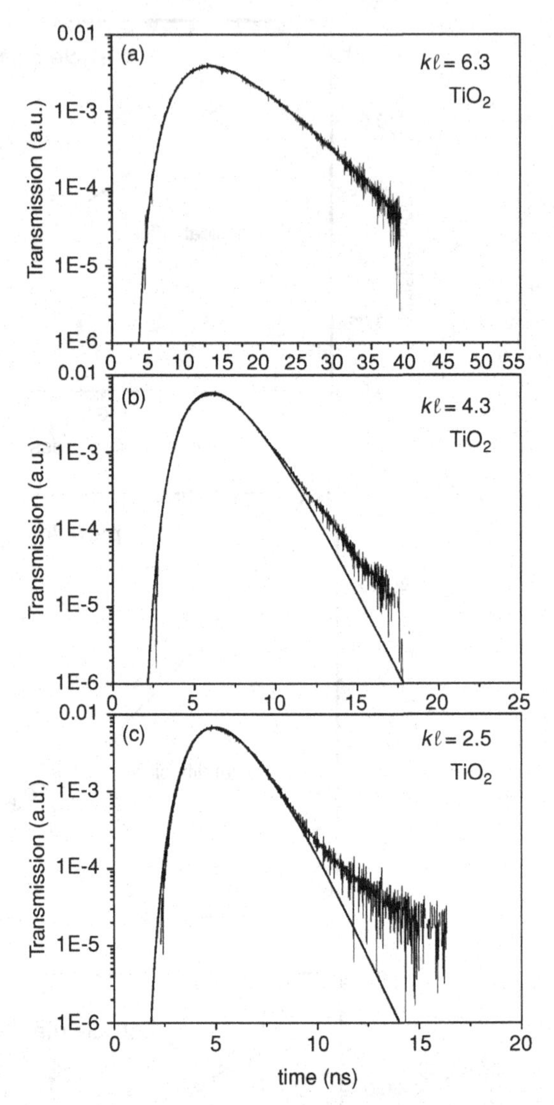

Fig. 8.8 Time-resolved transmission data for 20 ps laser pulse for three samples consisting of dense TiO_2 spherical particles with average diameter d in the range of 250–550 nm with 25% polydispersity and sample length L [16]. The experimental results are compared to diffusion theory including absorption. (a) $d = 550$ nm, $L = 2.5$ mm, $D = 22\text{m}^2/\text{s}$, $\ell_{\text{abs}} = 2600$ mm; (b) a sample with intermediate size of particles, $L = 1.51$ mm, $D = 13\ \text{m}^2/\text{s}$, $\ell_{\text{abs}} = 380$ mm); (c) $d = 250$ nm, $L = 1.48$ mm, $D = 15\ \text{m}^2/\text{s}$, $\ell_{\text{abs}} = 340$ mm). The $k\ell$ values indicated in the figures are obtained from backscattering cone measurements. The laser wavelength is 590 nm.

from the backscattering data and from the $T \propto \ell/L$ dependence has the same value $\ell = 9.8\ \mu\text{m}$. A deviation from this law for thicker samples ($L > 500\ \mu\text{m}$, $L^{-1} < 2\ \text{mm}^{-1}$) is a signature of absorption which has been estimated to have a characteristic length $\ell_{\text{abs}} \equiv \alpha^{-1}$, corresponding to the absorption coefficient value $\alpha < 0.13\ \text{cm}^{-1}$. When the particle mean diameter goes down to 1 μm, the $T \propto \ell/L$ law is no longer the case. Instead, the quadratic dependence $T \propto L^{-2}$ emerges (Fig. 8.7b). This type of behavior was predicted by the scaling theory of localization at the localization transition [1, 18]. For a comparison, the dashed line shows the $T \propto \ell/L$ law with $\ell = 0.17$ nm used for fitting which has been evaluated from backscattering cone analysis. Finally, for smaller particle diameter 0.3 μm the exponential transmission versus length law is observed,

$$T(L) \propto \exp\left(-\frac{L}{\ell_{\text{loc}}}\right), \tag{8.19}$$

which is a distinctive manifestation of the light localization regime. The characteristic scaling length parameter ℓ_{loc} for exponential decay is called the localization length, which in the case under consideration was found to be $\ell_{\text{loc}} = 4.3$ μm.

The exponential law expressed by Eq. (8.19) formally coincides with the Bouger law (8.4) inherent in inelastic (absorptive) losses. Sometimes, Eq. (8.4) is used in the form,

$$T(L) = \exp(-L/\ell_{\text{abs}}), \quad \ell_{\text{abs}} \equiv \alpha^{-1}, \tag{8.20}$$

to emphasize this similarity.

Although for larger GaP particles examined the absorption coefficient was found to be rather small, an increase for submicrometer powder entities can not be excluded. This controversial issue was raised in the press after the first publication on exponential transmission behavior [14].

The general relation for transmission of a slab of complex media with absorptive losses and multiple scattering [19],

$$T(L) = \frac{\sinh(\ell/L) + (z_e/\ell_{\text{abs}})\cosh(\ell/\ell_{\text{abs}})}{(1 + z_e^2/\ell_{\text{abs}}^2)\sinh(L/\ell_{\text{abs}}) + (2z_e/\ell_{\text{abs}})\cosh(L/\ell_{\text{abs}})}, \tag{8.21}$$

reduces to exponential laws in each of the two extreme cases: Eq. (8.19) in the case of a purely dispersive medium for $L > \ell_{\text{loc}}, \lambda, \ell$; and Eq. (8.20) for a purely absorptive medium with negligible multiple scattering for $L \geq \ell_{\text{abs}}$.

In experiments with Ge powders (with $n = 4.1$ in the near infrared, Ge is the best candidate for light localization performance) it was found that it has non-negligible absorption which does influence the transmission of light and partially contributes to the exponential $T(L)$ law observed [15]. Therefore further signatures for light localization are to be searched for.

An important further experiment would be a time-resolved transmission experiment in which reduction of the diffusion constant at the Anderson localization transition is expected. Maret with co-workers [16, 17] considered possible alternative manifestations of the Anderson localization of light beyond the exponential transmission versus length dependence expressed by Eq. (8.19). Experiments were performed with dense TiO_2 ground beads of a size close to the optical wavelength packed in dense layers of 1.2–2.5 mm thickness. Increasing deviations were observed from the exponential time dependence predicted by the diffusive transport theory (see Eqs. (8.15) and (8.16) in Section 8.1). For a sample with larger $k\ell$ value (Fig. 8.8(a)) the temporal profile of the output light pulse features good agreement with the theory of diffusive transport based on Eqs. (8.15) and (8.16),

$$T(t) \propto \exp\left[-\left(\frac{\pi^2 D(t)}{(L + 2z_e)^2} + \frac{1}{\tau_{\text{abs}}}\right)t\right], \tag{8.22}$$

and exhibits an exponential tail at longer times. For smaller $k\ell$ values, a discrepancy with the diffusive transport theory is visible which increases for smaller $k\ell$ (Fig. 8.8b,c). The transmission tail was found to be reasonably described by a modified Eq. (8.22) with time-dependent diffusion coefficient $D(t)$,

$$T(t) \propto \left(\frac{D(t)}{D_0}\right)^2 \exp\left[-\left(\frac{\pi^2 D(t)}{(L+2z_e)^2} + \frac{1}{\tau_{abs}}\right)t\right]. \tag{8.23}$$

It was found that $D(t)$ bears witness to a decrease with time as $1/t$, as expected from the theory for the localization regime. The above results indicate that time-resolved light flight measurements offer further insight towards discrimination of light propagation regimes near the localization threshold. In the near future, new experiments can be foreseen to bring about a wealth of interesting phenomena related to the Anderson localization of classical waves.

Anderson localization is possible not only in optics but also in other fields. It has been reported for electromagnetic waves in the microwave range [20] as well as for ultrasonic waves [21].

8.4 Light in fractal structures

I conceived and developed a new geometry of nature and implemented its use in a number of diverse fields. It describes many of the irregular and fragmented patterns around us, and leads to full-fledged theories, by identifying a family of shapes I call fractals.

Benoit B. Mandelbrot [22]

Optical fractal filters

The notion "fractals" was introduced in science to identify the unusual geometrical objects of complex shape whose properties were found to be characterized by means of fractional dimensionalities. The rigorous mathematical definition of fractals is based on the specific definition of dimensionalities and can be found in [22, 23]. Here, for clarity and simplicity, we shall emphasize the two principal features inherent in fractals. The first characteristic feature is geometrical *self-similarity*, which means that smaller portions of a fractal object exhibit the same shape as the bigger ones. Self-similarity can be accomplished by means of an algorithm that implies sequential repetition of a certain constructive procedure. Examples of one-dimensional fractal structures are given in Figure 8.9. The second characteristic feature is fractional dimensionality. This is always smaller for fractals than the dimensionality of the space in which fractals are embedded. For every structure presented in Figure 8.9 dimensionality D is less than unity as is indicated in the top line of the figure. Fractional dimensionality means that the amount of matter, contained in a fractal object, e.g. in terms of mass as $m = \rho^D$ where ρ is the material density, can not be expressed with integer D and typically contains a fractional power factor in the range $d - 1 < D < d$,

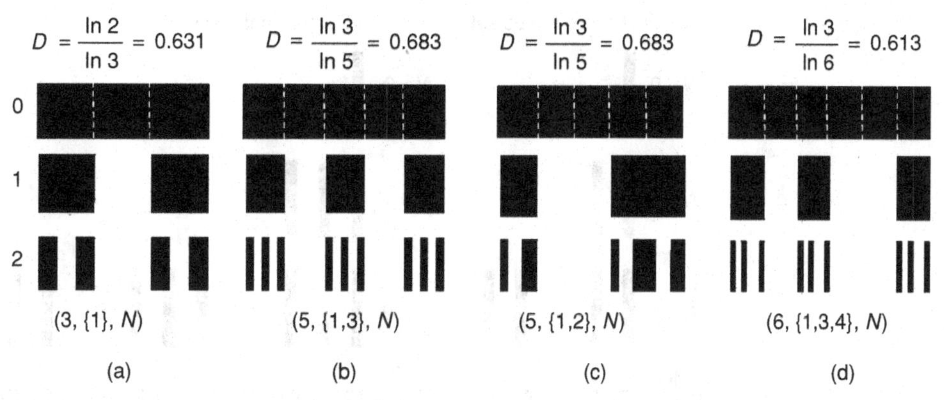

Fig. 8.9 **The first three generations (from the top to the bottom) of the four different one-dimensional fractal structures.**

with $d = 1, 2, 3$ being the integer dimensionality of the Euclidian space in which the fractal object is embedded. Notably, unlike continuous solid species, in fractal structures different physical properties (electrical and thermal conductivities, diffusion and others) may follow equations containing different dimensionalities D_i that do not coincide with the dimensionality D used to calculate the amount of matter $m = \rho^D$.

Figure 8.9(a) shows a fractal structure known as the triadic *Cantor set*.[1] It is generated by means of infinite iterative substitution of a middle 1/3 portion of a linear segment. The number 3 is referred to as the generator, $G = 3$. Figure 8.9(b) presents an algorithm for generating a pentadic Cantor set. In this case $G = 5$. Cantor structures comprise fractal sets with dimensionality,

$$D_{\text{Cantor}}(G) = \frac{\ln[(G+1)/2]}{\ln G}, \tag{8.24}$$

i.e. $D(3) = \ln 2/\ln 3$ for triadic and $D(5) = \ln 3/\ln 5$ for pentadic Cantor sets. Similarly, higher-order ($G = 7, 9, \ldots$) Cantor sets can be generated. More complex asymmetric fractal structures are shown in Figures 8.9(c) and 8.9(d). Their dimensionalities obey the law,

$$D = \frac{\ln g}{\ln G}, \tag{8.25}$$

where g is the integer defining the number of remaining portions. For representative asymmetric structures the dimensionality is given in the top line of Figure 8.9. Comparing Eq. (8.25) and Eq. (8.24) one can see that symmetric and asymmetric one-dimensional fractal structures obey the same formula with $(G + 1)/2 = g$ defining the number of remaining portions for the particular cases of symmetric Cantor structures.

It is convenient to use the following simple notation to define a given fractal structure as shown in the bottom line of Figure 8.9. First, the G value should be indicated. This gives the number of equal portions in which the original linear segment is partitioned. Second,

[1] This algorithm to construct a set with fractional dimensionality was proposed by a German mathematician Georg Cantor (1845–1918).

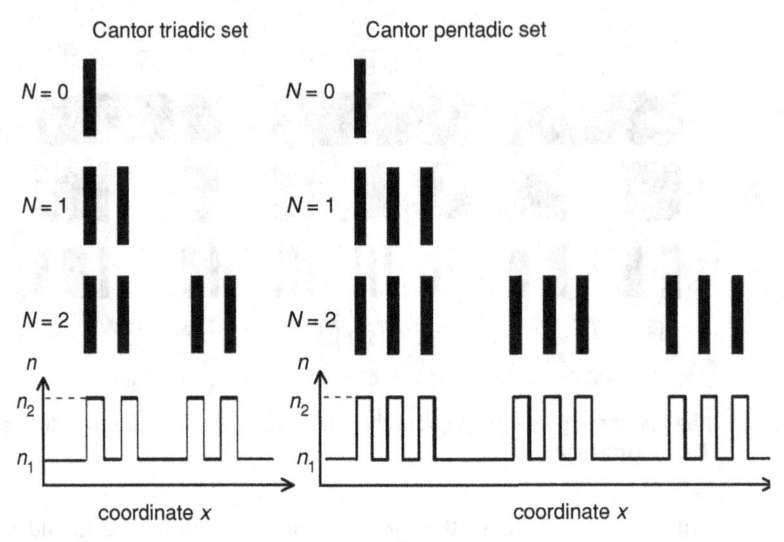

Fig. 8.10 **Triadic (left) and pentadic (right) multilayer optical Cantor filters and corresponding refractive index profiles for filters with $N = 3$. For these filters, the total number of layers equals $3^N = 9$ for triadic and $5^N = 25$.**

the portions to be removed in every iteration are to be indicated. To do this, portions are numbered sequentially starting from 0 and then the numbers of removing portions are written in {} brackets. Finally, the generation number N completes the unequivocal notation of a given fractal structure. Multilayer structures consisting of a pair of materials with different refractive indexes organized according to a fractal algorithm are referred to as *fractal filters*. Fractal multilayer filters were introduced into optics by Jaggard and Sun in 1990 [24, 25]. Their optical properties have been examined in detail in [26–31].

Sequential deposition of layers in multilayer optical films can be performed according to the fractal generation algorithm. Then fractal optical filters can be developed. It is of general physical interest to trace the regularities in their spectral properties with respect to electromagnetic wave propagation. Triadic and pentadic optical Cantor filters are shown in Figure 8.10. For any type of fractal filter composed of a material A with elementary layer thickness d_{A} (dark layers in Figure 8.10) and another material B with elementary layer thickness d_{B} (light layers), the self-similar construction algorithm gives rise to the following expression for the total geometrical thickness of the filter,

$$D_N = d_{\mathrm{B}}(G^N - g^N) + d_{\mathrm{A}} g^N, \tag{8.26}$$

which for Cantor structures ($g \equiv (G+1)/2$) reads,

$$D_N^{\mathrm{Cantor}} = d_{\mathrm{B}} \left[G^N - \left(\frac{G+1}{2} \right)^N \right] + d_{\mathrm{A}} \left(\frac{G+1}{2} \right)^N . \tag{8.27}$$

Self-similarity of fractal structure results in the following recurrent relation for the total geometrical thickness of fractal filters,

$$D_N = g D_{N-1} + (G-g) G^{N-1} d_{\mathrm{B}}, \tag{8.28}$$

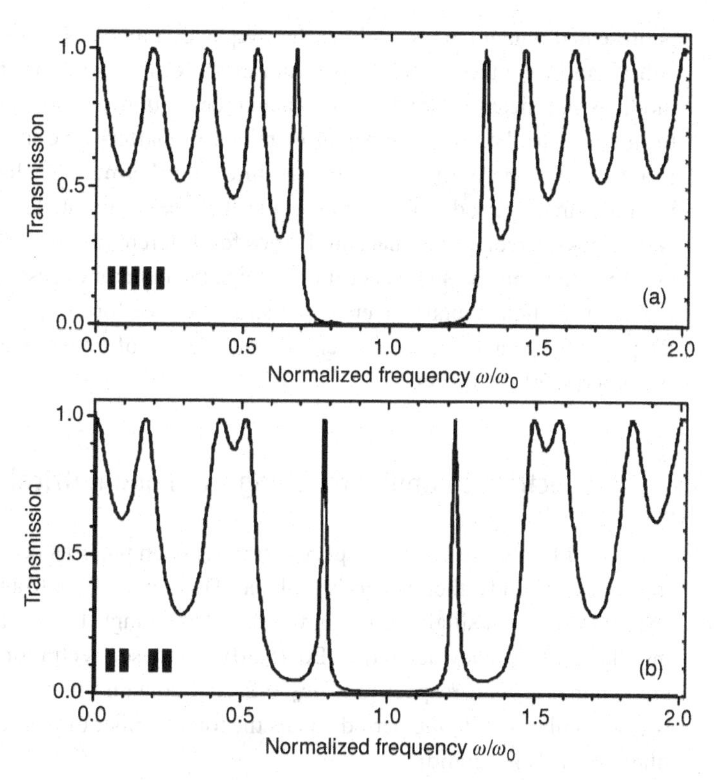

Fig. 8.11 Transmission spectra of (a) a nine-layer periodic quarter-wave stack and (b) a nine-layer Cantor triadic filter. The structures are shown in the left bottom corners of each graph. Adapted from [28].

which for Cantor filters with $g \equiv (G+1)/2$ reads,

$$D_N = \frac{G+1}{2} D_{N-1} + \frac{G-1}{2} G^{N-1} d_{\rm B}. \tag{8.29}$$

To analyze the spectral properties of optical fractal filters, for simplicity and clarity, consider filters consisting of elementary layers of a pair of materials A and B with refractive indexes $n_{\rm A}, n_{\rm B}$ and thicknesses $d_{\rm A}, d_{\rm B}$, satisfying an equality $n_{\rm A}d_{\rm A} = n_{\rm B}d_{\rm B}$. The latter condition was shown in Section 7.2 to give rise to the notion of "quarter-wave layers" and "quarter-wave stacks" since the middle stop-band wavelength λ_0 of a periodic quarter-wave stack obeys $\lambda_0 = 4n_{\rm A}d_{\rm A} = 4n_{\rm B}d_{\rm B}$ relation. This simplification offers convenient comparison of spectra for different generations, since the transmission spectrum of every (periodic and non-periodic) quarter-wave multilayer structure obeys periodicity on the frequency scale and symmetry within every period with respect to the middle-gap frequency,

$$\omega_0 = \frac{2\pi c}{\lambda_0} = \frac{\pi c}{2n_{\rm A}d_{\rm A}} = \frac{\pi c}{2n_{\rm B}d_{\rm B}}, \tag{8.30}$$

the period in the transmission spectrum being equal to $2\omega_0$.

A transmission spectrum of a simple nine-layer triadic Cantor filter is compared in Figure 8.11 with a nine-layer periodic quarter-wave stack. One full period in the spectrum is presented which corresponds to the frequency range $0 < \omega_0 < 2\omega_0$. Although there is still a gap in the transmission spectrum of the Cantor structure around ω_0, the spectrum modifies significantly. Notably, the total number of transmission peaks in both cases is equal to nine, i.e. to the total number of elementary quarter-wave layers in the structure. This is a general property of multilayer structures. Furthermore, each spectrum is symmetrical within a single period $2\omega_0$ with respect to the $\omega = \omega_0$ point. This is a property of quarter-wave stacks. In computational simulations for different fractal filters it was found that every period in the transmission spectrum contains the number of resolvable peaks which equals the number of elementary layers in the structure, i.e., for example 3^N peaks for triadic and 5^N peaks for pentadic Cantor filters. Further details of transmission spectra are defined by the ordering of the layers.

Spectral scalability resulting from geometrical self-similarity

For fractal filters, we consider spectral transmission patterns centered at the middle of the transmission band rather than the stop-band. This corresponds to analysis of the transmission spectrum within a single period in the frequency range $1 < \omega/\omega_0 < 3$ with ω_0 being the middle-gap frequency defined by Eq. (8.30). All these spectra for various fractal structures are symmetric with respect to $\omega/\omega_0 = 2$, and contain the resolvable transmission peaks whose number within the period equals the total number of elementary layers in the fractal filter under consideration.

Optical fractal filters were found to exhibit the principal spectral property resulting from their geometrical self-similarity. *For every type of fractal structure, the transmission spectrum of any given generation with number $N > 2$ contains embedded transmission spectra of all preceding generations with numbers $n = 1, 2, \ldots, N-1$, the spectrum of every preceding generation being squeezed along the frequency axis by a factor G^{N-n}.* This property has been called the *spectral scalability of fractal filters*.

The spectral scalability of fractal filters is illustrated and explained in Figures 8.12–8.14. Let us begin from triadic Cantor filters as the simplest example. Figure 8.12 shows the full periods for (a) 81-layer filter of 4th generation, (c) 27-layer filter of 3rd generation, and (f) 9-layer filter of 2nd generation. The three-fold (expanded along the frequency axis) central portion of the $N = 4$ filter coincides with the full period of the $N = 3$ filter (compare patterns in (b) and (c)). The nine-fold ($9 = 3^2$) expanded central portion of the $N = 4$ filter fits the full spectrum of the $N = 2$ filter (compare patterns in (d) and (f)). The central portion of the $N = 3$ filter, in its turn, fits the full spectrum of the $N = 2$ filter when expanded by a factor of three. Scalability means reproducible spectral patterns in a "physical" rather than rigorous mathematical sense. Comparing graphs (d), (e), (f) one can see that patterns are rather similar, although sharpness of resonant peaks shows a tendency to lower finesse with growing N.

The spectral scalability is inherent in all types of multilayer fractal filters, both symmetric (Cantor and non-Cantor) and asymmetric. This is shown for representative structures in Figures 8.13 and 8.14. Symmetric fractal structures in the very central portion of transmission spectra around $\omega/\omega_0 = 2$ exhibit G number of resolvable peaks. For symmetric

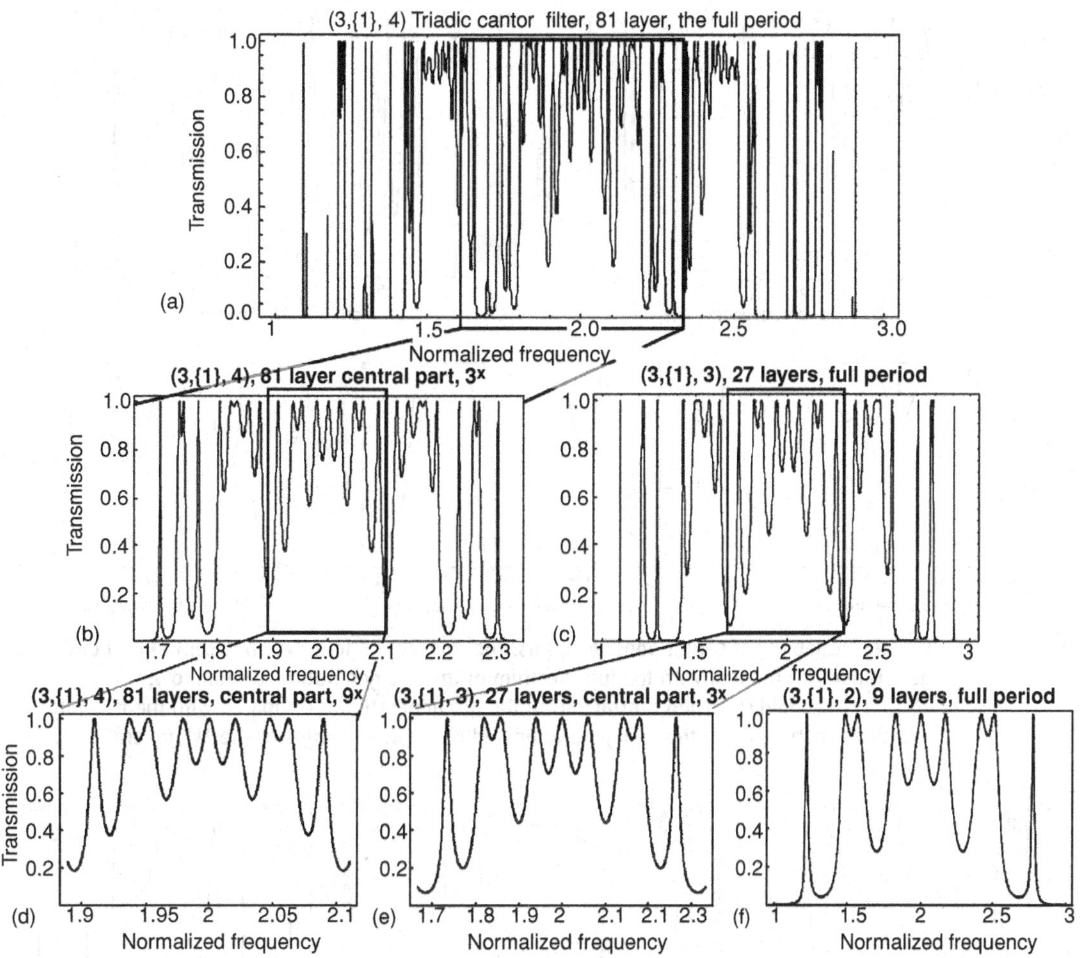

Fig. 8.12 **Scalability of optical spectra for triadic Cantor filters $(3, \{1\}, N)$: (a) the full period for $N = 4$; (b) the central part magnified in the frequency scale by 3 versus (c) the full period for $N = 3$; (d) the central part for $N = 4$ magnified by $9 = 3^2$; (e) the central part for $N = 3$ magnified by 3; (f) the full period for $N = 2$. Compare patterns in (b) versus (c) and in (d) versus (e) and (f). All resonance peaks in transmission have the maximal value of unity. Somewhat lower values apparent for a few peaks in (a) are artifacts due to high finesse of resonances and finite-step numerical calculations, as well as finite spatial resolution of the used software and hardware. Adapted from [28].**

fractal structures every transmission peak features unity maximal transmission whereas for asymmetric structures this is not the case.

Experimental performance of fractal filters needs precise deposition or growth of multiple subwavelength layers. A representative example is given in Figure 8.15 where the measured and calculated spectra are presented for triadic filters of second and third generation. Good agreement of the measured and calculated data is seen. Experimentally feasible spectra differ from the ideal picture given in Figure 8.12 for a number of reasons. First, dependence of

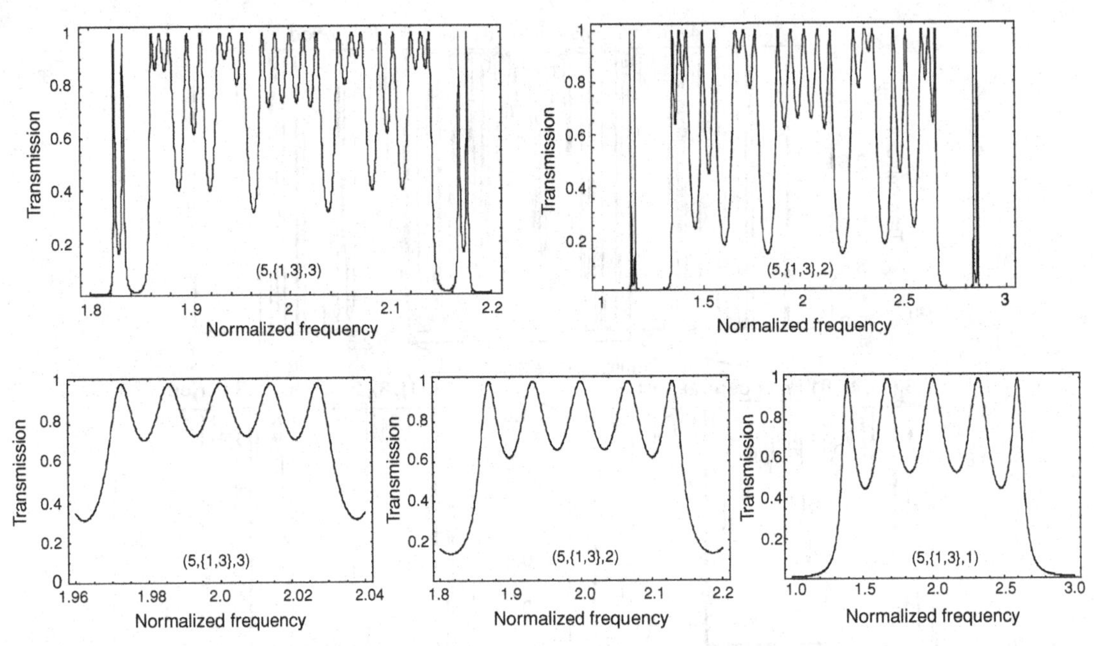

Fig. 8.13 Spectral scalability of Cantor pentadic filters. Top row: $G = 5$-fold stretched central part of the 3rd generation is compared with the full spectrum of the 2nd generation. Bottom row: $G^2 = 5^2 = 25$-fold stretched central part of the 3rd generation is compared with the five-fold extended central part of the 2nd generation and with the full period of the 1st generation [28].

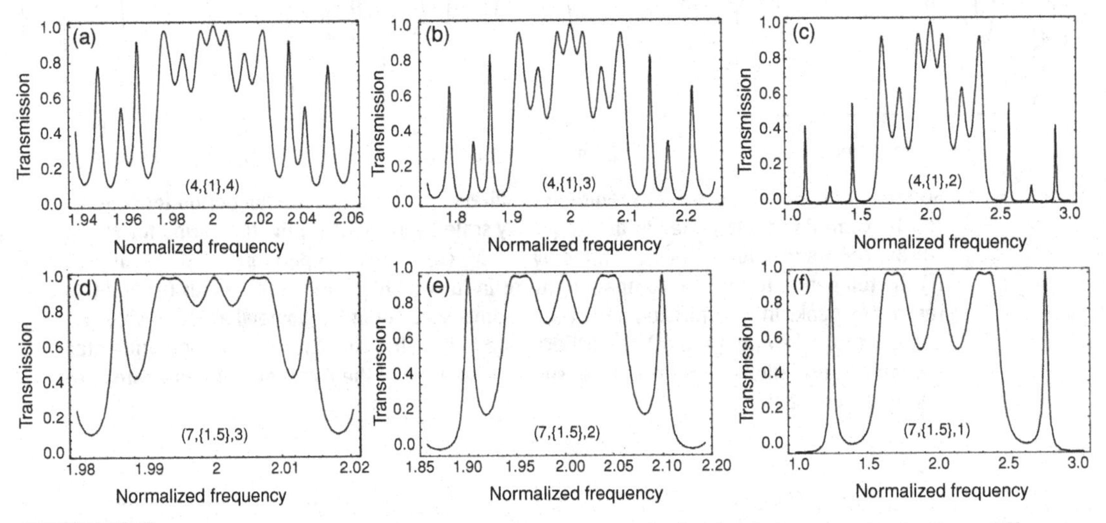

Fig. 8.14 Spectral scalability of non-Cantor fractal structures [31]. (a)–(c) show spectra for three different generations of the asymmetric structure $(4, \{1\}, N)$. (c) presents the full spectrum whereas (a) and (b) represent the central portions with (a) $G^2 = 4^2 = 16$-fold stretching and (b) $G^1 = 4$-fold stretching with respect to (c). (d)–(f) show spectra for three different generations of the symmetric $(7, \{1, 5\}, N)$ fractal structure. (f) presents the full spectrum whereas (d) and (e) represent the central portions with (d) $G^2 = 7^2 = 49$-fold stretching and (e) $G^1 = 7$-fold stretching with respect to (f).

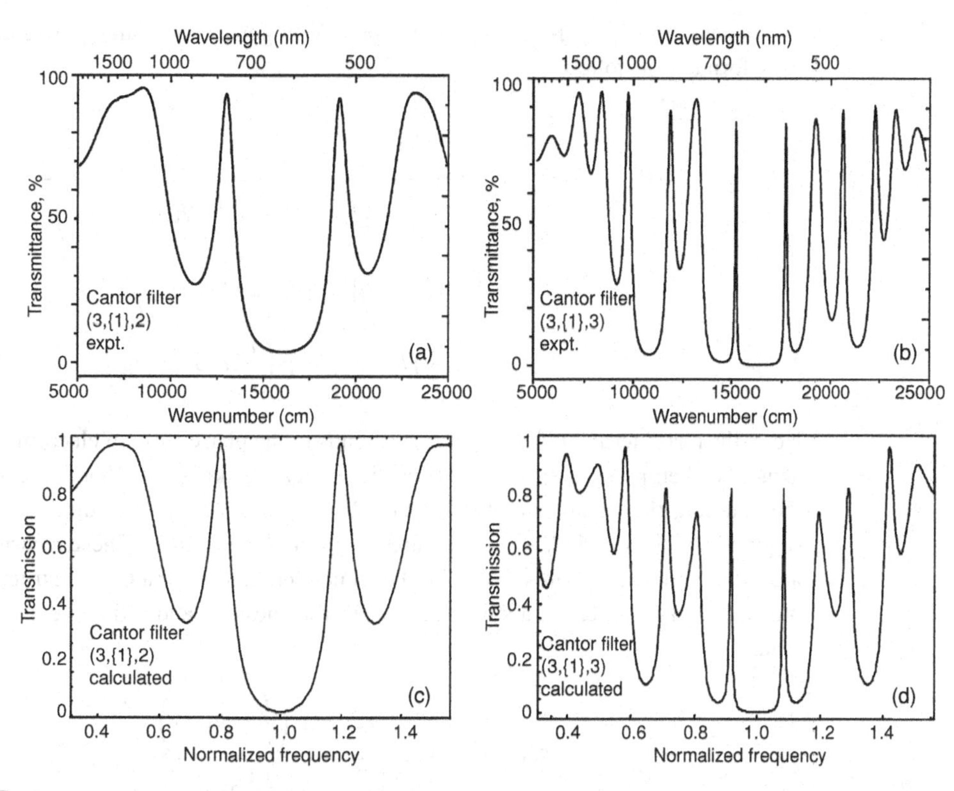

Fig. 8.15 **Transmission spectra of triadic Cantor filters of the second (9 layers) and the third (27 layers) generations consisting of ZnSe and Na_3AlF_6 [28]. (a) $N = 2$, experiment; (b) $N = 3$, experiment; (c) $N = 2$, calculations, (d) $N = 3$, calculations.**

refractive index versus wavelength gives rise to an asymmetry of the transmission spectra on the frequency axis. Second, finite absorption by dielectric materials of short-wave radiation does not allow observations in the wide frequency range. Third, finite deviations in the thickness and internal inhomogeneity of polycrystalline layers give rise to partial scattering, resulting in transmission peaks of height lower than unity. The finite angular aperture of incident light additionally contributes to lowering of maximal transmission since it means a deviation from the ideal one-dimensional propagation problem. Finally, a glass substrate from one side and ambient air from the other side result in a deviation of experimentally designed structures contrary to the ideal ones examined in theory. All theoretically examined structures discussed so far have included only two media, implying the ambient environment coincides with the B-material constituting a fractal filter. Vertical multilayer structures obtained by templated etching, like that shown in Figure 7.16(b) in Chapter 7, offer better options for development of fractal filters closely reproducing the ideal ones. However this approach has not been reported in the context of fractal filters so far.

Zhukovsky *et al.* proved analytically the spectral scalability of fractal filters [29, 31]. Consider quarter-wave Cantor filters. For triadic Cantor structures. Jaggard and Sun [24, 25] derived the recurrent relations between intensity reflection $R(\omega)$ and transmission $T(\omega)$

coefficients for $(N+1)$-th generation versus N-th generation using generalization of the Airy formulas as follows,

$$R_{N+1}(\omega) = R_N(\omega) + \frac{R_N(\omega)[T_N(\omega)]^2 \exp\left(\frac{i}{c}2\omega G^N d_A n_A\right)}{1-[R_N(\omega)]^2 \exp\left(\frac{i}{c}2\omega G^N d_A n_A\right)}, \tag{8.31}$$

$$T_{N+1}(\omega) = \frac{[T_N(\omega)]^2 \exp\left(\frac{i}{c}\omega G^N d_A n_A\right)}{1-[R_N(\omega)]^2 \exp\left(\frac{i}{c}2\omega G^N d_A n_A\right)}. \tag{8.32}$$

Note, the phase term $\omega G^N d_A n_A/c$ corresponds to the phase shift an electromagnetic wave acquires when passing through a layer with optical thickness $G^N d_A n_A$. It corresponds to the addition, the Cantor structure of the $(N+1)$-th generation acquires as compared to sequentially displaced $(N+1)$ structures of the N-th generation. These recurrent relations are valid until $N > 0$ holds. For $N = 0$ the known relations for a Cantor structure building block should be used, i.e. a single layer with thickness d_B and refractive index n_B. These read,

$$R_0(\omega) = -r + \frac{rtt' \exp\left(\frac{i}{c}2\omega n_B d_B\right)}{1-r^2 \exp\left(\frac{i}{c}2\omega n_B d_B\right)}, \tag{8.33}$$

$$T_0(\omega) = \frac{tt' \exp\left(\frac{i}{c}\omega n_B d_B\right)}{1-r^2 \exp\left(\frac{i}{c}2\omega n_B d_B\right)}, \tag{8.34}$$

with elementary amplitude reflection and transmission coefficients at the borders of materials A and B,

$$r \equiv \frac{n_B - n_A}{n_B + n_A}, \quad t \equiv \frac{2n_A}{n_B + n_A}, \quad t' \equiv \frac{2n_B}{n_B + n_A}. \tag{8.35}$$

Spectral scalability comes from the invariance of the phase term in Eqs. (8.31) and (8.32) with respect to simultaneous transformations $N \to N+1$ and $\omega \to \omega/G$ because of the evident relation,

$$\frac{i}{c}\omega G^N d_A n_A \equiv \frac{i}{c}\frac{\omega}{G} G^{N+1} d_A n_A. \tag{8.36}$$

This means the phase term favors spectral scalability with the G value as the scaling factor. Let us have a closer look at the remaining terms. The two functions are to be compared,

$$T_{N+1}\left(\frac{\omega}{G}\right) \text{ versus } T_N(\omega). \tag{8.37}$$

Using Eq. (8.32) with $G = 3$ these functions read,

$$T_{N+1}\left(\frac{\omega}{3}\right) = \frac{\left[T_N\left(\frac{\omega}{3}\right)\right]^2 \exp\left(\frac{i}{c}\frac{\omega}{3}3^N d_B n_B\right)}{1-\left[R_N\left(\frac{\omega}{3}\right)\right]^2 \exp\left(\frac{i}{c}\frac{\omega}{3}2\cdot 3^N d_B n_B\right)}, \tag{8.38}$$

$$T_N(\omega) = \frac{[T_{N-1}(\omega)]^2 \exp\left(\frac{i}{c}\omega\cdot 3^{N-1} d_B n_B\right)}{1-[R_{N-1}(\omega)]^2 \exp\left(\frac{i}{c}2\omega\cdot 3^{N-1} d_B n_B\right)}, \tag{8.39}$$

Since the phase exponents in Eqs. (8.36) and (8.37) are exactly equal, the sole difference between $T_{N+1}\left(\frac{\omega}{3}\right)$ and $T_N(\omega)$ is contained in the coefficients, $T_N\left(\frac{\omega}{3}\right)$, $R_N\left(\frac{\omega}{3}\right)$ and $T_{N-1}(\omega)$, $R_{N-1}(\omega)$, respectively. Expanding these coefficients in the same way, using Eqs. (8.31) and (8.32), one can see the difference will again present only in the coefficients, this time, $T_{N-1}\left(\frac{\omega}{3}\right)$, $R_{N-1}\left(\frac{\omega}{3}\right)$ and $T_{N-2}\ (\omega)$, $R_{N-2}\ (\omega)$, respectively. Tracing this procedure down sequentially and seeing that all frequency-dependent exponents that appear along the way are equal for both terms in Eq. (8.36), we finally arrive at $N = 0$. At this point, the factors to be compared are $T_1\left(\frac{\omega}{3}\right)$, $R_1\left(\frac{\omega}{3}\right)$ and $T_0\ (\omega)$, $R_0\ (\omega)$. The corresponding phase terms are $i\omega d_A n_A/c$ and $i\omega d_B n_B/c$. These are equal if the quarter-wave condition is met. The difference in coefficients is smaller as r decreases, and the agreement is total if $r^2 \approx 0$. As we have seen, all frequency-dependent exponents in the expression (8.37) are equal at any stage of decomposition. So it can be written that the quantities in Eq. (8.37) have *identical phase structure*, with a minor difference in the coefficients. Since the characteristic spectral features (transmission resonances and local band gaps) are essentially phase phenomena (resulting from constructive and destructive interference, respectively), similar phase structure results in similar appearance of the spectral portraits as confirmed by Figure 8.12. It is the phase structure of the recurrent relations which determines the apparent scalability of spectral transmission profiles, although the absolute transmission coefficient values in the dips between resonant peaks show certain deviations from precise scalability. As can be seen in Figures 8.12–8.14 lower generations obey a higher finesse of the transmission bands.

More complete formulation of scalability can be made in terms of the amplitude scaling parameter, γ, which alters the absolute value of transmission without modification of its extrema, i.e. the scalability relation now reads [29],

$$\left[T_{N+1}\left(\frac{\omega}{G}\right)\right]^\gamma = T_N(\omega). \tag{8.40}$$

Computational and analytical estimates give the γ value,

$$\gamma \approx \left(\frac{G}{g}\right)^2, \tag{8.41}$$

that is, it is determined by the filling factor g/G of the fractal structure under consideration, i.e. by the fraction of remaining matter in the fractal construction algorithm. Amplitude scaling using the above factor provides more perfect fitting of spectra for different generations (Fig. 8.16).

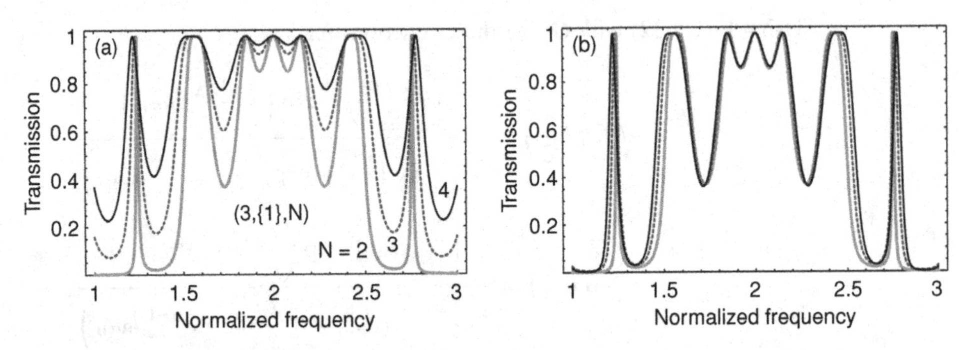

Fig. 8.16 **Exact comparison of the scaled spectra for triadic Cantor structures: (a) unchanged, (b) raised to a power γ [29].**

Recurrent relations for triadic Cantor structures Eqs. (8.31) and (8.32) can be generalized for all self-similar multilayer structures by means of the multiple-reflection formalism presented by Mandatori *et al.* [32]. Based on such generalization, Zhukovsky proved analytically scalability for every family of self-similar multilayer structures in the form of Eq. (8.40) [31].

Splitting of transmission bands in symmetrical fractal filters

Symmetrical fractal filters similar to those shown in Figure 8.10 exhibit splitting of the resonant transmission bands with the multiplicity growing with generation number. Narrow transmission bands result from localization of light-wave energy in certain cavities presenting in the filter as defects with respect to periodic structure. Looking at the junior triadic and pentadic Cantor filters in Figure 8.10, one can see for the second generation the triadic filter can be treated as a cavity BBB in the center surrounded by a pair of "mirrors" formed by a triple ABA-structure, whereas the pentadic filter contains two BBBBB cavities separated by the three ABABA multilayer mirrors. For higher generations, the number of existing cavities grows by the factor of $g \equiv (G+1)/2$, i.e. it doubles for triadic and triples for pentadic structures. Transmission resonances split accordingly. This is a manifestation of the general behavior of coupled oscillators. It can be traced in classical mechanics for coupled pendulums and in radiophysics for coupled *LC*-circuits. Optically coupled cavities exhibit the same behavior, resulting from interference of waves confined in identical cavities coupled via a finite size separation of a certain length that favors coupling of waves in distant cavities. The relevant quantum mechanical analog is splitting of particle energy levels in coupled quantum wells.

Multiple splitting is illustrated in Figures 8.17 and 8.18. Second-order triadic Cantor filters exhibit a transmission band at normalized frequency $\omega/\omega_0 = 0.78$. The third order has an emerging new transmission band near $\omega/\omega_0 = 0.705$ and the double-splitting near $\omega/\omega_0 = 0.78$. The corresponding field profiles clearly show that the new resonant peak corresponds to a new cavity whereas the split band corresponds to localization of light in two symmetric cavities. The new generation brings a quarterly split band at $\omega/\omega_0 = 0.78$,

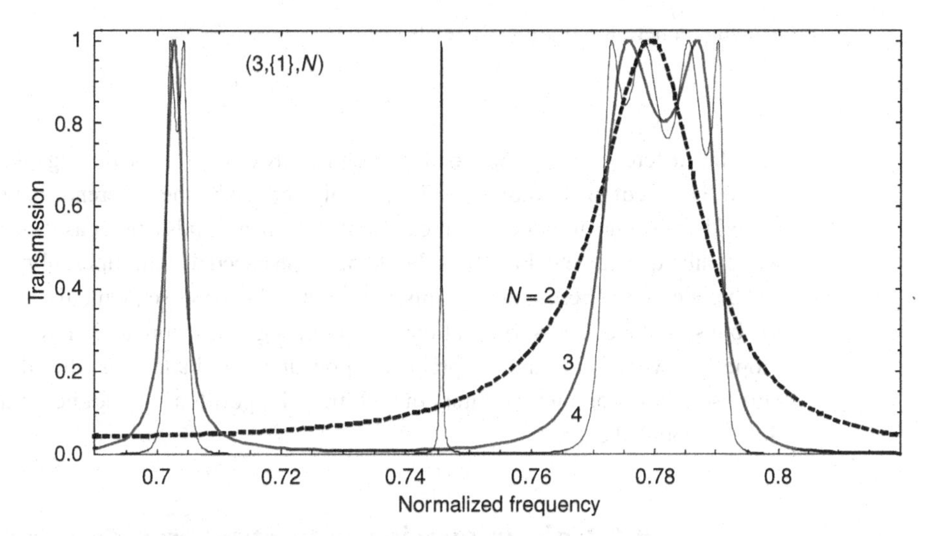

Fig. 8.17 **A portion of the frequency scale with transmission bands of triadic Cantor filters of second, third and fourth generations [28].**

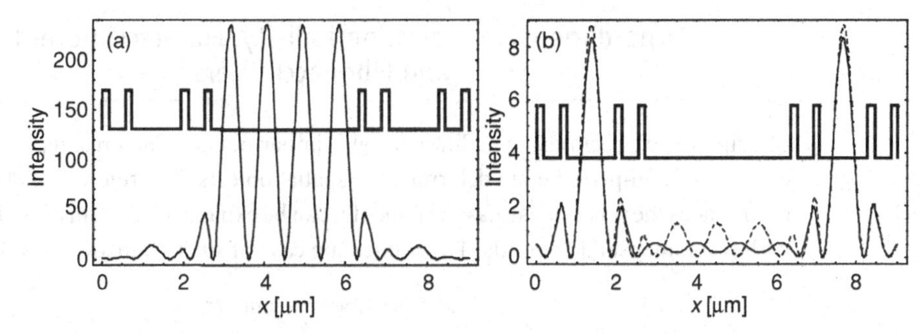

Fig. 8.18 **Electric field profiles superimposed with the refraction index profile for a third-order Cantor filter [28]. (a) Normalized frequency is 0.705; (b) normalized frequency is 0.775 and 0.785.**

double splitting at $\omega/\omega_0 = 0.705$ and the new emerging single band at $\omega/\omega_0 = 0.745$. A further generation (containing $3^5 = 243$ layers) gives rise to eight-fold splitting of the original resonance (not presented in the figure).

The following regularities can be seen. First, every new resonant band appears to be sharper than the previous ones. This is because of a higher Q-factor of resonance resulting from larger numbers of layers confining light in a given cavity. Second, every new resonant band has the splitting factor,

$$S_{\text{new}} \equiv G - g = \frac{G-1}{2}. \tag{8.42}$$

We have seen for triadic structures this splitting is absent ($S_{\text{new}} = 1$) whereas for pentadic $S_{\text{new}} = 2$, i.e. every new transmission band appears in the form of a doublet. For semiadic structures triplets will appear etc. Third, sequential splitting of the existing bands with growing generation number N reads,

$$S_{\text{old}} \equiv g = \frac{G+1}{2}. \tag{8.43}$$

Note the difference $S_{\text{old}} - S_{\text{new}} \equiv 1$, independently of G. This is nothing else but a straightforward result of the Cantor algorithm, simply defined by the difference between neighboring odd and even numbers. This means that if the new bands appear as triplets, older bands will exhibit quartets etc. Finally, it should be emphasized that multiple splitting occurs without broadening of resonant transmission bands. The band sequentially goes wider at the top but simultaneously shrinks at the bottom thus getting more and more rectangular. This property is very important for practical applications in band-pass filter design. It is used purposefully in commercial filters though fractal algorithms are not necessarily applied to develop coupled cavities.

8.5 Light in quasiperiodic structures: Fibonacci and Penrose structures

One-dimensional quasiperiodicity: Fibonacci potentials and Fibonacci filters

Quasiperiodic structures, in addition to self-similar fractal structures, represent yet another example of non-periodic but deterministic spatial objects. The rigorous mathematical definition states the function is quasiperiodic if it can be expanded in a sum of periodic functions with incommensurate periods. The immediate case of such a function is, for example,

$$f(x) = \sin(x) + \sin(x/\pi). \tag{8.44}$$

This is plotted in Figure 8.19 to give an idea of how quasiperiodic functions can look. For multilayer structures, the *Fibonacci sequence* offers an algorithm to develop quasiperiodic lattices in one dimension.

The Fibonacci sequence is based on the algorithm obeying regularities of Fibonacci numbers F_N.[2] The Fibonacci numbers form the following set. The first two numbers are $F_0 = 1$ and $F_1 = 1$. Then every next number is equal to the sum of the two previous numbers, $F_N = F_{N-2} + F_{N-1}$:

$$1, 1, 2, 3, 5, 8, 13, 21, 34, \ldots. \tag{8.45}$$

The important property of the Fibonacci sequence is the limit of the ratio of the two neighboring terms which is referred to as the "golden mean",

$$\lim_{N\to\infty} \frac{F_N}{F_{N-1}} = \frac{\sqrt{5}-1}{2} = 0.6180339887\ldots \tag{8.46}$$

[2] Leonardo Pisano Fibonacci (1180–1240) was an Italian mathematician.

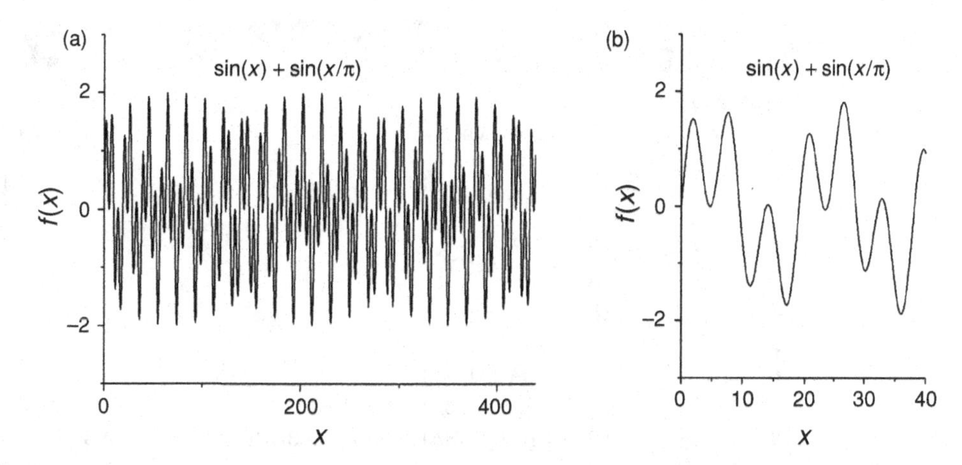

Fig. 8.19 An example of a quasiperiodic function (8.44) plotted for (a) larger and (b) smaller range in the x scale.

and expressed as an irrational number. Geometrically this corresponds to the case when, for two segments with lengths a and b the relation holds,

$$\frac{a}{b} = \frac{b}{a+b}. \tag{8.47}$$

When solving Eq. (8.47) one arrives at,

$$\tau = \frac{a}{b} = \frac{\sqrt{5}-1}{2}. \tag{8.48}$$

A multilayer Fibonacci sequence is built as follows. A layer of material A and another layer of material B form the first two generations. Then every next term in the set represents the sum of the two previous terms, i.e.

$$\text{A, B, AB, BAB, ABBAB,} \ldots \tag{8.49}$$

This is shown in Table 8.1. The total number of layers in every sequence forms the Fibonacci set (8.45).

A one-dimensional Schrödinger equation has been examined with a stepwise quasiperiodic potential whose value is equal to either of the two fixed values sequentially in accordance with the Fibonacci set [33–36]. For such a potential, an electron was found to have the *critical wave function* with strong spatial fluctuations. Such a function, being principally extended does not look like a Bloch wave and, for example, does not allow the notion of an envelope function to be used, which has been most fruitful in the study of both extended and localized states. An example of such a function is presented in Figure 8.20.

This type of wave function correlates with the properties of the energy spectrum. Typically, extended wave functions are indicative of states belonging to continuous bands, whereas localized wave functions with exponential decay describe states belonging to the discrete portion of energy spectra. In a quasiperiodic potential, a quantum particle has

Table 8.1. Fibonacci multilayer structures for 0-8 generations

Generation number	Structure sequence	Total number of layers
0	A	1
1	B	1
2	AB	2
3	BAB	3
4	ABBAB	5
5	BABABBAB	8
6	ABBABBABABBAB	13
7	BABABBABABBABBABABBAB	21
8	ABBABBABABBABBABABBABABBABBABABBAB	34

Note. For clarity, a part of every $(N+1)$-th sequence corresponding to the $(N-1)$-th generation is underlined.

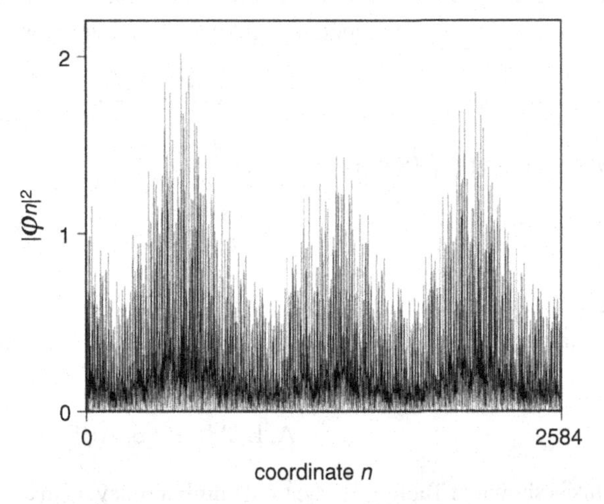

Fig. 8.20 Electron charge profile for a Fibonacci stepwise potential in a finite slab containing $F_{17} = 2584$ cells [36]. The graph presents the probability of finding an electron at the site with number n.

extended singular wave function. Such states belong neither to the discrete nor to the continuous portion of the energy spectrum. The energy spectrum for a quasiperiodic potential cannot be characterized in terms of discrete or continuous portions of the spectrum. Instead, energy values form a Cantor set with an infinite number of discrete values but without continuous bands developed. The spectrum as a whole can be described as singular continuous. The peculiar feature of the problem is the extended character of the wave function which means that electron transport is possible in spite of the absence of a continuous energy spectrum.

Unusual electron properties in quasiperiodic potentials have stimulated extensive research of the optical counterparts. The propagation and localization of light waves in

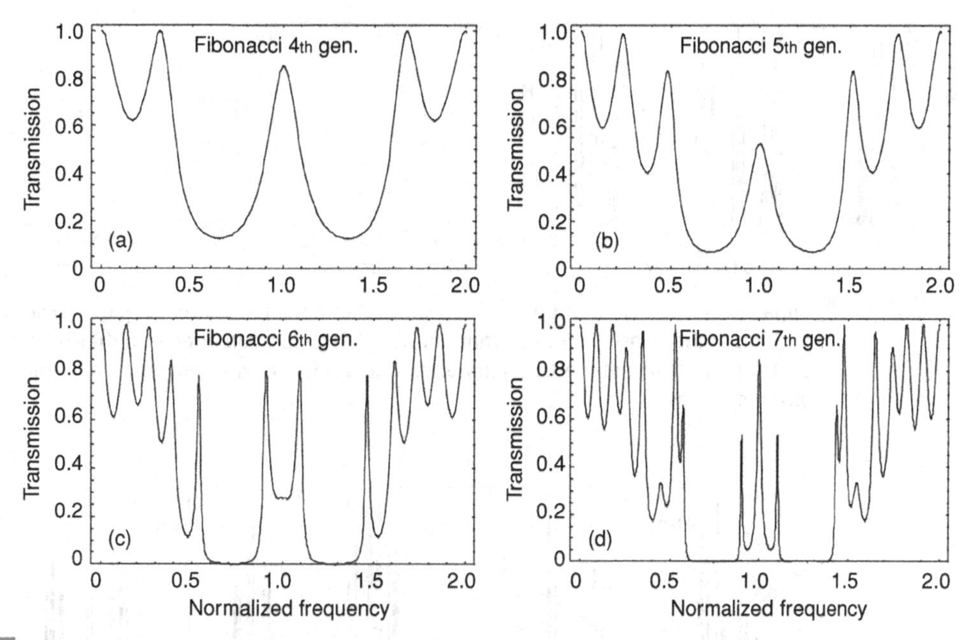

Fig. 8.21 **Calculated transmission spectra of quarter-wave Fibonacci filters of (a) 4th, (b) 5th, (c) 6th and (d) 7th generations [31].**

Fibonacci optical filters, i.e. multilayer structures organized according to the Fibonacci algorithm have become the subject of thorough analysis by many authors. The experimental performance of Fibonacci structures and their studies have confirmed that, similar to other types of potentials, the quasiperiodic ones give rise to classical counterparts with peculiar features found for the quantum mechanical Schrödinger problem.

Spectral regularities inherent in Fibonacci structures have been derived in analytic form based on the transfer matrix approach and relevant algebraic considerations [37–39]. Because of the restricted scope of the book these cumbersome derivations are not reproduced here and only the numerical results in visual form will be used to illustrate the basic optical properties of quasiperiodic structures.

Junior generations of Fibonacci filters feature a transmission band in the center which first splits into two and then into three sub-bands (Fig. 8.21). This central triplet is indicative for the Cantor triadic set and for higher generations definite self-similar and scaling features develop which are inherent in fractals. Self-similarity is shown in Figure 8.22. Different portions of transmission spectra for the same high-order generation of Fibonacci filter do exhibit similar spectral shape which becomes apparent when using the 5.11-fold stretch of the frequency axis. Self-similarity is apparent already for the 10th generation. For higher generations it becomes more and more pronounced.

Transmission spectra of Fibonacci filters also show spectral scalability which means that spectra of different generations have a similar shape when the frequency axis is properly scaled. This property is shown in Figure 8.23. Notably, the scaling factor 5.11 remains the same as in Figure 8.23, used to demonstrate internal self-similarity of the spectrum.

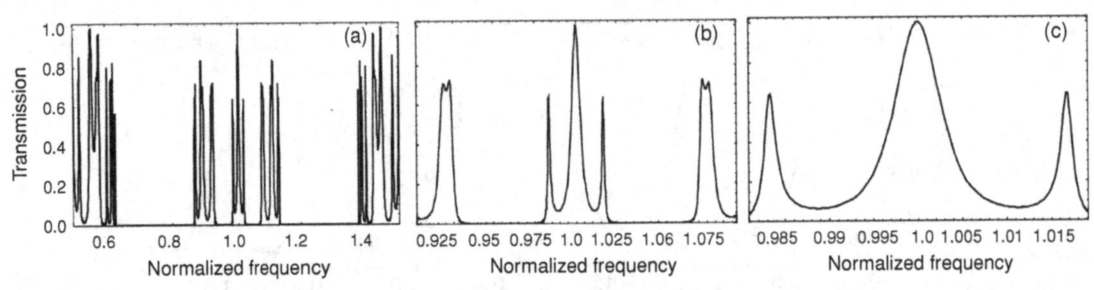

Fig. 8.22 Apparent self-similarity of transmission spectra for a Fibonacci filter [31]. (a)–(c) show the central portions of the transmission spectrum for the 10th generation filter with (b) single and (c) double 5.11-fold stretching of the frequency axis. Note a self-similar ternary pattern of the spectrum at different scales.

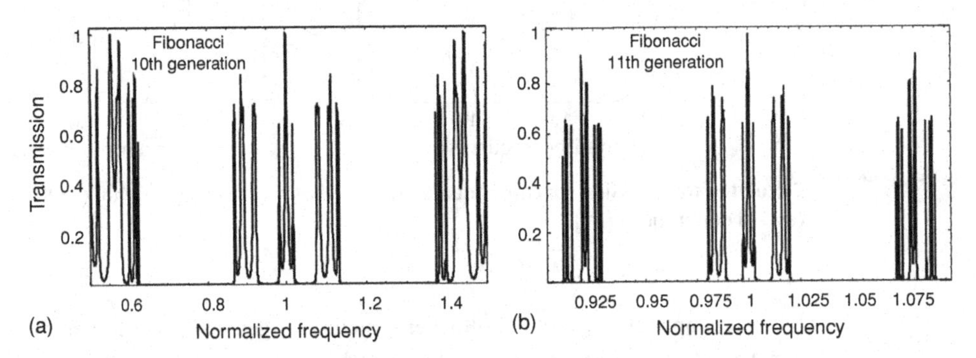

Fig. 8.23 Scalability of transmission spectra of Fibonacci multilayer structures [31]. (a) The central portion of the spectrum for the 10th generation; (b) the central portion of the spectrum for the 11th generation with 5.11-fold stretching of the frequency axis.

Refractive indexes of constituent materials define the absolute value of the scaling factor describing spectral self-similarity and scalability.

Notably, transmission spectra of generic Fibonacci structures do not exhibit perfect transmission bands. Many among well-resolved peaks have a transmission coefficient much lower than unity. As was emphasized in Section 8.4 when discussing Figure 8.14, this is rather a property of asymmetric multilayer structures than a property inherent in a certain algorithm of the filter design. Huang *et al.* [40, 41] considered in detail spectral properties of symmetric structures consisting of the two counterwise Fibonacci filters. They found that the symmetric structures do exhibit unity transmission in every peak along with remaining self-similarity of the spectrum.

Unlike fractal multilayer structures, quasiperiodic structures should contain larger numbers of layers (higher generation), for their scaling properties to manifest themselves noticeably. Gellerman *et al.* reported on the experimental realization of a multilayer Fibonacci filter up to the ninth order (55 layers) by means of thin film vacuum deposition using SiO_2/TiO_2 dielectric layers [42], with a reasonable agreement of the scalable transmission spectra with calculated ones.

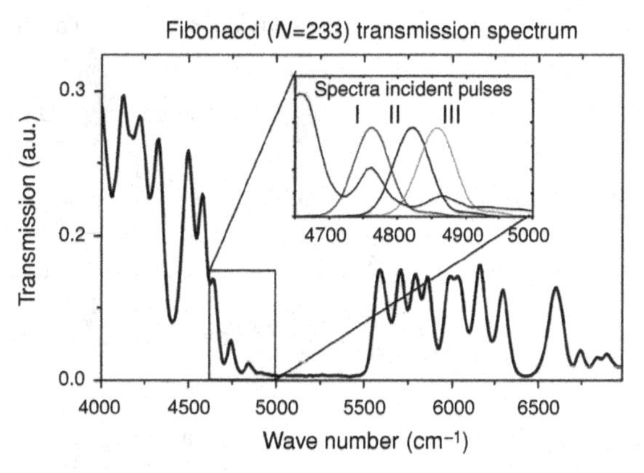

Fig. 8.24 **Measured transmission spectrum of a 12th order Fibonacci ($N = 233$) optical filter [43]. Around wave numbers 5000–5500 cm^{-1} the system exhibits a pseudo gap. The inset shows three examples of the power spectrum of the incoming laser pulses in a time-resolved experiment reported in Figure 8.25. The spectrum was recorded with a standard absorption spectrometer, hence the transmission peaks are broadened due to the angular spread of the incoming light on the sample, and lateral sample inhomogeneities.**

Laser pulse shaping with Fibonacci filters

An interesting approach to fabrication of one-dimensional quasiperiodic structures of higher generations has been proposed by Dal Negro *et al.* [43]. The technique is based on electrochemical etching of silicon to get randomly arranged porous silicon whose porosity is defined by the electrical current value. Finite porosity in turn defines a modified refractive index of the composite effective medium silicon/air. Therefore, an alternating current gives rise to alternating refractive index, thus providing a route towards the manifold regular alteration of refractive index along the current flow. This approach was first proposed for dielectric Bragg mirrors and planar microcavities fabrication by Pavesi *et al.* in 1995 [44]. In this way, a 12th-order Fibonacci filter has been developed consisting of 233 layers exhibiting transmission bands and gaps in the infrared (Fig. 8.24).

These Fibonacci filters were found to modify significantly the temporal shape of the ultrashort laser pulses (Fig. 8.25). In agreement with theoretical predictions, pulse stretching and oscillations were observed. When the incoming pulse is resonant with one transmission peak, the pulse is significantly delayed and stretched. This stretching becomes surprisingly strong close to the band edge (curve IIIa). In addition to the delay and stretching, when the spectrum of the laser pulse overlaps with two adjacent narrow transmission modes a strongly oscillatory behavior is observed (curve IIa). These oscillations can be interpreted as due to beating between individual band-edge modes. Indeed the frequency of the oscillations corresponds to the frequency difference between the peaks in the transmission spectrum.

Another example of the application of Fibonacci filters is laser pulse compression which becomes more efficient in quasiperiodic structures as compared to periodic ones because of enhanced dispersion of group velocity. Akhmanov *et al.* [45] proposed to use a linear dispersive medium for optical pulse compression provided that the pulse phase is modulated.

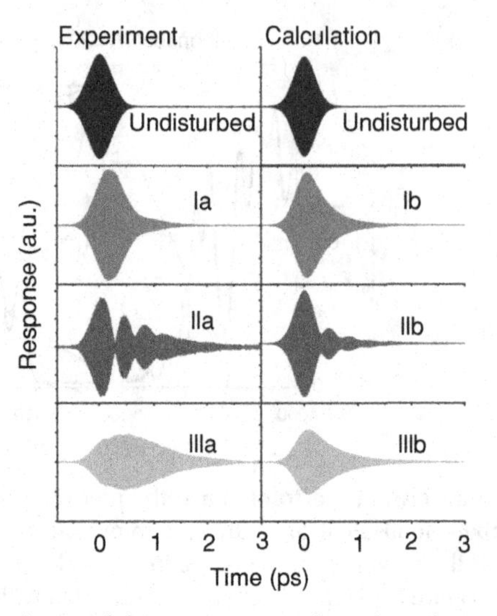

Fig. 8.25 **Experimental data and calculation of the transmission through a Fibonacci filter at four different frequencies [43]. The reference pulse is also plotted for comparison. The time offset corresponding to the total optical thickness of the sample has been subtracted in all cases. When the laser pulse is resonant with one band-edge state the transmitted intensity is strongly delayed and stretched. When two band-edge states are excited, mode beating is observed.**

Consider a phase-modulated laser pulse of Gaussian type,

$$E(t)|_{x=0} = A \exp\left(-\frac{1}{2}(\tau_0^{-2} + i\alpha_0)t^2\right) \exp(i\omega_0 t), \tag{8.50}$$

where τ_0 is the characteristic duration of the pulse (a pulse that is not a transform-limited pulse and thus can be compressed), α_0 is the rate of frequency modulation, which is commonly referred to as a *chirp*,[3] and ω_0 is the central frequency. In the slowly varying envelope approximation with second-order dispersion taken into account, the spatial evolution of the pulse duration reads,

$$\tau(x) = \tau_0 \sqrt{(1 - \alpha_0 \kappa_2 x)^2 + \left(\frac{\kappa_2 x}{\tau_0^2}\right)^2}, \tag{8.51}$$

where $\kappa_2 = \partial^2\kappa/\partial\omega^2$ is the dispersion of the group velocity, $v_g = \partial\omega/\partial\kappa$. In the case $\alpha_0\kappa_2 < 0$ the chirped light pulse is stretched while propagating through a structure, whereas in the case $\alpha_0\kappa_2 > 0$ the chirped pulse experiences compression upon propagation down to the minimal pulse duration, and then increases again. The minimal duration of such a phase-modulated Gaussian light pulse reads [45],

$$\tau_{\min} = \frac{\tau_0}{\sqrt{1 + \left(\alpha_0 \tau_0^2\right)^2}}. \tag{8.52}$$

[3] The term “chirp”, according to the *Encyclopaedia Britannica*, describes the ‘characteristic short sharp sound specially of a small bird or insect’ and implies that the frequency of the sound varies during the call.

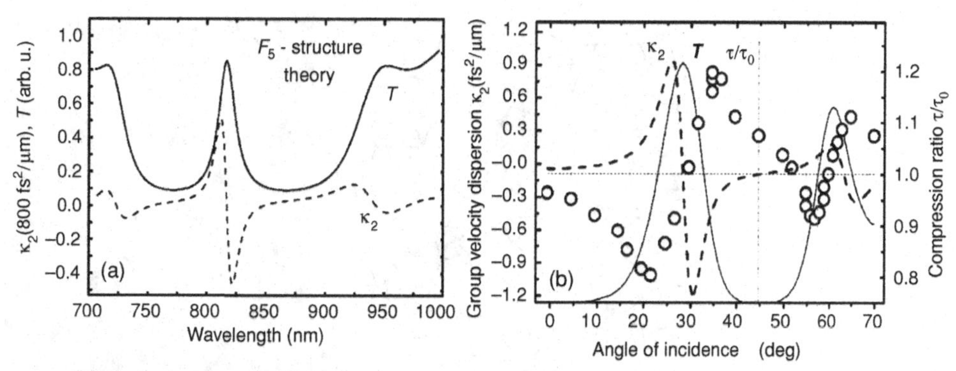

Fig. 8.26 **Application of a Fibonacci filter for chirped pulse compression (adapted from [46]). (a) Calculated transmission T and group velocity dispersion κ_2 for a 5th order Fibonacci ZnS/Na_3AlF_6 filter. (b) Measured compression ratio τ/τ_0 (circles) superimposed with calculated group velocity dispersion κ_2 and transmission T (not scaled) for that filter as a function of the angle of incidence. Compression corresponds to $\tau/\tau_0 < 0$.**

For this case of maximum compression, the pulse becomes transform limited and its duration $\tau_{\min}$ is related to the spectral width $\Delta\omega$ as $\tau_{\min}\Delta\omega = 0.44$ for a Gaussian shape pulse. The duration $\tau_{\min}$ occurs at a distance,

$$L_{\text{compr}} = \frac{\left(\alpha_0\tau_0^2\right)^2}{\alpha_0\kappa_2[1+\left(\alpha_0\tau_0^2\right)^2]}. \tag{8.53}$$

Experiments on the application of Fibonacci filters for chirped laser pulses have been reported by Makarava *et al.* [46]. Strong group velocity dispersion develops in the middle of the gap where the resonant transmission peak occurs (Fig. 8.26). Accordingly, pulse compression is observed whenever $\kappa_2 > 0$, whereas for $\kappa_2 < 0$ pulse expansion is the case.

Two-dimensional quasiperiodicity: Penrose quasicrystals

To understand the key properties of aperiodic tilings the intuitive concepts of order and symmetry need to be suitably extended beyond some of the traditional views related to the concept of periodicity.

Enrique Maciá [47]

Two-dimensional quasiperiodic structures exist in the form of tilings, consisting of polygons covering a plane without gaps or overlaps in such a way that the resulting overall pattern lacks any translational symmetry. The existence of aperiodic sets of polygons capable of tiling a plane in an aperiodic fashion was proved in the 1960s. A representative 2D-quasiperiodic tiling proposed by R. Penrose in 1974 is shown in Figure 8.27. It can be obtained via the projection of a five-dimensional periodic lattice onto a planar surface. It is the basic feature of quasiperiodic structures that they can be seen as projections of periodic structures in space of higher dimensionalities [49, 50]. In quasiperiodic tilings,

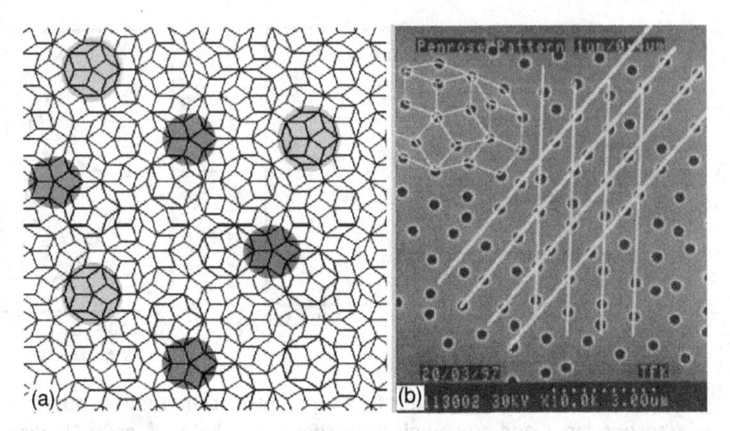

Fig. 8.27 (a) Penrose tiling illustrating the local isomorphism characteristic of Conway's theorem (adapted from [47]) and (b) its photonic crystal implementation [48]. Light gray and dark gray fields in the left panel indicate two types of repeated patterns. Copyright 1999 Elsevier Science Ltd. [47].

the notion of repetitiveness typical of translational symmetry is replaced by that of local isomorphism, which expresses the occurrence of any bounded region of the whole tiling infinitely, often across the tiling, irrespective of its size. For the particular case of *Penrose tiling*, the *Conway theorem* states that for any local pattern having a certain diameter, an identical pattern can be found within a distance of two diameters. This interesting statement is illustrated in Figure 8.27.

Penrose tiling features ten-fold rotational symmetry, a degree of symmetry that could not be achieved with any perfectly periodic structure. The highest degree of rotational symmetry achieved with a planar periodic structure is six-fold, inherent in a triangular or hexagonal close-packed lattice. It is important that higher-order rotational symmetry is favorable for omnidirectional photonic band gaps to develop. A two-dimensional photonic quasicrystal can be developed, e.g. based on Penrose tiling by putting air holes or vice versa dielectric rods in every point corresponding to intersections. Such a crystal fabricated by Krauss *et al.* is shown in Figure 8.27. Zoorob *et al.* fabricated a planar quasiperiodic structure featuring even higher, 12-fold rotational symmetry by means of an appropriate arrangement of etched holes in silicon nitride ($n = 2.02$) and silicon oxide ($n = 1.45$) crystalline material [51]. In spite of relatively low refractive index contrast, an omnidirectional two-dimensional photonic band gap was observed. Note that for an omnidirectional gap to develop in a typical periodic planar structure much higher refractive index contrast (>3) is mandatory, such as air/silicon or air/GaAs cases (see Chapter 7 for properties of periodic structures.)

8.6 Surface states in optics: analog to quantum Tamm states

In 1932, very soon after publication of the pioneering paper by Kronig and Penney on electron properties in a periodic stepwise potential [52], I. E. Tamm considered the case

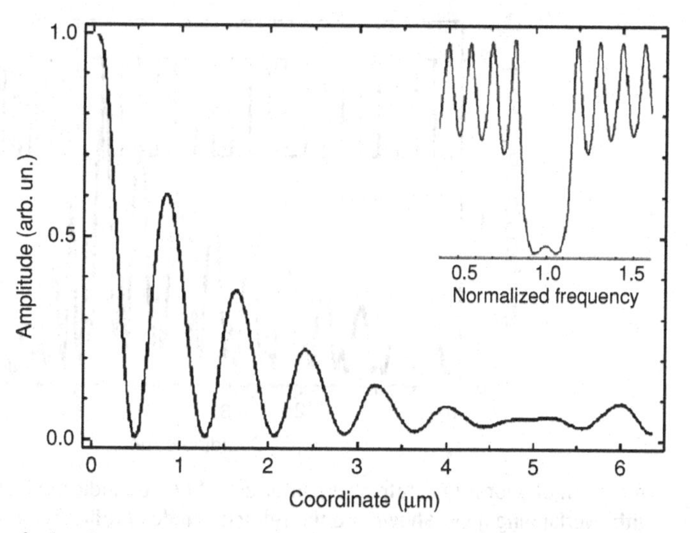

Fig. 8.28 **Development of a localized state within the band gap of a seven-period quarter-wave multilayer structure bordering at 6 μm a dielectric medium with very high dielectric constant [55]. The plot shows the electric field amplitude profile at the mid-gap frequency (normalized frequency = 1). The insert shows the central portion of the transmission spectrum.**

of a high-potential barrier at the edge of a periodic potential [53]. This case corresponds to the real situation at a crystal surface where finite energy is necessary for an electron to leave a crystal for the vacuum. He found there is a state of an electron which develops at the border whose wave function exponentially vanishes outside a crystal and the energy eigenvalue lies inside the band-gap region. These surface states in quantum mechanics are referred to as Tamm states.

Tamm states result from the analysis of a single-particle Schrödinger equation and therefore have an electromagnetic analog, provided the relevant dielectric interface is developed, as discussed in Chapter 3 for a number of model problems. Probably, Kossel was the first to suggest that localized states could exist in optics near the boundary between a homogeneous and a layered medium [54].

The formation of the optical analog of the Tamm state is shown in Figure 8.28. At the border of a seven-period quarter-wave multilayer with very highly-refractive adjacent medium (near 6 μm), a pronounced increase in the field amplitude is apparent. In the transmission spectrum (insert in Figure 8.28) this state manifests itself as enhanced transmission in the centre of the band gap. It is instructive to compare this figure with similar data for a periodic structure in Chapter 7 (Figures 7.5 and 7.7).

Another type of surface state in optics is presented in Figure 8.29. It is as interface mode which develops at the boundary between two periodic dielectric structures, one having a period close to the wavelength of light and the other close to double the wavelength. The interface state features the electric field distribution decaying exponentially in the surrounding media. The frequency of this state lies inside the overlapping gaps (stop-bands, reflections bands) of the two periodic media. The state manifests itself as a sharp transmission peak within the reflection bands of both periodic structures. Interestingly, the

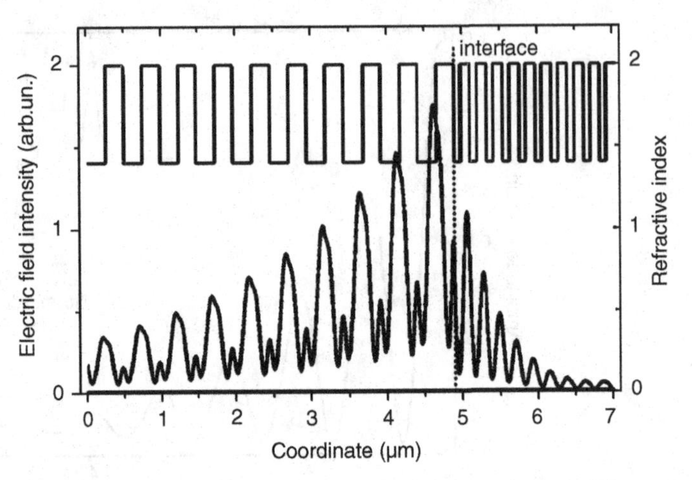

Fig. 8.29 **Formation of a localized state at the interface of two periodic media with different periods but with overlapping gaps. Shown are the refractive index profile (upper curve) and electric field profile (lower curve). Adapted from [56].**

in-plane dispersion law for this interface state is a continuous parabolic $\omega(k)$ law, which by means of the substitutions $E = \hbar\omega$, $p = \hbar k$, leads to a parabolic $E(p)$ relation inherent in a particle with mass equal to approximately 10^{-5} of the free electron mass [56]. In such a situation, it is typical to ascribe the appropriate effective mass to a photon with $E = \hbar\omega$, $p = \hbar k$. We must bear in mind, however, that no photon has been introduced into consideration so far, and introducing photons into this problem is not justified. One should rather speak about the mass-like behavior of electromagnetic waves in a complex medium. A discussion in terms of the mass-like dispersion law as an indication of light energy storing is rather instructive with possible application of Einstein's $E = mc^2$ relation to estimate the energy stored. Such a consideration of parabolic dispersion laws in the electrodynamics of complex media belongs to Rivlin [57].

Generally, the properties of optical Tamm states are only at the initial stage of theoretical understanding and experimental investigation. These states of light will be the subject of close consideration in the near future. Surface and interface modes can be purposefully used in various practical devices and components as desirable lossless modes for light generation, emission and propagation.

8.7 General constraints on wave propagation in multilayer structures: transmission bands, phase time, density of modes and energy localization

Consider a one-dimensional dielectric multilayer structure consisting of N sequential layers along the x axis, each having a thickness d_j and a refractive index n_j, infinite in the transverse directions in the y–z plane and surrounded on both sides by free space with

$n_0 = 1$. Consider a normally incident plane monochromatic wave propagating through such a structure. Our goal is to evaluate the general constraints on light propagation in terms of certain conserving quantities that are invariant with respect to the spatial arrangement of layers in the structure under consideration [58]. We deal with the one-dimensional Helmholtz equation,

$$\frac{\partial^2}{\partial x^2}E(x) + [n(x)]^2\frac{\omega^2}{c^2}E(x) = 0, \tag{8.54}$$

to evaluate the transmission $T(\omega)$ and reflection $R(\omega)$ functions which are complex, taking into account the phase shift a wave acquires in the structure in question.

Transmission spectra and phase shifts

Assume all the layers have parameters such that the optical path $n_j d_j$ is the same for any j, so that,

$$n_1 d_1 = n_2 d_2 = \cdots = n_j d_j = \cdots = n_N d_N \equiv \frac{\pi c}{2\omega_0}, \tag{8.55}$$

where ω_0 is defined as the *central frequency*. So far such structures have been referred to as *quarter-wave structures* (see, e.g. Eq. (8.30) in this chapter, and Eq. (7.12) in Chapter 7).

For any even multiple of ω_0 the propagating wave passes each constituent layer without reflection and therefore gains the phase shift,

$$\Delta\varphi = \frac{\omega}{c}n_j d_j, \tag{8.56}$$

which is the same for all layers in view of the quarter-wave condition (Eq. 8.55). As a result, the structure becomes fully transparent ($|T(2m\omega_0)| = 1, m = 1, 2, 3\ldots$) for any value of number of layers, N and for any arrangement of constituent layers, the total phase shift being a sum of the shifts for all the layers. That is, the transmission reads as follows,

$$T(\omega)|_{\omega=2m\omega_0} = \exp\left(i\frac{\omega}{c}\sum_{j=1}^{N} n_j d_j\right) = \exp(iNm\pi). \tag{8.57}$$

Equation (8.57) defines equidistant frequency points where the phase and the wave number are linear functions of ω independent of the arrangement of the layers. The wave number and frequency are related as,

$$k_m(\omega_m) = k(2m\omega_0) = \frac{Nm\pi}{d^*} = \frac{N\lambda_0}{2d^*}\frac{\omega_m}{c}, \tag{8.58}$$

with $d^* \equiv \sum_{j=1}^{N} d_j$ being the total length of the structure under consideration. One can see Eq. (8.58) describes a linear wave number versus frequency function as is the case for a homogeneous medium.

Quarter-wave multilayer structures possess periodic transmission spectra,

$$T(\omega + 2m\omega_0) = T(\omega), \quad m = 1, 2, \ldots \tag{8.59}$$

Moreover, within every period, the transmission spectrum is symmetrical with respect to every odd multiple of ω_0 , i.e. $\omega = (2m-1)\omega_0$ $(m = 1, 2, \ldots)$ points. These properties have been discussed in Chapter 7 (Section 7.2) for quarter-wave *periodic* multilayer structures, but they are generally inherent in every quarter-wave structure of arbitrary geometry. Symmetry can be expressed as,

$$T(m\omega_0 + \delta\omega) = T(m\omega_0 - \delta\omega), \quad 0 < \delta\omega < \omega_0. \tag{8.60}$$

Spectral periodicity and symmetry of quarter-wave multilayer structures are direct consequences of the formulas for transmission and reflection of a single dielectric layer (Chapter 3, Eq. (3.21)).

Let us reduce further consideration for visual clarity to binary structures consisting of the two types of materials A and B with refractive indexes n_A and n_B embedded in an ambient medium B. Figure 8.30 shows transmission spectra calculated for a number of nine-layer structures with arbitrary alteration of the A and B layer. Let us consider it in more detail. In case (a) a nine-layer structure of material A exhibits just equidistant Fabry–Perot modes in the transmission spectrum. It is a trivial case of transmittivity of a dielectric layer (see Chapter 3, Figs. 3.7, 3.8 and Eq. (3.42)). Case (b) shows a triple structure with corresponding triple Fabry–Perot modes in the transmission spectrum. Case (c) presents the familiar periodic structure, a one-dimensional photonic crystal or a multilayer dielectric Bragg mirror. This has been discussed in detail in Chapter 7 (Section 7.2 and Fig. 7.7). The transmission spectrum has a stop-band (reflection band) around ω_0 with (nearly) unity reflection and (nearly) zero transmission, surrounded by two symmetrical transmission bands each containing distinct transmission peaks. Panels (d), (e) and (f) present variations to the periodic structure which can be treated as (d) a defect (extra B in the middle) inside a periodic structure, (e) two spaced defects (extra As) in a periodic structure and (f) single defect (B instead of A in the middle) forming a long BBB cavity surrounded by a couple of ABA stacks forming mirrors. Accordingly, with respect to the periodic arrangement, one or two sharp transmission bands develop inside the stop band, with other transmission peaks being shifted and reorganized. Notably, every structure contains exactly nine transmission peaks within the period $[0, 2\omega_0]$ (the two half-bands in every spectrum at $\omega = 0$ and $\omega = 2\omega_0$ together give a single peak). The only exclusion is case (d) where the very last layer B merges with the ambient medium thus reducing the actual number of layers in the structure in question to eight. One can consider the total number of transmission peaks discussed for fractal structures, where within the range $[0, 2\omega_0]$ the number of peaks was always equal to the number of constituent elementary layers (see Figs. 8.11–8.14). Such "number of peaks conservation" behavior leads to the straightforward analog of energy level splitting in a sequence of identical quantum wells with potential barriers in between. A particular case of a periodic arrangement of wells has been discussed in Chapter 3 (Section 3.5). This analogy is essentially the result of Helmholtz and Schrödinger equation similarities, which were thoroughly examined in Chapter 3.

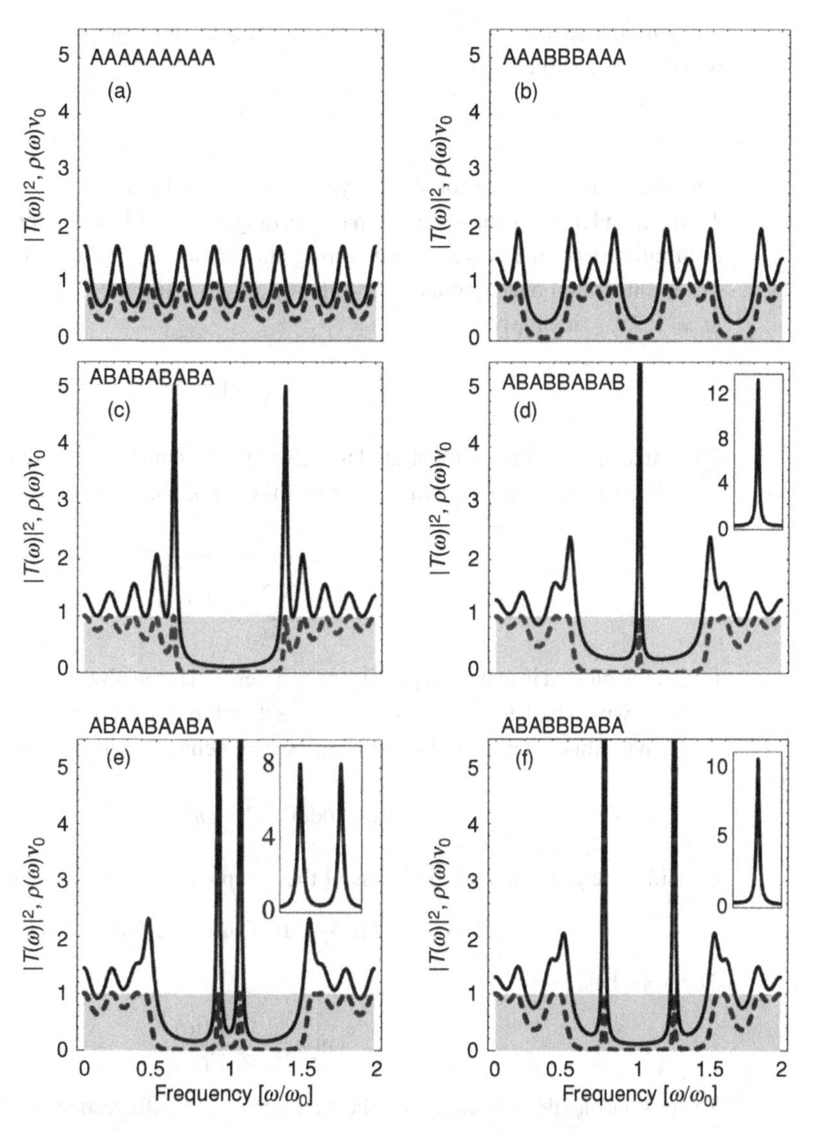

Fig. 8.30 **Spectral properties of binary quarter-wave nine-layer stacks with various arrangements of A and B layers within the spectral range $[0, 2\omega_0]$. Dashed lines show transmission spectra. Solid line shows $\rho(\omega)v_0$ function defined by Eqs. (8.60), (8.63). The square under this function equals the square of the gray band with height 1 and length $2\omega_0$. Adapted from [58].**

The conservation law

The apparent number of transmission peaks conserved is believed to indicate a certain integral relationship governing the limits and restrictions of modification of electromagnetic wave propagation in an inhomogeneous medium. Such a relationship does exist and is most observable for quarter-wave multilayer stacks because of the existing periodicity in the transmission spectrum. Therefore, only the frequency range $[0, 2\omega_0]$ is to be examined. It was shown [58] the value

$$\rho(\omega) = \frac{1}{d^*}\frac{d\varphi}{d\omega}, \tag{8.61}$$

does integrally conserve for all N-layer stacks defined only by the total geometrical length D, independently of the specific spatial arrangement of layers. Here φ is the total phase accumulated by the wave packet during its transmission through the multilayer stack. When integrated over frequency within any single full period $[2m\omega_0, 2(m+1)\omega_0]$ where $m = 1, 2, \ldots$ it simply reads,

$$\int_{2m\omega_0}^{2(m+1)\omega_0} \rho(\omega)d\omega = \frac{\pi N}{d^*}. \tag{8.62}$$

This amazing property is illustrated in Figure 8.30 where the $\rho(\omega)$ function is presented for a number of multilayer structures. For visual clarity, the value $\rho(\omega)v_0$ with

$$v_0 \equiv \frac{d^*}{\sum_{j=1}^{N}(n_j d_j / c)} \tag{8.63}$$

is actually plotted in every graph. The parameter v_0 has the physical meaning of the maximal velocity, with which light can traverse the stack without reflections at the interlayer refraction steps. With this notation and dimensionless frequency $\tilde{\omega} = \omega/\omega_0$ Eq. (8.62) reads,

$$\int_{2m}^{2m+2} v_0 \rho(\tilde{\omega})d\tilde{\omega} = 2, \quad m = 0, 1, 2, \ldots \tag{8.64}$$

Consider the $\rho(\omega)$ function in terms of the complex transmission $T(\omega)$,

$$T(\omega) \equiv \mathrm{Re}T(\omega) + i\mathrm{Im}T(\omega) \equiv |T(\omega)| \exp(i\varphi). \tag{8.65}$$

Then one has,

$$\tan\varphi = \frac{\mathrm{Im}T(\omega)}{\mathrm{Re}T(\omega)}. \tag{8.66}$$

To find out $d\varphi/d\omega$ we shall take the first derivative with respect to ω from both sides of Eq. (8.66), i.e.

$$\frac{d}{d\omega}(\tan\varphi) = \frac{d}{d\omega}\left(\frac{\mathrm{Im}T(\omega)}{\mathrm{Re}T(\omega)}\right). \tag{8.67}$$

This gives,

$$\sec^2(\varphi)\frac{d\varphi}{d\omega} = \frac{[\mathrm{Im}T(\omega)]'\mathrm{Re}T(\omega) - \mathrm{Im}T(\omega)[\mathrm{Re}T(\omega)]'}{[\mathrm{Re}T(\omega)]^2}, \tag{8.68}$$

whence using the identity $\sec^2\varphi \equiv 1 + \tan^2\varphi$ and solving Eq. (8.68) with respect to $d\varphi/d\omega$, we finally arrive at the expression for $\rho(\omega)$,

$$\rho(\omega) = \frac{1}{d^*}\frac{d\varphi}{d\omega} = \frac{[\mathrm{Im}T(\omega)]'\mathrm{Re}T(\omega) - \mathrm{Im}T(\omega)[\mathrm{Re}T(\omega)]'}{d^*|T(\omega)|^2}, \tag{8.69}$$

with the derivation taken with respect to ω. Taking into account the symmetry of the transmission spectrum Eq. (8.60) one can write instead of Eq. (8.64) the relation,

$$\int_{m}^{m+1} v_0\rho(\tilde{\omega})\mathrm{d}\tilde{\omega} = 1, \quad m = 0, 1, 2, \ldots. \tag{8.70}$$

One can see that, indeed, for every case presented in Figure 8.30 positive alterations of $\rho(\tilde{\omega})v_0$ with respect to unity in certain frequency intervals are compensated by negative ones in the another intervals. The gray band indicates the square of 1×2 on the $\rho(\tilde{\omega})v_0 - \tilde{\omega}$ plane which appears to be equal to the total square under the curve $\rho(\tilde{\omega})v_0$ in every panel.

Does the above conservation law hold for other multilayer structures beyond quarter-wave stacks? Because there is inverse proportionality between ω_0 and the optical path of the constituent layers, then even if the quarter-wave condition (8.55) is broken, but the quantities $n_j d_j$ all remain commensurate, the same reasoning can be applied. Equations (8.62) and (8.70) can then be obtained by subdivision of the constituent layers, accompanied by the consequent increase in the central frequency $\omega_0 \to N\omega_0$. In the limiting case of mathematically incommensurate layers, N goes to infinity, and the structure appears to possess the same freedom as a continuously inhomogeneous medium would, retaining only the asymptotic relation,

$$\lim_{N\to\infty} \frac{1}{N}\int_{0}^{N\omega_0} \rho(\omega)\mathrm{d}\omega = \frac{\pi}{d^*}. \tag{8.71}$$

Phase time and traversal velocity

Let us now discuss in more detail the physical content of the observed conservation law. The $\rho(\omega)$ function, as defined by Eq. (8.61), is proportional to the value,

$$\tau = \frac{\mathrm{d}\varphi}{\mathrm{d}\omega}, \tag{8.72}$$

which has dimensions of time and is referred to as the *phase time, group delay or Wigner time*, since it was introduced by E. Wigner in 1955 when considering the problem of scattering in quantum mechanics [59]. This notion is extensively used when analyzing light traversal in tunneling processes (see Chapter 10, Section 10.2). Therefore the $\rho(\omega)v_0$ function plotted in Figure 8.30 for various structures gives a direct insight into the time of light flight through a given structure. A straightforward conclusion is apparent: *every peak in transmission remarkably coincides with a peak in Wigner time.* That is, *high transmission always occurs at the expense of slowing down in propagation.* For a complex medium, the traditional group velocity notion $v_g = \mathrm{d}\omega/dk$ can not be directly applied since the wave number k is undefined. However, based on the phase time notion Eq. (8.72), traversal velocity can be defined as,

$$v_\tau = \frac{d^*}{\tau} = \frac{1}{\rho(\omega)}, \tag{8.73}$$

and the $\rho(\omega)$ function appears to be inversely proportional to the traversal time. Therefore, the statement holds: *"total modification of phase time (and traversal velocity) is limited by the conservation law: lower times in one portion of the frequency are compensated by higher times in some other portions"*. For quarter-wave stacks, conservation occurs exactly within any range of frequency with the width $m\omega_0$, $m = 1, 2, 3, \ldots$. For multilayer stacks with commensurable layer thicknesses, this range is finite and is determined by the thickness scalability. For incommensurate stacks conservation occurs in the limit of infinite frequency range, and holds approximately for finite ranges becoming more accurate for wider ranges.

Notably, the conclusion on the slowing down of propagation in transmission peaks remarkably coincides with the $v_g = \mathrm{d}\omega/\mathrm{d}k$ velocity dependence on frequency derived by D'Aguanno *et al.* [60] and discussed in detail in Chapter 7 (Fig. 7.32) for periodic structures where the notion of wave number with respect to Bloch waves is possible.

Longer traversal time for transmission peaks means it is light energy concentration (storage, localization) which enables light to pass through a complex structure with minor (sometimes zero) reflection and high (sometimes unity) transmission.[4]

Density of modes conservation

The $\rho(\omega)$ function has straightforward relation with the *density of modes* $D(k)$ with respect to wave number k and $D(\omega)$ with respect to frequency ω

$$D(\omega) = D(k)\frac{\mathrm{d}k}{\mathrm{d}\omega} \tag{8.74}$$

considered in Chapter 2 (Section 2.2). For one-dimensional space, the density of modes D_1 with respect to wave number is simply the constant value (see Eq. (2.23) in Chapter 2)

$$D_1(k) = \frac{1}{\pi}, \tag{8.75}$$

whence the density of modes with respect to frequency ω is entirely determined by the dispersion law relating k and ω, namely

$$D_1(\omega) = \frac{1}{\pi}\frac{dk}{d\omega}. \tag{8.76}$$

Notably, this appears to be inversely proportional to the group velocity $v_g = \mathrm{d}\omega/\mathrm{d}k$, i.e.

$$D_1(\omega) = \frac{1}{\pi}\frac{\mathrm{d}k}{\mathrm{d}\omega} = \frac{1}{\pi}\frac{1}{v_g}. \tag{8.77}$$

This relation coupling group velocity and the one-dimensional density of electromagnetic modes is so amazing that sometimes $1/\pi$ is omitted and the one-dimensional density of

[4] Not rigorously but meaningfully, "sometimes" here means "for symmetric structures". For asymmetric structures the conservation law certainly holds but not every peak in transmission equals unity. Examples of transmission spectra for asymmetric structures can be found in Figure 8.14, and Figures 8.21–8.23.

modes is treated simply as v_g^{-1} (see, e.g. [26, 58, 61]). By no means should this relation be overemphasized, since it actually holds exclusively for one-dimensional space and proportionality to v_g^{-1} simply comes from the derivation rule for a complicated function, as is clearly seen from Eq. (8.74).

For a continuous medium with refractive index n where $k = \omega n/c$, the density of modes reads,

$$D_1(\omega) = \frac{n}{\pi c}. \tag{8.78}$$

For a periodic medium, the Bloch wave number is defined and the group velocity notion $v_g = \mathrm{d}\omega/\mathrm{d}k$ remains valid, although it no longer defines the energy velocity as has been discussed in detail in Chapter 7 (Section 7.12). The anti-correlation between transmission and group velocity in a periodic multilayer structure (Fig. 7.32 and [60]) along with the inverse proportionality of the density of modes to group velocity allows for the following statement to be formulated: *transmission maxima for a periodic multilayer structure correspond to the density of modes maxima and vice versa.*

In an arbitrary multilayer structure with no periodic arrangement implied, the notion of wave number can not be introduced as for a plane wave in a continuous medium or for a Bloch wave in a periodic medium. However, we can speak about a certain "efficient" wave number $\tilde{k}$, defining this based on the phase shift a wave acquires when passing through a finite stack as follows,

$$\tilde{k}(\omega) = \frac{\varphi(\omega)}{d^*} = \frac{\arg T(\omega)}{d^*}. \tag{8.79}$$

We can now introduce the "efficient" density of electromagnetic modes $\tilde{D}_1(\omega)$ defined as,

$$\tilde{D}_1(\omega) = \frac{1}{\pi}\frac{\mathrm{d}\tilde{k}}{\mathrm{d}\omega}, \tag{8.80}$$

which is essentially the *local* density of modes averaged over the finite stack in question.

With this notion we can write instead of Eq. (8.61) the relation,

$$\rho(\omega) = \frac{1}{(d^*)^2}\frac{\mathrm{d}\tilde{k}}{\mathrm{d}\omega} = \frac{\pi}{(d^*)^2}\tilde{D}_1(\omega). \tag{8.81}$$

Equation (8.81) together with Eq. (8.62) gives rise to the conservation law for the density of modes $\tilde{D}_1(\omega)$,

$$\int_{2m\omega_0}^{2(m+1)\omega_0} \tilde{D}_1(\omega)\mathrm{d}\omega = \frac{N}{d^*}. \tag{8.82}$$

Finally, the density of modes integrated over a stack under consideration simply reads,

$$\int_0^{d^*} \mathrm{d}x \int_{2m\omega_0}^{2(m+1)\omega_0} \tilde{D}_1(\omega)\mathrm{d}\omega = N, \tag{8.83}$$

i.e. *the number of modes in the frequency interval $2\omega_0$ simply equals the number of quarter-wave layers.* Note the same is valid for transmission peaks: the number of peaks does also equal the number of constituent layers. This coincidence of transmission peaks number, number of layers and modes number looks rather reasonable and intuitive. The above reasoning on extension of the conservation law from quarter-wave to arbitrary commensurate layers and incommensurate ones remains valid. Also valid is the statement on average deviation of the density of states from the vacuum value,

$$D_1^{\text{vac}}(\omega) = \frac{1}{\pi c}. \tag{8.84}$$

The total deviation equals zero when integrated over the period in the transmission spectrum, i.e.

$$\int_{2m\omega_0}^{2(m+1)\omega_0} [\tilde{D}_1(\omega) - D_1(\omega)]\mathrm{d}\omega = 0, \tag{8.85}$$

for quarter-wave stacks. For other stacks with commensurate thickness the frequency range of integration should be multiplied accordingly, whereas for arbitrary stacks integration over the whole frequency range should be performed, i.e.

$$\int_0^\infty [\tilde{D}_1(\omega) - D_1(\omega)]\mathrm{d}\omega = 0. \tag{8.86}$$

Running ahead, we emphasize that the latter statement holds in multi-dimensional spaces as well and can be formulated as follows: "*The density of electromagnetic modes can be redistributed over the frequency range but can not be totally modified.*" This statement is based on the Barnett–Loudon sum rule for radiative lifetimes modification [62] and will be evaluated in Chapter 13. Equation (8.85) is the one-dimensional precursor of this general law.

Summary and conclusion

For multilayer stacks with an arbitrary arrangements of layers, the conservation law for the density of modes holds, which states that total deviation of the mode density from the vacuum value equals zero when integrated over the whole frequency axis. For quarter-wave stacks, the transmission spectrum and density of mode spectrum are periodic with period $2\omega_0$ and symmetric with respect to points $\omega_m = m\omega_0, m = 1, 2, \ldots$ where ω_0 is defined by Eq. (8.55). Therefore, for quarter-wave structures the above conservation law holds for integration within any of the ranges of $[m\omega_0, (m+1)\omega_0]$. For quarter-wave stacks the number of transmission peaks and the number of modes (obtained by integration of the mode density over stack length) within every range $[2m\omega_0, 2(m+1)\omega_0]$ equals the number of constituent layers. For arbitrary stacks, transmission peaks correspond to maxima in the density of modes and to minima in the traversal velocity defined as the stack length divided by the phase time (Wigner time). Therefore, higher transmission of light through a multilayer stack (and higher density of modes) occurs at the expense of slower propagation of light and light energy accumulation.

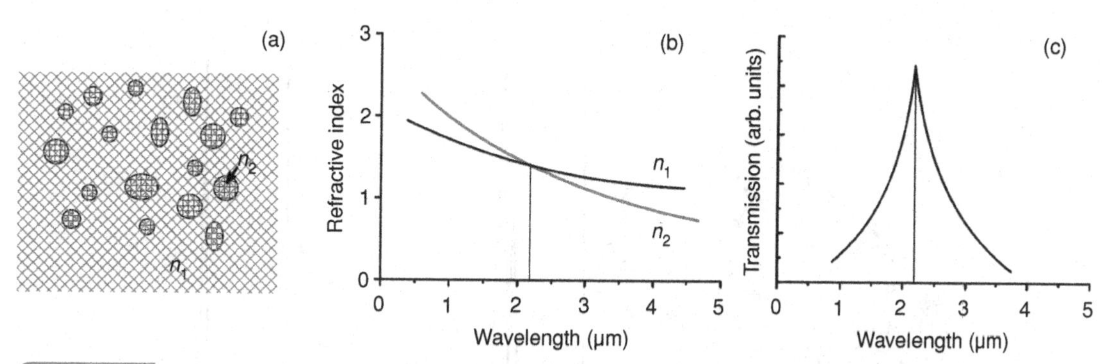

Fig. 8.31 Christiansen's filters: (a) a dielectric powder in an ambient medium with different refraction properties, (b) refraction spectra of two media, (c) transmission spectrum of a slab featuring a maximum at the wavelength where $n_1 = n_2$.

8.8 Applications of turbid structures: Christiansen's filters and Letokhov's lasers

In this concluding section we consider a couple of straightforward applications of random scattering media in optics. More than a century ago, in 1884, C. Christiansen[5] found that sometimes a turbid mixture of colloidal particles in solution becomes transparent at a certain wavelength [63]. This effect is referred to as the *Christiansen effect* and is explained in Figure 8.31. Transparency occurs at that wavelength where the refractive indexes of constituent substances obey the same value $n_1(\lambda^*) = n_2(\lambda^*)$. Then the otherwise turbid dispersive composite medium becomes homogenous and exhibits selective transparency in a narrow spectral range around the intersection point λ.

Christiansen's filters belong to the class of selective *dispersive filters*. They have found commercial application, mainly in infrared spectroscopy and techniques where standard interference filters based on multilayer films are hard to develop because of low durability and high friability of thicker ($\gg 1\,\mu$m) polycrystalline films. This needs the thorough technology of developing powders with desirable refractive indexes, their agglomeration and fritting [64].

Another application of random powder-like media is the fabrication of *random lasers*. Such a laser was proposed for the first time by V. S. Letokhov in 1968 [65]. Its action occurs by means of closed loops of light propagation in a medium with optical gain and scatterers (Fig. 8.32). Lasing develops for those loops where the gain exceeds the losses. Thus laser-like light sources can be developed as mesoscopic materials rather than as devices. Probably the first evidence of coherent radiation from a powder of active material with optical gain without any additional feedback was reported in 1986 by Markushev *et al.* [66]. These results are shown in Figure 8.32. A powder of neodymium-doped glass particles exhibits

[5] Christian Christiansen (1843–1917) was a Danish physicist, the doctoral advisor to Niels Bohr.

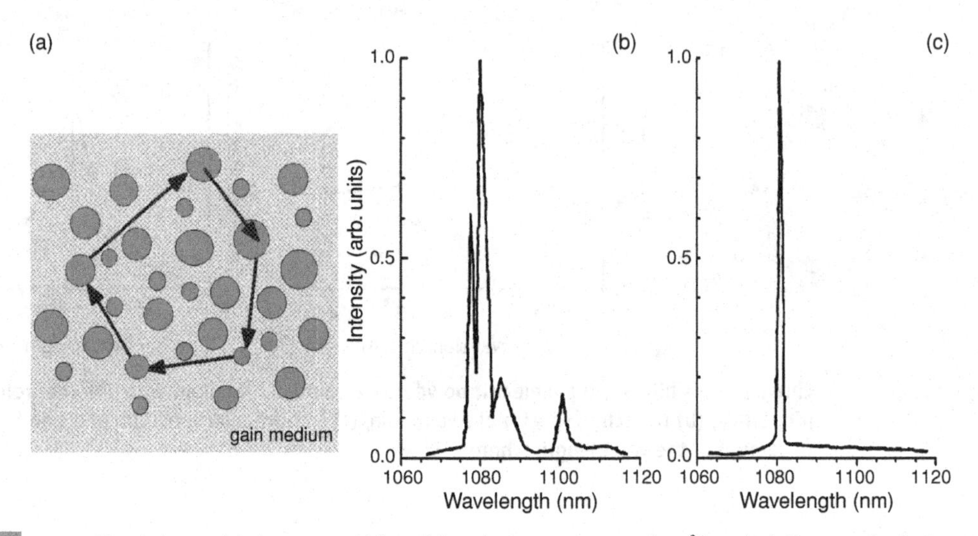

Fig. 8.32 **A Letokhov's laser: (a) the general idea, (b) emission spectrum of Nd^{3+}-containing powder below the threshold, and (c) above-threshold pumping [66].**

a threshold behavior with a single narrow-band emission above the threshold pumping, whereas at lower pumping below the threshold multiple emission lines are present.

Nowadays, experimental activity on random lasing is extensive [67–69]. This trend gained additional interest when novel active laser materials based on ceramics came about [70]. In ceramics, random lasing is readily observed because of higher active center concentrations as compared to traditional glass and crystal-based media.

Problems

1. What can you say about the reflection spectra of multiple structures whose calculated transmission spectra were shown in this chapter? Try to recover and discuss reflection spectra for a few selected structures. (Hint: note there were no dissipation losses involved in the calculations).

2. Calculate the total number of elementary layers in the non-Cantor fractal filters whose spectra are presented in Figure 8.14. Derive the general formula for the total number of layers and geometrical length of fractal filters.

3. Consider semiadic Cantor filters. Plot a few first generations, derive the relation for the total length and the total number of layers. Formulate the scaling factor for transmission spectra and the splitting factor.

4. Try to elaborate qualitative arguments as to why higher-order rotational symmetry helps to get an omnidirectional electromagnetic band gap.

5. Based on the issues discussed in Chapters 7 and 8 formulate general regularities of classical wave propagation in complex structures in terms of their correlation with, and dependence on, geometrical parameters.

6. Prove that any multilayer stack where layer thicknesses are commensurate can be treated as a quarter-wave one. Define the frequency scale for such a stack where the conservation law evaluated in Section 8.7 will hold.

References

[1] P. W. Anderson. The question of classical localization: a theory of white paint? *Philos. Mag. B*, **52** (1985), 505–510.

[2] V. L. Kuz'min and V. P. Romanov. Coherent phenomena in light scattering from disordered systems. *Physics – Uspekhi*, **39** (1996), 231–260.

[3] D. S. Wiersma, M. Colocci, R. Righini and F. Aliev. Temperature-controlled light diffusion in random media. *Phys. Rev. B*, **64** (2001), p. 144208.

[4] F. J. P. Schuurmans, D. Vanmaekelbergh, J. van de Lagemaat and A. Lagendijk. Strongly photonic macroporous gallium phosphide networks. *Science*, **284** (1999), 141–143.

[5] F. J. Poelwijk. Interference in random lasers. Ph. D. thesis, University of Amsterdam (2000).

[6] M. P. van Albada and A. Lagendijk, Observation of weak localization of light in random media. *Phys. Rev. Lett.*, **55** (1985), 2692–2694.

[7] P. E. Wolf and G. Maret, Weak localization and coherent backscattering of photons in disordered media. *Phys. Rev. Lett.*, **55** (1985), 2696–2699.

[8] M. Kaveh, M. Rosenbluh, I. Edrei and I. Freund. Weak localization and light scattering from disordered solids. *Phys. Rev. Lett.*, **57** (1986), 2049–2052.

[9] A. Z. Genack and J. M. Drake. Relationship between optical intensity, fluctuations and pulse propagation in random media. *Europhys. Lett.*, **11** (1990), 331–336.

[10] F. J. P. Schuurmans, M. Megens, D. Vanmaekelbergh and A. Lagendijk. Light scattering near the localization transition in macroporous GaP networks. *Phys. Rev. Lett.*, **83** (1999), 2183–2186.

[11] P. W. Anderson. Absence of diffusion in certain random lattices. *Phys. Rev.*, **109** (1958), 1492–1505.

[12] S. John. Electromagnetic absorption in a disordered medium near a photon mobility edge. *Phys. Rev. Lett.*, **53** (1984), 2169–2172 .

[13] D. S. Wiersma, P. Bartolini, A. Lagendijk and R. Righini. Localization of light in a disordered medium. *Nature*, **390** (1997), 671–675.

[14] F. Scheffold, R. Lenke, R. Tweer and G. Maret. Localization or classical diffusion of light? *Nature*, **398** (1999), 206–207; see also D. S. Wiersma, P. Bartolini, A. Lagendijk and R. Righini. Reply. *Nature*, **398** (1999), 207.

[15] J. Gómez Rivas, R. Sprik and A. Lagendijk. Optical transmission through very strong scattering media. *Ann. Phys., (Leipzig)* **8** (1999) Spec. Issue, p. I-77– I-80.

[16] M. Störzer, P. Gross, C. M. Aegerter and G. Maret. Observation of the critical regime near Anderson localization of light. *Phys. Rev. Lett.*, **96** (2006), p. 063904.

[17] C.M. Aegerter, M. Störzer, S. Fiebig, W. Bührer and G. Maret. Observation of Anderson localization of light in three dimensions. *J. Opt. Soc. Amer. A*, **24** (2007), A23–A27.

[18] E. Abrahams, P.W. Anderson, D.C. Licciardello and T.V. Ramakrishnan. Scaling theory of localization: absence of quantum diffusion in two dimensions. *Phys. Rev. Lett.*, **42** (1979), 673–676.

[19] N. Garcia, A. Z. Genack and A. A. Lisyansky. Measurement of the transport mean free path of diffusing photons. *Phys. Rev. B*, **46** (1992), 14475–14479.

[20] R. Dalichaouch, J. P. Armstrong, S. Schultz, P. M. Platzman and S. L. McCall. Microwave localization by 2-dimensional random scattering. *Nature*, **354** (1991), 53–55.

[21] R. L. Weaver. Anderson localization of ultrasound. *Wave Motion*, **12** (1990), 129–142.

[22] B. B. Mandelbrot. *The Fractal Geometry of Nature* (New York: Macmillan, 1982).

[23] J. Feder. *Fractals* (New York: Plenum Press, 1988; Moscow: Mir, 1991).

[24] D. L. Jaggard and X. Sun. Reflection from fractal multilayers. *Opt. Lett.*, **15** (1990), 1428–1430.

[25] X. Sun and D. L. Jaggard. Wave interactions with generalized Cantor bar fractal multilayers. *J. Appl. Phys.*, **70** (1991), 2500–2507.

[26] C. Sibilia, I. S. Nefedov, M. Scalora and M. Bertolotti. Electromagnetic mode density for finite quasi-periodic structures. *J. Opt. Soc. Amer. B*, **15** (1998), 1947–1952.

[27] S. V. Zhukovsky, S. V. Gaponenko and A. V. Lavrinenko. Spectral properties of fractal and quasi-periodic multilayered media. *Nonlinear Phenomena in Complex Systems*, **4** (2001), 383–389.

[28] A. V. Lavrinenko, S. V. Zhukovsky, K. S. Sandomirskii and S. V. Gaponenko. Propagation of classical waves in nonperiodic media: Scaling properties of an optical Cantor filter. *Phys. Rev. E*, **65** (2002), p. 036621.

[29] S. V. Zhukovsky, A. V. Lavrinenko and S. V. Gaponenko. Spectral scalability as a result of geometrical self-similarity of fractal multilayers. *Europhys. Lett.*, **66** (2004), 455–461.

[30] S. V. Zhukovsky, S. V. Gaponenko and A. V. Lavrinenko. Optical properties of fractal layered nanostructures. *Bulletin Russ. Acad. Sci. (Physics)*, **68** (2004), 29–32.

[31] S. V. Zhukovsky. Propagation of electromagnetic waves in fractal multilayers. Unpublished Ph.D. thesis, Institute of Molecular and Atomic Physics, Minsk (2004).

[32] A. Mandatori, C. Sibilia and M. Centini. Birefringence in one-dimensional finite photonic bandgap structure. *J. Opt. Soc. Amer. B*, **20** (2003), 504–513.

[33] M. Kohmoto, L. P. Kadanoff and C. Tang. Localization problem in one dimension: mapping and escape. *Phys. Rev. Lett.*, **50** (1983), 1870–1872.

[34] M. Kohmoto and B. Sutherland. Critical wave functions and a Cantor-set spectrum of a one-dimensional quasicrystal model. *Phys. Rev. B*, **35** (1987), 1020–1033.

[35] E. Macia and F. Dominguez-Adame. Physical nature of critical wave functions Fibonacci systems. *Phys. Rev. Lett.*, **60** (1996), 10032–10036.

[36] E. Macia. Physical nature of critical states in Fibonacci quasicrystals. *Phys. Rev. B*, **60** (1999), 10032–10036.

[37] M. Kohmoto, B. Sutherland and K. Iguchi. Localization in optics: Quasiperiodic media. *Phys. Rev. Lett.*, **58** (1987), 2436–2438.

[38] M. S. Vasconcelos, E. L. Albuquerquey and A. M. Mariz. Optical localization in quasiperiodic multilayers. *J. Phys.: Condens. Matter.*, **10** (1998), 5839–5849.

[39] E. Maciá. Exploiting quasiperiodic order in the design of optical devices. *Phys. Rev. B*, **63** (2001), p. 205421.

[40] X. Huang, Y. Wang and C. Gong. Numerical investigation of light-wave localization in optical Fibonacci superlattices with symmetric internal structure. *J. Phys. C: Cond. Matt.*, **11** (1999), 7645–7651.

[41] X. Q. Huang, S. S. Jiang, R. W. Peng and A. Hu. Perfect transmission and self-similar optical transmission spectra in symmetric Fibonacci-class multilayers. *Phys. Rev. B*, **63** (2001), p. 245104.

[42] W. Gellerman, M. Kohmoto, B. Sutherland and P. C. Taylor. Localization of light waves in Fibonacci dielectric multilayers. *Phys. Rev. Lett.*, **72** (1994), 633–6336.

[43] L. Dal Negro, C. J. Oton, Z. Gaburro, L. Pavesi, P. Johnson, A. Lagendijk, R. Righini, M. Colocci and D. S. Wiersma. Light transport through the band-edge states of Fibonacci quasicrystals. *Phys. Rev. Lett.*, **90** (2003), p. 055501.

[44] V. Pellegrini, A. Tredicucci, C. Mazzoleni and L. Pavesi. Enhanced optical properties in porous silicon microcavities. *Phys. Rev. B*, **52** (1995), R14328–R14331.

[45] S. A. Akhmanov, V. A. Vyslouch and A. S. Chirkin. *Optics of Femtosecond Laser Pulse* (New York: AIP, 1992).

[46] L. N. Makarava, M. M. Nazarov, I. A. Ozheredov, A. P. Shkurinov, A. G. Smirnov and S. V. Zhukovsky. Fibonacci-like photonic structure for femtosecond pulse compression. *Phys. Rev. E*, **75** (2007), p. 036609.

[47] E. Maciá. The role of aperiodic order in science and technology. *Rep. Progr. Phys.*, **69** (2006), 397–441.

[48] T. F. Krauss and R. M. De La Rue. Photonic crystals in the optical regime: past, present and future. *Progr. Quant. Electron.*, **23** (1999), 51–96.

[49] C. Janot. *Quasicrystals* (Oxford: Oxford University Press, 1997).

[50] T. Fujiwara and T. Ogawa. *Quasicrystals* (Berlin: Springer Verlag, 1990).

[51] M. E. Zoorob, M. D. B. Charlton, G. J. Parker, J. J. Baumberg and M. C. Netti. Complete photonic bandgaps in 12-fold symmetric quasicrystals. *Nature*, **404** (2000), 740–743.

[52] R. de Kronig and W. G. Penney. Quantum mechanics of electrons in crystal lattices. *Proc. Royal Soc. London*, **130A** (1931), 499–512.

[53] I. E. Tamm. Possible bound states of an electron on a crystal surface. *Phys. Z. Sowjetunion*, **1** (1932), 733–740.

[54] D. Kossel. Analogies between thin film optics and electron-band theory of solids. *J. Opt. Soc. Amer.*, **56** (1966), 1434–1439.

[55] S. V. Zhukovsky. Private communication. 2007.

[56] A. V. Kavokin, I. A. Shelykh and G. Malpuech. Lossless interface modes at the boundary between two periodic dielectric structures. *Phys. Rev. B*, **72** (2005), p. 233102.

[57] L. A. Rivlin. Photons in a waveguide (some thought experiments). *Physics – Uspekhi*, **40** (1997), 291–303.

[58] S. V. Zhukovsky and S. V. Gaponenko. Constraints on transmission, dispersion, and density of states in quarter-wave multilayers and stepwise potential barriers of arbitrary geometry. *Phys. Rev. E*, **77** (2008), p. 046602.
[59] E. P. Wigner. Lower limit for the energy derivative of the scattering phase shift. *Phys. Rev.*, **98** (1955), 145–147.
[60] G. D'Aguanno, M. Centini, M. Scalora, C. Sibilia, M. J. Bloemer, C. M. Bowden, J. W. Haus and M. Bertolotti. Group velocity, energy velocity, and superluminal propagation in finite photonic band-gap structures. *Phys. Rev. E*, **63** (2001), p. 036610.
[61] J. M. Bendickson, J. P. Dowling and M. Scalora. Analytic expressions for the electromagnetic mode density in finite, one-dimensional, photonic band-gap structures. *Phys. Rev. E*, **53** (1996) 4107–4121.
[62] S. M. Barnett and R. Loudon. Sum rule for modified spontaneous emission rates. *Phys. Rev. Lett.*, **77** (1996), 2444–2446.
[63] C. Christiansen. Untersuchungen über die optischen Eigenschaften von fein verteilten Körpern. *Ann. Phys. Chem.*, **23** (1884), 298–306.
[64] N. A. Borisevich, V. G. Vereschagin and M. A. Validov. *Infrared Filters* (Minsk: Nauka i Tekhnika, 1971 (*in Russian*); NASA 1974).
[65] V. S. Letokhov. Generation of light by a scattering medium with negative resonance absorption. *Sov. Phys. JETP*, **26** (1968), 835–842.
[66] V. M. Markushev, V. F. Zolin and Ch. M. Briskina. A powder laser. *J. Appl. Spectr.*, **45** (1986), 847–850.
[67] *Optical Properties of Nanostructured Random Media*, ed. by V. M. Shalaev (Berlin: Springer, 2002).
[68] D. S. Wiersma. The smallest random laser. *Nature*, **406** (2000), 132–133.
[69] H. Cao, Y. G. Zhao, S. T. Ho, E. W. Seelig, Q. H. Wang and R. P. H. Chang, Random laser action in semiconductor powder. *Phys. Rev. Lett.*, **82** (1999), 2278–2281.
[70] Y. Feng, J.-F. Bisson, J. Lu, S. Huang, K. Takaichi, A. Shirakawa, M. Musha and K. Ueda, Thermal effects in quasi-continuous-wave Nd^{3+}:$Y_3Al_5O_{12}$ nanocrystalline-powder random laser. *Appl. Phys. Lett.*, **84** (2004), 1040–1042.

9 Photonic circuitry

"The light should be confined gently in order to be confined strongly".
Y. Akahane et al. [1]

The complex propagation of light in periodic and inhomogeneous media considered in previous chapters forms a solid basis for purposeful controlling of light wave propagation and light energy accumulation. Coupling with light sources and light modulators gives rise to the well-defined concept of photonic circuitry. This recently emerged field is becoming more and more mature. The principal solutions in photonic circuitry are overviewed in this chapter. The discussion is kept at an introductory level to provide conceptual ideas and principal approaches without going too deeply into detail. The extensive list of references will partly compensate for the somewhat sketchy style in this chapter.

9.1 Microcavities and microlasers

In a sense, an optical microcavity, or a microresonator, can be treated as a wavelength-scale topological construction capable of accumulating and storing light. This can be implemented with respect to light impinging from the outside as well as with respect to light generated inside the cavity under consideration. The word "generated" here implies spontaneous emission, spontaneous Raman scattering rather than necessarily lasing. Light energy accumulation and storage becomes possible by the spatial confinement of light waves in a cavity. Primary examples of (micro)cavities and (micro)resonators were treated in Chapter 3 (Section 3.4) when the resonant tunneling of electromagnetic waves was analyzed; namely a one-dimensional problem of an electromagnetic wave impinging onto a pair of parallel metallic thin film layers serving as mirrors with dielectric spacing (Fig. 3.19). A series of pronounced resonant transmission peaks develop corresponding to standing waves in a cavity between mirrors. Resonances occur for every light wavelength corresponding to an integer number of half-waves over the cavity (optical) length. This statement is rigorous under the assumption of infinitely high "barriers" for light waves in terms of perfect unity reflection and zero skin effect. In reality this never happens. We have seen in Chapter 3 and in Chapter 6 that neither metal offers perfect reflection ($R < 100\%$ always holds) or negligible skin depth (Tables 3.4, 3.5 and 6.1). Additionally, metal mirrors possess dissipative losses because of the inevitable imaginary component of the complex dielectric function when a desirable negative real part of the dielectric permittivity is the case.

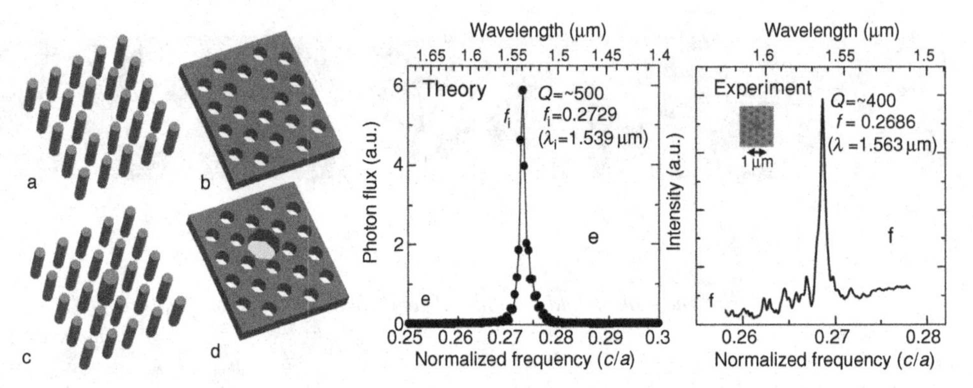

Fig. 9.1 **Microcavities inside a two-dimensional photonic crystal slab. (a)–(d) Possible geometries of a single defect. (e) Calculated and (f) measured response of a (d)-type cavity in an InGaAsP photonic crystal slab for a transverse-electric-like (TE) mode reported by S. Noda *et al.* [2]. Reprinted with permission from AAAS.**

In Chapter 7 we have seen that, since Rayleigh's prediction back in 1887, periodic dielectric structures offer a solution for nearly perfect mirrors where losses can actually become negligibly small (see Figs. 7.7 and 7.25). A defect introduced in a periodic dielectric stack comprises a planar all-dielectric Fabry–Perot cavity. Tracing the familiar trend of quantum–optical analogies, one can see that a defect in a photonic crystal should behave similarly to a defect in a periodic crystal lattice with respect to an electron. The latter case was the subject of Section 4.5 where resonant localized electron states were considered with energy values inside the band gap. The same occurs in optics and the planar Fabry–Perot cavity is a one-dimensional version of the more general consideration of defects inside photonic crystals.

For the band-gap region, an ideal photonic crystal represents a perfect mirror for the full solid angle when light impinges on its surface from a continuous dielectric medium.[1] A cavity made in a three-dimensional photonic crystal will confine certain modes in all three directions. This cavity should not necessarily be a vacuum or air void. It can be formed by any material disturbing the periodicity of a crystal. It can also be performed in a two-dimensional photonic crystal and this is shown in Figure 9.1. Not only it is much easier to draw such cavities in two dimensions. It is also much easier to fabricate a cavity in a two-dimensional slab as compared to the three-dimensional case, because in two dimensions submicron lithography and etching can be readily performed for many semiconductor materials with high refractive index. Figure 9.1(e and f) shows calculated and measured resonant response of a single cavity in a two-dimensional InGaAsP quarternary semiconductor compound. The topology of a system under consideration has been designed to meet resonant behavior near the 1.55 μm wavelength used for optical communication.

Every resonator can be characterized by the *Q-factor* which is the ratio of the energy stored in a cavity to the energy lost in a single period of oscillation. In other words it can be

[1] Or from another photonic crystal with different band structure that allows propagation of waves of the given frequency. That case would represent a kind of a "photonic heterostructure".

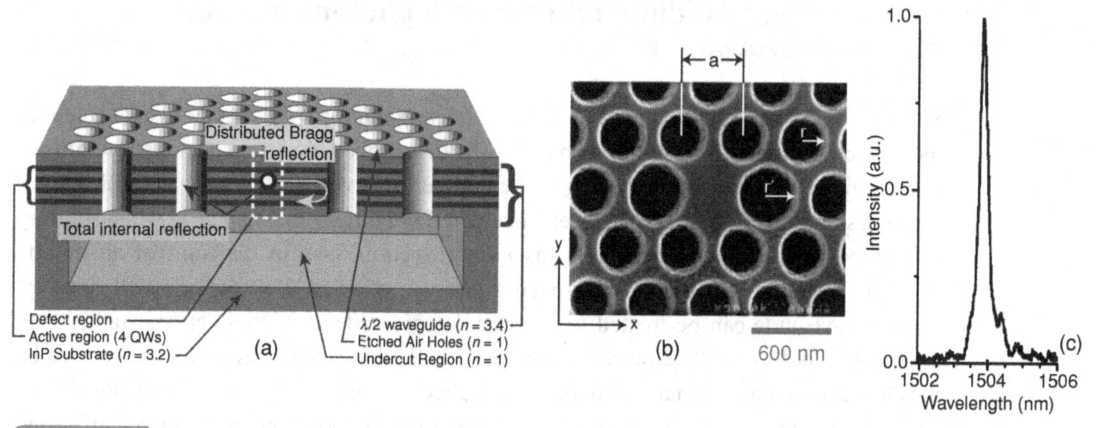

Fig. 9.2 **A microcavity photonic crystal laser: (a) design, (b) implementation and (c) performance. Semiconductor material is InGaAsP. Triangular lattice period $a = 515$ nm, pore radius $r = 180$ nm. The two enlarged holes have a radius $r' = 240$ nm. These are designed to split the dipole mode degeneracy. The InGaAsP membrane is 220 nm (half-wavelength) in thickness. Adapted from [3]. Reprinted with permission from AAAS.**

expressed as the number of cycles for its energy to decay by the factor of $e^{-2\pi}$. In spectral response, the Q-factor manifests itself as the ratio of the resonance frequency ω_0 to the resonance width $\Delta\omega$, i.e.

$$Q = \frac{\omega_0}{\Delta\omega}. \tag{9.1}$$

The Q-factor for the cavity shown in Figure 9.1(d–f) was predicted to be about 500 and was found to be about 400.

If a microcavity inside a photonic crystal is not an empty void but a gain medium, then a microlaser inside a photonic crystal can be performed. Painter *et al.* [3] demonstrated an extremely small laser which measures only 0.03 μm^3! Its design is clear from Figure 9.2(a,b). The same material was used as in Figure 9.1, namely an InGaAsP quarternary semiconductor. The cavity design corresponded to that in Figure 9.1(b). The active layer had half-wavelength thickness to provide additional confinement across the photonic crystal plane. In this plane confinement was provided by means of Bragg reflections because of the photonic band gap. The laser structure was optically pumped with a semiconductor laser of 830 nm wavelength, focused with a microscope objective to a spot size of 3 μm on the sample surface. Actually the authors reported on the active media volume rather than the laser volume. The reader is proposed to evaluate the real laser volume themselves (see Problem 3). The active medium consisted of four quantum wells which should not be mistreated as a photonic Bragg structure. Quantum well size was defined by the desired electron, rather than light-wave, confinement. The operating wavelength was close to 1.5 μm. Note, a quantum-well-based microcavity photonic crystal laser represents a class of photonic microdevices with simultaneously exploited electron and electromagnetic wave confinement, which provides a route to nanophotonic engineering of electronic and photonic bands and resonances.

9.2 Guiding light through photonic crystals

Since a single defect inside a photonic crystal forms a microcavity, regular displacement of those microcavities will result in formation of a delocalized electromagnetic state with the number of resonant sub-bands equal to the number of cavities. A hint to such splitting of transmission/reflection peaks has been considered previously in Chapter 8, when splitting in symmetric fractal structures was discussed (Section 8.4). In the limit of an infinite number of cavities and negligible spacing between cavities, a waveguide will develop. Such a waveguide can be treated as a linear defect inside a photonic crystal rather than a point-like defect in the case of a single cavity. Such a linear waveguide defect can be developed in a three-dimensional periodic lattice as well as in a two-dimensional one. The latter case is feasible by means of submicron photolithography and can be performed commercially.

A traditional waveguide operates using total internal reflection and therefore does not allow strong bending because of leakage of radiation when the total internal reflection condition is removed. A photonic crystal waveguide operates using multiple scattering on regularly displaced scatterers (hollow cylindrical voids) in a photonic crystal plane and by total internal reflection across that plane. A typical example of waveguide engineering using a silicon photonic crystal is shown in Figure 9.3. The waveguide has been designed for the range of wavelengths around 1.3 μm which is used in the optical communication industry and corresponds to the transparency region of silicon crystals. Notably, a simple introduction of a curvilinear defect does not guarantee high transmission because of losses in the bend areas (note, the cross-section of such a waveguide is sub-wavelength size). However bend losses can be minimized by means of adaptive optimization of the scatterer's topology which appears by no means to be periodic. This approach has become a standard trend in waveguide design for micro- and nanophotonic circuitry [4–13].

Considering more complex implementations of photonic crystal waveguiding. Figure 9.4a shows an optimized design of a silicon photonic crystal Y-type beam splitter. A single beam can be split into two equal portions as shown in Figure 9.4b. Two beams with different wavelengths λ_1, λ_2 can also be discriminated and sent to individual channels #1 and #2 provided the spectral spacing is approximately $|\lambda_1 - \lambda_2| > 50$ nm. For multiple wavelength division a more complicated demultiplexer can be designed and implemented as is seen in Figure 9.4(c) and (d).

Optical components presented in Figures 9.3 and 9.4 do demonstrate the emerging field of silicon photonics. The optical transparency of silicon in the strategic optical communication wavelength range of 1.3–1.5 μm, combined with the unprecedented advances in silicon sub-micron-scale technology, has inspired many research centres over the world to go into the field of silicon photonics. All the above nanostructures are made of silicon using silicon-on-insulator (SOI) technology combined with high resolution lithography. The latter is typically implemented in the form of the electron-beam- (e-beam) rather than photolithography. Silicon-on-insulator nano-imprint technology has been developed in recent years promising a cheaper route towards commercial fabrication of planar nanostructures. Imprinting is performed by means of reactive-ion etching using a master stamp structure made by

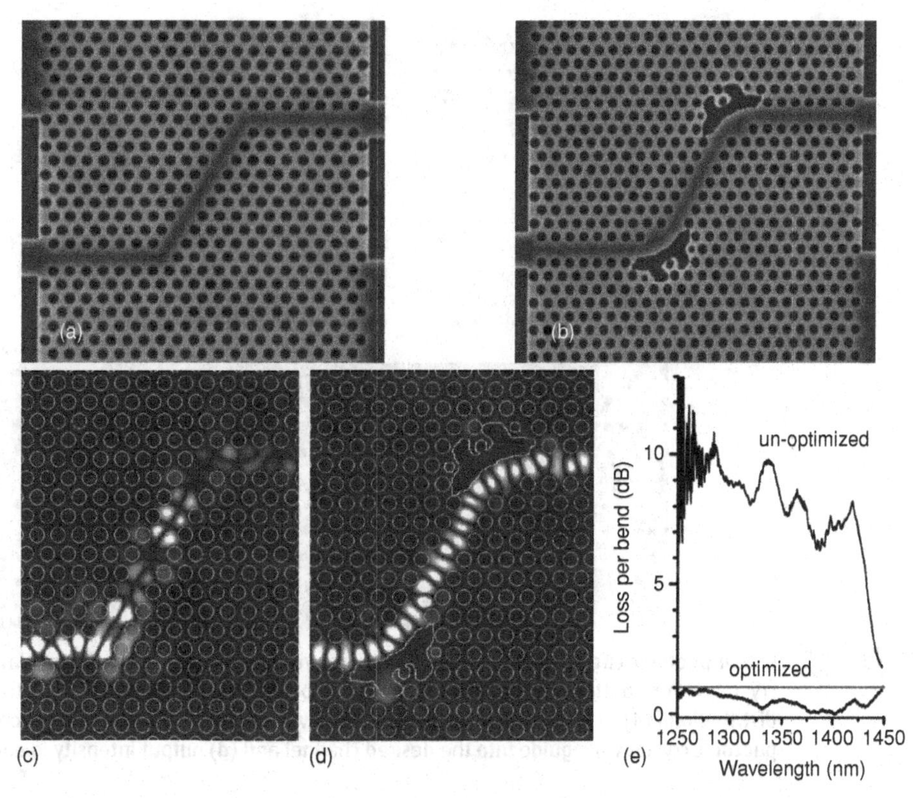

Fig. 9.3 (a), (b) Fabricated two-dimensional photonic crystal structures with a waveguide using a silicon wafer and submicron photolithography/etching technique. The hole diameter is 275 nm. (a) corresponds to a generic design whereas (b) represents the optimized design of bends areas. The contrast and brightness of the images have been changed for clarity. (c) and (d) Calculated steady-state magnetic field distribution for the fundamental photonic band gap using the two-dimensional finite difference time-domain method for (c) generic and (d) optimized design. (e) Measured loss per bend for the un-optimized 60° bends and the topology-optimized 60° bends. Both spectra have been normalized to transmission through a straight photonic crystal waveguide of the same length. A gray line marks a bend loss of 1dB. Adapted from [13].

e-beam lithography. The feasibility of spatial resolution of 30 nm in imprinting technology has been demonstrated [16]. Not only waveguides, splitters, multiplexers–demultiplexers but also optical and electro-optical modulators and switches become feasible. It is important for practical purposes that, based on SOI, these components can readily become CMOS-compatible, i.e. integrable in the modern microelectronics production cycle.[2]

It is important to note that propagation of a light wave through a photonic crystal waveguide structure is characterized by the complicated dispersion law and strong delay in group velocity. For such structures "slow light" and "large group delay index" are the relevant notions with the delay being so big that the group velocity can measure just a few

[2] CMOS stands for "complementary metal-oxide-semiconductor" and denotes the modern silicon microelectronics technology platform.

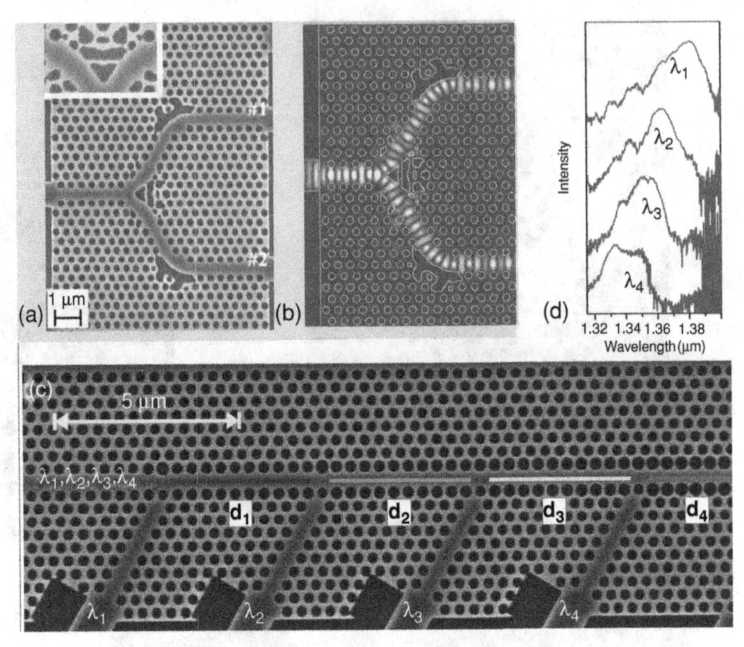

Fig. 9.4 **Planar photonic circuitry components based on two-dimensional silicon-on-insulator photonic crystals. (a) Y-splitter with optimized light-wave bending along with (b) calculated spatial field distribution [14]. (c) Demultiplexer providing delivery of every λ_i wavelength from a single photonic crystal waveguide into the desired channel and (d) output intensity in each channel [15].**

per cent or even less of the speed of light in a vacuum, c [17]. This time delay and slowing down become rather favorable for all non-linear optical components. The light propagation slowing down is equivalent either to an increase in the efficient path length or to an increase in the optical absorption coefficient or non-linear susceptibility in case of non-resonant non-linearity. Therefore it can be purposefully exploited for optical switching and modulation. The design known as a *Mach–Zehnder interferometer* has gained considerable attention in this context.

A photonic crystal waveguide design of a Mach–Zehnder interferometer is shown in Figure 9.5. The output signal is defined by the phase matching (mismatching) in a pair of guiding arms. External impact can be used to change the phase shift in either arm and thus an interferometer can be used as a modulator or a switch. Thermo-optical modulation with sub-microsecond time has been demonstrated [17]. Electro-optical modulation is feasible as well, in particular, using liquid crystal impregnation and exploiting the sensitivity of liquid crystal refractive index to an external electric field. All-optical switching can also be performed by means of strong optical pumping resulting in non-linear refraction. Note, under certain conditions semiconductor materials can offer a non-linear refraction index change of the order of $\Delta n = 10^{-2}$ [19] (see Section 7.13). For CMOS-compatible electro-optics, modification of silicon refraction in Si-based structures is crucial. This can be performed by means of injection of carriers to modify the refractive index depending on carrier density. For a Mach–Zehnder modulator design, switching times in the microsecond range at sub-milli-Ampere modulation current have been reported for an 80-μm-long device [20].

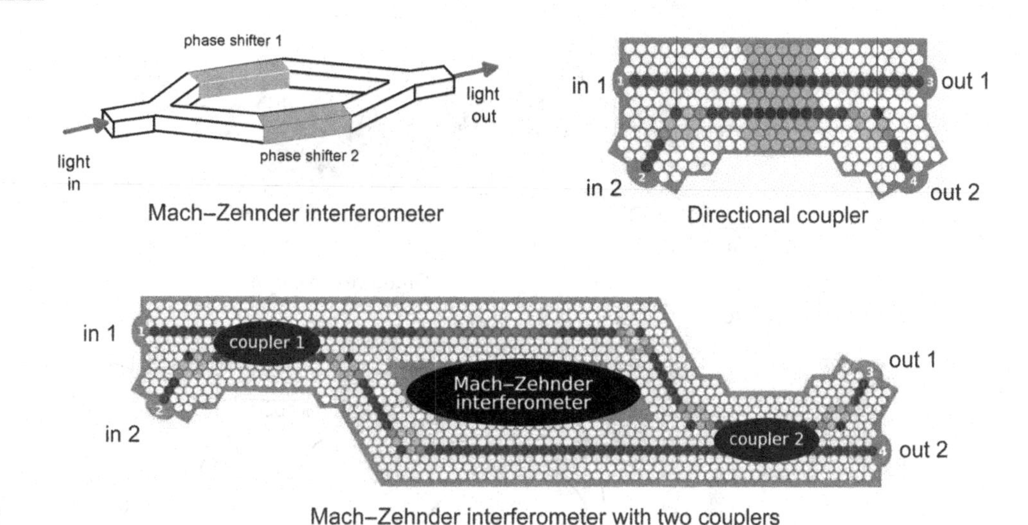

Fig. 9.5 **Photonic circuitry components: a Mach–Zehnder interferometer, a directional coupler and a combination of an interferometer with two couplers (adapted from [8]).**

In 2004 INTEL Corp. [21] reported news of a Si based Mach–Zehnder interferometer in which one branch experiences refraction modulation (and phase shift) without direct carrier injection/depletion using a capacitor formed by a metal-oxide-semiconductor (MOS) structure. Although the geometrical size was still rather large (about 1cm), not only had undesirable thermal and power consumption issues been eliminated, but also high speed operation was announced of the order of 1 GHz. This becomes possible with a capacitor effect since no finite carrier generation and recombination rates are involved in the operation. Two symmetrical MOS capacitors are made as waveguides with the length being equal to a few mm. The phase shift at 1.54 μm wavelength to get nearly 20 dB transmission modulation is affordable with a few volts applied to the structure. The large size is not the only drawback of the Mach–Zehnder modulator. Further problems arise, e.g. from high input losses because of the bent waveguide structure as well as high sensitivity to bias drift.

A further example of a basic photonic circuitry element is a *directional coupler* (Fig. 9.5). This consists of two waveguides that come close to each other so that the fields can couple in a section of certain length. The operation principle of the coupler is as follows [18]. In the coupling section, the two aligned single-mode waveguides form a dual-mode waveguide that supports an even and an odd propagating guided mode with respect to a certain symmetry of the system. These modes propagate with different wave vectors k_{even} and k_{odd}, respectively. Due to mode beating, the resulting field shifts, as a function of coupler length, periodically between the two waveguides with period $L_{\text{B}} = 2\pi/\left|k_{\text{odd}} - k_{\text{even}}\right|$. Thus, if the coupler length L is equal to L_{B} (or a multiple of it), light fed into one of the waveguides exits in the same waveguide (the "bar" state), while if the coupler length is half the beat length (or an odd multiple of it), the light is completely coupled over to the other waveguide (the "cross" state). A directional coupler with optional control of the refractive index of constituent materials by means of electric field effects or optical nonlinearities can be used for (electro)optical modulation and switching [22].

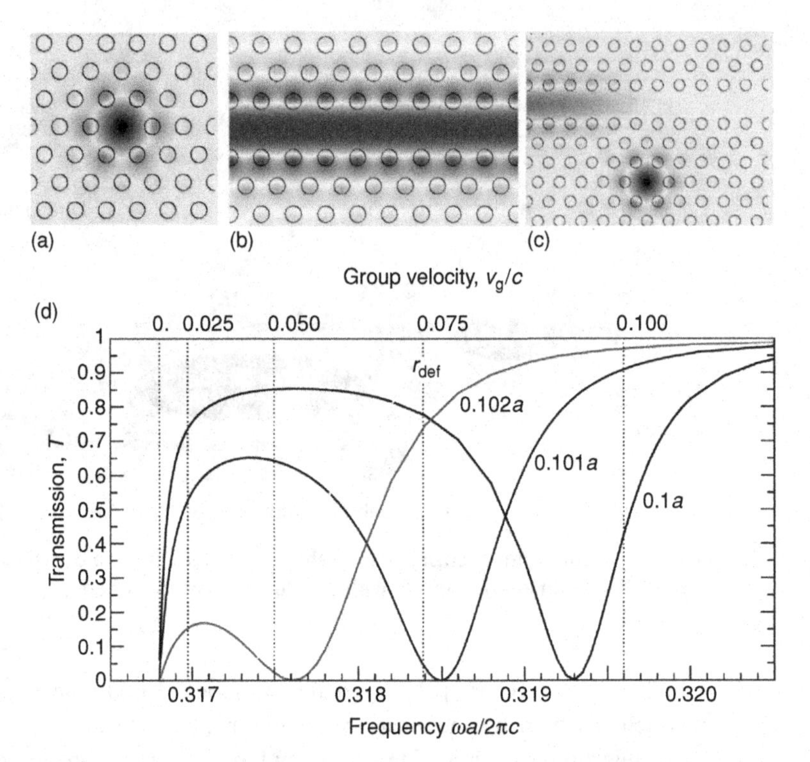

Fig. 9.6 **Optical properties of a photonic crystal waveguide coupled to a microcavity formed by a single rod [23]. The photonic crystal consists of Si rods in air with lattice period a and rods radius $0.25a$. (a) Field distribution for a cavity. (b) Field distribution for a waveguide. (c) Field distribution of a coupled waveguide–cavity system. (d) Transmission function versus normalized frequency. At frequency 0.3168 the band edge is located. Here the group velocity for light in the waveguide tends to zero. For a higher frequency group velocity monotonically rises with frequency but still remains a small fraction of c. Transmission features strong dependence on frequency as well as on the defect rod radius r_{def}.**

For a variety of photonic crystal-based components K. Busch and co-workers elaborated an elegant calculational approach based on magnetic field expansion into Wannier functions [10,18]. These functions form an orthonormal basis derived from the Bloch functions of the underlying photonic crystal. Interestingly, Wannier functions are extensively used in the electron theory of solids. Therefore, along with Bloch functions, Brillouin zones and band structure presentations considered in Chapter 7, the Wannier functions formalism represents one more example of fruitful transfer of theoretical notions from electron theory of solids to photonics. With these techniques the properties of the devices shown in Figure 9.5 were examined.

One more example of photonic circuitry is a photonic crystal waveguide coupled to a microcavity formed as a defect in a photonic crystal. Using the above Wannier function formalism, Mingaleev *et al.* [23] explored the properties of such a system based on a silicon-air photonic crystal formed by a lattice of rods with period a and rod radius $0.25a$ (Fig. 9.6). Optical bistability has been predicted for such a structure in the slow-light regime with lower switching intensity correlating with propagation slowing down. The Q-factor of the structure grows inversely proportional to the group velocity of light at the resonant frequency and can be as large as 10^3. Accordingly, the power threshold required for all-optical switching vanishes as a square of the group velocity.

Actually, a high-Q cavity buried in a periodic thin film multilayer structure (i.e. a Fabry–Perot interferometer) with a nonlinear-optical active medium providing intensity-dependent resonance have been known as bistable elements since the 1970s [24, 25]. This simple concept of an optical processing component is still in use and is under consideration as a possible component of silicon photonics [26]. Bistability arises due to the double-valued function of transmission versus incident intensity because of the cavity's capability to keep stored light. Going beyond the single dimension not only brings a new flavor to this old field, but extends its application area in terms of complicated micrometer-scale planar circuitry.

For all high-Q components, understanding the transient behavior is important. Not only is the operation intensity necessary, derived by examination of the steady-state equations, but the finite switching energy defined by the Q-factor and the finite switching time defined by the energy accumulation in a high-Q component should never be ignored. The importance of these issues can be illustrated by an amazing effect of critical switching slowing down. The switching time can become indefinitely long when the input intensity is kept negligibly above the threshold [25]. In many instances, this phenomenon is a characteristic of a variety of phase transitions including optical bistability, as an example of non-equilibrium phase transition.

Recalling the high-Q-based solutions in photonic circuitry beyond nanophotonics, a number of proposals for all-optical and electro-optical switching should be mentioned. These solutions are based on controllable light propagation through a linear fiber coupled to a ring fiber resonator or a microdisk resonator [27]. *Whispering gallery modes* first introduced by Rayleigh more than a century ago when considering extraordinary audibility in certain church buildings, are the essence of these solutions. These modes develop over the perimeter of a disk or a sphere and can exhibit a superior Q-factor of the order of 10^6 or even higher. Control of light propagation is performed by means of coupling modulation between a fiber and a high-Q cavity. The macroscopic size and high Q value form the basic limit in the switching time (see Problem 9).

9.3 Holey fibers

> *...we had received some "photonic crystal fiber" from the group of Philip Russell at the University of Bath in the UK. We had found out too late that these British researchers had actually pioneered micro-structured silica fibers some years earlier. Launching about 170 mW into a 30 cm length of photonic crystal fiber, we immediately produced a frequency comb spanning more than an octave.*
>
> *T. W. Hänsch, Nobel lecture (2005)*

Traditional optical fibers are based on the internal reflection phenomenon in optics and for this reason feature at least two principal drawbacks. First, the guiding inner part

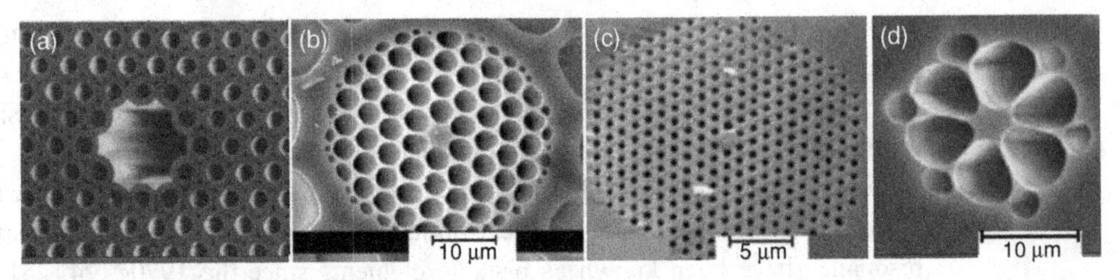

Fig. 9.7 **Various design of PC-fibers: (a) fiber with hollow core [29], and (b)–(d) fibers with solid cores and holey structures used for generation of white light, short-pulse compression, non-linear-optical interactions enhancement [30, 31].**

of fiber should possess higher refractive index than the outer part, which is referred to as the cladding layer. This results in unavoidable absorption losses. Second, bending a traditional optical fiber may break the condition of total internal reflection resulting in leakage of light energy outside the fiber, i.e. undesirable losses. In 1996 Philip Russell with co-workers pushed an ingenuous idea to provide reflection in optical fiber by means of regular microchannels forming a two-dimensional photonic crystal [28]. In this case the central core guide may be hollow (Fig. 9.7(a)). Later on the idea was extended to the concept of a photonic crystal fiber for which the principal feature is periodic microstructure with photonic band gap instead of a total reflection arrangement. Since the first publication in 1996, the field of photonic crystal fibers has become very wide and extensive.

Photonic crystal fibers with a hollow core like that shown in Figure 9.7(a) have distinguished transmission properties because no absorptive loss is introduced by the core itself. However, in many applications losses in the core is not the crucial issue and fibers with a solid core surrounded by the periodic structure have found considerable interest as well (Fig. 9.7(b) and (c)). Finally, even an aperiodic holey structure around the core can sometimes be useful as well (Fig. 9.7(d)). For this reason along with the original notation "photonic crystal fiber", the term "holey fibers" is widely used as a more general term.

The principal advantage of holey fibers as compared to their traditional counterparts is the wider frequency range of the single-mode propagation condition. It arises from the complicated dispersion of an electromagnetic wave propagating in such a fiber [32–34]. In such a fiber a multitude of nonlinear processes are enhanced because of the long interaction distance and the light slowing down. The cladding microstructure does not necessarily resemble a photonic crystal but may be treated in a certain range in terms of an effective medium with frequency-dependent refractive index. Higher harmonic generation, phase modulation, four-wave processes, parametric effects, stimulated Raman scattering and coherent anti-Stokes Raman scattering – all of these processes result in efficient broad-band continuum generation. Hollow fibers additionally offer the nonlinear processes of gases (and air) to be efficiently exploited (Kerr-type nonlinearity, plasma induced nonlinearities due to multi-photon ionization) with the advantage of higher break-down powers as compared to condensed matter.

Multiple applications of holey fibers in laser technology and non-linear optics have been discussed in reviews and books [30, 31, 35–39]. To summarize, advances in microstructured fiber technologies enabled the development of a new class of fiber-optic frequency converters, broad-band light sources, and short-pulse lasers. The frequency profile of dispersion and the spatial profile of electromagnetic field distribution in waveguide modes of microstructured fibers can be tailored by modifying the core and cladding design on a micro- and nanoscale. Pulse widths range from a few nanoseconds to a few optical cycles (several femtoseconds). Light power ranges from hundreds of Watts (owing to enhanced nonlinearities) to several gigaWatts (owing to the hollow core which allows the use of gases for nonlinear pulse transformation with optical breakdown powers higher than condensed matter analogs). Generation of higher harmonics (10th to 20th!) has been reported for gas-filled fiber pores. In new fiber lasers, microstructured fibers provide a precise balance of dispersion within a broad spectral range, allowing the creation of compact all-fiber sources of high-power ultrashort light pulses.

9.4 Whispering gallery modes: photonic dots, photonic molecules and chains

At the end of the nineteenth century, Rayleigh referred to the Whispering Gallery audibility in the dome of St. Paul's Cathedral in London (Fig. 9.8) as an excitingly amazing acoustic phenomenon with the origin being unclear for acoustics experts [40]. Some time later he managed to make it clear by considering the acoustic modes extending in a resonant manner along the inner circumference of the gallery [41]. Electromagnetic modes of this type are therefore also referred to as "*whispering gallery modes*" and have recently taken a noticeable place in microwave and optical circuitry.

In the optical range, whispering gallery modes are inherent in semiconductors (in the spectral range of low absorption) or dielectric microdisks and microspheres (including solid spherical particles and also shell-like dielectric envelopes). These modes develop owing to total internal reflection at the interface and therefore can only exist if the material has high enough refractive index as compared to the ambient medium. Examples of such modes in a microdisk or in a solid microsphere are given in Figure 9.9. In a similar manner such modes exist in microrings and in shell-like dielectric envelopes.

In a dielectric microsphere, whispering gallery modes correspond to light that is trapped in circular orbits just within the surface of the structure (Fig. 9.9(b)). The modes are most strongly coupled along the equatorial plane and they can be thought to propagate along a zig-zag path around the sphere. The modes are characterized by two polarizations (transversal electric, TE-modes and transversal magnetic, TM-modes) and three numbers which are the radial n, the angular l, and the azimuthal m mode numbers. The value of l is close to the number of wavelengths that fit into the optical length of the equator. The value of $l - m + 1$ equals the number of field maxima in the polar direction, i.e. perpendicular to the equatorial plane. Mode number n is equal to the number of field maxima in the direction

Fig. 9.8 The side view of St. Paul's Cathedral in London, England. The circular gallery which runs at the point where the vault of the dome starts to curve inwards, is called the Whispering Gallery. The name comes from the fact that a person who whispers facing the wall on one side, can be clearly heard on the other, since the sound is carried perfectly around the vast curve of the dome.

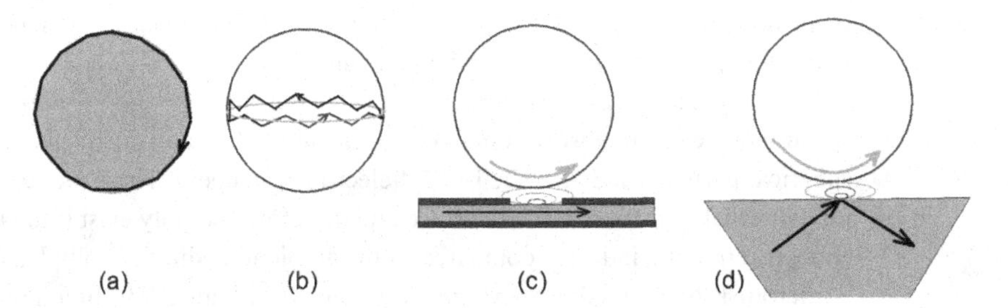

Fig. 9.9 Examples of whispering gallery modes (a) in a microdisk, (b) in a microsphere, and launching light into such modes using evanescent wave leakage (c) from a fiber and (d) from a prism with total internal reflection.

along the radius of the sphere and $2l$ is the number of maxima in the azimuthal variation of the resonant field around the equator. The resonant wavelength is determined by the values of n and l. The electromagnetic field is described by the spherical Bessel functions much similar to the problem of a quantum particle in a spherical potential well (see Section 2.6 in Chapter 2), whispering gallery modes being electromagnetic counterparts of electron states with very large l values.

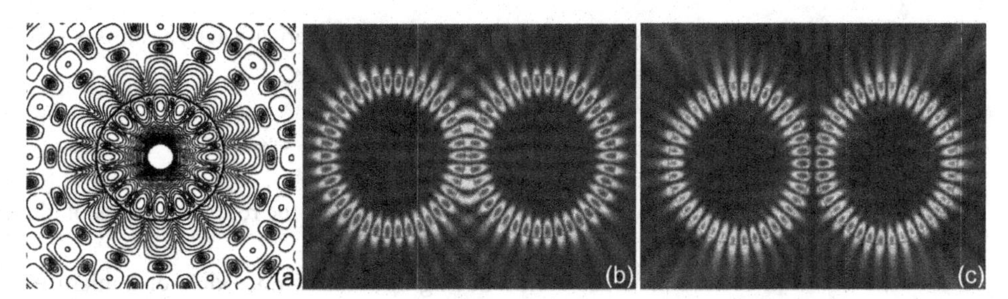

Fig. 9.10 **Light intensity distribution (a) in a single microdisk [42] and (b), (c) in coupled microspheres [43]. See text for detail.**

In Figure 9.10(a), light intensity distribution is presented for a semiconductor microdisk with diameter 2 μm. The disk border is marked by a black circle in the middle. The field distribution is shown within the area of 2 μm × 2 μm. The azimuthal number 8 defines $2 \times 8 = 16$ nodes over the inner perimeter of the disk under consideration. In the near vicinity outside the disk, light takes the form of an evanescent field with smooth build-up of a plane wave afterwards. In 1992 S. L. McCall with co-workers proposed a semiconductor microdisk laser as a promising design of micro-optical circuitry [42]. A single microdisk, a microsphere and even sometimes a dielectric or semiconductor micrometer-sized box are often referred to as *photonic dots* to emphasize the small number of high-Q optical modes by analogy to quantum dots (see Chapter 5) with respect to electrons. Photonic dots are considered as important components in photonic circuitry including lasers, incoherent light sources and optical switching elements. The principal feature is the extreme Q values of the whispering gallery modes. For example, an estimate for a fused high-quality silica with losses of 10 dB/km at a wavelength of 632 nm (He–Ne laser) promises Q of the order of 10^9 for micrometer-sized cavities if scattering losses are ignored. Scattering into the non-resonant modes is severely inhibited in high-Q cavities, whereas scattering into resonant modes does not mean losses.[3] This makes predictions of $Q = 10^{12}$ justified for millimeter-sized spheres in the optical range [44]. High Q-factors mean sharp optical resonance (Eq. (9.1)) and therefore microspheres can be used as ultra-narrow optical filters [45]. It is equally important for a low lasing threshold and for low switching levels. For example, a spherical cavity can be considered instead of a Fabry–Perot resonator in the known schemes of refractive optical bistability based on intensity-controlled resonance wavelength [24, 25]. Braginsky *et al.* proposed the following estimate for an optical bistable element based on tuneable resonance wavelength of a whispering gallery mode [46]. If a spherical cavity is made of a silicate glass with embedded semiconductor nanocrystals (like those discussed in Chapter 5), with typical nonlinear susceptibility $\chi^{(3)} \approx 10^{-9}$ ESU [47] and absorptive

[3] Light scattering probability is proportional to the local density of modes and therefore is strongly enhanced for resonant modes and strongly inhibited for off-resonant modes. The primary enhancement (inhibition) factor is the mode Q-value, i.e. a kind of mode "strength" as compared to the open space in terms of light energy concentration/storing. The additional factor comes from spatial localization of the mode under consideration and reads as λ^3/V where λ is the mode wavelength and V is the actual volume of confined mode in a cavity. Therefore for surface-like modes (which is the case for whispering gallery modes) the scattering can be additionally modified. This issue will be discussed in detail in Chapter 14.

losses 10^4 dB/km for such a material, a 4 μm spherical cavity promises $Q = 3 \cdot 10^4$. Then the bistability threshold,

$$W_{\text{bist}} \approx \frac{n^4 \omega V}{32\pi\chi^{(3)} Q^2} = 4 \times 10^{-5} W, \tag{9.2}$$

and switching time $\tau \approx Q/\omega = 10^{-11}$s. Here n is the refractive index of glass and V is the volume.

The serious problem of introducing light into a microdisk or a microsphere should always be kept in mind when speaking about applications. It is not crucial for lasing as well as for incoherent light emitters because in these cases pump light can be launched at a different wavelength for which total internal reflection can be overcome. However, for every switching or waveguiding application an adequate technique of launching light into a high-Q-disk or sphere is a problem. This can be solved by means of light tunneling using evanescent fields (Fig. 9.9(c) and (d)). For more detail on light tunneling see Chapter 10. Further examples of feeding light by means of tunneling will be discussed in Chapter 11 for surface plasmon polariton excitation at the metal–dielectric interface. Additionally, Chapter 14 will deal with the relation between tunneling and the local density of electromagnetic modes.

When two identical microdisks or microspheres are coupled, collective modes develop. Splitting of original resonances inherent in a single cavity occurs similar to splitting in coupled planar cavities, discussed in Chapter 3 (Fig. 3.24) and in Chapter 8 (Fig. 8.17). Spherical symmetry of cavities makes electromagnetic field distribution similar to that for electron wave function in diatomic molecules. Figure 9.10(b) shows *the bonding* mode and Figure 9.10(c) shows the *antibonding* mode in coupled spheres referred to by analogy with bonding and antibonding electron orbitals in molecules. Coupled disks or coupled spheres are often referred to as *photonic molecules* to complete the analogy with electrons in molecules. Note, the photonic bonding mode has high intensity in the spacing between spheres whereas the antibonding mode features negligible intensity between spheres. The mode structure of coupled spheres has been examined experimentally by several groups using calibrated polymer particles with internal light sources therein (like, e.g. semiconductor quantum dots) as probe light [48, 49]. Lasing in coupled bispheres has also been performed [50].

The next step towards more complicated components consisting of dielectric microspheres is sequential chains of identical particles. In such a chain collective modes can develop which in the limit of an ideally periodic infinitely long chain can be described as Bloch waves [51, 52]. These chains have been shown to offer transportation of light via sequential coupling of spheres including curvilinear displacement [53]. The distinctive feature of such waveguides is crucial sensitivity to coupling efficiency between neighboring spheres and extreme slowing down of light velocity [54]. For finite numbers of spheres multiple splitting of resonances occurs similar to splitting of energy levels in a sequential series of identical quantum wells [55]. Fine aspects of mode coupling in dielectric spheres have also recently been discussed [56]. This field gave rise to the special notation "coupled-resonator optical waveguide" (CROW) and is becoming more and more diverse. For this reason the reader is referred to the above cited works for more detail.

9.5 Propagation of waves and number coding/recognition

Propagation of waves in heterogeneous media which have topological inhomogeneities along the propagation direction with size comparable to the wavelength in a vacuum or in a continuous medium is characterized by multiple scattering and interference. In a medium with randomly distributed spatial inhomogeneities, wave propagation is diffusive and in many cases wave localization occurs. In deterministic non-continuous media, resonance propagation (tunneling) of waves is possible (see Chapters 7 and 8). If inelastic scattering (dissipation) of waves in a medium is negligible, propagation features complementary spectra of resonant transmission and reflection. These are determined by the spatial distribution of the material parameter which defines the speed of wave propagation. These parameters are potential energy for quantum particles, refractive index for electromagnetic waves and material density for acoustic waves. Multilayered media with stepwise variation of the parameter determining wave propagation velocity comprise a specific case of deterministic non-continuous medium.

A layered structure features a stepwise dependence of dielectric function or potential energy along the direction normal to the layer plane. These structures form a specific class of deterministic complex media. It is possible to treat multilayered spatial structures as numbers. Within the framework of such a consideration, propagation of classical waves and quantum particles can be treated as number recognition [57]. A structure containing N substances with different material parameters is treated as a number written to base N. Then propagation of classical waves (and quantum particles) through such a structure can be viewed as number recognition. Layered structures built of two substances with different values of a physical parameter which determines wave propagation velocity will correspond to binary numbers. For example, we ascribe for clarity “1” to a layer with higher value of potential energy, U in the case of quantum particles, refractive index n in the case of electromagnetic waves, density in the case of acoustic waves, and “0” to a layer with the lower value of the proper parameter (Fig. 9.11). Then a finite sequence of layers can be treated as the relevant number.

A representative set of spectral portraits for a few binary numbers is shown in Figure 9.12. Spectral portraits of many binary multilayer structures shown in Chapter 8 can also be added. Principally, a spectral portrait can be assigned to every number. Is this assignment unique? In other words, is it always possible to recover the original spatial distribution of refractive index based on a given spectral portrait? Most probably, in general formulation the problem is not solvable. It is essentially reduced to analysis of the inverse problems in the case of the Helmholtz equation.

For possible applications in data storage devices, a problem can be formulated of identifying classes of sequences possessing a unique relationship between the spatial structure (i.e. a number value) and the resonant frequency (energy) spectrum relevant to propagation of waves through such a structure. Numbers belonging to these classes feature unique spectral portraits and can be unambiguously recognized by means of wave propagation. In optics this property can be used in optical data recording and read-out, in nano- and opto-electronics it can be used in engineering nanostructures with a predefined energy

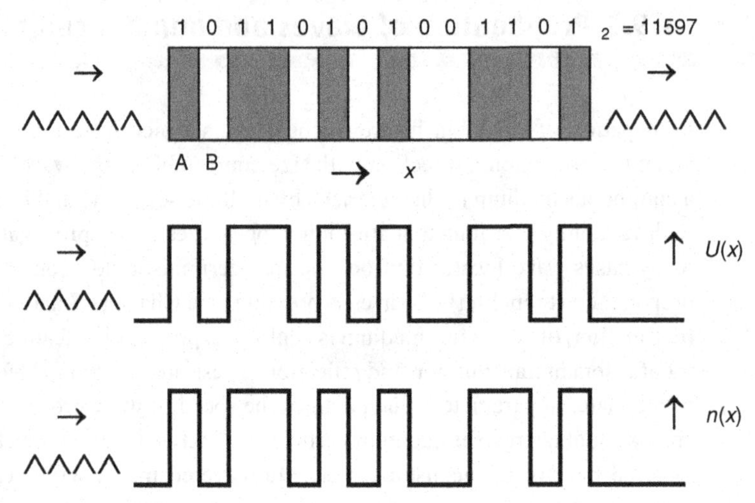

Fig. 9.11 A binary number $10110101001101_2 = 11597$, its representation in the form of a multilayer structure consisting of two substances, A and B, and the corresponding profiles of potential energy and refractive index [57].

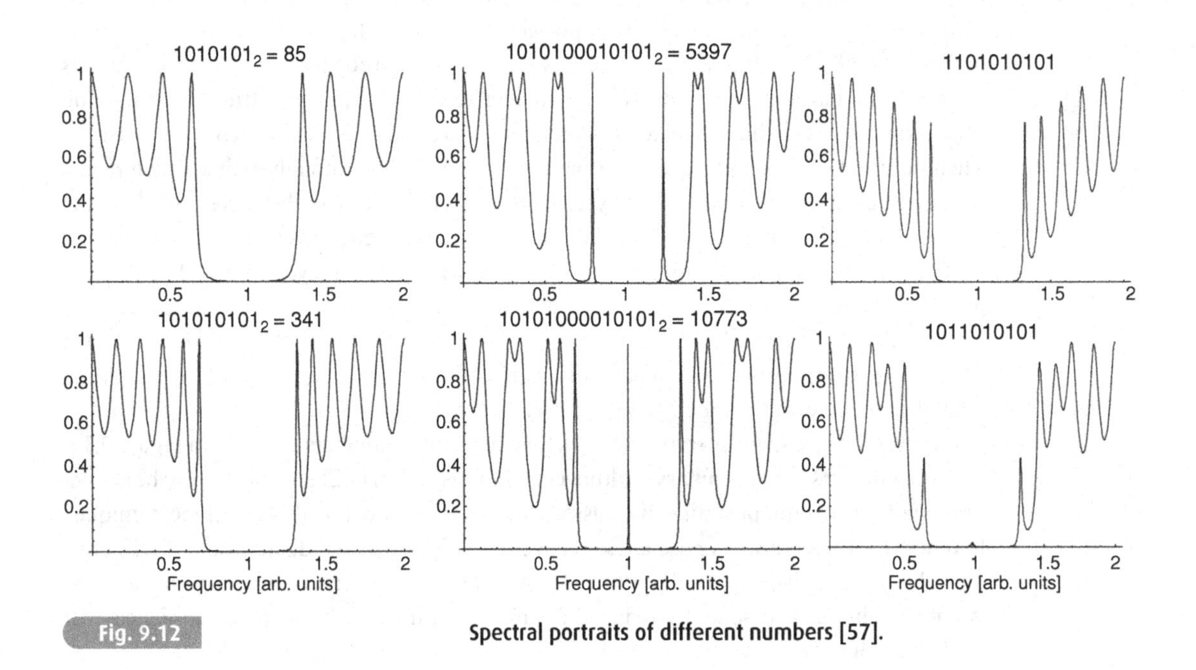

Fig. 9.12 Spectral portraits of different numbers [57].

spectrum of electrons. It is clear that not only identification of a specific number, but also the possibility to code and identify any given number is necessary for information coding. Most probably, an optimal solution of this problem can be gained by searching among all the sets allowing strict identification by spectral portraits, for at least one complete set of numbers [58], i.e. a set providing representation of any given number as a sum of a few different numbers belonging to the set. In case such a set cannot be found, coding of numbers can be performed using representation of a given number as a sum or a product of numbers with unique portraits. However, in this case the gain in information density will not be as significant as it might be in the case of a full set.

9.6 Outlook: current and future trends

Further progress in the research and development of photonic circuitry will no doubt bring a number of really serious solutions for embedding of micro-optical components into existing microelectronic data processing systems, as well as novel compact and efficient components for optical networks. These challenges have replaced the more massive breakthrough which had been expected at the end of the twentieth century and was related to all-optical computing. Insufficient economic investment was found to provide substantial support to all-optical processors, whereas steady progress in silicon microelectronics did push forward the demand for silicon-compatible (better to say "CMOS-compatible", see footnote in Section 9.2 for CMOS explanation) photonics. Therefore optical interconnects, all-optical and electro-optical ultrafast switches with CMOS compatibility are important. Silicon photonics prospects are reviewed in [59, 60].

Additionally, new phenomena in photonic components are still being sought. Representative examples are the *optical Hall effect* and dynamically tuned cavities. The standard Hall effect implies development of a transverse voltage with respect to the originally flowing electrical current in a conductive material when an external magnetic field is applied in the plane normal to current flow. The emerging transverse electric field has a direction normal to both current vector and magnetic field vector. It arises from the Lorenz force acting on electrons moving in the external electric field. Complex media that significantly slow down light propagation are believed to exhibit light scattering in the transverse direction when an external magnetic field is applied [61]. Dynamical tuned cavities offer a possibility to release the light energy stored for the spectrum defined by the original tuning of the cavity in the course of radiation decay from the cavity, with the output spectrum changing in the process of decay [62]. Additional reading on optical microcavities is recommended [63]. Microcavity effects on the spontaneous emission of light by atoms, molecules and quantum dots embedded therein will be considered in detail in Chapter 14.

Magneto-photonic crystals are also being investigated [65] but have not been considered in this chapter because of the restricted scope. These structures promise a giant optical Hall effect which should manifest itself as strong deflection of light passing through such a crystal because of the field effect on the photonic band structure [66].

To summarize, it is reasonable to present an overall expectation of photonic circuitry prospects which was discussed a few years ago by Noda and co-workers (Fig. 9.13). This implies a three-dimensional photonic crystal with embedded array of microlasers generating a set of desirable wavelengths. These wavelengths are then delivered to an optical switch providing data coding and afterwards are submitted to waveguide channels to individual destinations. The components discussed in this chapter do give hope for such circuitry to be developed with possible CMOS compatibility.

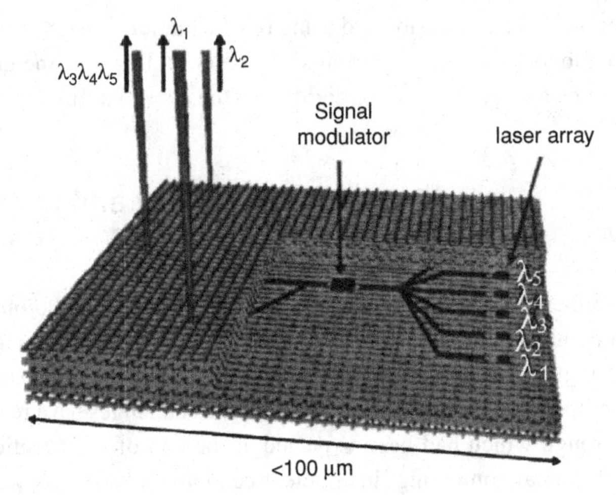

Fig. 9.13 The bright future of nanophotonic circuitry. Reprinted with permission from [64]. Copyright 1999 AIP.

Problems

1. Recalling the earlier example of an optical cavity in Figure 3.19 in Chapter 3, explain in terms of Q-factor why resonant transmission bands get narrower for higher frequencies.

2. Estimate the Q-factor in terms of $\lambda_{max}/\Delta\lambda$ for a cavity formed by two metal mirrors whose transmission spectrum was shown in Figure 3.20.

3. For a microlaser shown in Figure 9.2 explain why it is not correct to measure its volume as 0.03 μm^3. Estimate the actual volume of this laser. Use the epigraph to this chapter as a hint.

4. Find out the electronic analog to an infinitely large number of identical cavities in a photonic crystal. Look for the relevant figure in Chapter 4.

5. Explain radiation leakage in a severely bent traditional dielectric optical fiber waveguide.

6. Based on data on the electric field effect on the absorption spectrum of a Ge quantum well elaborate an electro-optical switching device using photonic crystal architecture. Hints: consider the electric field effect on refractive index implying that absorption and refraction spectra are related via Kramers–Kronig relations in the form similar to that in Figure 6.10. That is, change in absorption will necessarily give rise to change in refraction.

7. Explain the slowing down of an optical bistable element switching at an input intensity slightly exceeding the threshold.

8. Consider lasing as an example of non-equilibrium phase transition in optics.

9. Estimate the basic limit for the switching time of a system consisting of a macroscopic fiber coupled to a microdisk with radius 10 μm and refractive index of the disk material 3.4. Imply $Q = 10^6$ for a microdisk. Hint: consider the time that is necessary for steady-state light intensity in a cavity to build up in terms of Q roundtrips of the light wave therein.

10. Along with the circular whispering galleries, there are elliptical galleries in many cathedrals providing perfect audibility when a speaker and a listener are located in certain positions. Explain the phenomenon. Extend it to three-dimensional ellipsoidal constructions. Find out the optical analog. Look at the quantum analog [67] and explain it. Hint: recall the ellipse definition in geometry based on equal sums of distances from any point on an ellipse to its focuses.

11. Find the wrong light beam in the cover image.

References

[1] Y. Akahane, T. I. Asano, B.-S. Sond and S. Noda. High-Q photonic nanocavity in a two-dimensional photonic crystal. *Nature*, **425** (2003), 944–947.

[2] S. Noda, A. Chutinan and M. Imada. Trapping and emission of photons by a single defect in a photonic bandgap structure. *Nature*, **407** (2000), 608–610.

[3] O. Painter, R. K. Lee, A. Scherer, A. Yariv, J. D. O'Brien, P. D. Dapkus and I. Kim. Two-dimensional photonic band-gap defect mode laser. *Science*, **284** (1999), 1819–1821.

[4] A. Chutinan and S. Noda. Waveguides and waveguide bends in two-dimensional photonic crystal slabs. *Phys. Rev. B*, **62** (2000), 4488–4492.

[5] K. Busch, G. von Freymann, S. Linden, S. F. Mingaleev, L. Tkeshelashvili and M. Wegener. Periodic nanostructures for photonics. *Phys. Rep.*, **444** (2007), 101–202.

[6] T. Sondergaard and A. Lavrinenko. Large–bandwidth single–mode photonic crystal waveguides. *Optics Commun.*, **203** (2002), 263–270.

[7] A. A. Green, E. Istrate and E. H. Sargent. Efficient design and optimization of photonic crystal waveguides and couplers: The interface diffraction method. *Opt. Express*, **13** (2005), 7304–7311.

[8] D. Hermann, M. Schillinger, S. F. Mingaleev and K. Busch. Wannier-function based scattering-matrix formalism for photonic crystal circuitry. *J. Opt. Soc. Amer. B*, **25** (2008), 202–209.

[9] J. S. Jensen and O. Sigmund. Systematic design of photonic crystal structures using topology optimization: low-loss waveguide bends. *Appl. Phys. Lett.*, **84** (2004), 2022–2024.

[10] K. Busch, S. F. Mingaleev, A. Garcia-Martin, M. Schillinger and D. Hermann. The Wannier function approach to photonic crystal circuits. *J. Phys.: Cond. Mat.*, **15** (2003), R1233–R1250.

[11] A. Lavrinenko, P. I. Borel, L. H. Frandsen, M. Thorhauge, A. Harpøth, M. Kristensen, T. Niemi and H. Chong. Comprehensive FDTD modelling of photonic crystal waveguide components. *Opt. Express*, **12** (2004), 234–248.

[12] J. S. Jensen, O. Sigmund, L. H. Frandsen, P. I. Borel, A. Harpøth and M. Kristensen. Topology design and fabrication of an efficient double 90° photonic crystal waveguide bend. *IEEE Photon. Technol. Lett.*, **17** (2005), 1202–1204.

[13] L. H. Frandsen, A. Harpøth, P. I. Borel, M. Kristensen, J. S. Jensen and O. Sigmund. Broadband photonic crystal waveguide 60° bend obtained utilizing topology optimization. *Optics Express*, **12** (2004), 5916–5921.
[14] P. I. Borel, L. H. Frandsen, A. Harpøth, M. Kristensen, J. S. Jensen and O. Sigmund. Topology optimised broadband photonic crystal Y-splitter. *Electron. Lett.*, **41** (2005), 69–71.
[15] T. Niemi, L. H. Frandsen, K. K. Hede, A. Harpøth, P. I. Borel and M. Kristensen. Wavelength division de-multiplexing using photonic crystal waveguides. *IEEE Photon. Technol. Lett.*, **11** (2006), 226–228.
[16] P. I. Borel, B. Bilenberg, L. H. Frandsen, Th. Nielsen, J. Fage-Pedersen, A. V. Lavrinenko, J. S. Jensen, O. Sigmund and A. Kristensen. Imprinted silicon-based nanophotonics. *Optics Express*, **15** (2007), 1261–1266.
[17] Y. A. Vlasov, M. O'Boyle, H. F. Hamann and S. J. McNab. Active control of slow light on a chip with photonic crystal waveguides. *Nature*, **438** (2005), 65–69.
[18] K. Busch, S. F. Mingaleev, A. Garcia-Martin, M. Schillinger and D. Hermann. The Wannier function approach to photonic crystal circuits. *J. Phys.: Cond. Mat.*, **15** (2003), R1233–R1250.
[19] M. V. Ermolenko, S. A. Tikhomirov, V. V. Stankevich, O. V. Buganov, S. V. Gaponenko, P. I. Kuznetsov and G. G. Yakushcheva. Ultrafast nonlinear absorption and reflection of ZnS/ZnSe periodic nanostructures. *Photonics and Nanostructures*, **5** (2007), 101–105.
[20] Y. Jiang, W. Jiang, L. Gu, X. Chen and R. T. Chen. 80-micron interaction length silicon photonic crystal waveguide modulator. *Appl. Phys. Lett.*, **87** (2005), p. 221105.
[21] A. Liu, R. Jones, L. Liao, D. Samara-Rubio, D. Rubin, O. Cohen, R. Nicolaescu and M. Pannicio. A high-speed silicon optical modulator based on a metal-oxide-semiconductor capacitor. *Nature*, **427** (2004), 615–618.
[22] F. Cuesta-Soto, A. Martínez, J. García, F. Ramos, P. Sanchis, J. Blasco and J. Martí. All-optical switching structure based on a photonic crystal directional coupler. *Opt. Express*, **12** (2004), 161–167.
[23] S. F. Mingaleev, A. E. Miroshnichenko and Yu. S. Kivshar. Low-threshold bistability of slow light in photonic-crystal waveguides. *Opt. Express*, **15** (2007), 12380–12385.
[24] F. V. Karpushko and G. V. Sinitsyn. Bistable optical element for integrated optics based on nonlinear semiconductor interferometer. *J. Appl. Spectr.*, **29** (1978), 1323–1326.
[25] H. M. Gibbs. *Optical Bistability: Controlling Light with Light* (Orlando: Academic Press, 1985; Moscow: Mir, 1988).
[26] C. A. Barrios, V. R. Almeida and M. Lipson. Low-power-consumption short-length and high-modulation-depth silicon electrooptic modulator. *J. Ligthwave Technol.*, **21** (2003), 1089–1099.
[27] Q. Xu, B. Schmidt, S. Pradhan and M. Lipson. Micrometer-scale silicon electro-optic modulator. *Nature*, **435** (2005), 325–328.
[28] J. C. Knight, T. A. Birks, P. St. J. Russell and D. M. Atkin. All-silica single-mode fiber with photonic crystal cladding. *Opt. Lett.*, **21** (1996), 1547–1549.
[29] R. F. Cregan, B. J. Mangan, J. C. Knight, T. A. Birks, P. St. J. Russell, P. J. Roberts and D. C. Allan. Single-mode photonic band gap guidance of light in air. *Science*, **285** (1999), 1537–1539.

[30] A. M. Zheltikov. Let there be white light: super continuum generation by ultrashort laser pulses. *Physics – Uspekhi*, **49** (2006), 605–628.

[31] A. M. Zheltikov. Microstructure optical fibers for a new generation of fiber-optic sources and converters of light pulses. *Physics – Uspekhi*, **50** (2007), 705–728.

[32] J. Broeng, D. Mogilevtsev, S. E. Barkou and A. Bjarklev. Photonic crystal fibers: A new class of optical waveguides. *Opt. Fiber Technol.*, **5** (1999), 305–330.

[33] D. Mogilevtsev, T. A. Birks and P. St. J. Russell. Group velocity dispersion in photonic crystal fibres. *Opt. Lett.*, **23** (1998), 1662–1664.

[34] T. A. Birks, J. C. Knight and P. St. J. Russell. Endlessly single-mode photonic crystal fiber. *Opt. Lett.*, **22** (1997), 961–963.

[35] A. Bjarklev, J. Broeng and A. S. Bjarklev. *Photonic Crystal Fibres* (Boston, MA: Kluwer Academic Publishers, 2003).

[36] J. M. Dudley, G. Genty and S. Coen. Supercontinuum generation in photonic crystal fiber. *Rev. Mod. Phys.*, **78** (2006), 1135–1180.

[37] A. M. Zheltikov. *Optics of Microstructured Fibers* (Moscow: Nauka, 2004) – in Russian.

[38] P. St. J. Russell. Photonic-crystal fibers. *J. Lightwave Technol.*, **24** (2006), 4729–4749.

[39] K. V. Dukel'skii, Y. N. Kondrat'ev, A. V. Khokhlov, V. S. Shevandin, A. M. Zheltikov, S. O. Konorov, E. E. Serebryannikov, D. A. Sidorov-Biryukov, A. B. Fedotov and S. L. Semenov. Microstructured lightguides with a quartz core for obtaining a spectral supercontinuum in the femtosecond range. *J. Opt. Technol.*, **72** (2005), 548–550.

[40] Lord Rayleigh. *The Theory of Sound.* Second Edition. 1894–1896 (Ch. XIV, sec. 287).

[41] Lord Rayleigh. The problem of the whispering gallery. In *Scientific Papers*, vol. **5**, pp. 617–620. (Cambridge, England: Cambridge University, 1912).

[42] S. L. McCall, A. F. J. Levi, R. E. Slusher, S. J. Pearton and R. A. Logan. Whispering-gallery mode microdisk lasers. *Appl. Phys. Lett.*, **60** (1992), 289–291.

[43] S. Preu, H. G. L. Schwefel, S. Malzer, G. H. Döhler, L. J. Wang, M. Hanson, J. D. Zimmerman and A. C. Gossard. Coupled whispering gallery mode resonators in the terahertz frequency range. *Opt. Express*, **16** (2008), 7336–7343.

[44] M. L. Gorodetsky, A. D. Pryamikov and V. S. Ilchenko. Rayleigh scattering in high-Q microspheres. *J. Opt. Soc. Amer. B*, **17** (2000), 1051–1057.

[45] S. P. Vyatchanin, M. L. Gorodetskii and V. S. Il'chenko. Tunable narrow-band optical filters with modes of the whispering gallery type. *J. Appl. Spectr.*, **56** (1992), 182–187.

[46] V. B. Braginsky, V. S. Ilchenko and M. L. Gorodetsky. Optical microcavity with whispering-gallery modes. *Physics – Uspekhi*, **160** (1990), 157–159.

[47] S. V. Gaponenko, H. Kalt and U. Woggon. *Semiconductor Quantum Structures. Part 2. Optical Properties* (Berlin: Springer Verlag, 2004).

[48] Yu. P. Rakovich, J. F. Donegan, M. Gerlach, A. L. Bradley, T. M. Connolly, J. J. Boland, N. Gaponik and A. Rogach. Fine structure of coupled optical modes in photonic molecules. *Phys. Rev. A*, **70** (2004), 051801(R).

[49] B. Möller, U. Woggon and M. V. Artemyev. Photons in coupled microsphere resonators. *J. Optics A: Pure and Appl. Opt.*, **8** (2006), 113–121.

[50] Y. Hara, T. Mukaiyama, K. Takeda and M. Kuwata-Gonokami. Photonic molecule lasing. *Opt. Lett.*, **28** (2003), 2437–2439.

[51] N. Stefanou and A. Modinos. Impurity bands in photonic insulators. *Phys. Rev. B*, **57** (1998), 12127–12133.

[52] A. Yariv, Y. Xu, R. K. Lee and A. Scherer. Coupled-resonator optical waveguide: a proposal and analysis. *Opt. Lett.*, **24** (1999), 711–713.

[53] B. M. Möller, U. Woggon and M. V. Artemyev. Coupled-resonator optical waveguides doped with nanocrystals. *Opt. Lett.*, **30** (2005), 2116–2118.

[54] Y. Hara, T. Mukaiyama, K. Takeda and M. Kuwata-Gonokami. Heavy photon states in photonic chains of resonantly coupled cavities with supermonodispersive microspheres. *Phys. Rev. Lett.*, **94** (2005), 203905.

[55] B. M. Möller and U. Woggon. Band formation in coupled-resonator slow-wave structures. *Opt. Express*, **15** (2007), 17362–17368.

[56] L. I. Deych, C. Schmidt, A. Chipouline, T. Pertsch and A. Tünnermann. Optical coupling of fundamental whispering-gallery modes in bispheres, *Phys. Rev. A*, **77** (2008), 051801(R).

[57] S. V. Gaponenko, S. V. Zhukovskii, A. V. Lavrinenko and K. S. Sandomirskii. Propagation of waves in multilayered structures viewed as number recognition. *Opt. Commun.*, **205** (2002), 49–57.

[58] R. L. Graham, D. E. Knuth and O. Patashnik. *Concrete Mathematics: A Foundation for Computer Science* (Addison-Wesley: Reading, MA, 1994, Moscow: Mir, 1998).

[59] R. Soref. The past, present, and future of silicon photonics. *IEEE J. Select. Topics Quant. Electron.*, **12** (2006), 1678–1687.

[60] M. Haurylau, G. Chen, H. Chen, J. Zhang, N. A. Nelson, D. H. Albonesi, E. G. Friedman and P. M. Fauchet. On-chip optical interconnect roadmap: challenges and critical directions. *IEEE J. Select. Topics Quant. Electron.*, **12** (2006), 1699–1707.

[61] C. Koerdt, G. L. J. A. Rikken and E. P. Petrov. Faraday effect of photonic crystals. *Appl. Phys. Lett.*, **82** (2003), 1538–1540.

[62] S. F. Preble, Q. Xu and M. Lipson. Changing the color of light in a silicon resonator. *Nature Photonics*, **1** (2007), 293–296.

[63] K. Vahala (Ed.). *Optical Microcavities* (Singapore: World Scientific, 2004).

[64] A. Chutinan and S. Noda. Highly confined waveguides and waveguide bends in three-dimensional photonic crystal. *Appl. Phys. Lett.*, **75** (1999), 3739–3741.

[65] A. K. Zvezdin and V. A. Kotov. *Modern Magnetooptics and Magnetooptical Materials* (Bristol: IOP Publishing, 1997).

[66] A. M. Merzlikin, A. P. Vinogradov, M. Inoue and A. B. Granovsky. Giant photonic Hall effect in magnetophotonic crystals. *Phys. Rev. E*, **72** (2005), 029902.

[67] H. C. Manoharan, C. P. Lutz and D. M. Eigler. Quantum mirages formed by coherent projection of electronic structure. *Nature*, **403** (2000), 512–515.

10 Tunneling of light

Oh, Kitty! how nice it would be if we could only get through into Looking-glass House! I'm sure it's got, oh! such beautiful things in it! Let's pretend there's a way of getting through into it, somehow, Kitty. . . . In another moment Alice was through the glass, and had jumped lightly down into the Looking-glass room.

Lewis Carroll, "Through the Looking Glass (And What Alice Found There)"

Is it possible for light to "jump lightly" through the looking glass as Alice did? Actually there is no absolutely reflecting border for electromagnetic waves provided that the material forming the reflecting border is restricted in space along the light propagation direction. A variety of light-through-the-looking-glass tunneling phenomena will be the subject of this chapter with the intriguing and challenging issue of superluminal light propagation in tunneling events, as well as with parallel analogies to quantum mechanical counterparts in nanoelectronics. Before reading this chapter, it is advisable to read Sections 3.4 and 3.5 in Chapter 3 for an introduction to tunneling effects in quantum mechanics and optics, as well as Section 7.11 of Chapter 7 where the problem of speed of light evaluation in complex structures has been addressed.

10.1 Tunneling of light: getting through the looking glass

Probably, every known physical case of very high reflectivity of light at some material border or interface brings about an evanescent field which penetrates forward through the border or interface under consideration. Let us recall the cases of evanescent electromagnetic fields that have been identified in this book so far, emphasizing that all these phenomena exhibit the feasibility of making highly reflective barriers transparent in spatially restricted constructions.

The *first* example of tunneling in optics was given in Chapter 3 (Section 3.4, Figs. 3.14, 3.15). Propagation of light through a thin metal film placed in an ambient dielectric medium was considered in terms of the low but finite optical transparency of the film. It was shown to be the classical wave analogy to quantum particle tunneling. It has also been shown that the Fabry–Perot arrangement of a pair of metal films enables us to get very high transmittance at a certain resonant wavelength corresponding to the integer number of half-waves in the cavity between the mirrors.

The *second* example of an optical tunneling phenomenon was also mentioned in Section 3.4 (Fig. 3.16), namely, frustrated total internal reflection.

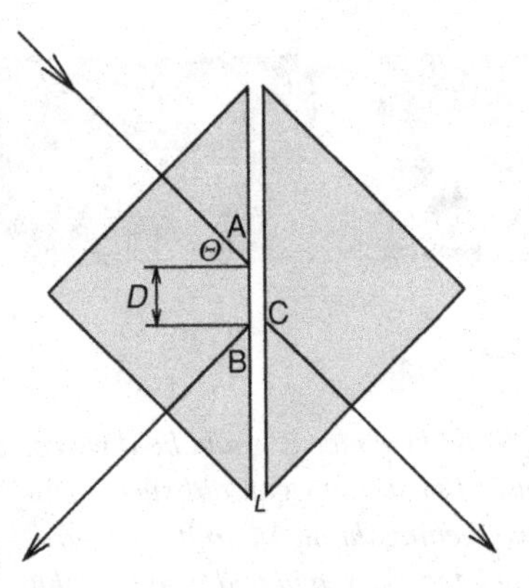

Fig. 10.1 **A two-prism geometry to observe frustrated total internal reflection.**

Let us have a closer look at frustrated total internal reflection (FTIR). Consider a typical FTIR experiment with a pair of prisms (Fig. 10.1).

When a light beam enters an interface between two different dielectrics with indexes of refraction n_1 and n_2 at an angle Θ, total reflection occurs if $n_1 > n_2$ and the angle Θ exceeds the critical angle value defined by the relation,

$$\Theta_{\text{crit}} = \arcsin \frac{n_2}{n_1}. \tag{10.1}$$

This conclusion is entirely based on geometrical optics. In reality (i.e. in wave optics) a number of events occur at and beyond the interface. The incoming light penetrates into the second medium near the point A and travels at a distance D along the interface to the point B before turning back into the first medium. The guesstimate about the finite shift of a light beam in the course of total reflection dates back to Isaac Newton. Its experimental discovery was made much later, in 1947, by F. Goos and H. Hänchen, and since then it has been referred to as the *Goos–Hänchen shift* [1]. In the second medium, the wave can be described by the wave number,

$$k_{\parallel} = k_0 n \sin \Theta, \quad k_0 = \omega / c \equiv 2\pi / \lambda_0, \quad n = n_1 / n_2, \tag{10.2}$$

for propagation from A to B along the interface and the imaginary value,

$$k_{\perp} = i k_0 \sqrt{n^2 \sin^2 \Theta - 1} \equiv i\kappa \tag{10.3}$$

relevant to traversal across the interface and an exponential decay in this direction as $\exp(-\kappa x)$ where κ is a real positive value. Here k_0, and λ_0 are the wave number and the wavelength of incident light in a vacuum, respectively, and ω is its frequency. If a third medium with refractive index $n_3 = n_1$ is displaced in close proximity leaving a narrow break L in the second medium, the reflection becomes "frustrated" in a sense that light (though rather low-intensity) recovers in the third medium and propagates outside. In Chapter 3 we remembered earlier work by L. I. Mandelstam in 1914 (Fig. 3.20) who applied a fluorescent dye rather than the second prism to trace the evanescent light. The two-prism displacement in frustrated total reflection measurements was introduced for the first time by J. S. Bose in 1927.

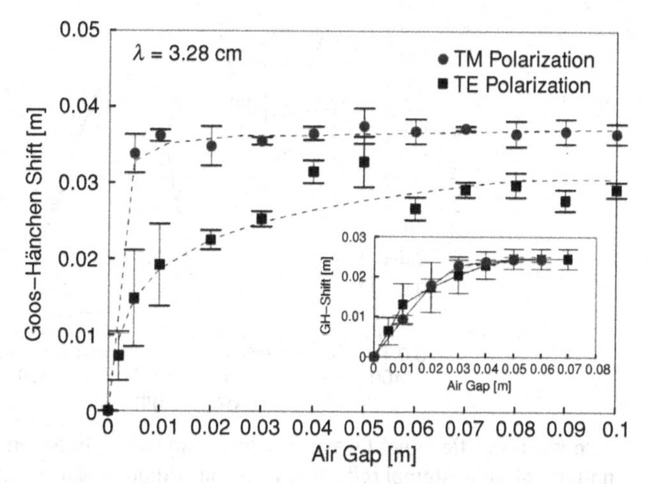

Fig. 10.2 **The Goos–Hänchen shift versus air gap width for TE and TM polarizations. The beam aperture is 80 mm. The insert shows the shift for a TM- and a TE-polarized beam for a larger aperture 190 mm [3].**

The Goos–Hänchen shift is expressed as [2],

$$D = \frac{\partial \varphi}{\partial k_{\parallel}}, \tag{10.4}$$

where φ is the phase shift. The Goos–Hänchen shift D depends on the spacing length between the prisms, on the angle of incidence Θ, on polarization and also on beam width. These regularities have been evaluated in experiments [3] with microwave electromagnetic radiation where measurements of shift value of the order of wavelength are essentially simplified because of a centimeter rather than a micrometer length scale.

Figure 10.2 shows the Goos–Hänsen shift value increase with the air gap width saturating at gap values close to the wavelength value of 3.28 cm. The absolute value of the shift is close to, but somewhat lower than, the wavelength value. Smaller beam widths facilitate difference for TE- and TM-polarized radiations whereas for larger beam widths this difference reduces (inset in Fig. 10.2).

It should be noted that not only the longtitudinal Goos–Hansen shift occurs under conditions of total reflection, but also a transverse shift takes place in the direction normal to the plane of incidence. This is referred to as the *Fedorov shift*. It was predicted by F. I. Fedorov in 1955 [4]. Later on N. N. Pun'ko and V. V. Filippov showed in experiments with microwaves that an incident light wave transforms into a pair of waves with elliptical polarization with symmetrical shift of the order of wavelength with respect to the plane of incidence [5].

Introducing thin layers of high-refractive material into the gap between the two prisms gives rise to multiple interference events and can result in resonant propagation of a wave through the air gap. This phenomenon, in essence, has the same origin as the resonant quantum and optical tunneling phenomena discussed in Chapter 3 (Sections 3.5 and 3.6). It is presented in Figure 10.3, where additional ZnS layers in the gap are seen to modify

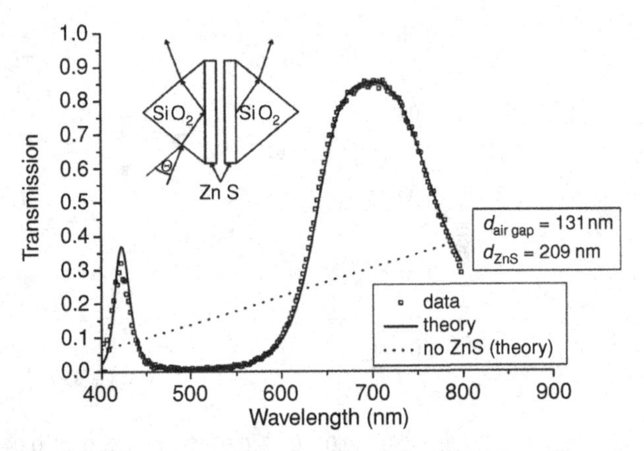

Fig. 10.3 **Transmission spectra of light waves traversing an air gap between two silica prisms under conditions of total internal reflection with and without a pair of thin ZnS layers. The experimental configuration is shown in the inset. Angle of incidence $\Theta = 68.2°$. Adapted from [6].**

the transmission, resulting in resonant enhancement of transmission along with additional inhibition of transmission otherwise.

The *third* example of evanescent light propagation was given in Chapter 7 when considering electromagnetic fields in periodic structures for frequencies corresponding to the band gap. Figure 7.5 shows the intensity profile exponentially vanishing from the front toward the rear end of the finite periodic structure. See also Figure 7.7(a) where optical transmission is shown for three different total numbers of layers in a periodic structure. A gradual decrease in transmission in the gap region with the length of the structure is obvious. The low transmission of a finite periodic structure in the gap region can be multiply enhanced by introducing defect layers therein, forming standing waves for gap wavelengths. This has been demonstrated for a multilayer structure comprising of coupled cavities buried inside a periodic multilayer slab (Fig. 3.24). Certainly, a combination of metal single-film mirror(s) and dielectric multilayer mirror(s) in a single or multiple cavity arrangement works as well. Only unavoidable absorption losses inherent in metals and possible absorption losses in dielectric materials restrict the maximal transmittance value affordable.

10.2 Light at the end of a tunnel: problem of superluminal propagation

What is the speed of light under the conditions of tunneling? What is the time delay between light entering and leaving the tunneling barrier? In Chapter 7 (Section 7.15) high transmission within the pass bands was shown to occur by means of slowing down of the light propagation: the group velocity $v_g = d\omega/dk$ was found to be an oscillating function whose minima perfectly correlate with the transmission maxima (see Fig. 7.32). What is going on inside the gap region where transmission crucially tends to zero with increasing number of periods in a multilayer periodic slab? Amazingly enough, the group velocity rises inside the gap and can become several times larger than the speed of light in a vacuum, c!

The group velocity v_g describes propagation of a pulse peak through the structure. Since in the gap region the dispersion curve $\omega(k)$ experiences a break, the traditional notion, $v_g = d\omega/dk$, is not applicable. Instead, the time τ of "light flight" through the structure under consideration should be calculated to get the group velocity, $v_g = L/\tau$, where L is the total length of the structure. The reasonable approach to traversal time definition is

$$\tau_\varphi := \frac{d\varphi}{d\omega}, \tag{10.5}$$

where φ is the phase of the transmission function. This time is referred to as the "*phase time*", group delay or *Wigner time*, since it was introduced by E. Wigner in 1955 when considering the problem of scattering in quantum mechanics [7].

The Hartman paradox in quantum mechanics

Using the mathematical analogy of the Schrödinger and Helmholtz equations, an approach can be used, developed in quantum mechanics for both quantum and classical tunneling problems, as was done in Section 3.4. From Eq. (3.66) for the wave function and standard conditions of $\psi(x)$ and $\psi'(x)$ continuity, one can derive the amplitude transmission coefficient t,

$$t = \frac{D}{A} = \frac{(1-r^2)\exp(-\kappa L)}{1-r^2\exp(-2\kappa L)}, \tag{10.6}$$

where r is the amplitude reflection coefficient at the potential step,

$$r = \frac{B}{A} = \frac{\kappa + ik}{\kappa - ik}, \tag{10.7}$$

and

$$\kappa = \sqrt{2m(U_0 - E)}/\hbar, \quad k = \sqrt{2mE}/\hbar, \tag{10.8}$$

for a quantum particle, and

$$\kappa_{EM} = \frac{\omega}{c}\sqrt{-\varepsilon_2}, \quad (\varepsilon_2 < 0), \qquad k_{EM} = \frac{\omega}{c}n_1, \quad n_1 = \sqrt{\varepsilon_1}, \quad (\varepsilon_1 > 0), \tag{10.9}$$

for an electromagnetic wave (see also Table 3.6). Here, as in Chapter 3, U_0 is the potential barrier height, ε_1, ε_2 are dielectric permittivities outside and inside the barrier, respectively.

In the limit of an opaque barrier $\kappa L \gg 1$, where L is the barrier width, Eq. (10.6) reduces to [8]

$$t \propto 2\left(1 - i\frac{k^2-\kappa^2}{2k\kappa}\right)^{-1} \exp(-\kappa L), \tag{10.10}$$

with the phase of t,

$$\varphi = \arctan \frac{\kappa^2 - k^2}{2\kappa k}, \tag{10.11}$$

which is relevant to the traversal time. Then the traversal time reads,

$$\tau_\varphi = \frac{\mathrm{d}\varphi}{\mathrm{d}\omega} = 2\left[1 + \left(\frac{k}{\kappa}\right)^2\right]^{-1} \frac{\mathrm{d}}{\mathrm{d}\omega}\frac{k}{\kappa}, \tag{10.12}$$

i.e. it depends on the ratio of k/κ. The expression (10.12) holds equally for quantum tunneling and electromagnetic tunneling phenomena. The peculiarities of a given experiment are accounted for via the dispersion relations $k(\omega)/\kappa(\omega)$. Remarkably, the traversal time is independent of the barrier length L provided that the condition $\kappa L \gg 1$ holds. This condition to a certain extent can be viewed as the condition $|t|^2 \ll 1$, i.e. low transmission coefficient. This means the group velocity $v_g = L/\tau_\varphi$ grows linearly with L, and seems to become an infinitely large value for wider barriers.

Consider the application of Eq. (10.12) to *electron tunneling time*. From Eq. (10.8) one gets,

$$\frac{k(\omega)}{\kappa(\omega)} = \sqrt{\frac{E}{U_0 - E}}, \tag{10.13}$$

which with substitutions $E = \hbar\omega$ and $\frac{\mathrm{d}}{\mathrm{d}\omega} = \hbar \frac{\mathrm{d}}{\mathrm{d}E}$ gives,

$$\tau_\varphi = 2\hbar\left(1 + \frac{E}{U_0 - E}\right)^{-1} \frac{\mathrm{d}}{\mathrm{d}E}\sqrt{\frac{E}{U_0 - E}},$$

whence,

$$\tau_\varphi = \frac{\hbar}{\sqrt{E(U_0 - E)}}. \tag{10.14}$$

Expression (10.13) is known as the "*Hartman paradox*". Since the original paper by T. Hartman in 1962 [9] it has stimulated vast activity in theoretical quantum mechanics and experimental nanoelectronics. It states that *the phase time of tunneling is independent of the barrier length*. In other words, for a very thick barrier there is a finite probability of finding an electron behind the barrier at a time that corresponds to faster-than-light propagation. Moreover, the expression has the feature that as the barrier height rises, the tunneling time decreases. That means the more repulsive the potential, the faster is particle penetration throughout. Note, there is no mass transfer with superluminal propagation because the probability of finding an electron behind the barrier is much smaller than unity.

To estimate the time scale under consideration, take typical, experimentally feasible E and U_0 values. For example, for $E = 1$ eV, $U_0 = 2$ eV, one has $\tau_\varphi = 0.66 \times 10^{-15}$ s.

Equation (10.14) can be written in the form,

$$\tau_\varphi = \frac{\hbar}{E}\frac{1}{\sqrt{(U_0 - E)/E}} = \frac{\hbar}{E}\sqrt{\frac{E}{U_0 - E}}, \tag{10.15}$$

which means it is the energy–time *uncertainty relation*,

$$E\tau > \hbar, \tag{10.16}$$

which gives a reasonable estimate of the traversal time τ_φ. The τ_φ value differs from the value $\tau = E/\hbar$ by the dimensionless factor of the square root of the inverse relative barrier height. The equality $\tau_\varphi = E/\hbar$ holds rigorously when $E = U_0/2$. A higher barrier gives shorter times, whereas for very low barriers the prerequisite condition may not be the case.

A reasonable rationale for uncertainty relation-defined time of particle traversal is as follows. To surmount the barrier a particle needs to "borrow" the energy deficiency of the amount, $U_0 - E$. This virtual borrowing may occur only for a period of time, $\tau_1 \le \hbar/(U_0 - E)$. On the other hand we can treat the situation as though an observer is losing a particle with energy E for the period it traverses under the barrier. Again, the uncertainty relation allows this loss of energy to exist for a period of time, $\tau_2 \le \hbar/E$. One can see Eqs. (10.14) and (10.15) give the harmonic average of the two times, i.e. $\tau_\varphi = \sqrt{\tau_1\tau_2}$. It is possible to show that the time given by Eq. (10.14) for the case of a long barrier defines the built-up time τ_{refl} of the reflected wave for a particle facing a potential step, $U_0 > E$. This built-up time reads,

$$\tau_{\text{refl}} = \frac{1}{\hbar}\frac{2m}{k\kappa} = \frac{\hbar}{\sqrt{E(U_0 - E)}} = \tau_\varphi. \tag{10.17}$$

Calculation of the reflection time and phase time for both cases of infinite and finite barrier length has been made by P. Davies [10]. Its compliance with the causality principle is discussed by A. Steinberg [11]. The key issue is that an observer behind a barrier cannot fix the absolute time when the particle starts traversing, i.e. the signal submission time. Only the difference between submission and reception of the signal (i.e. the particle traversal time) can be perceived. That is, faster-than-light signal transfer is not feasible.

Electromagnetic analog of the Hartman paradox

Consider now an electromagnetic implementation of the Hartman paradox. In Chapter 3, plasma properties were shown to give an immediate analogy to the quantum mechanical tunneling effect. Plasma possesses a negative dielectric permittivity, $\varepsilon < 0$, which corresponds to $E - U_0 < 0$ (see Chapter 3, Sections 3.3 and 3.4 for detail). Equation (10.11) now reads,

$$\tau_\varphi = \frac{d\varphi}{d\omega} = 2\left[1 + \left(\frac{k_{\text{EM}}}{\kappa_{\text{EM}}}\right)^2\right]^{-1} \frac{d}{d\omega}\frac{k_{\text{EM}}}{\kappa_{\text{EM}}}. \tag{10.18}$$

With substitutions according to Eqs. (10.9), using $\varepsilon_1 = 1$ (air) and,

$$\varepsilon_2 = 1 - \frac{\omega_p^2}{\omega^2}, \tag{10.19}$$

for plasma (see Eq. (3.54) in Section 3.3), Eq. (10.17) evolves to the expression,

$$\tau_\varphi = \frac{2}{\omega}\sqrt{\frac{\omega^2}{\omega_\mathrm{p}^2 - \omega^2}} = \frac{2}{\sqrt{\omega_\mathrm{p}^2 - \omega^2}}. \tag{10.20}$$

Bearing in mind that $\omega < \omega_\mathrm{p}$ (otherwise normal propagation rather than tunneling occurs) one can see in the low frequency limit ($\omega \ll \omega_\mathrm{p}$) the phase time tends to $2\omega_\mathrm{p}^{-1}$ which defines its lower limit, i.e.

$$\tau_\varphi > 2\omega_\mathrm{p}^{-1} \tag{10.21}$$

holds. Noteworthy, phase time is independent of the plasma layer thickness provided it is thick enough in terms of $\kappa_{\mathrm{EM}}L \gg 1$. Recalling the discussion in Section 3.4, $L = 1/\kappa$ corresponds to the skin layer thickness. To summarize, we can make the following statements:

(i) *The phase time for electromagnetic wave propagation in plasma when the plasma layer thickness is large compared to the skin layer thickness is thickness-independent, but can not be smaller than twice the inverse plasma frequency.*
(ii) *The phase shift occurs mainly within the skin layer remaining negligible otherwise.*
(iii) *The group velocity, defined as $v_\mathrm{g} = L/\tau_\varphi$ can exceed the speed of light in a vacuum, c.*

Plasma frequency entering the expressions for phase time (10.20) and (10.21) brings a reasonable physical insight to the underlying processes. As an electromagnetic wave whose frequency is below the plasma frequency approaches the air/plasma interface, the plasma reacts as a whole entity with maximal available rate (restricted by the inverse plasma frequency) independent of how thick it is. Its thickness defines the field amplitude at the output, whereas the exhausted replica of the incident wave emerges as fast as is allowed by the plasma frequency. No propagation process actually occurs. Instead, a splash-like motion of the plasma under the incoming impact of the electromagnetic radiation generates a wave of the same frequency at the rear end of the plasma layer. Therefore, the phase shift is negligible outside the skin layer. An instantaneous coherent shock-like splash does not bring a noticeable phase shift.

Phase time in frustrated total reflection

Phase time calculations have recently been applied to various tunneling phenomena in optics by A. Shvartsburg [12] and S. Esposito [8]. Calculations are made for complex amplitude transmission coefficient and then its phase is used to get the traversal time Eq. (10.5). Because of the cumbersome derivation, we shall discuss only the final results.

In what follows we assume the air gap, i.e. $n_1 = n,\ n_2 = 1$. Without restrictions for the gap width L (Fig. 10.1), Shvartsburg obtained the width-dependant expressions for the absolute value $|t|$ and phase shift φ in the course of a frustrated total reflection experiment of the complex amplitude transmission coefficient $t = |t| \exp(i\varphi)$. These read,

$$|t| = \frac{2n\kappa \cos\Theta}{k_0 \cosh(\kappa L)\left[4n^2\cos^2\Theta\left(\frac{\kappa}{k_0}\right)^2 + G^2 th^2(\kappa L)\right]^{1/2}}, \tag{10.22}$$

and

$$\varphi = \arctan\left[\frac{\mathrm{th}(\kappa L)}{2n\cos\Theta}\frac{k_0}{\kappa}G\right], \tag{10.23}$$

where the G value is different for TE- and TM-modes, namely,

$$G_{\mathrm{TE}} = 1 + n^2\cos 2\Theta, \quad G_{\mathrm{TM}} = \cos^2\Theta - n^2\left(\frac{\kappa^2}{k_0^2}\right). \tag{10.24}$$

In these equations, k_0 is the wave number in a vacuum (Eq. 10.2) and κ is the amplitude decay factor (Eq. 10.3). For a large width (when $\kappa L \gg 1$) the energy transmission coefficient exhibits an exponential decrease as,

$$|t|^2 \propto \exp(-2\kappa L), \tag{10.25}$$

whereas phase φ tends to a constant value.

The phase time elucidated from Eq. (10.23) in accordance with Eq. (10.5) reads,

$$\frac{\tau_\varphi}{\tau_0} = \frac{n\cos\Theta G[1 - \mathrm{th}^2(\kappa L)]}{2n^2\cos^2\Theta + \left[\frac{k_0}{\kappa}G\mathrm{th}(\kappa L)\right]^2}, \tag{10.26}$$

where $\tau_0 = L/c$ is the traversal time for normal propagation of light in a vacuum over the distance L. One can see the tunneling time is width-dependent because of the terms $\mathrm{th}^2(\kappa L)$. Although Eq. (10.26) looks rather cumbersome, its width dependence is rather simple: it grows continuously with L when $\kappa L \ll 1$ and then decreases with L, readily taking the value less than unity. Not only is the traversal time shorter than the propagation time in a vacuum for the same distance, but moreover, traversal times are smaller for larger widths exhibiting an asymptotic exponential decrease with L as $\exp(-2\kappa L)$. For larger values of Θ the phase time derived by Shvartsburg even becomes negative, which can not be treated as a physically meaningful result.

Esposito arrived at an asymptotic width-independent traversal time under the assumption of a large gap width ($\kappa L \gg 1$),

$$\tau_\varphi = \frac{2}{\omega}\frac{\sin^2\Theta}{\cos\Theta}\frac{k_0}{\kappa}, \tag{10.27}$$

using Eq. (10.12) and considering FTIR-tunneling as a quasi-one-dimensional problem.

This brief analysis of selected theoretical results unambiguously shows that the theory of light tunneling in frustrated total reflection is far from complete.

Experiments on the time delay in frustrated total reflection have been performed in the microwave range where the characteristic air gap for the evanescent wave to traverse is of the order of a few centimeters [13]. The tunneling time was measured for vacuum wavelength $\lambda_0 = 36$ mm under conditions of an asymptotic Goos–Hänsen shift, $D = 31$ mm. The measured tunneling time was equal to 117 ± 10 ps. The same time delay was documented for reflection. This is in amazing concordance with the $\tau = 1/\omega_0$ value, which for the wavelength used in the experiment is equal to 120 ps. This result gives a definite hint that it is the uncertainty relation that governs the traversal rate in frustrated total reflection experiments. The expression (10.26) is not straightforwardly applicable since it gives frequency-independent time. Expression (10.27) gives a lower value for the experimental parameters used, namely 81 ps. When propagating in a vacuum with speed c, over the time interval, ω_0^{-1} light travels over a distance, $k_0^{-1} \equiv \lambda_0/2\pi$. For the experimental conditions under consideration the traversal length has not been reported and thus it is not possible to judge whether superluminal traversal was or was not the case.

Superluminal propagation and energy flow in a photonic crystal slab

Using analytical expressions for the complex amplitude transmission coefficient derived by J. Bendickson *et al.* [14], it is possible to take the expression for the transmission phase and then to arrive at the analytical expression for the phase time defined by Eq. (10.5). For the simple case of a quarter-wave stack consisting of N layers with equal optical thickness, $n_1 d_1 = n_2 d_2$ (see Section 7.2 in Chapter 7 for discussion of the properties of quarter-wave periodic structures) it reads [8],

$$\tau_\varphi = \frac{d\varphi}{d\omega} = \frac{1}{\omega_0} \frac{\pi c \sinh N\Theta}{\sinh(N-1)\Theta + \dfrac{r_{02}^4 - 1}{t_{02}t_{21}t_{12}} \sinh N\Theta}. \tag{10.28}$$

Here ω_0 is the midgap frequency defined by,

$$\omega_0 = \frac{2\pi c}{\lambda_0}, \quad \lambda_0/4 = n_1 d_1 = n_2 d_2, \tag{10.29}$$

r_{ij}, t_{ij} are the amplitude reflection and transmission coefficients at the border of the two materials whose refractive indices are n_i, n_j (see Eq. (3.21) in Chapter 3), and Θ is defined as,

$$\sinh \Theta = \frac{1}{2}\left(\frac{n_2}{n_1} - \frac{n_1}{n_2}\right). \tag{10.30}$$

The expression (10.28) is valid within the vicinity of the midgap frequency, far from the edges of the transmission bands. The principal factor in Eq. (10.28) is the inverse midgap frequency which is, again, manifesting the uncertainty relation influence, whereas the rest of the complicated fraction is the constant factor defined by the periodic slab design only.

G. D'Aguanno *et al.* [15] have examined the group velocity in the photonic gap of a finite periodic slab based on the *dwell time* concept that was introduced into quantum mechanical tunneling by F. Smith in 1960 [16]. A quantum particle spends a mean time inside a potential barrier which is proportional to the probability of finding a particle therein, i.e.

$$\tau_{\text{dwell}} \propto \int_0^L |\Psi(x)|^2 \, dx. \tag{10.31}$$

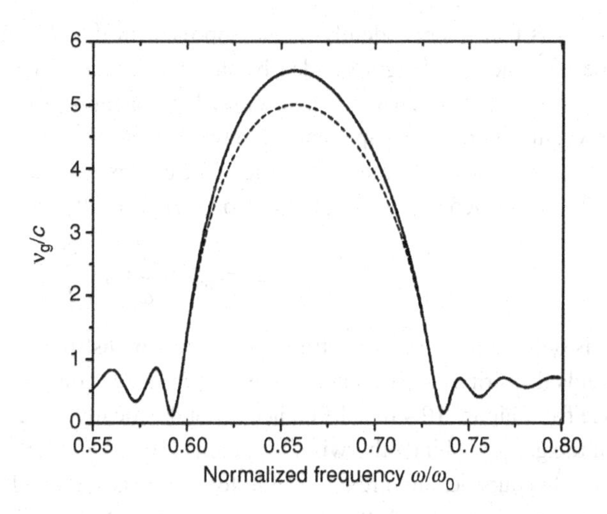

Fig. 10.4 The tunneling velocities defined as $v_g = L/\tau_{\text{dwell}}$ (solid line) and $v_g = L/\tau_\varphi$ (dashed line) in the photonic gap region and adjacent spectral ranges calculated for the 20-period structure discussed in Section 7.11 (see Fig. 7.32) [15].

For the optical problem, the probability density is replaced by the electromagnetic energy density to give,

$$\tau_{\text{dwell}} = \frac{1}{c}\int_0^L \varepsilon_\omega(x)\left|\frac{E_\omega(x)}{E_{I\omega}}\right|^2 \mathrm{d}x - \frac{1}{\omega}\mathrm{Im}\left(\frac{E_{R\omega}}{E_{I\omega}}\right). \tag{10.32}$$

Here $E_{I\omega}$, $E_{R\omega}$ are incident and reflected electric fields for a wave with frequency ω. Equation (10.32) states that the time the field spends inside the structure is proportional to the energy density integrated over the volume. The second term represents the difference in energy between the electric and magnetic components, and it has no quantum-mechanical counterpart. Equation (10.32) establishes a clear link between delay time and field localization.

Thus defined the time can be used to get the group velocity value as,

$$v_g = \frac{L}{\tau_{\text{dwell}}}, \tag{10.33}$$

bearing in mind this is not exactly the group velocity, since there is no propagating wave inside the barrier and the delay time, rather than velocity, remains the primary physical value to characterize the tunneling event.

In Figure 10.4 calculations are shown made for the same multilayer stack whose properties were presented in Figure 7.32. Unlike Figure 7.32 where light velocities in the transmission bands were examined, now the velocity Eq. (10.33) along with the velocity defined as $v_g = L/\tau_\varphi$ are presented. One can see perfect convergence of the two velocities in the transmission bands. Here Bloch waves with defined wave number and group velocity propagate (see Section 7.11), the velocity being lower than c everywhere with oscillations relevant to development of high transmission by means of slower propagation. In this section, the band-gap region is under examination. The visible divergence of the two velocities within the gap region is not physically valuable. The most important physical result is that, independently of the approach in defining the tunneling time, the ratio of the periodic slab length divided by the time exceeds the speed of light in a vacuum, c, by several times. That is, superluminal propagation of the electromagnetic field when traversing a photonic gap comes into the discussion.

Further analysis of tunneling time in the context of phase time and dwell time definitions [17, 18] showed that in fact, these two times basically converge,

$$\tau_\varphi = \tau_{\mathrm{dwell}} + \frac{1}{\omega}\mathrm{Im}\left(\frac{E_{R\omega}}{E_{I\omega}}\right). \tag{10.34}$$

It is only asymmetry in electromagnetic energy distribution between electric and magnetic fields that brings a difference in the phase time as compared to the dwell time. As one can see from Figure 10.4 this difference is not drastically large and occurs mainly in the middle of the gap vanishing otherwise. Equations (10.32) and (10.34) hold not only within the gap but also outside, and are applicable for transmission bands of periodic structures as well. For transmission bands the second term in Eq. (10.34) vanishes and the following statement can be made:

Phase time = dwell time (outside the photonic band gap).

Combining Eqs. (10.32) and (10.34), an alternative expression for the phase time can be written in the form,

$$\tau_\varphi = \frac{1}{c}\int_0^L \varepsilon_\omega(x)\left|\frac{E_\omega(x)}{E_{I\omega}}\right|^2 \mathrm{d}x. \tag{10.35}$$

This gives further insight into the meaning of phase time. In structures with a periodic dielectric function, the phase time is determined by the integrated product of the dielectric function and electric field spatial redistribution.

Equation (10.34) holds strictly for symmetric structures in symmetric ambient environments on both sides of the structure. For asymmetrical structures it should be replaced by a more complex expression [17]. There are, e.g. short periodic structures consisting of a few periods. Short symmetrical structures should contain half-integer numbers of periods. Long structures ($L \gg \lambda_0$) asymptotically can be treated as symmetrical ones.

It should be emphasized that $v_g > c$ occurs under conditions of vanishing transmission coefficients (see, e.g. Figures 7.7 and 7.25 for reference) of a periodic slab. The energy transport velocity under this condition is rather slow. The *energy velocity* v_E can be defined in a very clear and straightforward manner as

$$\textit{Energy velocity} = \frac{\langle \textit{Poynting vector} \rangle}{\langle \textit{Energy density} \rangle} \tag{10.36}$$

where $\langle \ldots \rangle$ means averaging over the volume. Thus defined velocity gives the energy flow rate through the slab as was shown in Section 7.12 (Eqs. (7.26) – (7.30)). It reads as the product of the dwell-time velocity and the intensity transmission coefficient [15],

$$v_E = \frac{L}{\tau_{\mathrm{dwell}}}|t|^2. \tag{10.37}$$

This relation converges perfectly with the energy velocity discussed in Section 7.12 for Bloch waves in transmission bands (Eq. 7.30) under conditions that are not severely rigorous and allow notation of velocity in terms of Eq. (10.33), even though one should bear in mind the delay time in tunneling is actually what is physically meaningful. It is rather instructive to observe the energy velocity given by Eq. (10.37) as a universal relation for the whole frequency range when writing about photonic crystals. Thus elucidated, *the energy velocity never exceeds the speed of light in a vacuum* in complete conformity with the causality principle. It is the condition $v_E < c$ which determines the upper limit for the tunneling process and one can see that since the transmission coefficient is far from unity, then the dwell-time- and phase-time-velocities can be much higher than c, without any confusion with respect to causality.

Let us discuss the experimental evidence for tunneling rates in photonic band-gap structures. The available experiments include optical range as well as radiofrequency range.

Using a two-photon interferometer based on a UV-laser and a nonlinear crystal, Steinberg *et al.* [19] have measured the time delay of tunneling across a barrier consisting of a 1.1-μm-thick 1D photonic band-gap material. The photonic band-gap structure was a $(5 + 1/2)$ period of $SiO_2(n = 1.41)$ and $TiO_2(n = 2.22)$ layers. With total thickness 1.1 μm a traversal time is 3.6 fs for traveling at the speed of light in a vacuum, c. The band gap extended from 600 to 800 nm approximately, the minimal transmission being about 1% at 692 nm. The peak of the photon wave packet appears on the far side of the band-gap structure 1.47 ± 0.21 fs earlier than it would if it were to travel at the vacuum speed of light, c. The apparent tunneling velocity $(1.7 \pm 0.2)\,c$ is superluminal. However, the authors note that this is not a genuine signal velocity, and Einstein causality is not violated.

Spielmann *et al.* [20] studied the propagation of electromagnetic wave packets through 1D photonic band-gap materials using 12 fs optical pulses. The measured transit time was found to be paradoxically short (implying superluminal tunneling) and independent of the barrier thickness for opaque barriers. These authors registered shortening of Fourier-limited incident wave packets upon propagation through these linear systems. They emphasized that, although in apparent conflict with causality and the uncertainty principle, neither of these general principles is violated because of the strong attenuation suffered by the transmitted signals.

Mojahedi *et al.* [21] used single microwave pulses centered at 9.68 GHz with 100 MHz (full width at half maximum) bandwidth and studied their evanescent tunneling through a one-dimensional photonic crystal. In a direct time-domain measurement, it was observed that the peak of the tunneling wave packets arrives (440 ± 20) ps earlier than the companion free space (air) wave packets.

Conclusions and outlook

Similarity of tunneling in quantum mechanics and in electromagnetism occurs, not only in the steady-state transmittance formulation, but in time-resolved tunneling processes as well. Plenty of unclear issues related to tunneling phenomena in both fields of physics exist. In theory, the complexity of time-resolved analysis comes from the time-dependent

Schrödinger and electromagnetic wave equations. To avoid obstacles from time-dependent equations, analysis is performed based on time-independent steady-state Schrödinger and Helmholtz equations by means of introducing the dwell time and phase time notions. The latter acquired a definite spread in the analysis of optical tunneling but the problem reduces to the evaluation of phase with respect to evanescent fields within a barrier. The range of obstacles extends from elucidation of the physically meaningless negative transit time in tunneling [12] to claims that tunneling is by no means a propagation process and the notion of velocity is not applicable to tunneling at all [22].

The lack of experiments where tunneling time is elucidated is also evident. For quantum particles, experimental data seem not to be available at all for tunneling times at the time of writing (2007). For optical experiments, superluminal traversal of photonic band gap materials seems to be reliably registered. In the microwave region, superluminal transit governed by the uncertainty relation was documented for frustrated total reflection experiments. The latter seems not to be the best experimental choice for testing theoretical predictions since even the theory of linear light propagation in this case is far from being complete.

It is reasonable to note that in the case that a consistent analysis of time-dependent equations in optics and quantum mechanics is performed, the concordance of quantum and classical tunneling may diverge because of the second-order time derivative inherent in the electromagnetic wave equation and the first-order time derivative inherent in the Schrödinger equation. Only in the slowly varying envelope function approximation can these two fields of physics further merge, since in this case the second time derivative can be approximated by the first one. In optics, such an approximation corresponds to pulses whose length is much larger than the inverse frequency, i.e. pulse duration should measure at least tens of femtoseconds. Superluminal transit times in tunneling do not violate the causality and relativity principles. Signal transmission cannot be performed at a speed determined by the tunneling traversal times. The energy velocity in every case remains subliminal and can be defined as the product of intensity transmission coefficient and dwell- or phase-time velocity.

10.3 Scanning near-field optical microscopy

In a traditional far-field optical microscope, the object is ideally illuminated by a monochromatic plane wave. The light, scattered by the object, is collected by a lens and imaged onto the retina or a photosensitive matrix detector. The lens is placed at least several wavelengths away from the object surface, i.e., in the far field. The spatial resolution is restricted by the *diffraction limit*,

$$\Delta x = \frac{\lambda}{2\pi \mathbf{NA}}, \tag{10.38}$$

which was first evaluated by M. Abbé in 1873. Here **NA** stands for *numerical aperture* which is an important characteristic of the imaging system,

$$\mathbf{NA} = n \sin(\alpha/2), \tag{10.39}$$

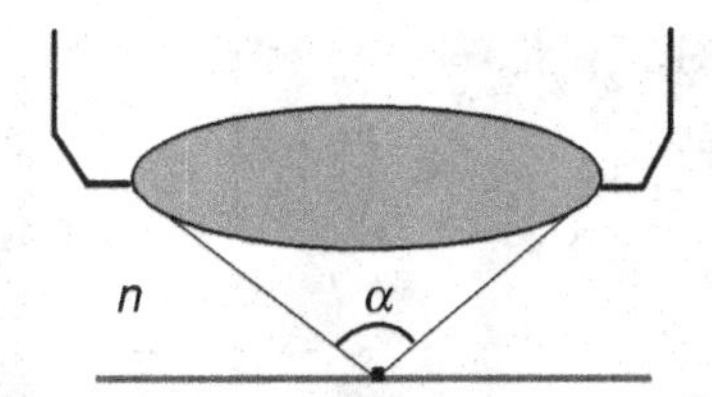

Fig. 10.5 The object and the lens in a traditional optical microscope.

where n is the refractive index of the ambient medium and α is the angle indicated in Figure 10.5.

With high-refractive ambient liquids **NA** $= 1.6$ is feasible. This number along with the wavelength form the basic limit of resolution in a microscope.

To negate this limit, the scanning *near-field* optical microscope (SNOM) was invented. In this device, imaging develops by means of in-plane point-to-point scanning of the surface under investigation with a sub-wavelength optical fiber. It is now roughly the size of a fiber tip that determines the resolution limit. However, a subwavelength tip allows only evanescent waves to be developed at the output and therefore light intensity falls exponentially outside the tip. Fine submicrometer-scale mechanics is mandatory to precisely control the tip–surface distance and its in-plane position.

The idea was first formulated by Irish physicist E. Synge in 1928. He proposed to use a sub-wavelength hole in a thin metal film as a primary light source that should be scanned over the surface at a sub-wavelength distance. At that time technical performance of micro-optics and micro-mechanics gave no chance for experimental implementation. In 1982 it was re-invented by D. Pohl at the IBM Rüschlikon Research Laboratory [23, 24]. This happened very soon after the quantum counterpart, the scanning tunneling microscope has been invented by G. Binnig and H. Rohrer at the same laboratory. It is reasonable to reproduce an abstract of Pohl's patent [24]:

"This optical near-field scanning microscope comprises an "objective" (aperture) attached to the conventional vertical adjustment appliance and consisting of an optically transparent crystal having a metal coating with an aperture at its tip with a diameter of less than one wavelength of the light used for illuminating the object. Connected to the aperture-far end of the "objective" is a photodetector via an optical filter and an optical fiber glass cable. Scanning the object is done by appropriately moving the support along x/y-coordinates. The resolution obtainable with this microscope is about 10 times that of state-of-the-art microscopes."

The invention became possible owing to the fabrication of a sub-wavelength optical aperture at the apex of a sharply tapered fiber probe tip that was coated with a metal, and implementation of a feedback loop to keep a constant spacing of only a few nanometers while raster scanning the sample in close proximity to the fixed probe. Nano-positioning mechanics became available owing to the development of the electron tunneling microscope.

A representative tip appearance is shown in Figure 10.6. Light is fed from the top and evanesces from the apex in the bottom end. The sample is placed on top of a hemispherical substrate which allows efficient harvesting of light emerging from the probe–sample interaction zone into the far field.

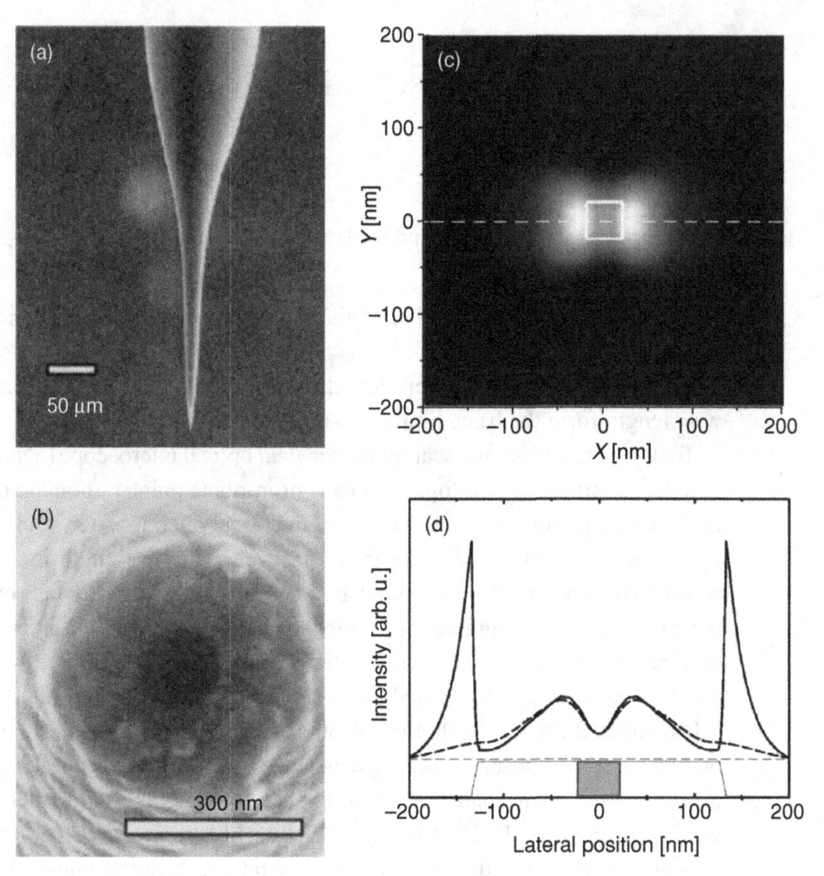

Fig. 10.6 Scanning near-field optical microscopy. (a) side view and (b) end view of a tapered fiber tip covered with aluminium. (c), (d) show computer simulation of signal obtained by means of a tapered fiber for a glass box $40 \times 40 \times 40$ nm^3 whose cross-sections are shown by a white square in (c) and gray square in (d). Light is polarized along the x-direction. To reproduce an experimental image, each pixel in (c) corresponds to a different calculation with a particular tip–sample position, the tip apex being kept at a constant height 2 nm above the top sample surface. (d) Comparison of different scanning modes: constant height (dashed line) and constant gap (solid line) measured along the dashed line in (c). The corresponding tip motion is depicted by gray lines at the bottom of the figure. Reprinted with permission from [25]. Copyright 2000 AIP.

Taking into account the subject of this book we shall concentrate on the tip properties rather than positioning techniques. It is clear that the tip should optimally meet two controversial requirements: (i) the spot size determined by the aperture diameter should be as small as possible, whereas (ii) the light intensity at the aperture should be as high as possible. Light waves fed into the thick origin of the tip reduce to a single mode in a thinner part of the tip (where the diameter is close to the light wavelength), which in turn evolves to an evanescent field at the output. The transmission efficiency is very small, 10^{-6}–10^{-8}, the rest of the radiation either reflects back into the fiber or is absorbed by the metal envelope of the tip. The theory of these processes is far from complete. Interpretation of signals collected in the course of scanning needs a thorough theoretical description of light

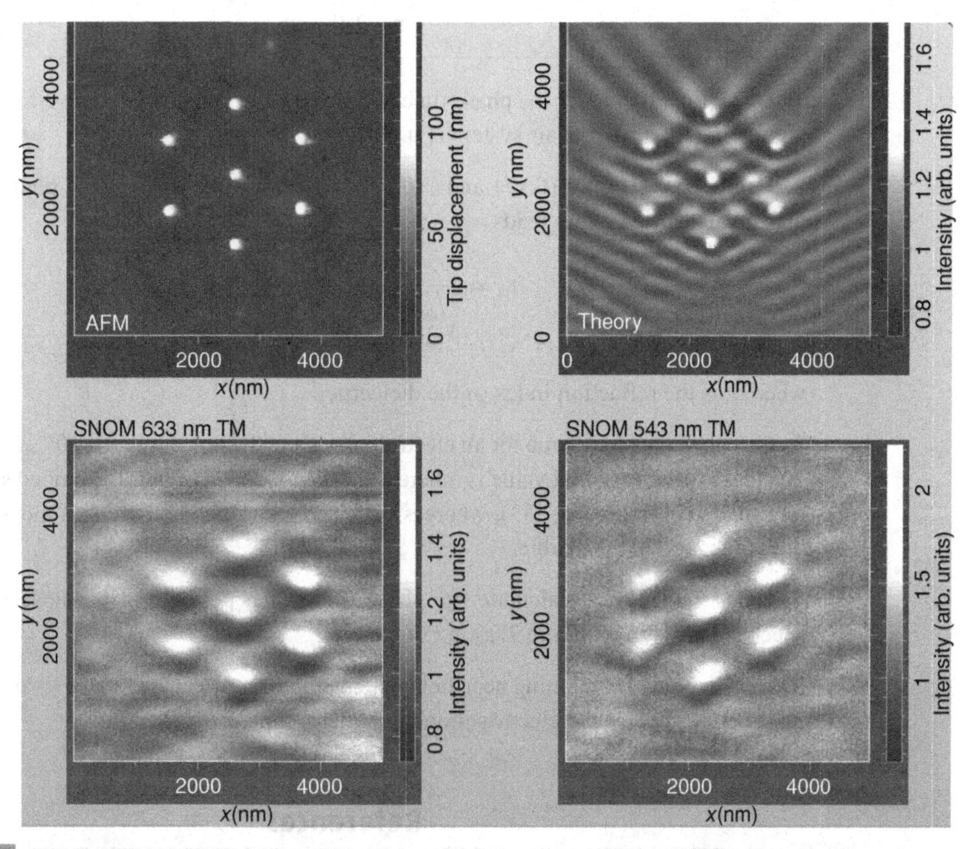

Fig. 10.7 **Imaging of the same portion of a surface containing seven arranged dots with approximately** 100 nm **height obtained by atomic force microscopy (AFM, top left), theoretical simulations of electric field distribution with no tip (top right), and with SNOM imaging using TM-field at** 633 nm **(bottom left) and** 543 nm **(bottom left).** ***Courtesy of A. Dereux.***

scattering, ideally, a rigorous solution of the inverse wave scattering problem. Figure 10.6(c) and (d) show numerical simulations of the light intensity collected in scanning a tip over a glass cube. The cube contour is shown as the white square in Figure 10.6 (c) and as a gray square in Figure 10.6 (d). Figure 10.6 (c) shows there is definite correlation between topographical structure and SNOM image, but the cube contour is rather smeared and the light distribution differs considerably along the x- and y-axes. Figure 10.6 (d) shows that the signal depends on whether scanning at a constant height above the surface, or at a constant gap is performed.

Figure 10.7 illustrates the real sub-wavelength resolution capability of a practical SNOM-device. Not that SNOM images differ from topographical data and from the theory with no tip involved. Typically resolution is still far from its ultimate limit. The image is also dependent on the wavelength of electromagnetic radiation used. We shall see in Chapter 14 (Section 14.8) that this is local density of electromagnetic modes (photon states) which defines the SMON image rather than simple topographical relief. For further reading on SNOM theory and practice, the following books and reviews are recommended [25–28].

Problems

1. Using Eq. (10.17) for the phase time and Eqs. (10.9) and (10.19) derive for yourself the phase time for a plasma/air system (Eq. 10.20).

2. Using Eqs. (10.9), (10.17) and (10.19) prove for yourself that the phase time for a plasma/dielectric system reads,

$$\tau_{\varphi} = \frac{2}{\sqrt{\omega_{\mathrm{p}}^2 - \omega^2}} \frac{n}{1 + (n^2 - 1)\dfrac{\omega^2}{\omega_{\mathrm{p}}^2}},$$

where n is the refraction index of the dielectric.

3. Calculate the phase time for an electron plasma with concentrations 10^{19}, 10^{20}, 10^{21} and 10^{22} cm^{-3}. The very first value is inherent in heavily doped or hardly pumped semiconductors (like, e.g. in Figure 7.33), whereas the very last value is inherent in metals providing high reflection in the visible.

4. Using Eq. (10.28) calculate the phase time for a periodic slab whose transmission spectrum is given in Figure 7.25 in Chapter 7.

5. Explain why the scanning near-field optical microscope can be treated as an electromagnetic counterpart of the electron tunneling microscope.

References

[1] F. Goos and H. Hänchen, Ein neuer und fundamentaler Versuch zur Totalreflexion. *Ann. Phys. (Leipzig)*, **1** (1947), 333–346.

[2] C. K. Carniglia and L. Mandel. Phase-shift measurement of evanescent electromagnetic waves. *J. Opt. Soc. Amer.*, **61** (1971), 1035–1043.

[3] A. Haibel, G. Nimtz and A. A. Stahlhofen. Frustrated total reflection: The double-prism revisited. *Phys. Rev. B*, **63** (2001), 047601.

[4] F. I. Fedorov. On the theory of total reflection. *Proc. Acad. Sci. USSR.*, **105** (1955), 465–470.

[5] N. N. Pun'ko and V. V. Filippov. Splitting of incident light beam under condition of total reflection into two elliptically polarized beams. *Opt. Spectr.*, **58** (1985), 125–129.

[6] I. R. Hooper, T. W. Preist and J. R. Sambles. Making tunnel barriers (including metals) transparent. *Phys. Rev. Lett.*, **97** (2006), 053902.

[7] E. P. Wigner. Lower limit for the energy derivative of the scattering phase shift. *Phys. Rev.*, **98** (1955), 145–147.

[8] S. Esposito. Universal photonic tunneling time. *Phys. Rev. E*, **64** (2001), 026609.

[9] T. E. Hartman Tunneling of a wave packet. *J. Appl. Phys.*, **33** (1962), 3427–3432.

[10] P. C. W. Davies. Quantum tunneling time. *Amer. J. Phys.*, **73** (2004), 23–27.

[11] A. M. Steinberg. Clear message for causality. *Phys. World*, **16**, issue 12 (2003), 19–20.

[12] A. B. Shvartsburg. Tunneling of electromagnetic waves: paradoxes and prospects. *Physics – Uspekhi*, **50** (2007), 43–58.

[13] A. Haibel and G. Nimtz. Universal tunneling time. *Ann. Phys. (Leipzig)*, **8** (2001), 707–712.

[14] J. M. Bendickson, J. P. Dowling and M. Scalora. Analytic expressions for the electromagnetic mode density in finite, one-dimensional, photonic band-gap structures. *Phys. Rev. E*, **53** (1996), 4107–4121.

[15] G. D'Aguanno, M. Centini, M. Scalora, C. Sibilia, M. J. Bloemer, C. M. Bowden, J. W. Haus and M. Bertolotti. Group velocity, energy velocity, and superluminal propagation in finite photonic band-gap structures. *Phys. Rev. E*, **63** (2001), 036610.

[16] F. T. Smith. Lifetime matrix in collision theory. *Phys. Rev.*, **118** (1960), 349–356.

[17] H. G. Winful. Group delay, stored energy, and the tunneling of evanescent electromagnetic waves. *Phys. Rev. E*, **68** (2003), 016615.

[18] G. D'Aguanno, N. Mattiucci, M. Scalora, M. J. Bloemer and A. M. Zheltikov. Density of modes and tunneling times in finite one-dimensional photonic crystals: A comprehensive analysis. *Phys. Rev. E*, **12** (2004), 016612.

[19] A. M. Steinberg, P. G. Kwiat and R. Y. Chiao. Measurement of the single-photon tunneling time. *Phys. Rev. Lett.*, **71** (1993), 708–711.

[20] Ch. Spielmann, R. Szipöcs, A. Stingl and F. Krausz. Tunneling of optical pulses through photonic band gaps. *Phys. Rev. Lett.*, **73** (1994), 2308–2311.

[21] M. Mojahedi, E. Schamiloglu, F. Hegeler and K. J. Malloy. Time-domain detection of superluminal group velocity for single microwave pulses. *Phys. Rev. E*, **62** (2000), 5758–5766.

[22] H. G. Winful. Energy storage in superluminal barrier tunneling: Origin of the "Hartman effect". *Opt. Express*, **10** (2002), 1491–1496.

[23] W. D. Pohl. Optical near-field scanning microscope. US Patent 4,604,520 (1986).

[24] D. Pohl, W. Denk and M. Lanz. Optical stethoscopy: Image recording with resolution $\lambda/20$. *Appl. Phys. Lett.*, **44** (1984), 651–653.

[25] B. Hecht, B. Sick, U. P. Wild, V. Deckert, R. Zenobi, O. J. F. Martin and D. W. Pohl. Scanning near-field optical microscopy with aperture probes: Fundamentals and applications. *J. Chem. Phys.*, **112** (2000), 7761–7774.

[26] D. Courjon. *Near-field Microscopy and Near-field Optics* (London: Imperial College Press, 2003).

[27] L. Novotny and B. Hecht. *Principles of Nano-Optics* (Cambridge: Cambridge University Press, 2007).

[28] C. Girard. Near-fields in nanostructures. *Rep. Prog. Phys.*, **68** (2005), 1883–1933.

11 Nanoplasmonics II: metal–dielectric nanostructures

Nanoplasmonics is a modern extensively expanding field of optics. It basically develops from the rather old field of metal optics including properties of dielectrics with subwavelength nanometer-size metal inclusions. For this reason, reading of Chapter 6 is strongly recommended prior to proceeding to the content of this chapter.

This chapter is organized as follows. First, local electromagnetic field enhancement of near metal singularities in a dielectric environment is considered. Further, multiple scattering phenomena are discussed including extraordinary transmission of a hole array in a metal film. Then, spatially organized metal–dielectric structures are discussed and for that section to be well understood, Chapter 7 will be helpful as will Chapter 10, where the tunneling of light in metal structures has been treated. Possible effects of metal nanoparticles on laser action and gain media properties are discussed as well. A separate section deals with general properties of negative refraction materials. This is included in this chapter since experimental realization of negative refraction in optics is based on periodic metal–dielectric metamaterials with specific subwavelength-scale architecture. Negative-refraction optics (and electrodynamics) emerged just a few years ago and the introductory style of this issue in the present chapter is believed will assist the interested reader in further self-learning through browsing in the many original papers on that matter.

As with all previous chapters in this book, the consideration in this chapter is organized rather like a guide into the field with emphasized physical phenomena than instructions for calculations and for fabrication of nanoplasmonic structures. The latter goals can be further perceived with reference to books and reviews [1–10].

11.1 Local electromagnetic fields near metal nanoparticles

Surface plasmon resonances inherent in metal nanoparticles embedded in a dielectric ambient medium promote development of high local light intensity in the close vicinity of a particle. In coupled nanoparticles and in the purposeful arrangement of particles, as well as with purposeful shaping of particles, electromagnetic field enhancement can be further magnified. This phenomenon forms the basic pre-requisite effect in linear and non-linear spectroscopies, including metal nanotextured surface-enhanced and metal tip-enhanced fluorescence and Raman scattering among linear optical phenomena, as well as nonlinear phenomena like, e.g. second harmonic generation, hyper-Raman scattering and others.

Surface plasmon oscillations give rise to a local increase in light intensity some time after the light enters the vicinity of a metal nanoparticle. The local areas of higher light intensity coexist with other areas where light intensity is not enhanced and may even be depleted as compared to the propagation of light in a continuous dielectric medium. Even for a single isolated particle the thorough electrodynamic calculations to be performed are very cumbersome and can only be realized numerically. It is the general opinion that the ideal case would be a small particle (preferably a prolate ellipsoid or spheroid as well as cone-like or other sharp tip) or, better, a pair of particles of material with low damping rate and minor interband transitions. Damping rate defines directly the sharpness of resonant response in terms of spectral width of the resulting real part of the dielectric function of the composite metal–dielectric medium. Interband transitions give rise to a finite imaginary part of the complex dielectric function and bring dissipative losses to the system in question. Because of the complexity of calculations we refer the reader to a recent book [2] where computational techniques are discussed in detail. In what follows only the final numerical results will be discussed.

In the list of metals promising superior local field enhancement, first place is given to silver, second to gold and third to copper, in accordance with damping rates and interband transitions. The crucial parameter is also the crossover point in the dielectric function where it passes through the zero value. For the above metals this point falls in the optical range. Alkali metals, even in spite of the minor contribution from interband transitions, are not suitable for local light intensity enhancement because their zero-crossover points go deep into the ultraviolet region.[1] These characteristics of metals were discussed in Chapter 6. Because of the above arguments, silver is a typical material for all model calculations.

Local fields near a single metal particle

For a single isolated silver particle local electric field enhancement factors $|\mathbf{E}|/|\mathbf{E}_0|$ of the order of 10^2 have been predicted for a 10–20 nm particle which corresponds to the intensity enhancement $I/I_0 \propto |\mathbf{E}|^2/|\mathbf{E}_0|^2 \approx 10^4$. The enhancement peaks at wavelength 370–400 nm, i.e. close to the extinction peak, are typically observed for dispersed silver nanoparticles (see Section 6.4). In Figure 11.1 the calculated field enhancement factor is plotted in the close vicinity of a spherical silver nanoparticle embedded in a medium with refractive index n_{out}. It is important to note that, for a given wavelength, field enhancement depends not only on particle size but on the refractive index of the ambient medium. Since local field enhancement manifests itself as superior Raman scattering and fluorescence enhancement for various probes, this effect can be purposefully exploited in various sensing applications where small deviations of ambient refractive index are to be monitored. Note that diminishing enhancement with growing radius is not as straightforward while the wavelength is fixed. For larger particles the optimal wavelength may shift to larger values because of the scattering contribution to extinction.

Along with the particle size and ambient environment control, there are two other ways to get superior field enhancement factors. These are particle shape design and multi-particle

[1] Additionally, alkali metals are not actually suitable for surface-enhanced spectroscopy experiments in air because of rapid oxidation.

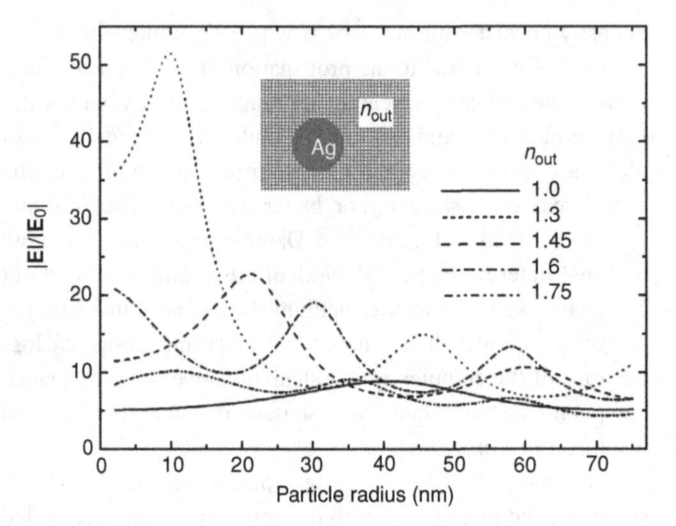

Fig. 11.1 **Local electric field enhancement factor $|\mathbf{E}|/|\mathbf{E}_0|$ near a spherical silver particle embedded in the medium with refractive index n_{out}. Wavelength is 400 nm. Adapted from [11].**

displacement. The enhancement strongly depends on a particle shape and gets higher for prolate particles. For example, for the bottom plane of a truncated tetrahedron with in-plane size 167 nm and 50 nm height the intensity enhancement was calculated to be as high as 47 000 [2]. Unlike smaller particles, in this case enhancement occurs at a wavelength of 646 nm. This red shift is from the scattering contribution to the extinction (see Fig. 6.13(b) in Section 6.4 for reference). For a prolate silver spheroid (length 120 nm, diameter 30 nm) intensity enhancement about 10^4 was obtained in calculations at a wavelength of 770 nm. Notably, the extreme enhancement numbers above are inherent in very restricted space portions, typically measuring 1–5 nm in one dimension. These small values might be questionable in the context of continuous electrodynamic applicability.

Spatially arranged particles

Coupled particles ("plasmonic dimers") show higher enhancement in the area between particles, the enhancement factors being strongly dependent on electric field orientation with respect to a dimer axis. Intensity enhancement of 10^4 was predicted between two silver spherical particles of 30 nm diameter and 2 nm spacing at wavelength 520 nm [2].

Figure 11.2 presents an electric field distribution between a pair of parallel silver cylinders for orientation of wave vector and electric field vector with respect to the axis connecting the centers of the cylinder cross-sections. This gives a reasonable picture of what is going on in the spacing between nanoparticles. Enhancement occurs in the middle between the cylinders and is higher for smaller diameters. For 5 nm spacing between 30 nm cylinders, the field amplitude enhancement is 24, i.e. intensity enhancement greater than 500 becomes feasible. This only happens for proper polarization with the electric field vector oriented along the axis between the cylinders. In the case of polarization across that axis,

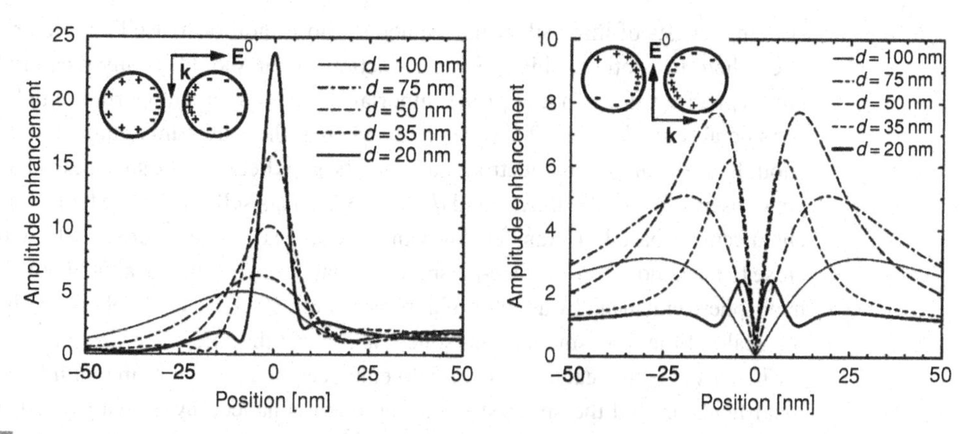

Fig. 11.2 Electric field amplitude enhancement for coupled silver cylinders. Diameter d for the two different orientations of the incident field $\mathbf{E}^0$ and wave vector $\mathbf{k}$. The separation between the cylinders is scaled with d and equals $0.2d$. Adapted from [12].

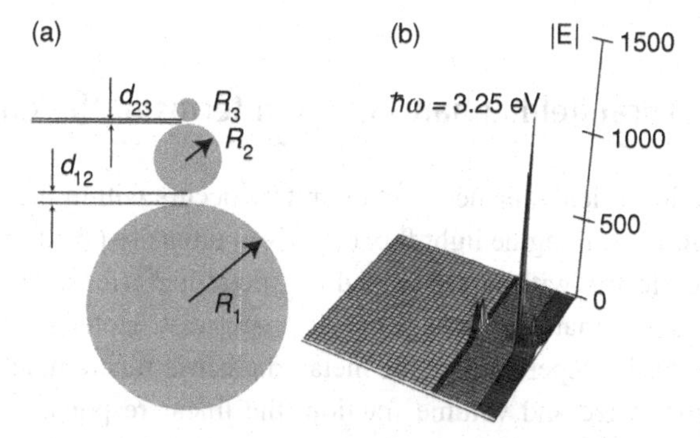

Fig. 11.3 Local field enhancement in a self-similar chain of metal nanoparticles. (a) Geometry of a nanolens consisting of three silver nanospheres is shown in cross-section through its plane of symmetry. Nanosphere radii are: $R_1 = 45$ nm, $R_2 = 15$ nm and $R_3 = 5$ nm. The separations between the nanosphere surfaces are $d_{12} = 4.5$ nm and $d_{23} = 1.5$ nm. (b) Local electric field enhancement coefficient as a function of the position in the cross-section through the plane of symmetry. Adapted from [13].

depletion of light in the middle occurs with minor enhancement in the near vicinity of each cylinder.

A transition from single particle to dimer provides a hint towards engineering of plasmonic nanostructures with enormous enhancement of the electromagnetic field. Further steps can be made based on more complex geometries. Sarychev and Shalaev computed the surface field distribution for an electromagnetic wave impinging the nanotextured surface formed by irregular particles with sufficiently dense concentration to develop percolation clusters [3]. They found in certain "hot spots" the intensity may rise by more than 10^4 times.

Stockman with co-workers [13] proposed an ingenious self-similar plasmonic "lens" using sequentially located nanoparticles with scalable size and spacing (Fig. 11.3). This system consists of three silver nanospheres whose radii decrease by a factor of $1/3$ from one sphere to another. This specific reduction factor should be significantly less than 1, but its precise value is not of prime importance. The larger nanoparticle's radius should be considerably smaller than the wavelength, whereas the minimum radius should still be large enough to use continuous electrodynamics. The gaps between the surfaces of the nanosphere are chosen to be, $d_{12} = 0.3R_2$, and $d_{23} = 0.3R_3$. This self-similar system exhibits superior enhancement based on transfer down the spatial scale. The external field with frequency close to the nanosphere surface-plasmon resonance excites the local field around the largest nanosphere enhanced by a factor of approximately 10. This local field is nearly uniform on the scale of the next smaller nanosphere and plays the role of an external excitation field for it. This in turn creates the local field enhanced by an order of magnitude. Similarly, the local fields around the smallest nanosphere are enhanced by a factor of 10^3 for the field amplitude. The problem was solved in the quasistatic approximation using a multipolar expansion for Green's function. The local field distribution shown in Figure 11.3(b) features the "hottest" spot in the smallest gap at the surface of the smallest nanosphere. This local field is enhanced by a factor of $|\mathbf{E}| \approx 1200$.

Local field enhancement in terms of the energy conservation law

The local field enhancement definitely occurs within the framework of the energy conservation law. Imagine light flux of a given intensity (Watts per unit square) impinging a slice of dielectric with rarely embedded metal inclusions. The optical transmission properties of such a material have been discussed in Chapter 6 in terms of the effective medium approach. Depending on the metal–dielectric function, ambient refractive index and metal particle size and volume fraction, the linear response is characterized by the extinction spectrum. Even in the case of zero absorption in a metal, the total flux outside the slice will be close to, but always less than, the input flux. Therefore, local field enhancement can be, in the first approximation, treated as redistribution of the incident field with light concentration near the metal bodies and light depletion otherwise. In that case the build-up of output flux will be instantaneous, controlled by the speed of light in a medium with a given refractive index. However, the local field enhancement seems to obey more complex regularities. Numerous experimental observations clearly indicate that luminescent probes (organic molecules, lanthanide ions, semiconductor quantum dots) dispersed randomly in a metal–dielectric composite medium enable harvesting of one order of magnitude more light as compared with similar samples without metal inclusions. For Raman scattering several orders of magnitude enhancement in far-field light harvesting can be readily performed. These phenomena are referred to as *Surface Enhanced Raman Scattering* (SERS) and *Surface Enhanced Fluorescence* (SEF). These phenomena will be the subject of close consideration in Part II where light–matter interaction in nanostructures will be considered. Here we want to emphasize that local field enhancement most probably occurs via light energy accumulation near metal nanobodies. This accumulation takes a certain time in accordance with energy conservation. Thus high secondary emission harvesting in the far field becomes possible. Therefore local field enhancement build-up most probably occurs by means of the time delay in light propagation, as happens in the case of interferometers, cavities and periodic structures. This issue will be the subject of thorough theoretical and experimental studies over the next decade. Note, the above consideration based on light accumulation in the near vicinity of a metal nanobody means that a metal nanobody with its close spatial vicinity acquires a finite Q-factor as does a cavity. Light energy accumulation and finite Q-factor for a single nanobody can be understood as a multiple reversible electromagnetic field – surface plasmon conversions. As a result, light spends more time near a nanobody. This gives rise to both energy accumulation and to slowing of propagation.

The Q-factor based interpretation becomes even more intuitive when spatially arranged nanobodies are discussed. Even in random ensembles multiple scattering and constructive interference of scattered waves are possible. For coupled nanobodies or spatial configurations like that in Figure 11.3, constructive interference effects are definitely more plausible.

11.2 Optical response of a metal-dielectric composite beyond Maxwell-Garnett theory

The simple Maxwell-Garnett theory based on electrostatic consideration of small spherical particles rarely dispersed in a host dielectric is valid only for low metal volume filling fraction $f \ll 1$. Beyond this limit (for larger particles, specific particle shape, arbitrary volume fraction) an analytical solution of the problem is not feasible and a thorough numerical computation within the framework of multiple scattering theory should be performed. While there are numerous data on light scattering behavior in complex dielectric media [14–18] only now are the first steps being made towards a consistent theory of metal–dielectric composites with arbitrary metal concentration.

In terms of modified scattering by a single particle with plasmonic resonance accounted for, recent calculations by R. Dynich and A. Ponyavina are instructive [11]. These authors showed more than three orders of magnitude enhancement in the near-field nanoparticle cross-section (Fig. 11.4). This enhancement in individual scattering, combined with interference of scattered waves results in complex properties of dense metal–dielectric composites. Additionally to multiple scattering and interference phenomena, percolation can develop at certain concentrations of the metal part of a composite. This brings about an additional multitude of effects since the electric conductivity further modifies optical response. Percolation phenomena are very important for clustered metal–dielectric structures. These are treated in detail by V. Shalaev [5]. Since fractal metal clusters typically occur in dense structures, inevitable dissipation losses prevent their use in optical filters, waveguides and other components, and they will not be discussed here.

Figure 11.5 shows transmission and reflection spectra for a single monolayer of silver spherical particles randomly located in a host medium with filling factor $f = 0.4$ (the main panel) and 0.6 (the inset) [19]. In these calculations size-dependent damping in the form of Eq. (6.73) has been involved. The inset shows the contribution from multiple scattering and interference. Without multiple scattering, as a starting approximation,

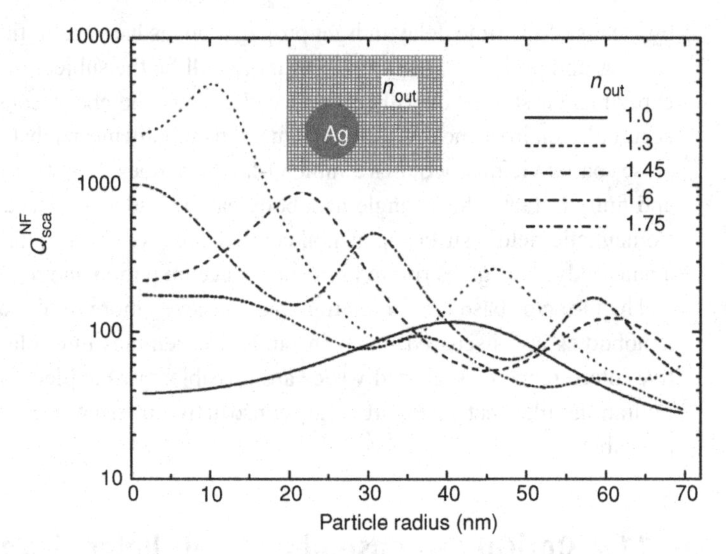

Fig. 11.4 Near-field scattering cross-section of a single silver particle depending on its radius for different values of ambient refractive index n_{out}. Wavelength is 400 nm. Adapted from [11].

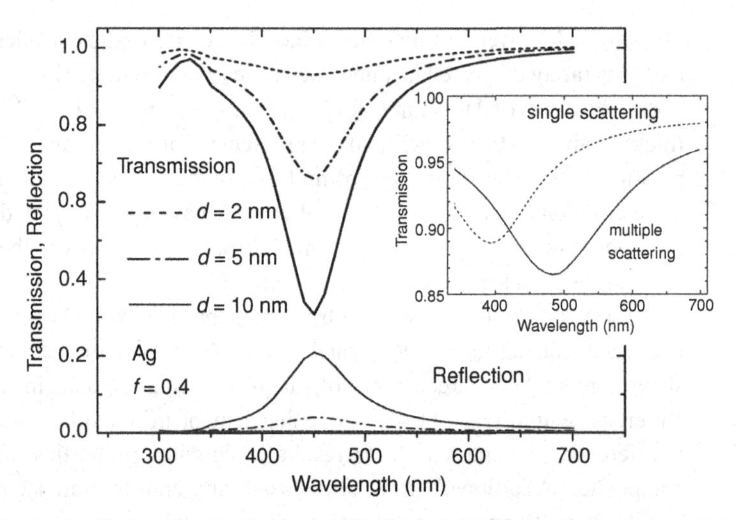

Fig. 11.5 Transmission and reflection spectra for a monolayer of identical silver spherical particles with various diameters $d = 2, 5, 10$ nm in a medium with refractive index $n_{out} = 1.4$. Silver volume filling factor $f = 0.4$. The inset shows transmission spectra calculated based on single scattering (dashes) and multiple scattering (solid line) approach (particle diameter $d = 2$ nm, volume filling fraction $f = 0.6$, ambient refractive index $n_{out} = 1.4$). Adapted from [19].

the straightforward Bouger law can be used based on a single particle extinction efficiency (Eq. 6.83 in Chapter 6). This result is shown in the inset by dashes, whereas the solid line represents the correct account for multiple scattering. One can see that multiple scattering shifts the transmission minimum (and the corresponding reflection maximum) to the red. This is clear from the inset in Figure 11.5. Note, single scattering

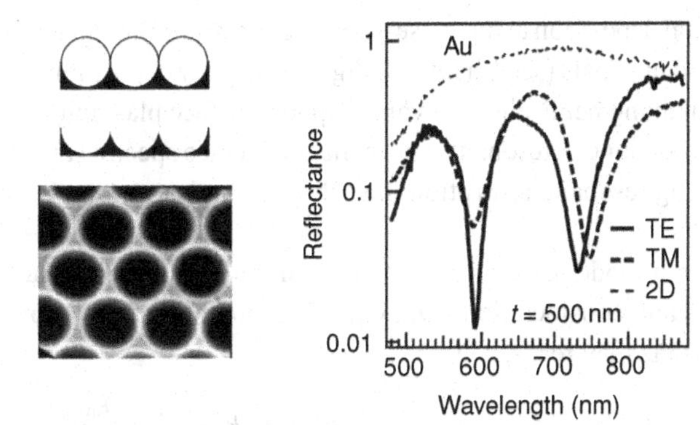

Fig. 11.6 A metal structure with hemispherical voids: (left) synthesis, and (right) reflectance spectra for a 500-nm-thick graded sample [21]. Void radius is 350 nm. Solid line corresponds to TE- and dashes correspond to TM-polarized incident light for an angle of incidence equal to 45°. The dotted curve is the reflectance of a reference unpatterned film electrodeposited under the same conditions.

calculations agree reasonably with the data on transmission spectra of a dielectric containing a small portion ($f \ll 1$) of silver particles of the same diameter, shown in Chapter 6 (Fig. 6.13a).

The above red shift from multiple scattering can be viewed as a result of self-consistent modification of the dielectric function of the composite medium. An electromagnetic wave experiencing a scattering event from a given nanoparticle further propagates in an effective medium whose dielectric function is modified by all other particles. Therefore the Fröhlich condition, expressed by Eq. (6.57), must be modified in a self-consistent manner to account for this effect (see Problem 2). As a starting approximation, the Maxwell-Garnett theory can be applied to get an estimate of the shift, at least in the sense of its direction on the wavelength scale. Such a self-consistent approach is in line with the single electron theory of solids, where an electron's motion in a self-consistent potential is considered where the potential comes from ion cores but is modified by the presence of other electrons.

The main panel shows development of a selective reflection band with the corresponding dip in transmission. Note, larger particles more readily form selective reflection. Smaller particles ($d < 10$ nm) show somewhat smeared reflection which is, most probably, because of enhanced damping (compare with Fig. 6.13(a), see also Eq. (6.73)).

Interestingly, for a single monolayer of spatially organized spherical metal particles, S. Kachan and A. Ponyavina found by means of the multiple scattering theory the blue shift of the transmission minima (and reflection maxima), as compared to random particles of the same volume fraction [20]. For silver particles with diameter 2 nm, filling fraction $f = 0.6$, and ambient refractive index $n_{\text{out}} = 1.4$ (i.e. the parameters used for the data shown in the inset in Fig. 11.5) a transition from random to a triangular or square arrangement shifts the spectra to the blue by approximately 50 nm. For a monolayer consisting of nanoparticle chains, strong dependence of transmission and reflection on incident light polarization has been revealed by calculations.

Another example of a spatially arranged metal dielectric structure is a regular metal sieve with subwavelength voids (Fig. 11.6, left). This can be fabricated by means of templated metal deposition using close-packed dielectric colloidal particles like those used to fabricate artificial opals (see Chapter 7, Fig. 7.22). After deposition the template can be removed. The remaining hemispherical voids support surface plasmon modes which manifest themselves as pronounced resonant dips in the reflectance spectra (Fig. 11.6, right). The theory predicts strong resonant absorption in such voids and, accordingly, resonantly inhibited reflectivity [22].

In a model case of a full spherical void in a solid metal, a rather straightforward electrostatic consideration can be applied. Equation (6.53) for polarizability α of a spherical particle now reads,

$$\alpha = 4\pi R^3 \frac{\varepsilon_{\text{void}} - \varepsilon_{\text{met}}}{\varepsilon_{\text{void}} + 2\varepsilon_{\text{met}}}, \tag{11.1}$$

where dielectric permittivities $\varepsilon_{\text{void}}$ of a void and ε_{met} of a metal stand instead of the dielectric permittivities of a metal particle and a host dielectric, respectively. Accordingly, the Fröhlich condition expressed by Eq. (6.57) now reads,

$$\text{Re}[\varepsilon_{\text{met}}(\omega)] = -\frac{1}{2}\varepsilon_{\text{void}}. \tag{11.2}$$

This "inverted" Fröhlich condition has been already used in Section 6.6 to define inner interface plasmon polaritons in a metal nanoshell.

11.3 Extraordinary transparency of perforated metal films

In 1998, T. Ebbesen with co-workers [23] reported on the extraordinary high transmittance of light through a perforated silver film with a regular array of subwavelength holes. These experimental data are presented in Figure 11.7 (left). The transmission spectrum exhibit well-known Rayleigh minima of Wood's anomaly which appears in any diffractive array roughly when an order of diffraction emerges tangent to the array.[2] The spectrum also features the extraordinary transmittance effect: at a wavelength $\lambda = 800$ nm, the intensity transmission coefficient is close to 15%. As the area covered by holes is only 11%, the normalized-to-area transmittance of light is 130%.

For a single aperture with diameter d, the intensity transmission coefficient for normal incidence reads [1],

$$T = \frac{64\pi}{27}\left(\frac{d}{\lambda_0}\right)^4, \tag{11.3}$$

and for the case under consideration the standard aperture theory for a single hole predicts a transmission coefficient of the order of 1%. Accounting for surface plasmon modes results

[2] In 1902 Robert Wood discovered sharp variations in diffracted light intensity for certain angles of incidence on a metal-grooved grating. The first attempt to explain Wood's anomalies was made by Lord Rayleigh in 1907. The theory of this phenomenon is based on surface plasmon excitation in metal grooves and was advanced in 1965 by A. Hessel and A. Oliner. See [7] for more detail.

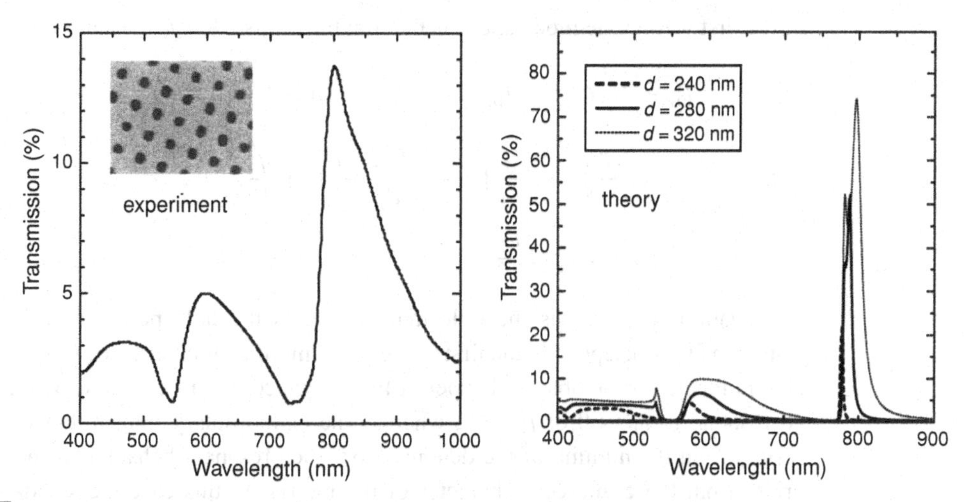

Fig. 11.7 **(*Left*) Experimental power transmittance at normal incidence for a square array of holes (lattice constant is 750 nm, average hole diameter is 280 nm) in a freestanding Ag film (thickness is 320 nm). *Inset*: electron micrograph of the perforated metal film. (*Right*) Calculated transmittance at normal incidence for an array of holes in a Ag film, defined by the lattice constant 750 nm, thickness 320 nm and three different hole diameters *d* indicated in the inset. Adapted from [24].**

in considerable modification of the standard aperture theory when a hole in a metal film is considered. Development of surface plasmon modes gives rise to efficient enlargement of an aperture in question and results in higher transmittivity for thinner, subwavelength-scale films [25].

However, properties of a single aperture, even modified to account for plasmons, can not explain the extraordinary transmission presented in Figure 11.7. A consistent explanation of the observed extraordinary transmittance of a hole array in a metal film is possible in terms of the multiple scattering theory with surface plasmon modes taken into account. The results of the relevant calculations for a silver film with perforated holes are shown in Figure 11.7, indicating that this theory does provide a reasonable rationale for the observed phenomenon. The transmission peak is predicted at the same position, near 800 nm, where it has been observed experimentally. To get a qualitative insight into the underlying physics, the simplified analytical model consideration is instructive, as has been proposed by L. Martín-Moreno and co-workers [24]. The complex transmission coefficient for electric field amplitude (light propagating along the z-axis) reads,

$$t_{010} = \frac{t_{01} t_{10} \exp(-|q_z|h)}{1 - \rho^2 \exp(-2|q_z|h)}, \tag{11.4}$$

formally coinciding with the formula for a triple dielectric–metal–dielectric structure, considered in Chapter 3 (Eq. 3.80). Here t_{01} and t_{10} are the amplitude transmission coefficients at interfaces, $\rho = \rho_{10} = \rho_{01}$ is the amplitude reflection coefficient, q_z is the z-component of the first waveguide mode and h is the film thickness. Now t_{01}, t_{10} and ρ read,

$$t_{01} = 2\frac{d}{L}G^{-1}, \quad t_{10} = \frac{2q_z}{k_0}G^{-1}, \quad \rho = \left(\frac{q_z}{k_0} - G_1\right)\left(\frac{q_z}{k_0} + G_1\right),$$

$$G = \left(\frac{d}{L}\right)^2 + 2\frac{k_0}{k_z}\left(\frac{d}{L}\frac{\sin(k_x d/2)}{k_x d/2}\right)^2 + \frac{q_z}{k_0}, \tag{11.5}$$

$$k_x^2 + k_z^2 = k_0^2 \equiv \frac{\omega^2}{c^2}.$$

In Equations (11.5) d is the hole diameter, L is the hole period ("lattice constant"). Equation (11.4) keeps only the first modes for simplification but gives a physically reasonable rationale for the observed superior transmittance. In this equation, Wood's anomalies correspond to $t_{01} = 0$ or $t_{10} = 0$, whereas the extraordinary transmittance near 800 nm results from the minima of the denominator. Such resonant behavior of the denominator arises from the evanescent character of the modes. In this case the standard restriction $\rho < 1$, which expresses flux conservation, relaxes. Instead, only the restriction $\mathrm{Im}[\rho] > 0$ holds, whereas the real part of ρ and its absolute value may be very large. Since $|\rho| \gg 1$ is possible, the denominator can take very small values providing a high transmission coefficient. It was shown, $|\rho|$ has a peak at a wavelength close to, but slightly larger than, the hole period L, with a maximum value that scales as $(L/d)^3$. Therefore it is the lattice period which determines the spectral position of the resonant transmission. This result does correlate remarkably with the observations. In Figure 11.7 the transmission maximum for a 750 nm period occurs at 800 nm wavelength, and in the original paper [23], the 900 nm period array showed peak transmission close to $\lambda = 1000$ nm.

Thus extraordinary transparency of a thin metal film with a regular array of subwavelength holes can be interpreted as surface-plasmon-assisted light tunneling. Plasmon development and coupling via periodic array needs a certain time and can not proceed instantaneously. The steady-state sustainable regime of high transmission occurs with the energy conservation law. A constant incoming flux will give constant output intensity after the finite build-up time necessary to gain a constructive result of multiple scattering. In this instance, the extraordinary transmission is similar to high transmission in a periodic dielectric medium or in an interferometer or coupled cavities, as has been considered in Chapter 7.

11.4 Metal–dielectric photonic crystals

Metal–dielectric composite materials can be further used for building complex metal–dielectric functional components including periodic structures with band gaps for the propagation of electromagnetic waves (so-called "photonic crystals"). Dielectric photonic crystals for the optical range have been thoroughly discussed in Chapter 7. In this section we consider the simplest periodic structure, namely a multilayer structure with periodicity in a single direction coinciding with the normal layers. Every odd layer in the stack will be the metal–dielectric composite whose dielectric function is assumed to obey the Maxwell-Garnett relation (Eq. 6.61). Let it consist of a dielectric material ($\varepsilon = 2.56$) with silver

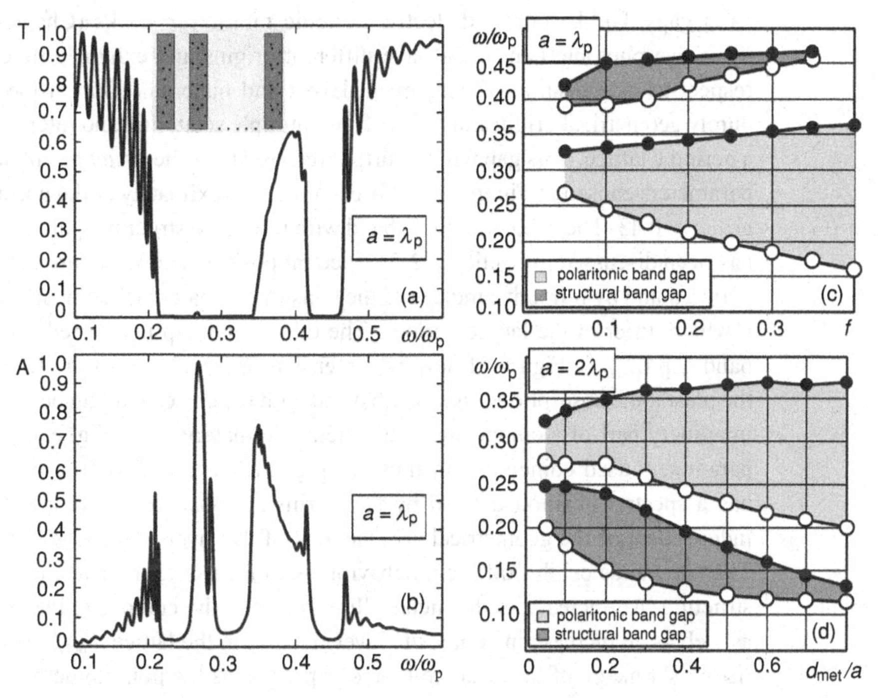

Fig. 11.8 **Properties of a metal–dielectric photonic crystal consisting of periodic layers with dispersed silver spherical nanoparticles [26]. (a) Transmission and (b) absorption spectra of a 16-period stack consisting of composite/air binary cells, the composite being a dielectric ($\varepsilon = 2.56$) material with silver nanoparticles (volume filling fraction $f = 0.2$), composite layer thickness d_{met} and air spacing thickness d_{air} obeying a condition $d_{met} = d_{air}$, $a = d_{met} + d_{air} = \lambda_p$ where λ_p is the light wavelength corresponding to the silver plasma frequency ω_p. (c) and (d) show positions of the upper (black dots) and lower (circles) band gap frequencies versus (c) relative fraction of the composite layer $d_{met}/a = d_{met}/(d_{met} + d_{air})$ in the periodic structure and (d) metal volume fraction in the composite material f. Parameters used are: (c) $a = \lambda_p$, $d_{met}/a = 0.5$, number of periods is $N = 8$, (d) $a = 2\lambda_p$, $f = 0.2$, $N = 16$.**

nanoparticles (the complex dielectric function obeys Eq. (6.30) with parameters $\varepsilon_\infty = 5$, $\hbar\omega_p \approx 9$ eV, $\hbar\Gamma = 0.02$ eV). The real and imaginary parts of the dielectric function for such a medium with the particular metal volume fraction 0.2 has been presented in Chapter 6 (Fig. 6.10). We shall look for the transmission and reflection spectra at normal incidence of a stack consisting of odd metal–dielectric composite layers of thickness d_{met}, and even dielectric layers with thickness d_{air}. Let us take for simplicity the air interlayer spacing ($\varepsilon = 1$). Thus, along the propagation direction one has a periodic medium with the period $a = d_{met} + d_{air}$. This problem has been examined numerically by P. Dyachenko and Yu. Miklyaev [26]. The results are presented in Figure 11.8. The other parameters of the system in question are: total number of periods in the stack $N = 16$, metal filling factor in composite layers $f = 0.2$, $d_{met} = d_{air} = a/2 = \lambda_p/2$, where $\lambda_p = 2\pi c/\omega_p$ is the light wavelength corresponding to the silver plasma frequency.

Figure 11.8(a) and (b) show transmission and absorption spectra, respectively. One can see the two pronounced dips in the transmission spectra which are referred to as photonic band gaps. Unlike purely dielectric periodic multilayer stacks (Chapter 7, Section 7.2) these two photonic band gaps have different origins and exhibit different behavior with respect to modification of constituent layers and materials. One of the band gaps has a purely geometrical origin and results from multiple scattering and interference of waves on a periodic lattice. This gap will be further referred to as the *structural* photonic gap. For the parameters chosen in Figure 11.8 it is centered approximately at the normalized frequency $\omega/\omega_{\mathrm{p}} = 0.43$. The behavior of this band with respect to structural parameters is the same as has been discussed in Section 7.2. Its spectral position and width are defined by the lattice period a and by refractive indices of the constituent materials (compare Figure 11.8(c) and d which differ in the lattice period). The other band gap is referred to as the *polaritonic* band gap [27]. In Figure 11.8(a) its center is near $\omega/\omega_{\mathrm{p}} = 0.28$. It clearly correlates with the plasmonic absorption (Fig. 11.8(b)) and with the corresponding behavior of the real and imaginary part of a composite metal–dielectric material (see Fig. 6.10, note the material parameters used coincide with those in Fig. 11.8(a) and (b)). The polaritonic band gap has a spectral position defined by the intrinsic properties of the metal and therefore is independent of the geometrical arrangement of the layers (compare Fig. 11.8(c) and (d)). The two bands exhibit different behaviors with respect to the content of metal within the structure. An increase in the metal filling factor in the composite layers f (Fig. 11.8(c)), as well as an increase in composite layer fraction in the lattice d_{met}/a (Fig. 11.8(d)), gives rise to shrinkage of the structural band gap whereas the polaritonic band gap widens. By means of lattice period a, variation, the structural gap can be tuned over the frequency and can get a location on either side with respect to the polaritonic gap (compare Fig. 11.8(c) and (d)). It is also possible to get both gaps overlapping within the same spectral range.

Therefore, metal–dielectric periodic structures offer definite prospects in photonic engineering. This subfield of photonics is only at the very beginning of its development and new interesting results and achievements are foreseen in the very near future. The example of a one-dimensionally periodic metal–dielectric structure given in Figure 11.8 is only the primary illustration of a new class of photonic materials. Two- and three-dimensional metal–dielectric photonic crystals are also under investigation [7, 27].

11.5 Nonlinear optics with surface plasmons

If a continuous (on a light wavelength scale) material medium exhibits an instantaneous reaction to external electromagnetic radiation, i.e. its dielectric susceptibility χ depends on the instantaneous field amplitude $\mathbf{E}$,

$$\chi(\omega, \mathbf{E}) = \varepsilon(\omega, \mathbf{E}) - 1, \tag{11.6}$$

then the polarization vector components can be expanded into a power series of the field amplitude as,

$$\frac{1}{\varepsilon_0} P_i = \sum_j \chi_{ij}^{(1)} E_j + \sum_{j,k} \chi_{ijk}^{(2)} E_j E_k + \sum_{j,k,l} \chi_{ijkl}^{(3)} E_j E_k E_l + \cdots, \tag{11.7}$$

implying that the susceptibility is a tensor value. In a simpler scalar form, polarization versus the incident field reads,

$$P = \varepsilon_0[\chi^{(1)}E + \chi^{(2)}E^2 + \chi^{(3)}E^3 + \cdots]. \tag{11.8}$$

In these equations $\chi^{(1)}$ describes the linear response of the material in question to the electromagnetic field, whereas $\chi^{(2)}$, $\chi^{(3)}$, etc. are referred to as second-order, third-order, etc. nonlinear susceptibilities, respectively. Expansions (11.7) and (11.8) mean that with a growing electric field value in electromagnetic radiation the material response is no longer linear. Nonlinear susceptibilities define the character and the value of the nonlinear response. The finite second-order susceptibility $\chi^{(2)}$ gives rise to second harmonic generation, sum- and difference-frequency generation. The finite third-order susceptibility $\chi^{(3)}$ gives rise to nonlinear refraction (Kerr nonlinearity) and self-focusing, four-wave mixing, hyper-Raman scattering and coherent anti-Stokes Raman scattering. For isotropic media, the second-order susceptibility equals zero. Therefore, the nonlinear response of isotropic media is typically characterized by the third-order susceptibility since higher-order nonlinearities require extreme radiation densities. In this case from the relation,

$$P \approx \varepsilon_0[\chi^{(1)}E + \chi^{(3)}E^3], \tag{11.9}$$

the simple expression for intensity-dependent refractive index $n(I)$ follows,

$$n(I) = n_0 + n_2 I, \tag{11.10}$$

where n_0 is the linear refractive index and n_2 is directly proportional to $\chi^{(3)}$.

In metal–dielectric nanostructures modified local electric field distribution of the electromagnetic radiation results in polarizability of the adjacent medium in the near vicinity of the nano-shaped metal surface (single metal particle, rod, tip, many-particle cluster or fractal aggregate) induced by the local field which can be multiply enhanced as compared to the free space, since the local electric field experiences strong enhancement. With respect to the electric field of the incident radiation $\mathbf{E}_0$ such a response will appear as enhancement of the corresponding susceptibilities. Such enhancement of optical nonlinearities of molecules or thin films adsorbed on a nanotextured metal surface has been known about since the 1970s. Reviews of the theory and experiments in this field can be found in [3, 5]. The theory is essentially reduced to the calculation of the local field distribution near a complex metal surface and to calculation of the modified nonlinear response as in [5]. The enhancement factor of second harmonic generation efficiency reads,

$$G_{\text{SHG}} = \left| \left\langle \left[\frac{E_\omega(\mathbf{r})}{E_\omega^{(0)}} \right]^2 \left[\frac{E_{2\omega}(\mathbf{r})}{E_{2\omega}^{(0)}} \right] \right\rangle \right|^2, \tag{11.11}$$

and accounts for possible electric field modification at the frequency ω of the incident field and of its second harmonics 2ω. In this and in the forthcoming formulas, the angular brackets $\langle \ldots \rangle$ mean averaging over an ensemble of molecules near metal nanobodies. For the N-th harmonic generation the enhancement factor is then evident, i.e.

$$G_{\text{NHG}} = \left| \left\langle \left[\frac{E_{\omega}(\mathbf{r})}{E_{\omega}^{(0)}} \right]^N \left[\frac{E_{N\omega}(\mathbf{r})}{E_{N\omega}^{(0)}} \right] \right\rangle \right|^2 . \tag{11.12}$$

For Kerr-type nonlinearity the enhancement factor is,

$$G_{\text{Kerr}} = \frac{\langle |\mathbf{E}(\mathbf{r})|^2 [\mathbf{E}(\mathbf{r})]^2 \rangle}{[\mathbf{E}^{(0)}]^4}, \tag{11.13}$$

and for a four-wave mixing enhancement factor one gets the relation,

$$G_{\text{FWM}} = |G_{\text{Kerr}}|^2 = \left| \frac{\langle |\mathbf{E}(\mathbf{r})|^2 [\mathbf{E}(\mathbf{r})]^2 \rangle}{[\mathbf{E}^{(0)}]^4} \right|^2 . \tag{11.14}$$

Expansions (11.7) and (11.8) do not imply any resonant optical transition with the population of electron excited states. If such transitions are involved, then high local field enhancement will result in modified linear response of molecules, atoms or other probes attached to the nanotextured metal surface. These are, first of all, optical absorption and photoluminescence. Resonance Raman scattering will also experience enhancement. These processes will be the subject of Chapters 13–16. However, optical transitions can also occur by means of two- and multi-photon absorption. These are nonlinear optical processes with the cross-section directly proportional to I^N, where N equals the number of photons absorbed simultaneously. In accordance with local field $E(\mathbf{r})$ enhancement and intensity $I \propto |E(\mathbf{r})|^2$ enhancement, many-photon absorption cross-sections will rise and multi-photon transitions will become more feasible. Typically, in most experimental conditions, even with high-intensity bench-top lasers of nano- and picosecond light pulses with repetition rates 10^1–10^3 Hz, multi-photon absorption essentially reduces to a two-photon process. Remarkably, modern bench-top femtosecond cw-like lasers with ultrashort pulses and high repetition rates in the MHz range do allow for two-photon absorption spectroscopy in routine experiments. This is very important for high resolution bio-imaging using fluorescent labels attached to biomolecules, or using the intrinsic photoluminescence inherent in a few biomolecules (e.g. green fluorescent protein). Higher resolution is achieved since a typically Gaussian (or other type) intensity distribution across an incident light beam under conditions of two-photon absorption generates a narrower photoluminescence spot. Therefore, metal nanoparticles can enhance photoluminescence imaging by means of two-photon absorption and enable high resolution fluorescence microscopy. A representative example has been reported by Cheng *et al.* [28]. Further details can be found in the book by P. N. Prasad [29].

11.6 Metal nanoparticles in a medium with optical gain

In Chapter 6 optical properties of dielectric media containing metal nanoparticles were discussed and the manifestation of surface plasmon resonance in the optical extinction of such composite materials has been emphasized. An intriguing question arises: what happens

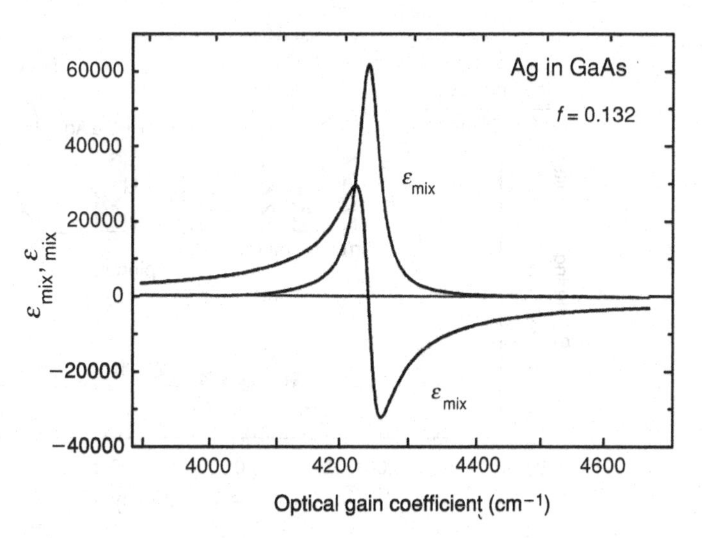

Fig. 11.9 **Real (ε'_{mix}) and imaginary (ε''_{mix}) parts of the dielectric function for GaAs gain medium ($n = 3.6$) with silver nanoparticles (adapted from [30]). Metal volume fraction is $f = 0.132$, wavelength $\lambda = 800$ nm.**

if metal nanoparticles are dispersed in a medium with optical gain? First, interplay of optical gain of the host medium and optical losses in inclusions, along with the complicated dependence of the real dielectric function of a metal on frequency may result in unusual optical properties of the composite mixture. Second, compensation for the absorption contribution to extinction, if any, will give superior light scattering because of strong enhancement of the scattering cross-section (Fig. 11.4). Third, high local fields may result in a lower lasing threshold. However, the latter effect can be readily masked by the above mentioned enhanced scattering. Fourth, enhanced scattering with compensated dissipative losses may result in Anderson localization of light (discussed in detail in Chapter 8). Studies in this direction only began a few years ago. In what follows the first results will be briefly overviewed.

Oraevsky and Protsenko [30] considered the possibility of compensating for the intrinsic metal losses in nanoparticles by means of gain adjustment of the ambient medium. The dielectric function ε_{gain} of a gain medium in the form,

$$\varepsilon_{gain}(\lambda) = [n(\lambda)]^2 - \left[g(\lambda)\frac{\lambda}{2\pi}\right]^2 - 2i\left[n(\lambda)g(\lambda)\frac{\lambda}{2\pi}\right], \qquad (11.15)$$

has been inserted in the Maxwell-Garnett relation (Eq. 6.61) and the real ε'_{mix} and imaginary ε''_{mix} parts of the complex dielectric function of the mixture have been elucidated. The results for silver nanoparticles in GaAs are presented in Figure 11.9. One can see, at a certain gain ($g > 4250\ \text{cm}^{-1}$) the imaginary part of the dielectric permittivity can take a zero or negative value, whereas the real part acquires very large values. Note, for nanoparticles of silver, where sharp plasmonic features are present, the gain value of the host medium was found to be rather high and attainable in a bulk solid medium like a semiconductor. Probably,

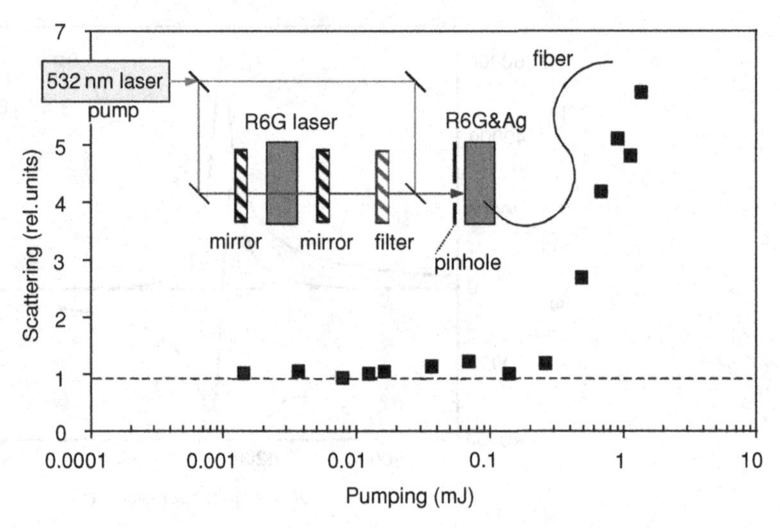

Fig. 11.10 **Relative intensity of Rayleigh scattering as a function of the pumping energy for a solution of Rhodamine 6G dye with Ag nanoparticles. Adapted from [32]. Inset: experimental setup. Copyright 2007 Elsevier Ltd.**

similar calculations for gold and copper nanoparticles will give more affordable values of gain to obtain compensation for metal losses.

With a similar idea, Lawandy [31] has predicted a singularity for the localized surface plasmon resonance in spherical metallic nanoparticles when the host dielectric medium has a critical value of optical gain. This singularity, resulting from canceling both real and imaginary terms in the denominator $[\varepsilon_{\text{metal}}(\omega) + 2\varepsilon_{\text{diel}}(\omega)]$ of a metal nanoparticle polarizability (see Eq. (6.53)) manifests itself as a sharp increase in Rayleigh scattering within the plasmon band and could lead to low-threshold random laser action and light localization effects.[3] Here $\varepsilon_{\text{metal}}$ and $\varepsilon_{\text{diel}}$ are complex dielectric functions of the metal and the host dielectric medium, respectively.

Experiments by Noginov with co-workers [32] did reveal pronounced threshold behavior of Rayleigh scattering of a mixture of the organic laser dye Rhodamine 6G (R6G) and silver particles (Fig. 11.10). In the pump–probe experiment, a fraction of the pumping beam was split off and used to pump a laser consisting of the cuvette with R6G dye placed between two mirrors. The emission line of the R6G laser (558 nm) corresponded to the maximum of the gain spectrum of the R6G dye in the mixtures studied. The beam of the R6G laser, which was used as a probe in the Rayleigh scattering, was aligned with the pumping beam in the beamsplitter and sent to the sample through a pinhole. The scattered probe light was seen in the spectrum as a narrow line on top of a much broader spontaneous emission band. The sixfold increase in Rayleigh scattering was observed in the dye–Ag aggregate mixture with the increase in pumping energy. The same authors have also examined the effect of silver particles on the operation of an R6G dye laser and found a decrease in laser efficiency instead of the expected improvement in laser performance. Addition of 1% silver particles to the laser dye resulted in a two-fold decrease in laser

[3] See Chapter 8 for explanations on Anderson localization (Section 8.3) and random lasers (Section 8.7).

energy output. The results are understood in terms of energy losses due to (i) incomplete compensation of metal losses by gain, and (ii) additional pump energy losses since a portion of the pump energy is a priori destined to produce a gain for metal loss compensation, rather than for compensation of other cavity losses. Local-field enhancement, discussed in Section 11.1, might not be very helpful in dye lasers since (i) it needs a certain time for build-up and (ii) it affects the properties of dye molecules in extreme proximity to a metal particle only, whereas losses are introduced throughout the dye solution modifying the properties of the gain medium in an "effective-medium" manner. Finally, the enhanced scattering along with random displacement of metal particles makes the gain medium strongly inhomogeneous and destroys development of a coherent wave in the laser cavity.

Strong enhancement of scattering can, however, become advantageous in resonatorless random lasers (see Section 8.7 for the general principle of random laser operation). Dice *et al.* reported on surface-plasmon-enhanced random laser emission from a suspension of silver nanoparticles in a laser dye operating at diffusive and subdiffusive scattering strengths [33]. The metal-nanoparticle-based random laser yields a larger linewidth narrowing at lower pump threshold than a dielectric-scatterer-based random laser under equivalent conditions.

Avrutsky has predicted the modification of surface plasmon polaritons [34] at the interface between metal and a dielectric with optical gain. The proper choice of optical indexes of the metal and the dielectric can result in an infinitely large effective refractive index of surface waves. Such resonant plasmons have extremely low group velocity and are localized in very close proximity to the interface.

11.7 Metamaterials with negative refractive index

"*This is such an esoteric invention, it will never, ever be of any practical importance*" – with these words the *maser* inventor Charles Townes was refused a patent for his invention by Bell Research Labs. Later on, in 1964, he received the Nobel award for that invention. This is a clear example of a novel idea which, at the beginning, is thought to be useless and idle. The maser was a precursor of lasers and nowadays everybody knows that printing (including this book) and data storage devices (including that in the author's laptop computer) are possible entirely due to laser developments. This section deals with another rather exotic idea concerning materials featuring negative refraction. Electrodynamic and optical effects for such materials were considered for the first time by V. G. Veselago in 1967 in his seminal paper [35]. Several decades later, at the very end of the last century, it did stimulate active research towards artificial metamaterials with that exotic property. At the time of publication, such an exotic property had not been anticipated and the paper was treated as a kind of "gedanken" experiment only. In the last decade, so-called "left-handed" metamaterials in the optical range of electromagnetic waves have become an active research field. Many details on the theory and many interesting experimental results in this field are omitted from this book because of the restricted scope. For further reading the review [6] and the book [36] are recommended.

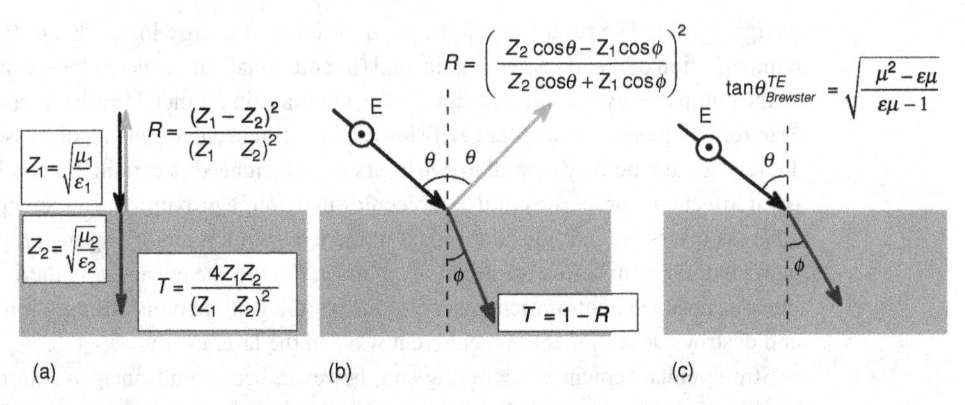

Fig. 11.11 **A few examples of electromagnetic wave reflection/transmission at an interface of two media with different magnetic permeabilities. (a) Normal incidence transmission and reflection are governed by impedances (mis)matching. (b) Modification of the Fresnel formula in terms of impedances. (c) Existence of the "magnetic" Brewster angle for TE-wave under condition** $\theta + \phi = \pi/2$.

Optics with $\mu \neq 1$

In the optical range, the magnetic permeability of materials typically equals unity, and for this reason the very notation of "μ" is not present at all in many formulas we use in our routine calculations in optics and optical engineering. This is a reasonable consequence of the properties of all existing natural and artificial materials which indeed have $\mu = 1$ in the optical frequency range. Let us forget for a moment about this property and remember the basic formulas where μ enters but has not been analyzed because of the "$\mu = 1$" convention.

Whenever $\mu \neq 1$ holds, the notion of *impedance* $Z(\omega)$ as a property of a medium becomes essential. It reads,

$$Z(\omega) = \sqrt{\frac{\mu_0\mu(\omega)}{\varepsilon_0\varepsilon(\omega)}} \equiv Z_0\sqrt{\frac{\mu(\omega)}{\varepsilon(\omega)}}, \tag{11.16}$$

and gains importance in electrodynamics with $\mu \neq 1$, which is the typical case in radiophysics. In optics ($\mu = 1$) reflection and transmission of electromagnetic waves at an interface of two media is governed by the relative refractive index n. In electrodynamics with $\mu \neq 1$ transmission and reflection at an interface are governed by the impedances of the interfacing media (Fig. 11.11). For example, the textbook formulas for reflection and transmission of light intensity at normal incidence (see Eqs. (3.22) and (3.26)),

$$R = \frac{(n-1)^2}{(n+1)^2}, \quad T = \frac{4n}{(n+1)^2}, \quad n = \frac{n_2}{n_1} = \frac{\sqrt{\varepsilon_2}}{\sqrt{\varepsilon_1}}, \tag{11.17}$$

must be modified by replacing (see, e.g. Section 1.5 in [37]),

$$\frac{n_2}{n_1} = \frac{\sqrt{\varepsilon_2}}{\sqrt{\varepsilon_1}} \to \frac{Z_1}{Z_2} = \frac{\sqrt{\varepsilon_2/\mu_2}}{\sqrt{\varepsilon_1/\mu_1}}, \tag{11.18}$$

to give,

$$R = \frac{(Z_1 - Z_2)^2}{(Z_1 + Z_2)^2}, \quad T = 1 - R = \frac{4Z_1 Z_2}{(Z_1 + Z_2)^2}. \tag{11.19}$$

Reflection at the interface vanishes when $Z_1 = Z_2$ and develops otherwise. Reflectionless propagation requires impedance matching, rather than refractive indexes matching, at the interface. Therefore reflection vanishes, not when $\varepsilon_1 = \varepsilon_2$, but when $\varepsilon_1/\mu_1 = \varepsilon_2/\mu_2$. Considering a vacuum (or air, $\varepsilon_1 = \mu_1 = 1$) interface with a material with finite ε and μ, one can see matching occurs for every material with $\varepsilon = \mu$.

Consider now the oblique incidence of an electromagnetic wave on an interface from medium 1 to medium 2. Let it be a TE-wave (s-polarization), i.e. the **E** vector is oriented normally to the plane of incidence (Fig. 11.11b). The reflection angle equals the incidence angle, as usual, and Snell's law $n_2 \sin\theta = n_1 \sin\phi$, $n_i = \sqrt{\varepsilon_i \mu_i}$ holds. The intensity reflection and transmission coefficients read [38],

$$R = \left(\frac{Z_2 \cos\theta - Z_1 \cos\phi}{Z_2 \cos\theta + Z_1 \cos\phi}\right)^2, \quad T = 1 - R, \tag{11.20}$$

differing from the familiar Fresnel formula by using impedances instead of refractive indices.

For a TM-wave (p-polarization) in traditional "$\mu = 1$"-optics at the *Brewster angle*,

$$\tan\theta_{\text{Brewster}}^{\text{TM}} = \frac{n_2}{n_1} = \sqrt{\varepsilon_2/\varepsilon_1}, \tag{11.21}$$

corresponding to the case $\theta + \phi = \pi/2$ reflection vanishes ($R = 0$). It can be intuitively understood that electric dipoles do not emit along their oscillation direction. What happens if $\mu_2 \neq 1$? In this case Eq. (11.21) modifies [39],

$$\tan\theta_{\text{Brewster}}^{\text{TM}} = \sqrt{\frac{(\varepsilon_2/\varepsilon_1)^2 - (\varepsilon_2/\varepsilon_1)(\mu_2/\mu_1)}{(\varepsilon_2/\varepsilon_1)(\mu_2/\mu_1) - 1}}. \tag{11.22}$$

Furthermore, coming back to the TE-wave (s-polarization), we now have in the case where $\theta + \phi = \pi/2$ the "magnetic" Brewster angle defined by the relation [39],

$$\tan\theta_{\text{Brewster}}^{\text{TE}} = \sqrt{\frac{(\mu_2/\mu_1)^2 - (\varepsilon_2/\varepsilon_1)(\mu_2/\mu_1)}{(\varepsilon_2/\varepsilon_1)(\mu_2/\mu_1) - 1}}. \tag{11.23}$$

For the particular case of $\varepsilon_1 = \varepsilon_2$ this reduces to,

$$\tan\theta_{\text{Brewster}}^{\text{TE}} = \sqrt{\mu_2/\mu_1} \equiv \frac{n_2}{n_1}, \tag{11.24}$$

supporting the notation of a "magnetic" Brewster angle. The intuitive explanation of vanishing reflection is that magnetic dipoles do not emit along their oscillation direction. These selected examples show clearly a wealth of effects and phenomena which are brought about in electrodynamics when media with $\mu \neq 1$ are involved.

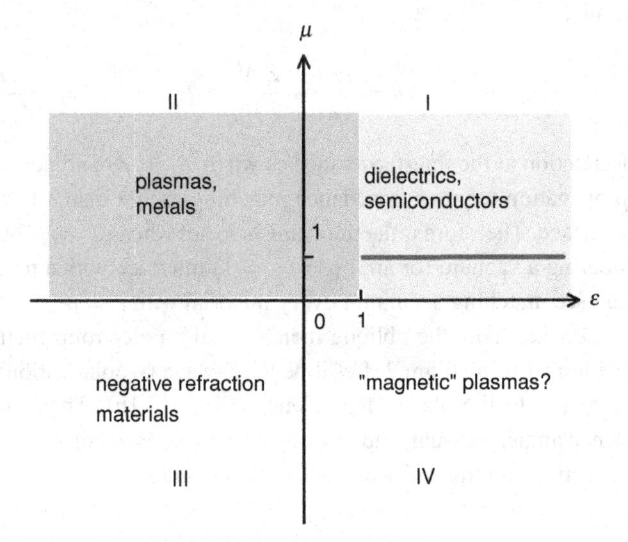

Fig. 11.12 **Possible combinations of the dielectric permittivity and the magnetic permeability of materials.**

Optics with "left-handed" materials

The whole multitude of material dielectric and magnetic properties can be placed in a $\varepsilon - \mu$ – plane (Fig. 11.12), as has been suggested by V. G. Veselago [35]. Segment I (positive ε, positive μ) corresponds to traditional dielectric and magnetic materials. As we mentioned, in the optical range, $\mu = 1$ typically holds. Furthermore, $\varepsilon > 1$ holds for dielectrics and semiconductors. This is indicated by a gray line in the figure. Segment II (negative ε, positive μ) corresponds to plasma (electrons in metal, ion plasma, etc.). Strictly, plasma extends up to $\varepsilon = 1$ (for frequencies over the plasma frequency). This is shown by the gray square in Figure 11.12. Negative dielectric permittivity results in tunneling of electromagnetic waves, skin-effect, volume and surface plasmon resonances, complex properties of composites comprising a host dielectric with positive permittivity with subwavelength-scale inclusions. These media are well identified, understood and attainable both as natural and man-made materials. What about the remaining two segments forming a semiplane with negative magnetic permeability? Such materials are not identified in nature but their existence is not forbidden in principle. Segment IV with positive dielectric permittivity but negative magnetic permeability can be assigned, by analogy with segment II, to hypothetical "magnetic" plasmas. The subject of our consideration in this section is segment III, which implies simultaneously negative dielectric permittivity and magnetic permeability. Such a combination has not been known up to very recently, but as V. G. Veselago noticed, since there is no basic restriction for such a combination it is worthwhile to examine what would happen if such materials were to be found or developed.

The immediate surprise of simultaneous $\varepsilon < 0$, $\mu < 0$ is the negative index of refraction n. In the relation,

$$n = \pm\sqrt{\varepsilon\mu}, \tag{11.25}$$

with typically positive values of permittivity and permeability, no doubt arises concerning the sign of refractive index. It is positive and takes the meaning of the factor defining the speed of wave propagation decrease with respect to the vacuum. In the case of both negative ε and μ, formally in terms of a purely mathematical treatment, the positive square root of the positive $\varepsilon\mu$ product seems reasonable as well. However this is not the case. The peculiarity of electrodynamics with both negative permittivity and permeability is the inverse direction of the wave vector **k** with respect to the vectorial product of **E** and **H**, $[\mathbf{E} \times \mathbf{H}]$. For a plane monochromatic wave, the Maxwell equations reduce to a pair of equations (SI units),

$$[\mathbf{k} \times \mathbf{E}] = \omega\mu_0\mu\mathbf{H}, \quad [\mathbf{k} \times \mathbf{H}] = -\omega\varepsilon_0\varepsilon\mathbf{E}, \tag{11.26}$$

whence for both positive ε and μ, the three vectors **E**, **H**, **k** form the "right-hand" set of vectors, i.e. the **k** direction coincides with the $[\mathbf{E} \times \mathbf{H}]$ defined direction. For both negative ε and μ, these three vectors do form a "left-hand" set. The **k** vector direction appears to be opposite to that defined by $[\mathbf{E} \times \mathbf{H}]$. On the other hand, the Poynting vector, which defines the energy flux density transfer, reads,

$$\mathbf{S} = [\mathbf{E} \times \mathbf{H}], \tag{11.27}$$

and appears to have the opposite direction with respect to **k**. Therefore, when an electromagnetic wave propagates in a "left-handed" material, the "left-hand" orientation of the **E**, **H**, **k** vector set gives rise to counterpropagating phase and group velocities. The opposite direction of the group velocity with respect to phase velocity is not the exclusive property of media with negative ε and μ, but can also be met in certain anisotropic media as well as in media with spatial dispersion.

To account for the opposite directions of phase and group velocity as well as "left-handedness" of a material with negative ε and μ, the "−" sign in the square root of Eq. (11.25) should be taken. Therefore, such materials are often referred to as "negative refraction materials". V. G. Veselago proposed to introduce "handedness", p, as a material property with $p = +1$ for "right-handed" and $p = -1$ for "left-handed" ones. With this notation, the generalized Snell's law can be formulated as,

$$\frac{\sin\theta}{\sin\phi} = \frac{n_2}{n_1} = \frac{p_2}{p_1}\left|\sqrt{\frac{\varepsilon_2\mu_2}{\varepsilon_1\mu_1}}\right|, \tag{11.28}$$

with notations for angles θ, ϕ shown in Figure 11.11(b). The reflection angle remains equal to the incidence angle, independently of material "handedness". One can see, when light is coming from a right-handed to a left-handed material, the refraction angle changes sign as compared to traditional optics.[4]

Consider refraction at the border of a vacuum (or air), $\varepsilon = \mu = 1$ and a left-handed medium (Fig. 11.13(a)) with $\varepsilon = \mu = -1$. For this case reflected intensity is zero because impedances are matched and only the refracted wave presents. One can see, the electric

[4] The notions "handedness", "left-, right-handed" used in this section should not be mistreated as the same notations used to characterize material chirality. The latter remains beyond the scope of this book.

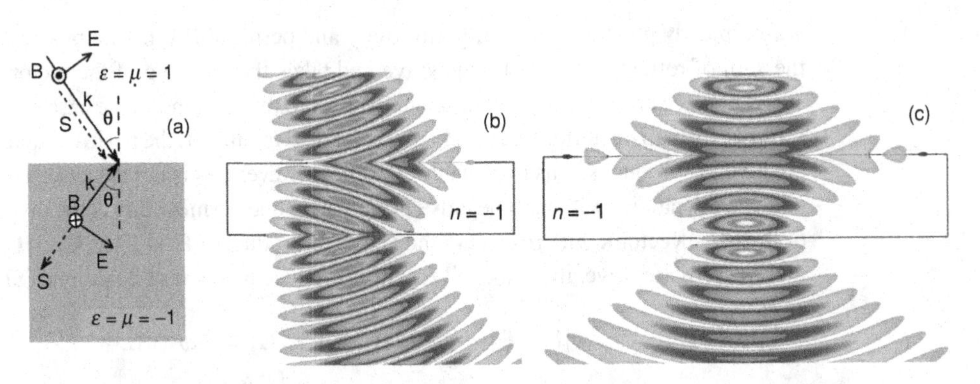

Fig. 11.13 **Propagation of electromagnetic waves impinging a slab of negative refractive index material from air. (a) Negative refraction at a matched interface; (b) calculated Gaussian beam propagation for oblique incidence [38]; (c) calculated Gaussian beam propagation at normal incidence with focusing effect [38].**

vector, the magnetic vector and the wave vector change symmetrically with respect to interface border. The refraction angle equals the angle of incidence in the absolute value but has the opposite sign. The phase velocity direction, defined by the wave vector, in a negative refraction medium has the opposite direction with respect to the energy flow defined by the Poynting vector. Figure 11.13(b) shows the results of numerical modeling for this case for a Gaussian incident beam, obtained by means of solution of the Maxwell equations [38].

Among other interesting phenomena, a few are the most important. A plane-parallel slab of a negative refraction material can serve as a lens at normal incidence. This notable property was highlighted by Veselago [35] and is illustrated in Figure 11.13(c), where the numerical solution for a Gaussian light beam is presented. Further effects include inverse Doppler shifts, modified Cherenkov radiation, unusual light tension instead of usual light pressure and the modified Fermat's principle [35, 40]. For the latter, it is shown that the traditional formulation in terms of the minimum (extremum) of the wave propagation time between two points is not correct in general. The right formulation involves the extremum of the total optical length, with the optical length for propagation through a negative refraction index material taken to be negative. Many further instructive examples of unusual properties of negative refraction materials (not necessarily from optics) can be found in articles [6, 38, 39] and in recent books [36, 41].

Although existence of materials with simultaneously negative ε and μ is not restricted by any general law, there is the basic restriction which dictates necessarily the frequency dependence of both negative $\varepsilon(\omega)$ and $\mu(\omega)$[35]. It comes from the expression for the total energy of an electromagnetic wave,

$$W = \frac{1}{2}(\varepsilon\varepsilon_0 E^2 + \mu\mu_0 H^2), \tag{11.29}$$

for the specific case of frequency-independent ε and μ. One can see for both negative ε and μ the sum in the right-hand part of Eq. (11.29) is negative, which is evidently meaningless.

The solution of the problem comes from the more general expression for energy,

$$W = \frac{1}{2}\left(\frac{\partial(\varepsilon\omega)}{\partial\omega}E^2 + \frac{\partial(\mu\omega)}{\partial\omega}H^2\right), \tag{11.30}$$

whence positive energy $W > 0$ necessarily requires that inequalities,

$$\frac{\partial(\varepsilon\omega)}{\partial\omega} > 0, \quad \frac{\partial(\mu\omega)}{\partial\omega} > 0 \tag{11.31}$$

hold. Hence, *functional dependencies $\varepsilon(\omega)$ and $\mu(\omega)$ for negative refraction materials are mandatory.*

The latter statement immediately brings further constraints into play. Note, the above consideration did imply the lossless case of purely real permittivity and permeability. Once we have arrived at the conclusion of their dispersive properties, the losses expressed by the imaginary permittivity and imaginary permeability should be necessarily inherent in the material in question. This is prescribed by the Kramers–Kronig relations, resulting in turn from the causality principle. In other words, it is the causality principle that allows frequency-dependent permittivity and permeability only at the expense of frequency-dependent losses.[5] Therefore, the search for negative refraction and other related phenomena can only be performed in media with finite losses. Thus, $\varepsilon(\omega)$ and $\mu(\omega)$ should necessarily be the complex functions,

$$\varepsilon(\omega) = \varepsilon'(\omega) + i\varepsilon''(\omega), \quad \mu(\omega) = \mu'(\omega) + i\mu''(\omega). \tag{11.32}$$

This issue has been discussed by Bush *et al.* [6]. The complex refractive index,

$$n(\omega) = n'(\omega) + in''(\omega), \tag{11.33}$$

is related to $\varepsilon(\omega)$ and $\mu(\omega)$ via the evident expression,

$$[n'(\omega) + in''(\omega)]^2 = [\varepsilon'(\omega) + i\varepsilon''(\omega)][\mu'(\omega) + i\mu''(\omega)], \tag{11.34}$$

whence,

$$n'(\omega) = \frac{\varepsilon'(\omega)\mu''(\omega) + \varepsilon''(\omega)\mu'(\omega)}{2n''(\omega)}. \tag{11.35}$$

Therefore, for positive $n''(\omega)$, the negative $n'(\omega)$ can become possible if and only if,

$$\varepsilon'(\omega)\mu''(\omega) + \varepsilon''(\omega)\mu'(\omega) < 0 \tag{11.36}$$

holds [6].

[5] The condition of inevitable losses still allows very low losses in the range of smooth frequency-dependent real parts of permeability and permittivity. Recall the low losses inherent in dielectrics at frequencies well below the band gap.

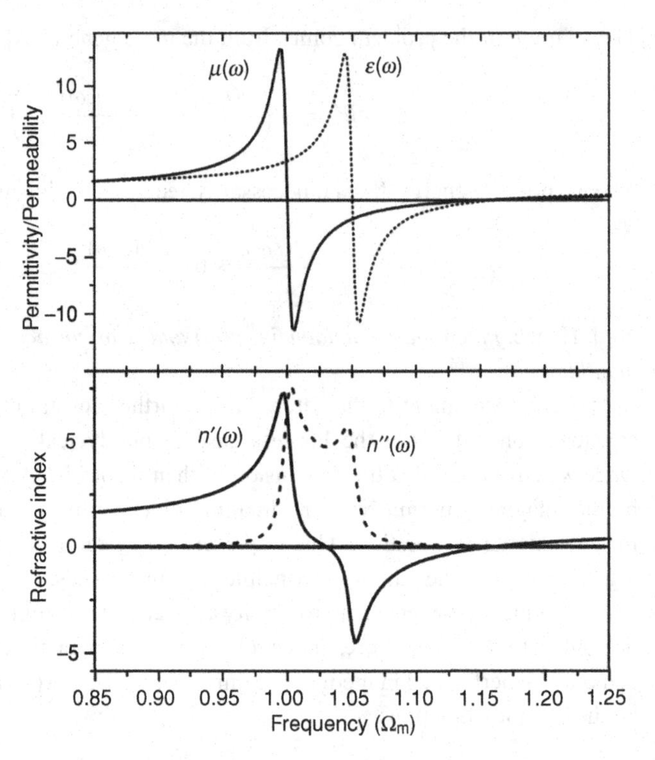

Fig. 11.14 (a) Example for the real part of permittivity $\varepsilon(\omega)$ (dotted) and permeability $\mu(\omega)$ (solid) in accordance with Eqs. (11.37). (b) Resulting real $n'(\omega)$ (solid curve) and imaginary $n''(\omega)$ (dashes) parts of refractive index. Parameters are: $\Omega_e/\Omega_m = 1.05$, $\omega_p/\Omega_m = 0.5$, $f = 0.25$, $\gamma_e/\Omega_m = 0.01$, $\gamma_m/\Omega_m = 0.01$. Reprinted with permission from [6]. Copyright 2007, Elsevier B.V.

Figure 11.14 shows an example of a hypothetical material with dielectric permittivity and magnetic permeability,

$$\varepsilon(\omega) = 1 + \frac{\omega_p^2}{\Omega_e^2 - \omega^2 - i\gamma_e\omega}, \quad \mu(\omega) = 1 + \frac{f\omega^2}{\Omega_m^2 - \omega^2 - i\gamma_m\omega} \tag{11.37}$$

along with the corresponding refractive index.

No natural material has been identified so far that features negative magnetic permeability in the optical range. In 1999, J. Pendry [42] proposed a design of a metamaterial with negative magnetic permeability. At a sub-wavelength scale, such a material should resemble a broken conductive ring (single or concentric coupled) or even a pair of plain metal plates separated by a dielectric with dimensions much lower than the wavelength (Fig. 11.15, left panel). Such units represent an LC-circuit, the inductance coming from a ring and capacitance coming from an air spacing. Negative permeability develops in a regular array of such LC-circuits. When such a material is combined with another periodic structure (Fig. 11.15, right panel), the overall response of the whole system acquires negative refraction features. This idea has been followed by many research groups, mainly in microwaves where the length scale allows for complicated sub-wavelength structures to be developed. For the optical region fine techniques on the nanometer scale using high-resolution lithography, etching, vacuum deposition etc. are necessary. Such an approach has been performed

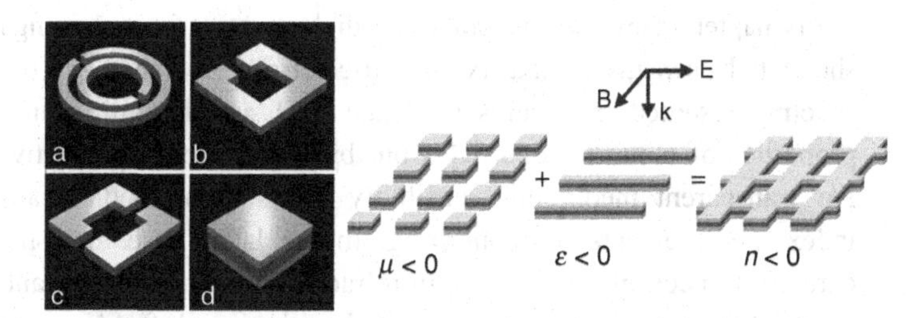

Fig. 11.15 Examples of elementary building units to get magnetic permeability in a regular array (a–d) and design of a negative refraction material by means of combined arrays with negative permeability and negative permittivity. Reprinted with permission from [6]. Copyright 2007, Elsevier B.V.

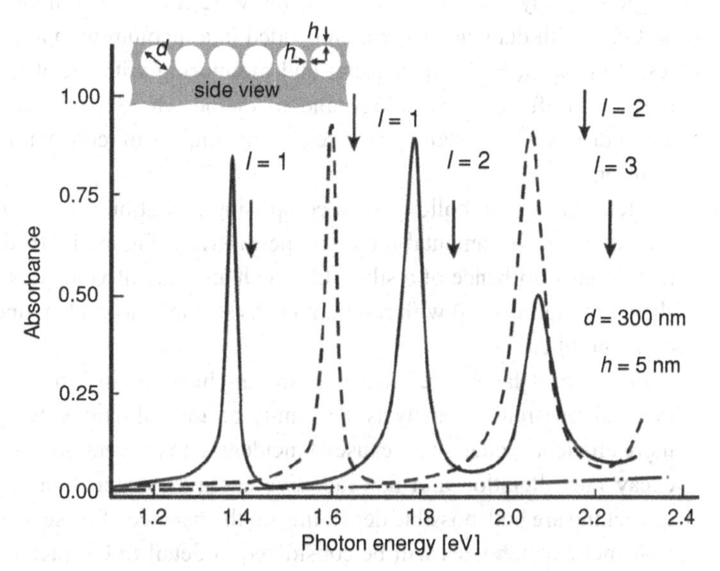

Fig. 11.16 Absorption spectra of light incident normally onto a silver surface with spherical inclusions of a material with dielectric constant $\varepsilon = 4.5$ (solid curve) and $\varepsilon = 3.3$ (dashed curve) [46]. The absorption of light on the surface of bulk silver is shown by dash-dotted curve. Vertical arrows mark the energies of the fundamental ($l = 1$), second ($l = 2$) and third ($l = 3$) plasmon modes of a single void in bulk silver.

recently by Grigorenko *et al.* [43] and Shalaev *et al.* [44]. These authors did manage to develop the proper geometry on the sub-wavelength scale and have reported on refractive index with the real part $n' = -0.3$.

11.8 Plasmonic sensors

Nanostructures with plasmonic response can be used in various sensor applications. These include sensors for refractive index determination and/or monitoring, fluorescent labels and fluorescent sensor enhancers and Raman detection systems.

In Chapter 6 the optical response of a dielectric matrix containing metal nanoparticles was shown to be sensitive to matrix dielectric permittivity (see Eqs. (6.61) – (6.63)). Extinction spectra presented in Figures 6.10 and 6.11 are determined not only by the intrinsic properties of metal nanoparticles but by the dielectric permittivity of the host as well. For transparent media this sensitivity offers an immediate application for refractive index measurements or monitoring. In the latter case, the spectral sensitivity of the extinction spectrum to variation in refractive index n is important. It can be expressed in terms of the maximum extinction wavelength λ_{max} shift $\Delta\lambda_{max}$ versus Δn, i.e. $S = \frac{\Delta\lambda_{max}}{\Delta n}$. A. D. Zamkovets *et al.* [45] examined, both theoretical and experimental, S values for different metals and n values. The S value was found to grow in the series Cu, Au, Ag, and for all above metals it was found to grow with increasing density of nanoparticles in a single monolayer. Values of $S \approx 200\,\text{nm}$ were found for a monolayer of close-packed Ag particles with diameter 10 nm, embedded in a medium with $n = 1.5$, i.e. $\Delta n = 0.01$ will result in $\Delta\lambda_{max} = 2$ nm. Sensing and monitoring with metal nanoparticles of refractive index is useful e.g. in biology and medicine where small changes of n of a solution are indicative of certain processes going on, or of concentration change in complex solutions.

Metal films with hollow voids comprising a continuous sub-space can also be used as sensors for environmental dielectric permittivity. Figure 11.16 demonstrates pronounced resonance absorbance of a silver film with an array of voids (a similar gold structure was shown in Figure 11.6) with resonant peaks strongly dependent upon dielectric permittivity of a void filler.

Fluorescent labels are used in many applications in biological, pharmacological and medical research and analysis. Proximity of a metal nanobody to a fluorophore results in more efficient excitation (because of incident field concentration and localization), radiative decay rate alteration and non-radiative energy transfer. Fluorescence enhancement and quenching are both possible depending on the balance of these three effects. Light emission near metal nanobodies will be considered in detail in Chapter 16. Here we only mention that fluorescence enhancement by metal tips, nanoparticles and nanotextured surfaces can exceed one order of magnitude and is viewed as a promising way towards high-sensitivity nano-bio-sensors in medicine. Enhanced fluorescence near metal nanobodies can also be exploited in other sensors, e.g. temperature sensors in case the luminescence spectrum or its intensity is sensitive to temperature.

Raman scattering spectroscopy is routinely used in molecular analysis in chemistry, biology, medicine and ecology. Raman scattering can be enormously enhanced by means of plasmonic effects when a molecule is adsorbed at a nanotextured metal surface [2]. The primary enhancement factor comes from incident-field enhancement (see Section 11.1). Local-field enhancement at the excitation wavelength provides more efficient virtual excitation of a molecule. Additionally, modification of photon density of states at the wavelength of scattered light provides further enhancement of the Raman signal by means of an increase in the scattering rate [47]. Thus detection of a single molecule becomes possible. Surface-enhanced Raman scattering (SERS) will be a subject for close consideration in Chapter 16. In spite of the impressive records of sensitivity reported for research data, SERS has not found either routine or commercial application to date. The main obstacles are poor reproducibility of results because of the multiple enhancement factors involved, and non-reproducible combinations of topology/adsorption parameters. Additionally, absence of a consistent theory as well as the circumspection of possible chemical modifications of adsorbed molecules makes industrial and commercial researchers wary of wide SERS application. It can be helpful in the detection of certain molecules but not in quantitative analyses.

Strong local enhancement of an incident electromagnetic field in plasmonic nanostructures can also be exploited to promote various photochemical processes by means of enhanced excitation efficiency [48], provided that non-radiative bypass in the form of energy transfer from an optically excited molecule (or other absorption complex or center) to the metal is damped.

11.9 The outlook

Optical properties of metal–dielectric structures influenced by volume and surface plasmonic resonances are a very active and expanding modern field of research. Because of the restricted scope of this book many interesting results and ideas are left for self-learning. These are properties of island metal films, where hot electrons are readily developed owing to discrete energy levels and inhibited electron–phonon interactions because of restricted mean free path [49]; fractal metal clusters where local fields are essentially influenced by percolation processes [3, 5]; the theory and implementation of three-dimensional metal–dielectric photonic crystals [7]; high-resolution imaging with metal–dielectric structures [42]; properties of sub-wavelength slits and apertures in metal films [50–53]; modified radiative decay of plasmons in periodic structures [54, 55]. For the actively developing field of modified emission and scattering of light in metal–dielectric nanostructures Chapters 14 and 16 are recommended.

Problems

1. Try to explain the difference in the field profile between a pair of cylinders (Fig. 11.2) for different light polarization.

2. Derive the modified Fröhlich condition for a composite medium with noticeable fraction of metal particles. Use the Maxwell-Garnett approximation for the effective dielectric function of the composite medium. Prove that the red shift of the transmission minimum (and the reflection maximum) will arise.

3. Try to elaborate an application proposal for enhanced nonlinearities in plasmonic structures.

4. Prove that the nonlinear response of two-photon-excited photoluminescence will actually increase spatial resolution of luminescent imaging.

5. Evaluate the dimensionality of impedance (Eq. (11.16)) and prove that it coincides with the dimensionality of electric resistance. Calculate vacuum impedance $Z_0 = \sqrt{\mu_0/\varepsilon_0}$. Compare it with the impedance of a home TV-cable (typically written on every cable on the outer dielectric envelope).

6. Recalling the Doppler effect discussed in many textbooks, consider its modification for light waves propagating in a negative refraction medium. In a gas of atoms at finite temperature the so-called *Doppler broadening* of spectral lines occurs, dominating the natural atomic linewidths. Consider the Doppler effect on atomic lines for the case of a negative refraction medium.

7. Find out the formulation of the *Kramers–Kronig relations* in a good textbook on optics and, based on these relations, discuss in more detail the restrictions on negative dielectric permittivity and negative magnetic permeability.

8. Find out and discuss the correlation of data presented in Figures 11.6 and 11.16.

9. Evaluate the spectral shift versus refractive index $S = \dfrac{\Delta\lambda_{\max}}{\Delta n}$ for the structure presented in Figure 11.16.

10. In Chapter 7 (Fig. 7.33) an example is given of nonlinear response of a periodic multilayer structure under conditions of strong optical pumping of a ZnSe/ZnS Bragg mirror. The response is supposed to result from the negative contribution to refractive index $\Delta n \approx -0.05$ of a ZnSe sublattice arising from a non-equilibrium electron–hole plasma. Estimate the plasma concentration neglecting damping. Consider only the electron contribution taking electron effective mass $m_e^* \approx 0.1 m_0$.

11. When an intense laser beam propagates in the atmosphere, a self-focusing phenomenon occurs because of positive change in refractive index with intensity. However, in the case of a very intense laser beam attainable with femtosecond lasers, instead of self-focusing, a filament similar to lightning develops. Explain the phenomenon implying multiphoton absorption resulting in ionization of atoms. Hint: recall the plasma contribution to refractive index from Problem 10.

References

[1] S. A. Maier. *Plasmonics: Fundamentals and Applications* (Berlin: Springer Verlag, 2007).

[2] K. Kneipp, M. Moskovits and H. Kneipp (Eds.). *Surface-Enhanced Raman Scattering. Physics and Applications* (Berlin: Springer, 2006).

[3] A. K. Sarychev and V. M. Shalaev. Electromagnetic field fluctuations and optical nonlinearities in metal-dielectric composites. *Phys. Rep.*, **335** (2000), 275–371.

[4] A. V. Zayats and I. I. Smolyaninov. Near-field photonics: surface plasmon polaritons and localized surface plasmons. *J. Opt. A: Pure Appl. Opt.*, **5** (2003), 816–850.

[5] V. M. Shalaev: *Nonlinear Optics of Random Media: Fractal Composites and Metal-Dielectric Films* (Berlin: Springer, 2000).

[6] K. Busch, G. von Freymann, S. Linden, S. F. Mingaleev, L. Tkeshelashvili and M. Wegener. Periodic nanostructures for photonics. *Phys. Rep.*, **444** (2007), 101–202.

[7] J. M. Lourtioz *et al. Photonic Crystals: Towards Nanoscale Photonic Devices* (Berlin: Springer, 2005).

[8] *Nanophotonics with Surface Plasmons.* Eds. V. M. Shalaev and S. Kawata (Amsterdam: Elsevier 2007).

[9] U. Kreibig and M. Vollmer. *Optical Properties of Metal Clusters* (Berlin: Springer, 1995).

[10] Yu. P. Petrov. *Clusters and Small Particles* (Moscow: Nauka, 1975) (in Russian).

[11] R. A. Dynich and A. N. Ponyavina. Metal particle size effect on local field near its surface. *J. Appl. Spectr.*, **75** (2008), 831–837.

[12] J. P. Kottmann and O. J. F. Martin. Retardation-induced plasmon resonances in coupled nanoparticles. *Opt. Lett.*, **26** (2001), 1096–1098.

[13] K. Li, M. I. Stockman and D. J. Bergman. Self-similar chains of metal nanospheres as an efficient nanolens. *Phys. Rev. Lett.*, **91** (2003), 227402.

[14] H. C. van de Hulst. *Light Scattering by Small Particles* (New York: John Wiley & Sons, 1957).

[15] M. Kerker. *The Scattering of Light and Other Electromagnetic Radiation* (New York: Academic Press, 1969).

[16] C. F. Bohren and D. R. Huffman. *Absorption and Scattering of Light by Small Particles* (New York: John Wiley & Sons, 1985).

[17] L. G. Astaf'eva, V. A. Babenko and V. N. Kuzmin. *Electromagnetic Scattering in Disperse Media: Inhomogeneous and Anisotropic Particles* (Berlin: Springer, 2003).

[18] M. I. Mishchenko, L. D. Travis and A. Lacis. *Scattering, Absorption, and Emission of Light by Small Particles* (Cambridge: Cambridge University Press, 2002).

[19] S. M. Kachan and A. N. Ponyavina. Spectral properties of close-packed monolayers consisted of metal nanospheres. *J. Phys.: Cond. Matt.*, **14** (2002), 103–111.

[20] S. M. Kachan and A. N. Ponyavina. The spatial ordering effect on spectral properties of close-packed metallic nanoparticle monolayers. *Surface Science*, **507–510** (2002), 603–608.

[21] S. Coyle, M. C. Netti, J. J. Baumberg, M. A. Ghanem, P. R. Birkin, P. N. Bartlett and D. M. Whittaker. Confined plasmons in metallic nanocavities. *Phys. Rev. Lett.*, **87** (2001), 176801.

[22] T. V. Teperik, V. V. Popov and F. J. Garcia de Abajo. Void plasmons and total absorption of light in nanoporous metallic films. *Phys. Rev. B*, **71** (2005), 085408.

[23] T. W. Ebbesen, H. J. Lezec, H. F. Ghaemi, T. Thio and P. A. Wolff. Extraordinary optical transmission through sub-wavelength hole arrays. *Nature*, **931** (1998), 667–669.

[24] L. Martín-Moreno, F. J. García-Vidal, H. J. Lezec, K. M. Pellerin, T. Thio, J. B. Pendry and T. W. Ebbesen. Theory of extraordinary optical transmission through subwavelength hole arrays. *Phys. Rev. Lett.*, **86** (2001), 1114–1117.

[25] A. Degiron, H. J. Lezec, N. Yamamoto and T. W. Ebbesen. Optical transmission properties of a single subwavelength aperture in a real metal. *Opt. Commun.*, **239** (2004), 61–66.

[26] P. N. Dyachenko and Yu. V. Miklyaev. One-dimensional photonic crystal based on a nanocomposite "metal nanoparticles – dielectric". *Kompyuternaya Optika*, **31** (2007), 31–34 (in Russian).

[27] A. Runs and C. G. Ribbing. Polaritonic and photonic gap interactions in a two-dimensional photonic crystals. *Phys. Rev. Lett.*, **92** (2004), 123901.

[28] Y. Cheng, J. Swiatkievich, T.-C. Lin, P. Markowicz and P. N. Prasad. Near-field probing surface plasmon enhancement effect on two-photon emission. *J. Phys. Chem. B*, **106** (2002), 4040–4042.

[29] P. Prasad. *Nanophotonics* (New York: John Wiley & Sons, 2004).

[30] A. N. Oraevsky and I. E. Protsenko. Optical properties of heterogeneous media. *Quantum Electronics*, **31** (2001), 252–256.

[31] L. M. Lawandy. Localized surface plasmon singularities in amplifying media. *Appl. Phys. Lett.*, **85** (2004), 5040–5042.

[32] M. A. Noginov, G. Zhu, V. P. Drachev and V. M. Shalaev. Surface plasmons and gain media. In: *Nanophotonics with Surface Plasmons.* Eds. V. M. Shalaev and S. Kawata (Amsterdam: Elsevier, 2007), 141–169.

[33] G. D. Dice, S. Mujumdar and A. Y. Elezzabi, Plasmonically enhanced diffusive and subdiffusive metal nanoparticle-dye random laser. *Appl. Phys. Lett.*, **86** (2005), 131105.

[34] I. Avrutsky. Surface plasmons at nanoscale relief gratings between a metal and a dielectric medium with optical gain. *Phys. Rev. B*, **70** (2004), 155416.

[35] V. G. Veselago. The electrodynamics of substances with simultaneously negative values of ε and μ. *Soviet Physics Uspekhi*, **10** (1968), 509–514.

[36] A. K. Sarychev and V. M. Shalaev. *Electrodynamics of Metamaterials* (Singapore: World Scientific, 2007).

[37] M. Born and E. Wolf. *Principles of Optics* (New York: Macmillan, 1964).

[38] R. W. Ziolkowski. Pulsed and CW Gaussian beam interactions with double negative metamaterial slabs. *Optics Express*, **11** (2003), 662–681.

[39] L. V. Panina, A. N. Grigorenko and D. P. Makhnovskiy. Optomagnetic composite medium with conducting nanoelements. *Phys. Rev. B*, **66** (2002), 155411.

[40] V. G. Veselago. On the formulation of Fermat's principle for light propagation in negative refraction materials. *Physics – Uspekhi*, **172** (2002), 1215–1218.

[41] N. Engheta and R. W. Ziolkowski. *Metamaterials: Physics and Engineering Explorations* (Wiley-Interscience, 2006).

[42] J. B. Pendry. Negative refraction makes a perfect lens. *Phys. Rev. Lett.*, **85** (2000), 3966–3969.

[43] A. N. Grigorenko, A. K. Geim, H. F. Gleeson, Y. Zhang, A. A. Firsov, I. Y. Khrushchev and J. Petrovic. Nanofabricated media with negative permeability at visible frequencies. *Nature*, **438** (2005), 335–338.

[44] V. M. Shalaev, W. Cai, U. K. Chettiar, H. K. Yuan, A. K. Sarychev, V. P. Drachev and A. V. Kildishev. Negative index of refraction in optical metamaterials. *Opt. Lett.*, **30** (2005), 3356–3358.

[45] A. D. Zamkovets, S. M. Kachan and A. N. Ponyavina. Concentrational enhancement of surface plasmon resonance sensitivity of metal nanoparticles to characteristics of dielectric environment. *J. Appl. Spectr.*, **75** (2008), 568–574.

[46] T. V. Teperik, V. V. Popov and F. J. García de Abajo. Total resonant absorption of light by plasmons on the nanoporous surface of a metal. *Sol. St. Phys.*, **47** (2005), 172–175.

[47] S. V. Gaponenko. Photon density of states effects on Raman scattering in mesoscopic structures. *Physical Review B*, **65** (2002), 140303 (R).

[48] L. E. Brus and A. Nitzan. Chemical processing using electromagnetic field enhancement, U.S. Patent No.: 4,481,091 (21 October, 1983).

[49] R. D. Fedorovich, A. G. Naumovets and P. M. Tomchuk. Electron and light emission from island metal films and generation of hot electrons in nanoparticles. *Phys. Rep.*, **328** (2000), 73–179.

[50] N. Fang, H. Lee and X. Zhang. Sub-diffraction-limited optical imaging with a silver superlens. *Science*, **308** (2005), 534–537.

[51] D. O. S. Melville and R. J. Blaikie. Super-resolution imaging through a planar silver layer. *Opt. Expr.*, **13** (2005), 2127–2134.

[52] A. Degiron, H. J. Lezec, N. Yamamoto and T. W. Ebbesen. Optical transmission properties of a single subwavelength aperture in a real metal. *Opt. Commun.*, **239** (2004), 61–66.

[53] F. J. García-Vidal, H. J. Lezec, T. W. Ebbesen and L. Martín-Moreno. Multiple paths to enhance optical transmission through a single subwavelength slit. *Phys. Rev. Lett.*, **90** (2003), 213901.

[54] S. Linden, J. Kuhl and H. Giessen. Controlling the interaction between light and gold nanoparticles: Selective suppression of extinction. *Phys. Rev. Lett.*, **86** (2001), 4688–4691.

[55] T. V. Teperik and V. V. Popov (2004). Radiative decay of plasmons in a metallic nanoshell. *Phys. Rev. B*, **69** (2004), 155402.

12 Wave optics versus wave mechanics II

12.1 Transfer of concepts and ideas from quantum theory of solids to nanophotonics

In Chapter 3 we discussed that optics played an important role in the development of quantum mechanics at its very early stages. Wave mechanics with respect to classical mechanics has been developed by analogy to wave optics with respect to geometrical optics. A number of similarities were outlined in that chapter between quantum mechanical and electromagnetic phenomena. Many decades later the reverse process happened. The advances in single particle quantum theory of solids that dealt exclusively with analysis of the Schrödinger equation in complex potentials with no collective phenomena and spin effects included, were systematically transferred to electromagnetism, and first of all to wave optics. We have shown the bulk of these effects and phenomena in wave optics of complex structures in Chapters 7–9. The transfer of concepts and phenomena is presented in Table 12.1 with the principal dates indicated. This transfer is a remarkable event in modern science. It is indicative of the useful exchange of ideas between two large fields of physics. In a sense, quantum theory did pay back to optics with high "interest" for originally borrowing optical ideas in the 1920s. It is owing to this transfer that the writing of this very book has become topical.

Among the quantum phenomena listed in Table 12.1, the band theory of solids in terms of electron Bloch functions, conduction band and valence band concepts, Brillouin zones and electron and hole effective mass have been overviewed in detail in Chapter 4. Anderson localization of electrons in random potentials has also been discussed in Section 4.5 when considering the properties of disordered solids. There are two more quantum phenomena that have been replicated in optical research but have not been mentioned in previous chapters.

Weak localization of electrons was identified as the specific phenomenon when the problem of the theoretical description of conductivity in impurity semiconductors and in metals arose at the beginning of the 1980s of the last century. To describe the peculiar behavior of conductivity at low temperature when scattering at lattice oscillations (phonons) becomes negligible, loop-like trajectories were shown to be crucially important. An electron experiencing multiple scattering can exhibit paths visiting the same sites more than once. At first glance, random travel should lead to the same results independently whether or not the same sites are visited twice. That in fact would be the case in classical mass-point mechanics or in ray optics. However, the wavy properties of electrons result in a coherent interference

Table 12.1. Phenomena identified in quantum theory and then transferred to optics

Phenomenon	Quantum theory	Optics
Energy bands in crystals	Electron theory of solids, 1930s	1990s
Localization in random potential	Anderson localization of electrons, 1958	Anderson localization of electromagnetic waves, 1984
Surface states	Tamm states of electrons, 1931	Optical Tamm states, 1980s
Quantum interference in a complex potential	Weak localization of electrons, 1982	Coherent backscattering, 1984
Fractal spectrum in a quasi-periodic potential	Fractal energy spectrum of quasicrystals, 1984	Fractal transmission spectrum of Fibonacci filters, 1994

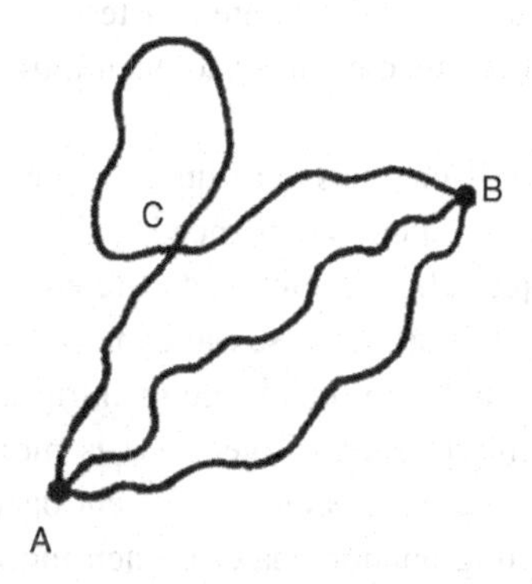

Fig. 12.1 Illustration of various possible paths for an electron to travel from A to B in a space with weak potential disorder. All paths without loops are on average equivalent and do not cause a new effect. However if a loop is present in the path, then at point C coherent addition of wave function amplitudes noticeably increases the net current between A and B.

effect in every node of such a loop-like portion of the whole path between the two contacts used to measure conductivity (Fig. 12.1). Constructive interference in the nodes of those loops makes the current higher. Representative references to this phenomenon in electronics are given in [1–5].

The quantum mechanical problem for an electron in a quasiperiodic potential became a subject of interest when a class of non-periodic solid-state materials was identified where atoms do not form a lattice with translational symmetry but, nevertheless the non-periodic arrangement features long-range order and exhibits regular scattering of X-ray radiation. Such systems feature a rotation symmetry of 5-, 8-, 10- and 12th order which are forbidden by the conventional Feodorov's groups of symmetry inherent in lattices with translational symmetry. These non-commensurate spatial atomic structures have been called *quasicrystals* [6–10]. Such structures can be treated in terms of six-dimensional space with translational symmetry, or can be constructed following the Penrose algorithm.

12.2 Why quantum physics is ahead

We can consider at least three reasons why after borrowing many ideas from wave optics, wave mechanics went ahead very rapidly and finally enabled transfer of its results back to optics:
1. Naturally existing systems have complicated spatial organization (crystals, glasses, disordered solids, quasicrystals). Therefore from the first steps, quantum mechanics faced problems of solving the Schrödinger equation with complex potentials. The concepts of Brillouin zones and Bloch waves were not a result of mathematical activity but were required in order to explain properties of real entities. It is therefore the structure of matter which in turn is governed by the Coulomb interaction of negatively charged electrons and positively charged nuclei and ions which determined the preceding formulation of tasks for the wave equation in complex potentials in quantum mechanics, rather than in optics. Since these tasks were formulated in terms of the single-particle Schrödinger equation immediately after this equation had been introduced in physics, the band theory of solids emerged.
2. Quantum mechanics has very quickly become a challenging physical theory with the flavor of unusual predictions enhanced by the general conception of the non-visual character of quantum objects. In this context it differed (and still does) drastically from the older field of wave optics where no surprise was expected and therefore none were searched for. Optical conceptions did not offer challenge or intrigue to researchers. It seems rather reasonable that quantum interference of electrons is more inviting for researchers as compared to its classical counterpart, namely, coherent back scattering in a turbid medium. Similarly, quantum tunneling phenomena are much more challenging for theorists as compared to glass transparency.
3. The development of electronics, fast emerging solid-state circuitry following invention of the transistor, growing integration of chips are other crucial factors that have pushed the quantum theory of solids ahead, leaving wave optics behind. It is important that electronics has attracted enormous financial resources, and accordingly, extensive research programs in the field of solid-state quantum theory.

12.3 Optical lessons of quantum intuition

Surprisingly enough, the organic consistency of optics and mechanics at the very beginning of the quantum era in physics, as well as the guiding role of optics in elaboration of its mechanical counterpart, seems to have been forgotten and regretfully is not included in the textbooks on quantum mechanics. Not only is the influence of quantum mechanics on nanophotonics important, but recalling the optical impact on quantum mechanics is rather instructive. Optical visual experience makes optical phenomena much more intuitive and understandable. Optical analogies should be used systematically to better understand the

Table 12.2. Wave optical and wave mechanical counterparts

Profile of potential (dielectric function)	Quantum particle	Electromagnetic wave
Upward/downward step	Transmission/reflection	Transmission/reflection (Fresnel laws)
Well with finite depth and width	Selective transmission/reflection	Selective transmission/ reflection (Fabry–Perot modes for dielectric films)
Barrier with finite depth and width	Selective reflection and transmission over a barrier	Selective transmission/ reflection, Fabry–Perot modes for an air slit between two dielectrics
	High reflection and tunneling throughout a barrier	High reflection and partial transmission of thin metal films, frustrated total reflection
A finite well between two barriers with finite depth and width	High off-resonant reflection and resonant tunneling throughout	High off-resonant reflection and high resonant transmission of a Fabry–Perot interferometer
Multiple identical barriers/wells	Multiple splitting of energy levels	Multiple splitting of transmission resonant bands
Periodic potential	Energy bands separated by band gaps	Transmission bands separated by band gaps (reflection bands, stop-bands)
	Effective mass near conduction band minimum	Effective medium approximation for smaller wavenumbers, effective mass consideration near edges of Brillouin zones
High barrier at the end of periodic potential	Surface (Tamm) states	Optical Tamm states
Random potential with weak disorder	Diffusive transport, Ohm's law	Diffusive transport, $T \sim 1/L$ law
	Weak localization	Coherent back scattering
Random potential with strong disorder	Anderson localization	Anderson localization of light
Quasiperiodic potential	Fractal energy spectra of natural quasicrystals	Fractal transmission/ reflection spectra of Fibonacci structures
Fractal potential	Not examined	Spectral scalability of transmission/reflection bands

quantum world of "mass-point waves". To assist in this understanding we summarize the whole list of counterparts identified to date in Table 12.2.

One more important note should be taken into account when considering the possible influence of optical analogies on our understanding of quantum-mechanical phenomena nowadays. Unlike electronic experiments, optical instrumentation offers today perfect coherent laser sources with controllable light wavelength and pulse shape. The electronic counterparts to lasers are not available. Additionally, purely wave phenomena like, e.g., weak localization, Anderson localization and tunneling, are seriously perturbed in electronic systems by electron–electron interactions. In optics there is no interaction between light waves, laser beams and photons. Therefore the above-mentioned phenomena, which are still under investigation in complex structures, can be better understood based on optical experiments.

At this point the consideration of electromagnetic waves in complex structures along with their analogies in quantum mechanics is complete. In what follows we concentrate on light–matter interactions in complex nanostructures. For these phenomena, the notions of photons and electromagnetic field quantization are important. Therefore, as promised in the Introduction (Chapter 1) it is the forthcoming Chapter 13 where photons eventually enter into the current *Introduction to Nanophotonics.*

Problems

1. Try to analyse why acoustics has not been used as the proper analogy to develop wave mechanics. Note, acoustics has been a rather well defined and developed field of research with a two-volume edition of Rayleigh's "Theory of Sound" [12] as a clear example of the state-of-the-art.

2. Make a transfer of phenomena listed in Table 12.2 towards acoustics.

3. All previous chapters from 2 to 11 have shown conclusively that wave phenomena in quantum mechanics have relevant counterparts in optics. Moreover, the quest for counterparts promotes new ideas and concepts in nanophotonics. Try to find out the new pair of counterparts that is not present in Tables 12.1 and 12.2. If you find it in the literature, discuss it with your classmates, colleagues and students. Sending a message to the author of this book will be greatly appreciated.

4. Try to find the quantum and optical counterparts which have not been indentified either in Tables 11.1 and 11.2 or in the other available sources. Then send the message to a good physics journal.

5. If you solved Problem 4 and found out that the relevant optical counterpart has not been observed, then write a research project and try to get a grant.

6. If your solved Problem 5 and got the grant then go ahead with the experiments and finally, send a report to a good physics journal.

References

[1] E. Abrahams, P. W. Anderson, D. C. Licciardello and T. V. Ramakrishnan. Scaling theory of localization: absence of quantum diffusion in two dimensions. *Phys. Rev. Lett.*, **42** (1979), 673–676.

[2] A. I. Larkin and D. E. Khmelnitskii. Anderson localization and anomalous magnetoresistance at low temperatures. *Sov. Physics – Uspekhi,* **136** (1982), 536–538.

[3] D. E. Khmelnitskii. Localization and coherent scattering of electrons. *Physica B+C (Amsterdam)*, **126** (1984), 235—241.

[4] G. Bergmann. Weak localization in thin films: a time-of-flight experiment with conduction electrons. *Physics Reports*, **107** (1984), 1–58.

[5] S. Chakravarty and A. Schmid. Weak localization: The quasiclassical theory of electrons in a random potential. *Physics Reports*, **140** (1986), 193–236.

[6] D. Shechtman. Metallic phase with long-range orientation order and no translational symmetry. *Phys. Rev. Lett.*, **53** (1984), 1951–1963.

[7] D. Levine and P. J. Steinhardt. Quasicrystals: a new class of ordered structures. *Phys. Rev. Lett.*, **53** (1984), 2477–2480.

[8] P. A. Kalugin, A. Yu. Kitaev and L. S. Levitov. $Al_{0.86}Mn_{0.14}$ – a six-dimensional crystal. *JETP Lett.*, **41** (1985), 119–123.

[9] C. Janot. *Quasicrystals* (Oxford: Oxford University Press, 1997).

[10] T. Fujiwara and T. Ogawa. *Quasicrystals* (Berlin: Springer Verlag, 1990).

[11] E. Maciá. The role of aperiodic order in science and technology. *Rep. Prog. Phys.*, **69** (2006), 397–441.

[12] J. W. Strutt (Lord Rayleigh). *The Theory of Sound* (London: Macmillan, 1877–1878).

[12] [illegible]

PART II

LIGHT–MATTER INTERACTION IN NANOSTRUCTURES

13 Light – matter interaction: introductory quantum electrodynamics

Generally speaking, fields and matter are the two entities which constitute the Universe. These entities continuously interact. The electromagnetic field is the specific type of field which contains the range of oscillation frequencies which human eyes are able to sense. After twelve chapters in this book, we are now approaching the point where photons enter nanophotonics. Photons are necessary to understand how matter emits light. This happens by means of quantum transitions where matter loses and the electromagnetic field gains a certain portion of energy and momentum. The emission of light, in a broad sense, includes all types of processes where the electromagnetic field gains a portion of energy and momentum from matter. This can be classified as different types of secondary radiation which include emission of photons and scattering of photons. In nanophotonics, these elementary processes of field–matter interaction experience modification because of light-wave confinement, which is typically explained in a rather elegant way as a consequence of the photon density of states modification. The main purpose of the present chapter is to explain the notions of field quantization, photons, emission and scattering rates in terms of quantum transitions and density of electromagnetic modes.

13.1 Photons

> *... the theory of light operating with continuous spatial functions will lead to contradictions with experiment when being applied to events of creation and transformation of light.*
>
> *Albert Einstein, 1905* [1]

Basic statements

Photons are elementary quanta of electromagnetic radiation. Radiation itself is viewed as an infinite set of harmonic oscillators. Every radiation mode defined by the wave vector **k** is treated as an oscillator with frequency $\omega_{\mathbf{k}}$. The energy spectrum of a harmonic oscillator in accordance with the solution of the Schrödinger equation with the parabolic potential (see Eq. 2.85 in Section 2.6) reads,

$$E(\omega_{\mathbf{k}}) = \hbar\omega_{\mathbf{k}}\left(n_{\mathbf{k}} + \frac{1}{2}\right),\ n_{\mathbf{k}} = 0, 1, 2, 3, \ldots \qquad (13.1)$$

The integer number $n_{\mathbf{k}}$ in Eq. (13.1) is treated as the *number of photons* in the **k**-mode of radiation. Every **k**-mode is characterized not only by wave vector **k** but also by the polarization state. The energy $\hbar\omega_{\mathbf{k}}$ is the photon energy. In what follows the "**k**" subscript will often be omitted.

Recalling the properties of electromagnetic waves we can make immediate statements about certain photon properties. It has momentum $\mathbf{p} = \hbar\mathbf{k}$. The dispersion law in a vacuum $\omega = ck$ (Eq. 2.9) evolves to the photon dispersion law,

$$\hbar\omega = \hbar ck \Rightarrow E = pc. \tag{13.2}$$

The linear dispersion law means the photon mass defining its inertia is zero in accordance with the definition of inertial mass,

$$m^{-1} = \frac{\mathrm{d}^2 E}{\mathrm{d}p^2}, \tag{13.3}$$

as has been discussed in Chapter 4 (Eq. (4.16)). It is a common treatise that since a photon never stops ("never has a rest") its rest mass m_0 is zero. This statement should, however, be taken rather as a convention than a real property. In terms of the general relation $E = m_0c^2$, the rest mass of a photon should read, $m_0 = \hbar\omega/c^2$. In accordance with the above mass–energy relation, a photon death offers energy release defined by its rest mass value.

The density of electromagnetic modes, derived in Chapter 2 (Eq. 2.25),

$$D(\omega) = \frac{\omega^2}{\pi^2 c^3}, \tag{13.4}$$

now aquires the meaning of the *density of photon states* in a vacuum and will be referred to as *photon DOS* in what follows. One more parameter of photons coming from classical electrodynamics is polarization.

Further properties of photons do not follow from the properties of classical waves. An arbitrary number of photons can exist in the same state. Photons belong to the class of elementary particles called *bosons*. In equilibrium, when radiation viewed as a gas of photons can be ascribed a temperature value, photons obey Bose–Einstein statistics,

$$N(\hbar\omega) = \frac{1}{\exp\left(\dfrac{\hbar\omega}{k_{\mathrm{B}}T}\right) - 1}. \tag{13.5}$$

Energy contained in every electromagnetic mode has no basic *upper* limit. However, it does have a basic *lower* limit defined by $\hbar\omega/2$ in Eq. (13.1). This gives rise to the concept of a physical vacuum and will be the subject of Section 13.2.

A further important property of photons is the non-conservation of their number. Because photon number does not conserve, their chemical potential, unlike atoms or molecules, equals zero and does not enter into Eq. (13.5). It means no free energy can be assigned per single photon.

Under equilibrium conditions, electromagnetic radiation is seen as an equilibrium gas of photons with the spectral distribution of energy density,

$$u(\omega) = \hbar\omega \frac{\omega^2}{\pi^2 c^3} \frac{1}{\exp\left(\frac{\hbar\omega}{k_B T}\right) - 1}, \tag{13.6}$$

which is referred to as the *black body radiation spectrum.*

Brief historical notes

The advent of the quantum theory of light into modern physics has occurred in parallel with development of the quantum theory of matter, i.e. wave mechanics. It was the black body radiation problem which challenged theorists over many years at the end of the nineteenth century. It resulted in 1900 in two, outstanding in essence but still preliminary, ideas advanced by Planck and Rayleigh. Planck introduced energy quanta into the empirically obtained formula for a blackbody spectrum, in the form of Eq. (13.6). However, he did not assume quantization of radiation. He quantized energy emitted by the cavity walls where radiation is confined. Rayleigh proposed to count discrete modes in a finite cavity and showed that density of modes per unit volume and unit wave number interval scales as k^2/π^2, and he explained the behavior of black body radiation in the long-wave limit (the second factor in the right-hand part of Eq. (13.6)).

In 1905, Einstein explained the regularities of the photoelectric effect implying quantization of light energy (he was awarded the Nobel Prize in 1921 for the laws of the photoelectric effect). His paper was entitled "About one possible heuristic point of view related to creation and transformation of light" [1]. The photoelectric effect is the emission of electrons from a metal into free space upon illumination by light. An emitted electron ("photoelectron") acquires kinetic energy E in accordance with the following balance,

$$\hbar\omega = E + \Phi, \tag{13.7}$$

where the threshold energy Φ is the work function of a metal under consideration. Einstein's idea of light quanta was the first solid stone in the basement of the quantum theory of light.[1]

This remarkable fact in the history of quantum physics is rather curious in its essence. Irony upon irony, the apparent explanation of the photoelectric effect provided by Eq. (13.7) did not necessarily require explanation based on light quanta. It can be explained in a semi-classical manner as follows. An electron performs a quantum transition from the ground state with energy E_g (bound state in an atom) to an excited state with energy E within a continuum of states inherent in infinite motion. This transition can be performed by an external perturbing electric field of incident light. According to the Bohr hypothesis, quantum transitions occur only if the light frequency meets the condition $\omega = (E - E_g)/\hbar$. One can see this consideration explains the behavior described by Eq. (13.7) if the energy an electron gains in the course of transition consists of work function Φ and kinetic energy E.

[1] Amazingly, the very name of Einstein in German can be loosely interpreted as "first stone".

However, Bohr's postulate appeared in 1913 only and it is exclusively owing to Einstein's ingeneous intuition that in 1905 the scientific community got the idea that light is absorbed by matter in portions called light quanta.

In 1916, i.e. three years after Bohr's formulation of his famous postulates, Einstein derived the black body radiation spectrum (Eq. 13.6), considering a set of quantum systems with discrete energy states and assuming precise balance in upward stimulated and downward spontaneous and stimulated transitions for every couple of energy levels [2]. Such balance is inherent if a set of elementary quantum systems (e.g. atoms) is in equilibrium with radiation. This outstanding paper will be considered in more detail in Section 13.5. Einstein worked within a paradigm that light is emitted in quanta as a consequence of quantum transitions between states with discrete energy values.

In 1924, Bose made the real breakthrough in the emerging quantum physics [3]. He consistently treated the black body radiation as an equilibrium gas of light quanta.[2] These light quanta were described in terms of a distribution function (13.5) derived upon assumption of quanta identity and their non-conserving number (the original Bose idea), density of states Eq. (13.4) in which quanta may exist (the original Rayleigh idea of counting modes in a cavity upgraded to a conceptual density of states notion) and the energy portion $\hbar\omega$ held by each quantum. Thus the concept of light quanta became more and more mature. Bose's approach to a distribution function was immediately transferred to particles of matter by Einstein (for atoms and molecules) in 1924–25 and by Dirac and Fermi in 1926 for electrons. Simultaneously, Rayleigh's mode density evolved to the density of states and was entered into many basic formulas at that time.

Very soon, in the fall of 1925, Born, Jordan and Heisenberg came up with their ingenious work on foundations of quantum mechanics for systems with multiple degrees of freedom [4]. Along with purely mechanical systems, they considered radiation confined in a cavity and proposed consideration of radiation in terms of elementary quantum oscillators, as expressed by Eq. (13.1), with $n_{\mathbf{k}}$ treated as the number of quanta in a certain state and the background additive defined by,

$$E_0 = \frac{1}{2}\hbar \sum_{\mathbf{k}} \omega_{\mathbf{k}}, \tag{13.8}$$

called "zero energy".

Formulation of the basic Schrödinger equation occurred in the same period in a paper published in 1926 [5] (see also Section 3.7 on the inspiring role of optics in formulation of wave mechanics).

Finally, in 1927, Dirac published the fundamentals of quantum electrodynamics under the title "The quantum theory of the emission and absorption of radiation" [6]. The prerequisite condition of his approach was clearly formulated as "*the number of light quanta per unit volume associated with a monochromatic light-wave equals the energy per unit volume of the wave divided by the energy $(2\pi h)\nu$*".

[2] The Bose consideration has been discussed in detail in Section 2.2.

The very notation "*photon*" had been introduced by G. N. Lewis in 1926 [7]. However, he did not consider radiation as an entity composed of light quanta. Photons according to Lewis were thought of as intermediate particles carrying energy for a certain period in the course of light absorption.

13.2 Wave–particle duality in optics

In modern physics wave–particle duality is fully appreciated. Many optical phenomena can be thoroughly interpreted in terms of classical wave optics without quantization of radiation. Many phenomena can be understood only assuming a discrete structure of electromagnetic radiation.

Many physicists deem that photons should only be involved in considerations if, and only if, there is no other way to explain the optical phenomenon in question. This paradigm was formulated clearly by W. Lamb in his distinguished paper entitled "Anti-photon" [8]. The author of this book supports this approach and it is exclusively because of this paradigm that photons only enter the content of this book in Chapter 13, but not in previous chapters. Light propagation phenomena seem not to need photon representation at all, probably down to the ultimate single photon level of detection. Most physicists do not adopt the concept of photons as corpuscular constituents or ingredients of light, as Einstein tended to outline in his pioneering paper [1]. Conception of a single-photon state of radiation still remains elusive even in multiple gedanken experiments. Particularly, the definite momentum inherent in a single photon leads to an image of an infinitely delocalized plane wave carrying energy $\hbar\omega$. Since we have successfully managed to understand the complex propagation of light in inhomogeneous media in previous chapters without photons, the problem of describing the field states in terms of single-photon parameters will not be touched upon at all. We shall involve the notion of photons when describing emission and scattering events resulting in releasing portions of energy from matter to the field, and ascribe the notion "photon" to these portions. Such a treatise meets reasonably minimal quantum electrodynamics without touching intimate features of light quanta.

The above example of a photoelectric effect looks rather intuitive in terms of photons but rather complicated in terms of a semi-classical approach. It is instructive to discuss briefly a few examples of the opposite type which can readily be intuitively understood in terms of wave optics, but appear to be completely counter-intuitive and cumbersome for description in photon language.

The first example is the Doppler shift of spectral lines and Doppler broadening of spectra. A gas of atoms at finite temperature exhibits atomic emission lines broadened because of the random motion of atoms with respect to a detector. This broadening dominates over the natural linewidth defined for a given atom by its radiative decay rate. The explanation originates from the acoustic counterpart for the source of sound moving towards and away from a listener. It is simple and clear in the context of wave optics. In terms of photons, one should suppose that an atom emits photons of the original frequency whereas a detector receives (another?) photons of shifted frequency. Every photon after being emitted seems to change its frequency by the time of detection. How did the photon learn that a detector is moving with respect to an atom while existing between the time of its creation in the course of downward transition of an atom and the time of its death in the course of the upward transition in a detector material? The explanation in terms of photons can be found only after cumbersome mathematics well beyond reasonable intuition. It takes several pages of formulas and can be found e.g. in the basic paper by Fermi [9].

The second example is change in frequency of light leaving a dynamically expanding (or shrinking) cavity. Light stored in a Fabry–Perot cavity changes its spectrum in the course of decay if the cavity expands or shrinks fast enough. This phenomenon has recently been reported [10]. Wave optics offers an immediate explanation which is again based on the acoustical counterpart. Sweeping sound occurs, e.g. in an electric guitar, for a vibrating string if the tension of the string is modulated by a guitar player's finger. It is a common way of getting higher expression in modern guitar playing. The above phenomenon in optics is explained as adjustment of light oscillations to the cavity modes. However, discussion in terms of photons is by no means straightforward. One has to explain why (and how) photons stored in the cavity have been replaced by other photons with adjusted frequency.

The intimate essence of a light quantum, in spite of its long centennial history in science, is still being actively discussed in terms of the real and gedanken experiments, where certain controversial issues in our understanding of radiation quantum properties became evident. A number of such examples can be found in [11–14]. The reader interested in deeper insight into the photon conception is referred to the topical books on quantum optics by R. Loudon [15], L. Mandel and E. Wolf [16], M. O. Scully and M. S. Zubairy [17].

13.3 Electromagnetic vacuum

Naturall reason abhorreth vacuum.[3]

Thomas Kranmer, Canterbury bishop, 1550 [18]

The electromagnetic vacuum is the state of an electromagnetic field in a free space where no photon exists in any mode, i.e. in Eq. (13.1), $n_{\mathbf{k}} = 0$ for all $\mathbf{k}$s. The vacuum state of field is characterized by the infinite energy defined by Eq. (13.8), or taking into account the continuous spectrum of electromagnetic waves in free space,

$$E_{\text{vacuum}} = \frac{1}{2}\int_0^{\infty} \hbar\omega \frac{\omega^2}{\pi^2 c^3} \mathrm{d}\omega = \infty. \tag{13.9}$$

Thus, decomposition of radiation in harmonic *quantum* oscillators gives rise to infinite minimal energy of radiation composed by finite zero energies $\hbar\omega/2$ of an infinite number

[3] Nature abhors vacuum.

of oscillators. The general approach of viewing fields as assemblies of quantum oscillators gives rise to a more general conception of a *physical vacuum* with the electromagnetic vacuum as the particular case.

Although the total energy contained in an electromagnetic vacuum is infinite, the energy density per unit volume and unit frequency (wavelength, wave number) range is finite and is defined by the product of the density of electromagnetic modes and the zero-energy inherent in every mode. For the frequency domain, the vacuum energy density $W(\omega)$ reads,

$$W(\omega) = \frac{1}{2}\hbar\omega\frac{\omega^2}{\pi^2 c^3}, \tag{13.10}$$

using Eq. (13.4) for electromagnetic mode density. We do not refer to photon density of states here since there are no photons in the vacuum state. Nor the factor $\frac{1}{2}\hbar\omega$ is treated as one half of the photon energy. Instead, as prescribed by Eq. (13.1), it is strictly the minimal energy every mode can carry on.[4]

In the energy scale, energy density per unit range of quantum energy reads,

$$W(E) = W(\omega)\frac{\mathrm{d}\omega}{\mathrm{d}E} = \frac{\omega^3}{2\pi^2 c^3} = \frac{4\pi(h\nu)^3}{h^3 c^3}. \tag{13.11}$$

Equations (13.10) and (13.11) define the finite energy contained in the vacuum state within a finite volume and a finite spectral range. The reader is requested to count how many Joules are contained within the visible in a unit volume (Problem 5). However, since the frequency range of existing electromagnetic waves has no upper limit, even every finite volume does contain an infinite amount of vacuum energy,

$$\int_V \mathrm{d}V \int_0^\infty W(\omega)\mathrm{d}\omega = \infty. \tag{13.12}$$

This infinity does manifest itself as a serious obstacle in many physical problems. In these cases it is simply ignored, relying on the assumption that it is only change of electromagnetic energy in the course of emission and absorption events that is meaningful and measurable.

From time to time speculative efforts to consider possible extraction of vacuum energy are published. Unfortunately, it is not possible since vacuum energy represents the minimal

[4] Although every person does most probably think by means of images, conceptions and associations rather than words (the author thinks he does ☺), the language we are using often has a serious influence on our mind. The extreme formulation of this problem "The spoken thought is getting to be a lie" belongs to Russian poet F. Tiutchev. In the specific case under consideration the statement "every mode carries energy" can be spoken in other words as "there is finite energy per mode" though it is hardly becoming clearer as to what is meant. The saying "energy contained in every mode" is inaccurate since a mode is by no means a real object, but is instead a way to describe the possible radiation state. And neither quantum system interacts with modes, independent of how many times this statement is found in the textbooks. "Materializing" modes is by no means relevant and has the same physical uselessness as "materializing" electric field lines in early electrodynamics. Similar, words like "a photon is a bundle of energy" are meaningless in spite of how much poetic emotion they can arouse.

energy the field can get and we can not create the field states with energy lower than the zero-energy value. One can see there is no photon in a vacuum so we cannot take a photon from a vacuum.

13.4 The Casimir effect

An electromagnetic vacuum is principally perceptible. If a pair of conductive plates (parallel for simplicity) confines a portion of space then the electromagnetic modes between and outside the plates are different. Semi-infinite space outside each plate allows a continuous set of modes whereas between the plates the mode set is cut off from the low-frequency side. This results in an attractive force known as the *Casimir effect.* This phenomenon was predicted by H. Casimir in 1948 [19].[5] Niels Bohr provided the principal hint about electromagnetic vacuum as the physical reason for intermolecular interactions to H. Casimir [20]. The arising forces are referred to as *Casimir forces*. For molecules and micro- and nanoparticles in colloidal solutions Casimir's approach explains *van der Waals forces*.

"*There exists an attractive force between two metal plates which is independent of the material of the plates as long as the distance is so large that for wave lengths comparable with that distance the penetration depth is small compared with the distance. This force may be interpreted as a zero point pressure of electromagnetic waves*"

– H. Casimir [19].

The outside and inside pressure difference gives rise to a compressive tension which reads, for spacing d, when counted per unit area,

$$F = \frac{\pi^2 \hbar c}{120 d^4} = 0.013 \frac{1}{d^4[\mu\mathrm{m}]} [\mathrm{dyne/cm^2}]. \tag{13.13}$$

This expression was derived [19] accounting for integration over all directions of wave vectors under the assumption of ideal reflecting walls and ignoring thermal fluctuations of electromagnetic radiation. Casimir forces were confirmed in many experiments and have attracted a lot of attention in the context of micro- and nanomechanics [20–23]. It has been known since 1960 that a thin wire before being soldered to a microchip hits the plate when close enough to it [24]. This is also taken into account in the theory of quantized fields and has become the subject of special books [25, 26]. The Casimir effect is often referred to as the most celebrated quantum electrodynamical effect. Also the hypothesis advanced by H. Puthoff on Casimir forces should be mentioned as the possible intrinsic origin of gravitation [27, 28]. In this hypothesis, gravitational attraction is a manifestation of the Casimir effect on a cosmological scale and is like van der Waals forces in a colloidal solution. Noteworthy, the Casimir effect has a clear classical analogy. This is the attractive

[5] Hendrik Casimir (1909–2000) was a Dutch physicist, the president of the European Physical Society from 1972 to 1975.

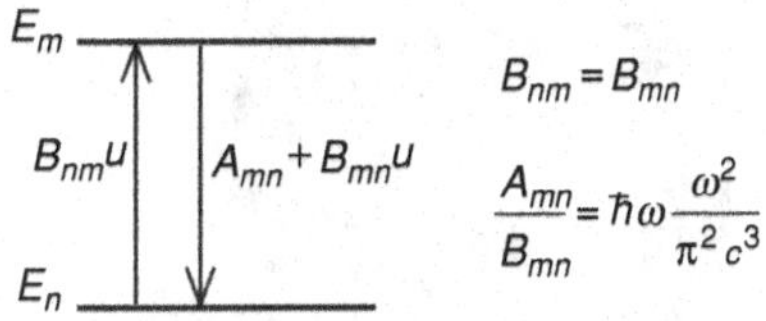

E_m E_n $B_{nm}u$ $A_{mn}+B_{mn}u$ $B_{nm}=B_{mn}$ $\frac{A_{mn}}{B_{mn}}=\hbar\omega\frac{\omega^2}{\pi^2c^3}$

Fig. 13.1 **A pair of energy levels of a quantum system and relations between Einstein coefficients *A*, *B*. *u* is the radiation energy density as in Eq. (13.6).**

force making two ships get closer to one another when moving on parallel courses at a relatively small distance [29].

When speaking about vacuum zero energy, one more reminiscence is worth mentioning. Oscillations of atoms in a crystal lattice are also treated in terms of quantum harmonic oscillators with a *phonon* being the vibrational counterpart of a photon. This conception dates back to 1925 [4]. Zero oscillations inherent in a quantum "vibrational vacuum" do prevent condensation of light (in the sense of low-weight, not luminous) atoms into solid matter. In particular, helium never exists as solid matter since the large amplitude of zero oscillations exceeds any possible value of the lattice constant.[6]

13.5 Probability of emission of photons by a quantum system

At the dawn of quantum mechanics, the Planck formula Eq. (13.6) for black body radiation, proposed in 1900 and remarkably fitting the experimental data, has often been chosen to test new ideas. After the formulation of Bohr's postulates in 1913, Albert Einstein considered the possibility of deriving the Planck formula in terms of balanced upward and downward transitions in an imaginary two-level system. In the works published in 1916–1917 entitled "Emission and absorption of light according to the quantum theory" [30] and "On the quantum theory of radiation" [31] he introduced coefficients A and B for spontaneous and stimulated transitions, respectively which have since then been referred to as *Einstein coefficients*. He arrived at the Planck formula in this representation and found it to be possible under two conditions. The first is the existence of downward stimulated transitions along with upward ones with the relation,

$$B_{nm} = B_{mn}. \tag{13.14}$$

Several decades later this prediction provided an understanding of optical gain in an ensemble of quantum systems with population inversion. The second condition is the following restriction for coefficients describing spontaneous and stimulated transitions,

$$\frac{A_{mn}}{B_{mn}} = \hbar\omega\frac{\omega^2}{\pi^2 c^3}. \tag{13.15}$$

[6] Note, it is the amplutide of atomic oscillations compared to the average interatomic distance that can serve as a reasonable criterion for discrimination of the liquid state versus the solid one.

Albert Einstein (1879–1955)

Paul Adrien Maurice Dirac (1902—1984)

The two prominent persons whose ideas essentially formed the subject of this and forthcoming chapters. Albert Einstein introduced in 1916 spontaneous emission probability and derived an expression for it to meet Planck's black body radiation formula. Paul Dirac in 1927 obtained spontaneous emission in the theory based on quantized electromagnetic radiation and showed that it is electromagnetic mode density that explains Einstein's formula for spontaneous emission probability.

The Einstein coefficient A has dimensionality s^{-1} and is equal to the inverse decay time of the excited state. It is the *spontaneous decay* rate of a quantum system. It describes the rate of spontaneous emission of photons by an excited atom, molecule, quantum dot or other simple quantum system.[7] It should be noted that Einstein did not pay attention that the right-hand part of Eq. (13.15) contained an electromagnetic mode density Eq. (13.4). It probably happened because at the time of writing that paper (1916), Rayleigh's approach of counting modes had not gained proper recognition in the physical community. It was Dirac in 1927 [32] and later on Fermi in 1932 [33] who introduced the concept of density of modes (and photon density of states) into quantum electrodynamics. Dirac introduced operators of creation and annihilation of photons and found that upward and downward transition rates for N photons in a given mode relate as $N/(N+1)$. The unity in the denominator was assigned to the spontaneous downward transition and it was the first time that spontaneous transitions were found but not introduced into the theory. Dirac understood that once he had found the occurrence of spontaneous transitions, the total emission rate should be proportional to the number of modes available, which reads as density of modes Eq. (13.4). Notably, he immediately made a comment that the scattering of photons should also obey the same proportionality to mode density. In what follows we reproduce a discussion on spontaneous emission in quantum electrodynamics according to R. Loudon [15].

[7] Very often the notations "spontaneous radiation" and "stimulated radiation" are used which are incorrect, since radiation can not be separated into spontaneous and stimulated forms.

First, radiation is seen in terms of quantum oscillators with energies expressed by Eq. (13.1) and photon numbers n_k and photon energies $\hbar\omega_{\mathbf{k}}$. Then photon creation $\hat{a}_{\mathbf{k}}^{+}$and annihilation $\hat{a}_{\mathbf{k}}$ operators are introduced along with the wave function $|n_{\mathbf{k}}\rangle$, describing the field state with n photons in a mode **k**. The operator $\hat{a}_{\mathbf{k}}^{+}$ when acting on a state with n photons (subscript "**k**" omitted for simplicity) converts it into the state with $(n+1)$ photons. The operator $\hat{a}_{\mathbf{k}}$ converts that state into the state with $(n-1)$ photons,

$$\begin{aligned}\hat{a}^{+}|n\rangle &= a^{+}|n+1\rangle,\\ \hat{a}|n\rangle &= a|n-1\rangle,\end{aligned} \tag{13.16}$$

where a^{+}, a are numbers.The normalization condition

$$\langle n-1|n-1\rangle = \langle n|n\rangle = \langle n+1|n+1\rangle = 1, \tag{13.17}$$

results in

$$\hat{a}|n\rangle = n^{1/2}|n-1\rangle, \tag{13.18}$$

$$\hat{a}^{+}|n\rangle = (n+1)^{1/2}|n+1\rangle. \tag{13.19}$$

When calculating the matrix elements defining probabilities of upward and downward transitions, the numerical coefficients in Eqs. (13.18 and 13.19) should be squared. Then upward and downward transition rates relate as $n_{\mathbf{k}}/(n_{\mathbf{k}}+1)$. Unity in the denominator means spontaneous transitions with a photon emitted into the k-mode are possible. To count the total probability that spontaneous decay of an excited state will happen, the probability for each mode should be summed over all the modes available within the full solid angle 4π and for the two possible light polarizations. Since in a free space modes form a continuous set, summing is replaced by integration. Consider the simplest case when an atom from an excited state with finite lifetime decays to the ground state with infinitely long lifetime. We can say then that the rate of spontaneous decay W_{10} of a system "atom+field" from the state $\langle E_{1,}0|$ (an atom is in the first excited state, the field is in the state with no photon in all modes with frequency $\omega = (E_1 - E_0)/\hbar)$ into the state $|E_0, 1\rangle$ (an atom is in the ground state whereas the field has a photon with frequency $\omega = (E_1 - E_0)/\hbar)$ is then given by the expression,

$$W_{10} = \frac{2\pi}{\hbar}D(\omega)|\langle E_1, 0|H_{\text{int}}|E_0, 1\rangle|^2, \tag{13.20}$$

with H_{int} being the interaction Hamiltonian.

Two assumptions should be emphasized which have been used in the above consideration. *The first assumption* is the very applicability of the notion "decay rate" to spontaneous transitions. This means that for a statistically large number of excited atoms (or other quantum systems) the number of detected transitions (i.e. the number of emitted photons) will be in direct proportion to the observation time. The spontaneous decay process is therefore treated as a *Markovian process*. This corresponds to the *Weisskopf–Wigner*

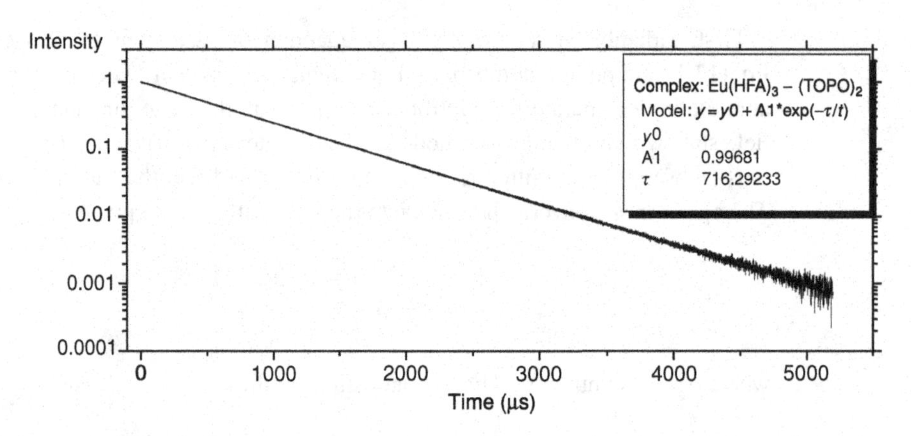

Fig. 13.2 **Decay of excited Eu^{3+} ions in a complex with an organic ligand in toluene solution. It exhibits with high accuracy a single-exponential behavior with lifetime $\tau = 716.3\,\mu s$. Intensity versus time dependence in a semi-logarithmic presentation obeys a straight line with high precision.**

approximation [34]. *The second assumption* is the implication that the field density of states (mode density, photon states density) $D(\omega)$ is a smooth function near the transition frequency ω.

To summarize, *in quantum electrodynamics spontaneous emission of photons arises without additional assumptions, the frequency dependence of the decay rate being defined by the summation of the transition rate over all the allowed final states of a field in a free space.*

Many physicists still anticipate the semiclassical view on spontaneous emission of light. In this context, once the electromagnetic field has been treated as a quantized one based on Eq. (13.1), then spontaneous transitions are believed to occur stimulated by vacuum zero-point radiation. In this presentation, spontaneous transitions are reduced to stimulated ones with vacuum field fluctuations being the source of stimulation. For example, V. Weisskopf [35], M. O. Scully and M. S. Zubairy [17] and many successors have adopted this consideration. Mandel and Wolf [16] refer to this treatise as follows:

"Sometime spontaneous transitions are viewed as being stimulated by vacuum fluctuations."

In all the above considerations the spontaneous decay rate is nevertheless seen to be proportional to the density of electromagnetic modes which we shall treat as density of photon states hereafter. Amazingly enough, spontaneous emission of light is a basic property of matter and in spite of the advances in quantum physics it is still the subject of debates. The reader is referred to several books and reviews on this issue [15–17, 36–38].

Figure 13.2 shows a representative example of the spontaneous decay law of excited atoms. Ions of lanthanides (Eu, Er, Tb, and other) emit light by means of quantum transitions from inner electron orbitals and therefore show very stable optical behavior even in solid matrices and solutions. A single-exponential decay,

$$N(t) = N(0)\,e^{-t/\tau}, \tag{13.21}$$

can be distinctly traced over several orders of magnitude resulting in a perfect straight line in a semi-logarithmic plot. The emission line has a peak at 613 nm and corresponds to the $^5D_0 \rightarrow ^7F_2$ electric dipole transition in Eu^{3+} ions.

For a model excited two-level system decaying into a free space by means of transition into the ground state the spontaneous decay rate reads [37],

$$W_{10} = \frac{4}{3}\frac{1}{4\pi\varepsilon_0}\mu^2\frac{1}{\hbar}\frac{\omega_0^3}{c^3} = \frac{1}{3\varepsilon_0}\mu^2\frac{\omega_0}{\hbar}\pi D(\omega_0), \tag{13.22}$$

where μ is the *dipole moment* for a given transition, ω_0 is the resonant frequency of transition $\omega_0 = (E_1 - E_0)/\hbar$, and density of photon states in free space $D(\omega_0)$ is defined by Eq. (13.4). The dipole moment value must be calculated explicitly by means of quantum electrodynamics for every specific system. The exponential decay of a two-level system results in a Lorentzian spectral line shape,

$$I(\omega) = I(\omega_0)\frac{\gamma_{10}^2}{\gamma_{10}^2 + (\omega - \omega_0)^2}, \tag{13.23}$$

with

$$\gamma_{10} = \frac{1}{2}W_{10}. \tag{13.24}$$

If transition from an excited state occurs not to the ground state of an atom (or another model quantum system), then the width of the emission spectrum will be the sum of the width inherent in the decay of the initial and the final states.

13.6 Does "Fermi's golden rule" help to understand spontaneous emission?

The statement that the spontaneous decay rate of an excited atom by means of photon emission is directly proportional to the electromagnetic mode density $D(\omega)$ is often referred to as *Fermi's golden rule.* Numerous books, reviews and popular texts cite Eq. (13.20) as Fermi's golden rule. This is a confusing mistake in scientific literature.

First, Fermi did not derive Eq. (13.20) for the photon emission rate. He did contribute considerably to earlier quantum electrodynamics. His basic paper "Quantum theory of radiation" published in 1932 [33] has had a strong impact on many researchers. It actually starts from evaluation of density of modes as a basic conception of the quantum theory of radiation and contains the derivation of Eq. (13.20). However, it had been published five years later after Dirac's original paper [6] and contained thorough citations to Dirac everywhere throughout the text. The first chapter of Fermi's paper is entitled "Dirac's theory of radiation".

Second, Fermi never called the expression for spontaneous emission of photons "the golden rule". Fermi did use this notation in the book "Nuclear Physics" published in 1950 [39] with respect to a quantum-mechanical expression for transition of a quantum system into the final state belonging to a continuous spectrum. Indeed, this is a common relation in quantum mechanics. It describes, e.g. scattering of an electron into a state with infinite motion. Notably, Fermi did refer to the classical textbook on quantum mechanics by Schiff [40], the first edition being published in 1949. However, this purely quantum-mechanical formula can not be applied to spontaneous decay of a quantum system with photon emission. Spontaneous emission itself can not occur within quantum mechanical theory. An excited atom will never emit a photon in quantum mechanics if there is no stimulating perturbation.

Thus one can see that Fermi never derived what he called himself a "golden rule" and never used the label "golden rule" for spontaneous emission of photons. Possibly, this confusing situation arose because of a certain misunderstanding of the density of states concept. Once appearing in quantum mechanics, density of particle (say, an electron) states seems to lead to the probability of photon emission, simply by replacing electron density of final states by photon density of states. It is however not the case since we need to quantize the field and find the spontaneous decay process itself, i.e. to make a principal step from quantum mechanics to quantum electrodynamics. The book by Berestetskii *et al.* [41] is among the textbooks on quantum electrodynamics that do not state a difference between Eq. (13.20) and its quantum-mechanical counterpart.

One more comment is reasonable in connection with using rules in science. Events and phenomena in Nature are driven by laws and by chance rather than by rules. This holds even when we can not evaluate the internal law and instead just observe the rule. The latter should simply be ruled out from basic science, although being useful in industrial and computational applications as a helpful recipe. The phrase "An atom emits light in accordance with the Bohr postulate and Fermi rule" is an analog of "Sun is rising" which we still use 500 years after Copernicus.

13.7 Spontaneous scattering of photons

Spontaneous scattering is a change of light properties, including either frequency, polarization or wave vector direction in the course of light–matter interaction without stimulating radiation involved. Otherwise it is referred to as stimulated scattering. In terms of the photon notion, spontaneous scattering can be treated as instantaneous virtual excitation of a quantum system with immediate emission of the photon, differing from the original in either of the following parameters: photon energy, photon momentum, polarization state. Spontaneous scattering can be separated into elastic and inelastic types. Elastic scattering implies that the only direction and polarization of light is modified whereas inelastic scattering means that the light frequency (photon energy) has been changed. A typical example of elastic scattering is the blue sky color on Earth. Elastic scattering can be thoroughly described in terms of wave optics as has been discussed in Chapter 8. Inelastic scattering implies energy exchange of radiation with a quantum system. In this event, the frequency of a scattered photon differs from the frequency of an incident photon by the characteristic value defined by the intrinsic vibrations of atoms in the matter. This phenomenon was discovered in 1928 by C. V. Raman and K. S. Krishnan [42] for molecular solutions and by L. I. Mandelstam and G. S. Landsberg [43] for solids. Since then it has gained the notation *Raman scattering*. Raman spectroscopy of molecules and solids is a vibrational spectroscopic method capable of identifying chemical bonds in the matter by the characteristic frequencies inherent in every pair of atoms involved in vibrations.

In terms of quantum electrodynamics, the Raman scattering rate can be written as,

$$W_{\mathrm{RS}} = n\frac{(2\pi)^2\omega\omega'}{\hbar^2}|S|^2\left[n' + \frac{\omega'^2}{(2\pi c)^3}\right], \tag{13.25}$$

where ω is the incident light frequency, ω' is the scattered light frequency, n is the incident photon number, n' is the scattered photon number, S is the matrix element of the transition under consideration, c is the speed of light in a vacuum. This consideration dates back to 1934 and belongs to G. Placzek [44]. The first term in the square brackets describes stimulated scattering, whereas the second term corresponds to spontaneous scattering. This equals the density of photon states at the scattered photon frequency per unit solid angle, i.e. $D(\omega')/4\pi$. Therefore the full rate of spontaneous Raman scattering in a vacuum within the full solid angle reads,

$$W_{\mathrm{RS}} = n\frac{(2\pi)^2\omega\omega'}{\hbar^2}|S|^2 D_0(\omega'). \tag{13.26}$$

There are now two comments to make. First, at $\omega = \omega'$ Raman scattering reduces to elastic scattering and one has $W_{\mathrm{RS}} \propto \omega^4$. This is a relationship for light scattering known since Rayleigh which explains the blue color of the sky on Earth and the reddish colors of sunrise and sunset. Second, photon scattering (both inelastic and elastic) is proportional to the density of photon states. The above treatise of the effect of photon density of states on light scattering can be found in [45].

It is instructive to consider possible convergence of the quantum and the classical descriptions of light scattering. In quantum optics, spontaneous photon scattering is viewed as a result of the virtual excitation of a quantum system with subsequent emission of another photon with different direction, polarization and (optionally) frequency as compared to the original photons. The classical picture for elastic scattering completely describes all scattering phenomena in terms of wave scattering and interference of scattered waves. The whole content of Chapter 8 confirmed this statement. In this context, the quantum presentation merges with classical wave optics perfectly. Inelastic scattering, including Raman scattering, is typically treated using the quantum-optical consideration. The classical wave-optical counterpart does exist but has not been elaborated in detail. It was proposed in the 1930s by L. I. Mandelstam. His elegant explanation reduces Raman scattering by a molecule to modulation of the incoming light wave by intrinsic molecular vibrations. The general result of amplitude modulation of harmonic oscillation with frequency ω by another harmonic oscillation with frequency Ω gives rise to the spectrum of resulting oscillations consisting of the 3 frequencies: ω, $\omega + \Omega$, $\omega - \Omega$. This is a well-known result in mechanics and radioengineering (see Problem 11).

To summarize, spontaneous emission and scattering rates are directly proportional to the density of photon states for emitted and scattered photons, respectively.

Problems

1. Based on Eq. (3.3) derive photon DOS for energy $D(E)$ and for momentum $D(p)$.

2. Calculate the photon rest mass for light of wavelength 500 nm.

3. Explain why and how the energy distribution function for photons differs from those for electrons and for atoms.

4. The dimensionality of energy density given by Eq. (13.10) is $\mathrm{J \cdot m^{-3} \cdot Hz^{-1}}$, whereas that given by Eq. (13.11) reads just $\mathrm{m^{-3}}$. Explain this seemingly confusing result.

5. Calculate the energy contained in an electromagnetic vacuum in one cubic centimeter within the visible range of the spectrum.

6. Calculate the contractive force between two ideally reflecting circular plates displaced parallel to each other for a few combinations of radius r and spacing d, e.g. $r = 100$ nm, $d = 100$ nm; $r = 1$ m, $d = 1$ mm, $r = d = 10^3$ m.

7. Explain the classical analogy of the Casimir effect for two ships in the sea.

8. Using the rate equations for a two-level system shown in Figure 13.1, show that in the limit of high radiation density populations of upper and lower levels will be equal to one half of the total number of systems under consideration. Explain why under this condition absorption saturation occurs.

9. Try to guess how the spontaneous emission rate will change if the final state of an atom belongs to a continuous spectrum. Hint: consider the number of ways a transition can occur and use the density of final states for the "atom+field" system.

10. Based on frequency-dependent density of photon state estimate the typical values of spontaneous decay rates in radiophysics with respect to optics. Consider a typical rate $10^8\ \mathrm{s^{-1}}$ inherent in the optical range and take the representative values of radio wavelength as 100 m (short-wave radio transmission range), 0.1 m (television and mobile phone transmission range).

11. To get more insight into the classical explanation of Raman scattering proposed by Mandelstam, consider an arbitrary harmonic oscillation with frequency ω whose amplitude is modulated with frequency $\Omega \ll \omega$ and show that the spectrum of the resulting oscillations consists of the three frequencies: ω, $\omega + \Omega$, $\omega - \Omega$.

References

[1] A. Einstein. Uber einen die erzeugung und Verwandlung des Lichtes betreffenden heuristischen Gesichtspunkt. *Ann. Physik*, **17** (1905), 132–148.

[2] A. Einstein. Strahlungs-Emission and – Absorption nach der Quantentheorie. *Verh. Deutsch. Phys. Ges.*, **18** (1916), 318–323.

[3] S. N. Bose. Planck's Gezets and Lichtquantenhypothese. *Zs. Physik*, **26** (1924), 178–181.
[4] M. Born, W. Heisenberg and P. Jordan, Zur Quantenmechanik. II, *Zs. Phys.*, **35** (1926), 557–615.
[5] E. Schrödinger. Quantisierung als Eigenwertproblem. *Ann. Physik*, **81** (1926), 109–139.
[6] P. A. M. Dirac. The quantum theory of the emission and absorption of radiation. *Proc. Royal Soc. London, Ser. A*, **114** (1927), 243–265.
[7] G. N. Lewis.The conservation of photons. *Nature*, **118** (1926), 874–875.
[8] W. E. Lamb. Anti-photon. *Appl. Phys. B*, **60** (1995), 77–80.
[9] E. Fermi. Quantum theory of radiation. *Rev. Mod. Phys.*, **4** (1931), 87–132.
[10] S. F. Preble, Q. Xu, and M. Lipson. Changing the color of light in a silicon resonator. *Nature Photonics*, **1** (2007), 293–296.
[11] A. Zajonc. Light reconsidered. *Opt. Phot. News Trends*, **3** (2003), S-2–S-5.
[12] R. Loudon. What is a photon? *Opt. Phot. News Trends*, **3** (2003), S-6–S-11.
[13] D. Finkelstein. What is a photon? *Opt. Phot. News Trends*, **3** (2003), S-12–S-17.
[14] A. Mathukrishnan, M. O. Scully, and M. S. Zubairy. The concept of photons – revisited. *Opt. Phot. News Trends*, **3** (2003), S-18–S-27.
[15] R. Loudon. *Quantum Theory of Light* (Oxford: Clarendon Press, 2000).
[16] L. Mandel and E. Wolf. *Optical Coherence and Quantum Optics* (Cambridge: Cambridge University Press, 1995).
[17] M. O. Scully and M. S. Zubairy. *Quantum Optics* (Cambridge: Cambridge University Press, 2001).
[18] *Oxford English Dictionary* (Oxford: Oxford University Press, 1987).
[19] H. B. V. Casimir. On the attraction between two perfectly conducting plates. Proc. Royal Netherland Acad. Arts Sci. **51** (1948), 793–795. Cited after *Gems from a Century of Science 1898–1997. Centenary Issue of the Proceedings of the Royal Netherlands Academy of Arts and Sciences*, Schoonhoven, L.M. (eds), Amsterdam, 1997, pp. 61–63.
[20] P. W. Milonni and Mei-Li Shih, Casimir forces. *Contemp. Phys.*, **33** (1992), 313–322.
[21] S. K. Lamoreaux. Demonstration of the Casimir force in the 0.6 to 6 mm range. *Phys. Rev. Lett.*, **78** (1997), 5–8.
[22] P. W. Milonni. *The Quantum Vacuum* (New York: Academic Press, 1994).
[23] G. L. Klimchitskaya and V. M. Mostepanenko. Experiment and theory in the Casimir effect. *Contemp. Phys.*, **47** (2006), 131–144.
[24] S. K. Lamoreaux. Hendrik Brugd Gerhard Casimir. *Proc. Amer. Philos. Soc.*, **146** (2002), 286–290.
[25] V. M. Mostepanenko and N. N. Trunov. *The Casimir Effect and Its Applications* (Oxford, Oxford University Press, 1997).
[26] K. A. Milton. *The Casimir Effect* (London: World Scientific, 2001).
[27] H. E. Puthoff. Gravity as a zero-point-fluctuation force. *Phys. Rev. A*, **39** (1989), 2333–2342.
[28] D. C. Cole, A. Rueda and K. Danley. Stochastic nonrelativistic approach to gravity as originating from vacuum zero-point field van der Waals forces. *Phys. Rev. A*, **63** (2001), 054101.

[29] S. L. Boersma. A maritime analogy of the Casimir effect. *Amer. J. Phys.*, **64** (1996), 539–542.

[30] A. Einstein. Strahlungs-Emission und -Absorption nach der Quantentheorie. *Verh. Deutsch. Phys. Ges.*, **18** (1916), 318–323.

[31] A. Einstein. Zur Quantentheorie der Strahlung. *Mitteilungen d. Phys. Ges. Zürich*, Nr.18 (1916); *Phys. Zs.*, **18** (1917), 121–128.

[32] P. A. M. Dirac The quantum theory of absorption and emission of light. *Proc. Royal Soc. London*, **112** (1927), 243–265.

[33] E. Fermi. Quantum theory of radiation. *Rev. Mod. Phys.*, **4** (1932), 87–132.

[34] V. Weisskopf and E. Wigner. Berechnung der naturlichen Linienbreite auf Grund der Diracschen Lichtheorie. *Zs. Phys.*, **63** (1930), 54–60.

[35] V. F. Weisskopf. Growing up with field theory (The development of quantum electrodynamics in half a century. Personal recollections). *Physics – Uspekhi*, **138** (1982), 455–470.

[36] G. S. Agarwal. *Quantum Statistical Theories of Spontaneous Emission and Their Relation to Other Approaches* (Berlin: Springer Verlag, 1974).

[37] A. N. Oraevskii. Spontaneous emission in a cavity. *Physics – Uspekhi*, **37** (1994), 393–406.

[38] D. S. Mogilevtsev and S. Ya. Kilin. *Quantum Optics Methods of Structured Reservoirs.* (Minsk: Belorusskaya nauka, 2007) – in Russian.

[39] E. Fermi. *Nuclear Physics* (Chicago: University of Chicago Press, 1950).

[40] L. I. Shiff. *Quantum Mechanics* (New York: McGraw-Hill, 1949, 1955).

[41] V. B. Berestetskii, E. M. Lifshitz and L. P. Pitaevskii. *Quantum Electrodynamics.* (Oxford: Pergamon, 1980).

[42] C. V. Raman and K. S. Krishnan. A new type of secondary radiation. *Nature*, **121** (1928), 501–501.

[43] G. S. Landsberg and L. I. Mandelstam. Uber die Lichtzerstrenung in Kristallen. *Zs. Physik.*, **50** (1928), 769–774.

[44] G. *Placzek*, Rayleight-Streuung und Raman-Effekt. *Handbuch der Radiologie VI, Teil II* (Leipzig: Akademische Verlagsgeselschaft, 1934).

[45] S. V. Gaponenko. Photon density of states effects on Raman scattering in mesoscopic structures. *Phys. Rev. B*, **65** (2002), 140303 (R).

14 Density of states effects on optical processes in mesoscopic structures

Begin by deciding how much of the universe needs to be brought into the discussion. Decide what normal modes are needed for an adequate treatment of the problem under consideration.

W. E. Lamb, 1995 [1]

In mesoscopic structures where the dielectric permittivity of space is inhomogeneous on a scale of light wavelength, the elementary processes of field–matter interaction experience modification which can be elegantly explained in terms of photon density of states modification. The main purpose of the present chapter is to provide certain instructive ideas on how spatial confinement will alter the spontaneous emission and the spontaneous scattering of light. Light–matter interaction in nanostructures with confinement of light waves and local singularities in dielectric function is a very active area of research that stimulates interesting experiments and device concepts. Control of the spontaneous emission of light in mesoscopic structures is the basic quantum electrodynamical effect which gains applications in many areas. Basic properties of model structures including microcavities, photonic crystals, layers and interfaces will be discussed. Emission rates and angular parameters of emitted radiation will be considered. General constraint in terms of spontaneous emission sum rules and local density of states redistribution will be outlined.

Edward Mills Purcell (1912–1997)

E. M. Purcell was an American physicist. In 1946 in a few lines he explained that vacuum mode density should be replaced by another term to offer unprecedented control over spontaneous emission rates by atoms in a cavity. In 1952 he shared the Nobel Prize with Felix Bloch for the discovery of nuclear magnetic resonance. Among his PhD students was N. Bloembergen (b. in 1920), the Nobel Prize winner in 1981 for laser spectroscopy.

14.1 The Purcell effect

Density of photon states which defines the probability of spontaneous decay of excited atoms and other quantum systems essentially depends on the properties of space around an emitter. If the space properties do not allow free propagation of certain modes the decay can be slowed down. If the space offers spatial concentration of light for certain modes, the decay will be promoted. If around an emitter there is no mode available in the space which could carry out the emitted radiation, then the decay rate will tend to zero. These qualitative considerations form a solid base for the broad trend in nanophotonics aimed at engineering the desirable decay rate and emission indicatrix of nanostructured materials as well as atoms and molecules in complex environments. E. M. Purcell was the first to outline the principal possibility of enhancing the spontaneous emission rate by putting an emitter in a cavity with resonance coinciding with the frequency of the quantum transition [2]. His pioneering paper was extremely short and clear. An abstract is reproduced in Figure 14.1. He proposed to enhance the probability of spontaneous transitions with emission of radiation by putting a quantum system in a cavity with resonant frequency coinciding with the frequency of the quantum transition. He suggested in this case the density of modes inherent in a free space should be replaced by the term describing electromagnetic radiation in a cavity.

There are two factors resulting in an increase of spontaneous transition rate. The first factor is proportional to the Q-factor of the resonant mode in a cavity because continuous modes are now to be replaced by discrete modes, every mode having spectral width $\Delta\omega = \omega/Q$. The second factor accounts for the actual volume occupied by a given mode as

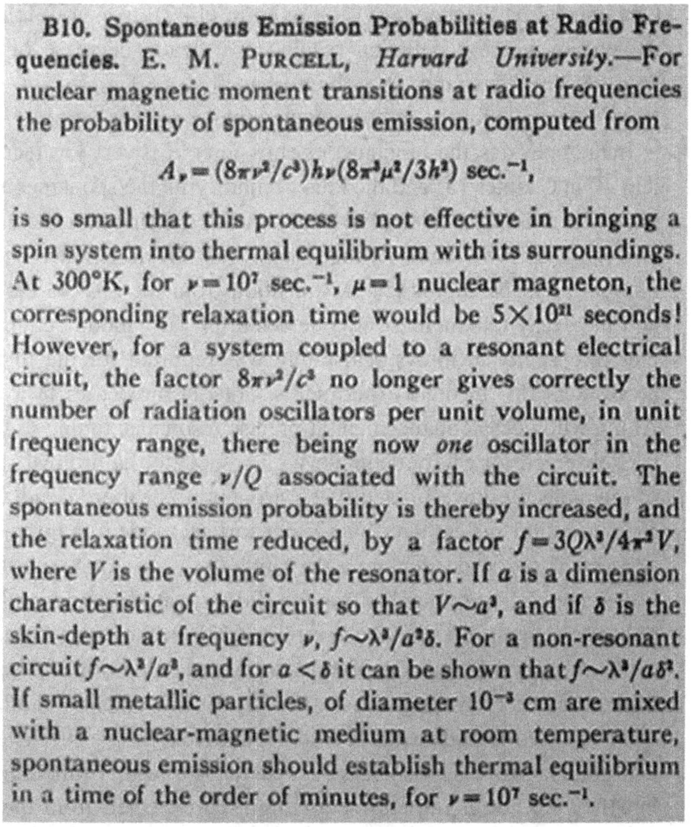

B10. Spontaneous Emission Probabilities at Radio Frequencies. E. M. PURCELL, *Harvard University*.—For nuclear magnetic moment transitions at radio frequencies the probability of spontaneous emission, computed from

$$A_\nu = (8\pi\nu^2/c^3)h\nu(8\pi^3\mu^2/3h^2) \text{ sec.}^{-1},$$

is so small that this process is not effective in bringing a spin system into thermal equilibrium with its surroundings. At 300°K, for $\nu = 10^7$ sec.$^{-1}$, $\mu = 1$ nuclear magneton, the corresponding relaxation time would be 5×10^{21} seconds! However, for a system coupled to a resonant electrical circuit, the factor $8\pi\nu^2/c^3$ no longer gives correctly the number of radiation oscillators per unit volume, in unit frequency range, there being now *one* oscillator in the frequency range ν/Q associated with the circuit. The spontaneous emission probability is thereby increased, and the relaxation time reduced, by a factor $f = 3Q\lambda^3/4\pi^2V$, where V is the volume of the resonator. If a is a dimension characteristic of the circuit so that $V\sim a^3$, and if δ is the skin-depth at frequency ν, $f\sim\lambda^3/a^2\delta$. For a non-resonant circuit $f\sim\lambda^3/a^3$, and for $a<\delta$ it can be shown that $f\sim\lambda^3/a\delta^2$. If small metallic particles, of diameter 10^{-3} cm are mixed with a nuclear-magnetic medium at room temperature, spontaneous emission should establish thermal equilibrium in a time of the order of minutes, for $\nu = 10^7$ sec.$^{-1}$.

Fig. 14.1 A photocopy of an abstract of Edward Purcell's paper published in 1946 (Physical Review, v. 69, p. 681). This is the abstract of his talk presented at the Material Research Society conference.

compared to the λ^3 value. Therefore for spontaneous emission rate in a cavity W_{cavity} versus rate in a vacuum W_{vacuum}, Purcell suggested the formula,

$$W_{\text{cavity}} = \frac{3}{4\pi^2}\frac{\lambda^3}{V}QW_{\text{vacuum}}. \tag{14.1}$$

An explicit derivation of this formula can be found in the review by A. N. Oraevsky [3]. Equation (14.1) corresponds to the precise resonance of a quantum system and a cavity and predicts strong enhancement of radiative decay. It is often referred to as the *Purcell effect* and the enhancement factor in this equation is referred to as the *Purcell factor*. Accordingly, strong detuning of a cavity from the transition frequency will offer poorer conditions for photon emission as compared to a free space, and one may expect inhibition of spontaneous decay. The complete expression for the arbitrary position of an atomic transition frequency ω_{a} with respect to the cavity resonant frequency ω_{c} was derived by Bunkin and Oraevsky in 1959 [3, 4]. It reads,

$$W_{\text{cavity}} = \frac{4\pi\mu^2}{\hbar V}\frac{Q}{\dfrac{(\omega_{\text{a}}-\omega_{\text{c}})^2}{\omega_{\text{c}}^2}Q^2+1}. \tag{14.2}$$

One can see at resonance ($\omega_a \to \omega_c$) Eq. (14.2) gives a Q-fold increase of spontaneous decay rate whereas detuning results in a steady decrease of W_{cavity}. For example, for $(\omega_a - \omega_c) = 0.5\omega_c$ and $Q \gg 1$ one has $W_{cavity}/W_{vacuum} = 4/Q$. The above predictions have been made for the radiofrequency range.

In radiophysics, the spontaneous emission rate is very low (see text in Fig. 14.1 and Problem 10 in Chapter 13) and the idea of modifying the spontaneous emission rate was aimed mainly at radiophysical applications. It was confirmed for the first time experimentally for that range in 1983 using Rydberg states (i.e. electron states with high principal quantum number n) in Na atoms [5]. A transition from the $23s$- to the $22p$- state was examined and the decay rate increase was detected. In 1987 modification of the spontaneous decay rate in a cavity in the optical range was reported for the first time by de Martini and co-workers [6]. In their experiments, organic molecules in a planar cavity were tested. Along with decay enhancement at precise resonance tuning of the cavity, decay inhibition was observed for cavity detuning with respect to the original spontaneous emission band of the molecules investigated. Very strong modification of radiative decay rate is expected for a quantum system decaying into a surface mode in a microcavity (whispering gallery modes). These modes feature a very high Q-factor along with the small volume occupied. For example, five-fold enhancement of decay rate for semiconductor quantum dots inside polymer microspheres has been reported [7]. In the case of a detuned cavity with respect to the transition frequency, 13-fold decay rate inhibition has been observed for Cs atoms in the infrared (wavelength 3.49 μm) [8]. Enhancement and inhibition of decay have also been reported for quantum dots in a planar cavity [9, 10]. These data confirm the basic approach to spontaneous decay in a cavity as a result of modified density of photon states but they are not as impressive as might be expected from the theory. There are several reasons for that. Among these are dependence of decay rate on the precise position of an emitter within a cavity, spread of radiative lifetimes over an ensemble, high spectral width (for molecules and quantum dots) with respect to cavity mode spectral width, contribution from non-radiative processes and others.

Along with decay rate modification, fluorescence line narrowing occurs for emitters in a cavity. This can be treated as squeezing of the originally wide but homogeneously broadened spectrum into the spectrum allowed by a cavity. An example of this phenomenon is given in Figure 14.2. The refractive index of a porous film depends on porosity which in the case of porous silicon is controlled by the etching conditions. An alternating current/time combination provides films of controllable thickness and porosity. Therefore, a microstructure can be developed on a silicon substrate in which alternating porous layers are used as an active medium containing buried silicon quantum dots and Bragg reflectors. This structure shows a very narrow emission spectrum as compared to that of a free porous film, the emission efficiency being enhanced by a factor of 16. These features are a consequence of the enhancement of spontaneous emission resonant with the optical modes of the Fabry–Perot resonator. Under these conditions no significant dependence of the linewidth on temperature was observed, confirming that the linewidth is determined only by the resonator finesse. The experimental findings are in good agreement with the calculated reflection and absorption spectra. In the case of coupled cavities, the emission spectrum condenses into a doublet line in accordance with intrinsic modes of a coupled cavity system [12].

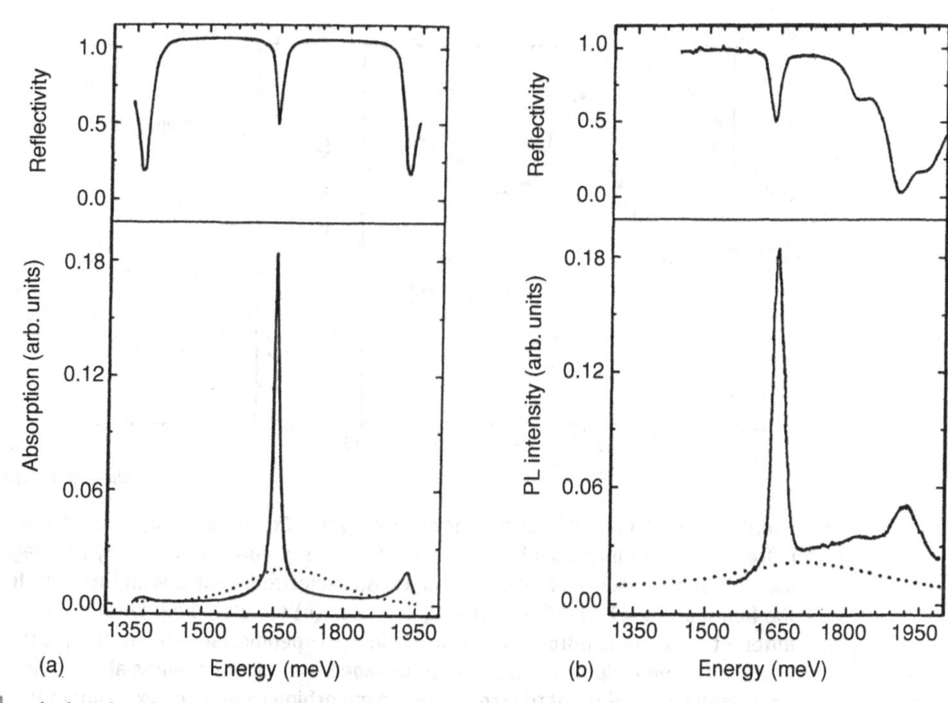

Fig. 14.2 Calculated and measured optical spectra of a planar porous silicon microcavity [11]. (a) Calculated reflectivity spectrum (upper panel) and absorption spectrum (lower panel) for a λ porous silicon microcavity. Also shown are the calculated absorption spectrum (dotted line) for a λ porous silicon layer without reflectors. (b) Room-temperature reflectivity spectrum of a λ porous silicon microcavity (upper panel) and photoluminescence spectra of a porous silicon layer (lower panel) without any reflectors (dotted line) and of a λ porous silicon microcavity (solid line).

Equations (14.1) and (14.2) imply that the very approach to spontaneous emission based on density of states is valid. Its applicability is restricted to weak coupling of a quantum system with an electromagnetic field in a cavity. The decay is supposed to be exponential but the rate is modified. Phenomena beyond this condition will also be discussed in the forthcoming sections. First, an intrinsically non-exponential decay in photonic crystals with incomplete gap will be discussed in Section 14.3. Second, oscillatory decay in a cavity in the strong coupling regime will be considered in Chapter 15 (Rabi oscillations). In the oscillatory regime the emitted radiation can survive long enough to re-excite the atom thus making spontaneous emission reversible and non-perturbative. This regime is inherent in high-Q cavities. The whole realm of cavity quantum electrodynamics has been thoroughly considered in books [13, 14] which are recommended for further reading. Another case of non-perturbative light–matter states occurs in photonic crystals with complete band gap and will also be discussed in Chapter 15.

For many experimentally feasible and practically essential situations, the perturbative approach in terms of density of states effects on light emission is valid. It is the dominating concept in explaining many experimental observations. To a large extent, engineering of photon density of states for practical purposes has become one of the key trends in nanophotonics. A few decades after Purell's prediction it was understood that spontaneous

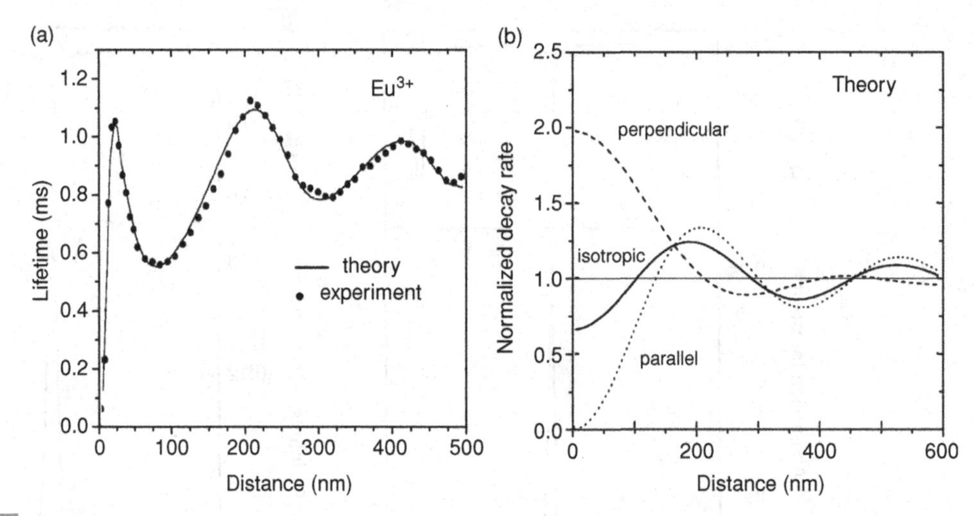

Fig. 14.3 Radative decay of an emitter in front of a mirror. (a) Experimental data for lifetime of Eu^{3+} ions in front of an Ag mirror as a function of ion/mirror separation reported by Drexhage [15] and the theoretical dependence obtained within classical electrodynamics with the dielectric constant of the film spacer 2.49, and that of the Ag $16 + 0.4i$. (b) Calculated decay rates for dipole emitters of different orientation in front of a perfect mirror, perpendicular to the mirror plane and parallel to it. The effect on a dipole emitter that is averaged over all directions is also shown and, importantly, is insufficient to account for the quenching seen in the experiment for close proximity of emitters to the mirror. Reprinted with permission from [16]. Copyright 1998 Taylor & Francis.

emission can be controlled and modified, not only in a cavity, but in many mesoscopic systems exhibiting modification of space properties on a wavelength scale. Further model systems will be discussed in the following subsections.

14.2 An emitter near a planar mirror

An emitter in front of a perfect mirror is the simplest system exhibiting modification of spontaneous emission rate. It was this system where modification of radiative lifetimes in optics was observed for the first time. In 1973 K. H. Drexhage published the pioneering paper [15] on systematic oscillatory behavior of decay time for Eu^{3+} ions in front of a silver mirror at a distance controllable by means of dielectric Langmuir–Blodgett film (Fig. 14.3a, circles). This simplest case can be understood in terms of classical electrodynamics if a dipole antennae is considered, emitting radiation into a half-space confined by a perfect conductive mirror from one side. Extensive discussion on the classical theory for this problem can be found in [13, 16]. Here we restrict ourselves to the net results only. In close proximity to the mirror, decay is promoted by surface plasmon polariton influence from a silver mirror. At distances more than 15–20 nm away its influence vanishes and emission is controlled by the mirror as a purely reflecting object with no conductivity or polaritons taken into effect. In this case oscillatory behavior is evident with damping upon distance. The model system of a dipole in front of a perfect mirror is found to explain this behavior (Fig. 14.3b). The mirror effect is essentially different for dipole orientation parallel and normal to the mirror plane. For parallel orientation decay is inhibited, whereas for perpendicular orientation decay is twice enhanced. In both cases oscillatory distance dependence with damping is inherent.

Since for this simple model structure classical electrodynamics explains the decay rate modification it is reasonable to put the question: is it possible to understand the Purcell effect for an atom in a cavity as a classical dipole emitter? The answer is positive. This problem is discussed in detail by E. A. Hinds in [13].

14.3 Spontaneous emission in a photonic crystal

Photonic crystals are believed to inhibit spontaneous emission completely if there is an omnidirectional band gap that offers no mode for an emitter to decay. Hence an excited state of an atom or a molecule can be "frozen" for an infinitely long time. The very idea of this physical phenomenon was proposed for the first time by V. P. Bykov in 1972 [17]. Much later, in 1987, E. Yablonovich [18] made a challenging proposal. In lasers, spontaneous transitions provide a bypass for an excited active medium to decay beyond the stimulated emission. If the cavity is perfect and the non-desirable dissipation losses in the active medium are negligibly small, then spontaneous decay remains the only process defining the finite threshold pumping to get lasing. Based on Bykov's original suggestion, Yablonovitch proposed to go towards thresholdless lasing by using an active medium in the form of a properly designed photonic crystal. Optical excitation of such a laser can be performed by higher-energy quanta where propagating modes exist.

These ideas turned hundreds of researchers across the world to look for the proper materials and techniques for making such lasers. Simultaneously, the basic science on spontaneous emission in band-gap structures had been developed. The basic theory of light waves in three- and two-dimensionally periodic dielectrics has been advanced and extended. The ingenious technological approaches were elaborated. Many issues related to light propagation properties and fabrication techniques of photonics crystals have been considered in Chapter 7. In this subsection we concentrate exclusively on spontaneous emission inside a photonic crystal. The theoretical issues on spontaneous emission of light in photonic band-gap structures are examined in detail in books and reviews [19–22].

Evaluation of radiative decay for a probe two-level system inside a photonic crystal gives not only inhibition of decay within the gap but also enhancement of decay at the gap boundary (Fig. 14.4). Well outside the gap decay rate obeys its value inherent in a vacuum. This property has a very fundamental origin. It means that strong modification of density of modes within a certain spectral range will be compensated by the opposite modification otherwise. This statement will be proved in Section 14.6. Consider experimental results on spontaneous decay of atoms and molecules in a photonic crystal.

The main problem in experimental observation of inhibited spontaneous decay is fabrication of a periodic structure offering an omnidirectional gap in the optical range. Most experiments have been performed with opal-based structures (bare silica colloidal crystals, their polymer and titania replicas). There is an incomplete gap in such structures which

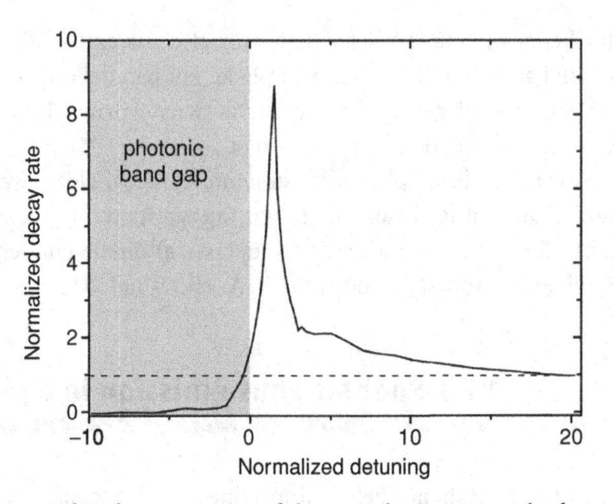

Fig. 14.4 **Calculated decay rate within the assumption of the perturbative approach of a two-level quantum system in a photonic crystal normalized with respect to vacuum rate versus normalized dimensionless detuning of the transition frequency with respect to photonic band gap edge. Dashed line shows vacuum rate. Adapted from [21].**

manifests itself as a reflection band with angular-dependent position. Opal-based photonic crystals do show a certain systematic modification of spontaneous decay. Probe emitters were organic molecules, lanthanide ions, and semiconductor quantum dots. Molecules show a systematic dip in the emission spectrum which is angular dependent and correlates with the spectral position of the reflection band (Fig. 14.5a) [23–25].

Luminescence decay if not single-exponential should preferably be characterized in terms of decay time distribution. To provide adequate model-independent data analysis, distributions of decay times $F(\tau)$ were recovered from measured luminescence kinetics according to the formula,

$$\int_{\tau_{\min}}^{\tau_{\max}} F(\tau)\exp(-t/\tau)\,\mathrm{d}\tau = I(t), \tag{14.3}$$

where $I(t)$ is the luminescence intensity. Elucidation of the decay time distribution in this ill-conditioned inverse problem was performed using the proper regularization method. One can see, when embedded in opals, molecules show a wider decay time distribution as in the reference polymeric film. Similar behavior was observed for Eu^{3+} ions embedded in opals using organic ligands (Fig. 14.6). Non-exponential decay with a pronounced spread of decay times is evident (compare with Fig. 13.2).

The problem of the spontaneous decay of a probe quantum system in a periodic structure with an incomplete band gap has become the subject of extensive analysis [27–30]. The key issues are the vectorial character of the electromagnetic field and the angular-dependent photonic band gap. In particular summation over angles for final states is to be made separating allowed and forbidden radiation modes. In this case non-exponential decay of the excited state was found to have an intrinsic origin which means that the Weisskopf–Wigner approximation fails. In other words, probability can not be assigned to the decay process. The notion of excited state lifetime vanishes. Instead, the decay features the properties of

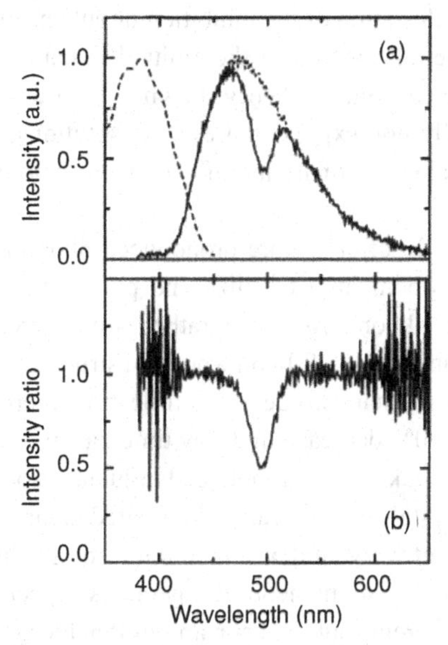

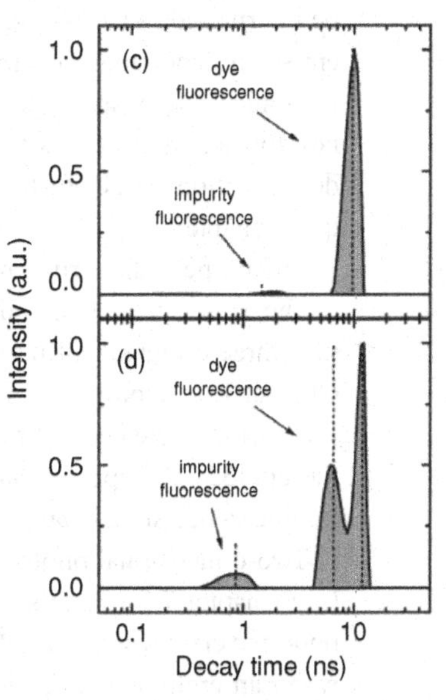

Fig. 14.5 **Spontaneous emission of 1,8-naphthoylene-1′,2′-benzimidazole (NBIA) dye molecules in poly(methylmethacrylate) (PMMA) impregnated into an artificial opal [25]. (a) Emission spectrum in opal (solid line), in a reference PMMA film (dots), and excitation spectrum (dashes). (b) Difference of emission spectra in an opal and in the reference film. (c) Decay time distribution in the reference film. (d) Decay time distribution in an opal.**

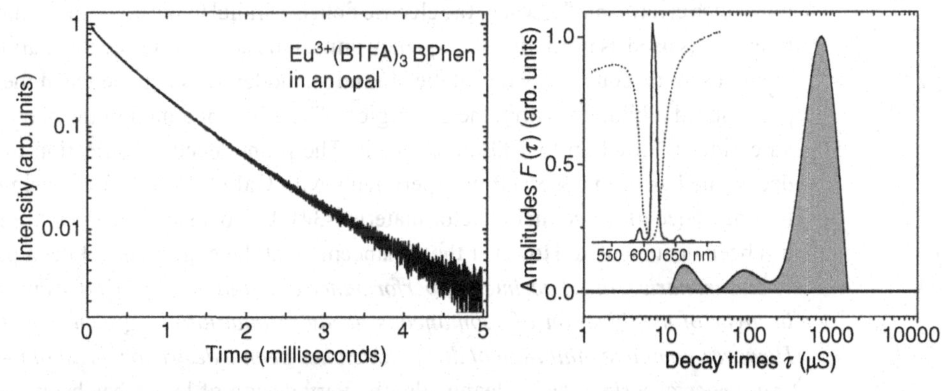

Fig. 14.6 **Luminescence decay of Eu^{3+} ions in Eu^{3+}-$(BTFA)_3$-BPhen complexes embedded in an opal by means of sol–gel reactions (left) and the elucidated decay time distribution (right). The insert shows the Eu^{3+} emission spectrum (solid line) superimposed with the reflection spectrum of an opal at normal incidence (dots). Adapted from [24, 26].**

a non-Markovian process. Therefore, non-exponential decay in Figures 14.5, 14.6 can be treated as a manifestation of the basic properties of spontaneous decay in media with a photonic pseudogap.

Experiments have been also performed with semiconductor quantum dots embedded in opals [31]. In this case inhomogeneous broadening of the emission spectrum resulting from size distribution gives rise to complicated modification of the emission spectrum. This can be interpreted as a possible manifestation of the inhibition or enhancement of spontaneous emission depending on the spectral position of the emitted radiation with respect to the photonic pseudogap (as is expected from the theory shown in Figure 14.4). Since the decay law for quantum dots is typically non-exponential even in solutions, and even in single-dot detection mode, further evaluation of its modification in photonic crystals is rather questionable.

Silicon periodic structures could offer a more pronounced photonic band gap because of the larger refractive index as compared to silica and polymers. However fabrication of a three-dimensional periodic silicon structure is rather complicated (see discussion in Chapter 7 on photonic crystal fabrication). Silicon woodpile structures consisting of only a few periods have been tested for spontaneous decay modification of erbium ions embedded therein [32]. An approximately 10% decrease in decay time has been reported though the luminescence signal was rather weak with a pronounced influence from background noise.

Two-dimensional photonic crystals can be readily fabricated using III–V solid solutions (see Chapter 7 for details) with refractive index $n > 3$. Interestingly, in a two-dimensional photonic crystal slab, modification of spontaneous decay rate is expected to occur, not only when an emitter is embedded therein, but also for an emitter located in the close (sub-wavelength) vicinity of a slab. This phenomenon has been examined in detail theoretically (Fig. 14.7) implying a probe dipole and a photonic crystal membrane with dielectric constant $\varepsilon = 11.76$ and thickness $d = 250$ nm, surrounded by up to 1 μm of air above and below. The membrane contained a hexagonal array of holes with radius $r = 0.3a$ at a lattice spacing of $a = 420$ nm. Such a structure possesses a band gap for a/λ in the range 0.25 to 0.33 for the transverse electric mode, where the electric field is parallel to the plane of the membrane. The ratio a/λ is used as normalized frequency units. One can clearly see the manifold enhancement of spontaneous decay rate at the blue edge, moderate enhancement at the red edge and pronounced inhibition within the gap region. The effect of a membrane persists even when an emitter is lifted up to 100 nm above it. The pronounced modification of spontaneous decay rate has been observed in experiments with GaInAsP ($n = 3.27$) using the intrinsic luminescence of the semiconductor material [34]. Up to a five-fold decrease in decay time has been documented. However the enhancement at the edges has not been pronounced.

To summarize, the experimental performance of photonic crystals does provide evidence in favor of modification of spontaneous decay in qualitative agreement with the theory. However, complete inhibition of decay which has been a desire for thresholdless lasing still challenges experimenters. Meanwhile, the very design of lasers has been changed during past decades. The most widespread semiconductor lasers have acquired vertical emitting geometry and operate at current densities several times exceeding the threshold. Therefore the development of materials with completely inhibited spontaneous emission will change the energetic efficacy only slightly.

14.4 Thin layers, interfaces and stratified dielectrics

Thin layers, interfaces of two different dielectrics and stratified dielectrics represent other sample structures where density of photon states differs from that in a continuous medium.

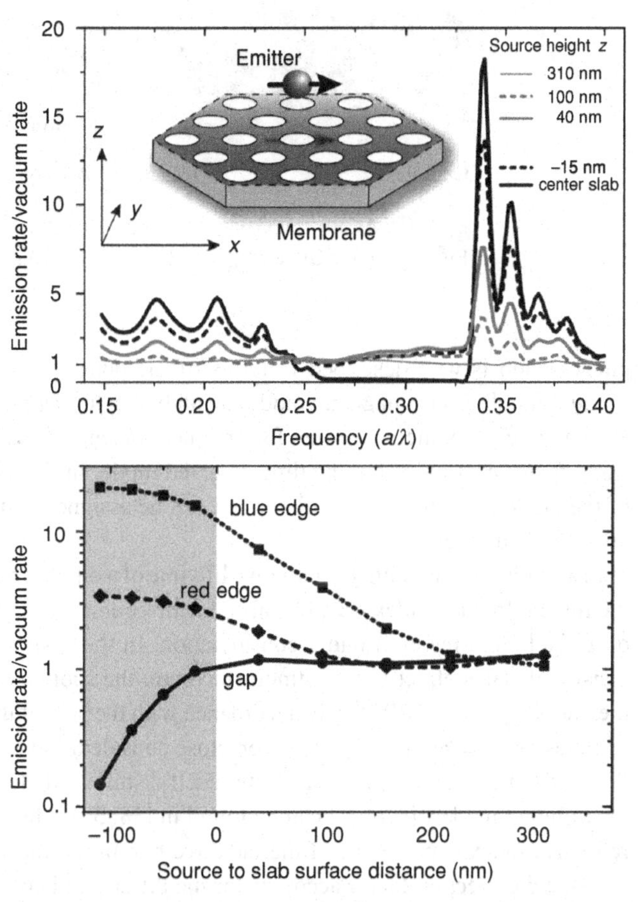

Fig. 14.7 Modified spontaneous emission rate of a model dipole emitter near the surface of a two-dimensional photonic crystal (PC) membrane [33]. Top: Emission rate normalized to the vacuum rate versus normalized frequency (a is the crystal period) for an x-oriented dipole at the central hole of a PC membrane. Black spectra correspond to dipoles *in* the slab and gray to dipoles *above* the slab. Bottom: Emission rate modification as a function of the height of a dipole above the PC membrane. Diamonds, circles and squares show data for frequencies below, in and above the gap, respectively. The shaded region shows the range of positions in the membrane.

In all the above structures the radiative lifetime of an excited quantum system modifies depending on the parameters of the structures and on the specific position of an emitter therein. This phenomenon is well established, examined, and reproducible, although precise calculation of the density of states is rather cumbersome. The radiative rate modification in these structures is much less pronounced as compared to microcavities and photonic crystals. However it is much more readily observed and reproducible because of simple experimental performance. For example, for Eu^{3+} complexes the systematic analysis of spontaneous decay rates in a thin submicrometer dielectric film surrounded by various dielectrics was performed by Urbach and Rikken [35]. For a film thickness more than 1000 nm the radiative rate obeys the value of 1.5 ms^{-1} inherent in a continuous medium. For thinner layers, decay rate monotonically falls to 1 ms^{-1} in a 10 nm thick layer. This

Table 14.1. Lifetimes of europium complexes in various environments

Complex	Environment		
	Toluene	Polystyrene	Dioxane
Eu^{3+}-$(HFAC)_3$-$(TOPO)_2$	0.713 ms	0.690 ms	0.570 ms
Eu^{3+}-$(BTFA)_3$-BPhen	0.642 ms	0.613 ms	0.520 ms

phenomenon is well described in terms of the density of states effect with a thorough account for all electromagnetic modes available including guided modes within a film.

The same phenomenon occurs for organic molecules in thin films. It is important when using fluorescent probes in the time-resolved mode. In this case even the precise distance of the probe molecule from an interface can be assigned. The representative experiments are described in [36, 37].

In a continuous medium the radiative lifetime of a dipole emitter can also be influenced by the finite refractive index n and local-field correction factor $f(n)$ describing polarizability of a dipole in a medium under consideration. In the first approximation, in terms of the density of states effect in a continous medium, the spontaneous decay rate should follow the law $W_{\rm rad}(n) = n^3 W_{\rm rad}^{\rm vacuum}$ in accordance with the substitution $c \to c/n$ into the density of states $D(\omega) = \omega^2/(\pi^2 c^3)$. However, close consideration requires a local field correction factor f to be involved since an emitter itself disturbs a dielectric medium. Two reviews of the different models involved can be found in [38, 39]. This effect is accounted for by the refractive-index dependence of the radiative rate in the form $W_{\rm rad} = nf^2(n)W_{\rm rad}^{\rm vacuum}$. The local-field correction factor accounts for the difference between the field in the vicinity of the emitter and the mean field in the medium, arising because a realistic emitter takes a finite volume and thus locally modifies the properties of the medium. The particular form of $f(n)$ depends on the model of the cavity formed in the dielectric around the emitter. For example $f(n) = (n^2 + 2)/3$ (the Lorentz model) and $f(n) = 3n^2/(2n^2 + 1)$ (the hollow cavity model) are used. However, in many experimental situations modification of lifetimes observed in routine experiments does not follow the models precisely. Representative examples of various lifetime data for two Eu^{3+} complexes in organic ligands are given in Table 14.1. This peculiarity has to be accounted for when interpreting various experiments on radiative lifetimes in nanostructured media. A thorough approach to adequate reference samples and careful comparison with data reported by various groups are mandatory.

The calculated radiative decay rate at the interface of two dielectric media are shown in Figure 14.8. From both sides of the interface at sufficient distance the rate tends to the constant value $W_{\rm rad}(n) = nW_{\rm rad}^{\rm vacuum}$, whereas within close proximity of the interface the rate oscillates around those values. Here the local-field correction factor was assumed to be equal to unity since the purpose of the calculations was to reveal interface effects on decay rate.

Semiconductor quantum dots represent an interesting example of artificial atom-like objects with size-dependent spectra of light absorption and emission as well as decay rates (see Chapter 5 for detail). There is an interesting option to further control their optical

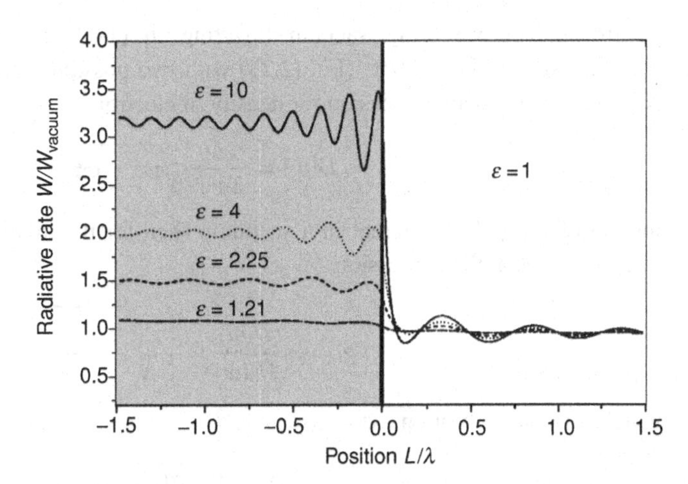

Fig. 14.8 **Calculated decay rate of an atom near a dielectric vacuum interface for four different values of dielectric permittivity. The dielectric area is shaded with gray. Adapted from [40].**

properties by means of the environmental dependence of the intrinsic radiative lifetime. In Chapter 3 (Table 3.1) the general trend was outlined for the dielectric properties of various materials. Namely, higher refractive indexes are inherent in narrow-band materials which are only transparent well into the infrared range. However, the same materials feature very small electron effective mass thus facilitating unprecedented short-wave shift of the absorption edge. Therefore it is possible starting from narrow-band original crystals by means of a nanometer-sized restriction to move their transparency range significantly towards the visible (defined by electron and hole confinement), while the refractive index is close to that of the bulk parent crystal. It becomes possible since sufficiently far from the sharp absorption in the spectrum, the refractive index is defined mainly by the crystal lattice. PbSe possesses the dedicated permittivity $\varepsilon = 23$ which gives the decay rate factor (using the so-called hollow cavity model) [41],

$$\frac{W_{\mathrm{rad}}(\varepsilon)}{W_{\mathrm{rad}}^{\mathrm{vacuum}}} = \sqrt{\varepsilon_{\mathrm{out}}} f^2(\varepsilon_{\mathrm{out}}) = \sqrt{\varepsilon_{\mathrm{out}}} \left[\frac{3\varepsilon_{\mathrm{out}}}{\varepsilon + 2\varepsilon_{\mathrm{out}}} \right]^2 .$$

Assuming $\varepsilon_{\mathrm{out}} = 1$ one has a lower limit for the decay rate $W_{\mathrm{rad}}(\varepsilon) = 0.014\, W_{\mathrm{rad}}^{\mathrm{vacuum}}$ without any purposeful structuring of the matter on the light wavelength scale! Assuming $\varepsilon_{\mathrm{out}} \to \varepsilon$ (quantum dot solids, see Section 5.7) one has an upper limit $W_{\mathrm{rad}}(\varepsilon) \to nW_{\mathrm{rad}}^{\mathrm{vacuum}}$.

14.5 Possible subnatural atomic linewidths in plasma

The dispersion law for electromagnetic waves in plasma reads (see Eq. (3.56)),

$$\omega^2 = \omega_{\mathrm{p}}^2 + c^2 k^2, \tag{14.4}$$

where ω_{p} is the plasma frequency defined by Eq. (3.54). This dispersion law was discussed in detail in Section 3.3. It differs substantially from the law, $\omega = ck$, inherent for electromagnetic waves in a vacuum. Inserting Eq. (14.4) into the general expression for mode density, $D(k) = k^2/\pi^2$ (Eq. (2.23) with two possible polarization states taken into account) results in an expression for density of electromagnetic modes in plasma,

$$D(\omega) = \frac{\omega}{2\pi^2 c^3}\sqrt{\omega^2 - \omega_p^2}, \tag{14.5}$$

instead of Eq. (14.3) inherent in a vacuum. Then for the modified linewidth (and decay rate) we arrive at the expression,

$$\gamma = \gamma_0 \frac{D(\omega)}{D_0(\omega)} = \gamma_0 \sqrt{1 - \frac{\omega_p^2}{\omega^2}}, \tag{14.6}$$

which can also be written as,

$$\gamma = \gamma_0 \sqrt{\varepsilon(\omega)}, \tag{14.7}$$

recalling the basic expression for dielectric function $\varepsilon(\omega)$ of plasma. One can see that atomic emission lines acquire sub-natural width in plasma. Is this effect observable? The serious obstacle is Doppler broadening of emission lines which typically dominate in the observed atomic spectra even at room temperature in gases. The standard approach to plasma preparation implies thermal ionization of atoms and needs extremely high temperatures (e.g. 10^4 K corresponds to so-called "cold" plasma). This leaves no chance to elucidate the fine effect of modified density of states on atomic linewidths. However an alternative means of plasma generation has been developed recently. This is laser ionization of atoms by means of resonant optical excitation of atoms using intense laser radiation. This approach can give ultracold plasma with temperature in the sub-Kelvin range [42, 43]. Under these conditions, observation of the predicted line narrowing becomes feasible.

14.6 Barnett–Loudon sum rule

In 1996 S. Barnett and R. Loudon proved the general theorem which formulates the sum rule for modified spontaneous emission probabilities [44]. It states that in a given point of space, modification of spontaneous emission rate of a dipole emitter in a certain frequency range will necessarily be compensated by the opposite modification. Precisely, modified rates are constrained by an integral relation for the relative emission rate modification $(W - W_0)/W_0$,

$$\int_0^\infty \frac{W(\mathbf{r},\omega) - W_0(\omega)}{W_0(\omega)} d\omega = 0. \tag{14.8}$$

Here $W_0(\omega)$ is the spontaneous emission rate in free space and $W(\mathbf{r},\omega)$ is the emission rate of an atom or molecule at position $\mathbf{r}$ as modified by the environment,whose effect is assumed to vary by a negligible amount across the extent of the emitting object. It follows that any reduction in spontaneous emission rate over some range of frequencies must necessarily be compensated by an increase over some other range of transition frequencies.

The proof is essentially based on the Weisskopf–Wigner probabilistic approach to spontaneous decay. The decay rate in the general case reads,

$$W(\mathbf{r},\omega_0)=\frac{1}{\hbar^2}\int_0^\infty d\omega\int_0^\infty d\omega'\mu_i\mu_j\langle 0|\hat{E}_i^+(\mathbf{r},\omega)\hat{E}_j^-(\mathbf{r},\omega')|0\rangle\delta(\omega'-\omega_0)$$
$$=\frac{1}{\hbar^2}\int_0^\infty d\omega\mu_i\mu_j\omega\omega_0\langle 0|\hat{A}_i^+(\mathbf{r},\omega)\hat{A}_j^-(\mathbf{r},\omega_0)|0\rangle,\ i,j=x,y,z, \tag{14.9}$$

where μ is the dipole matrix element for the transition and the relations were used between the transverse electric field $\hat{\mathbf{E}}$ and vector potential $\hat{\mathbf{A}}$ operators $\hat{\mathbf{E}}^{\pm}(\mathbf{r},\omega)=\pm i\omega\hat{\mathbf{A}}^{\pm}(\mathbf{r},\omega)$. Summation is implied for repeated indices i,j over three Cartesian coordinates. Further, the dissipation–fluctuation theorem is applied [45]. This gives the relation for vacuum expectation value in the form,

$$\langle 0|\hat{A}_i^+(\mathbf{r},\omega_0)\hat{A}_j^-(\mathbf{r},\omega)|0\rangle=2\hbar\mathrm{Im}G_{ij}^T(\mathbf{r},\mathbf{r},\omega_0)\delta(\omega-\omega_0), \tag{14.10}$$

where $G_{ij}^T(\mathbf{r},\mathbf{r},\omega_0)$ is the transverse *Green's function* defined by the solution of the partial differential equation,

$$-\left(\nabla^2+\frac{\omega^2}{c^2}\varepsilon(\mathbf{r},\omega)\right)G_{ij}^T(\mathbf{r},\mathbf{r}',\omega)=\frac{1}{\varepsilon_0c^2}\delta_{ij}^T(\mathbf{r}-\mathbf{r}'), \tag{14.11}$$

with $\varepsilon(\mathbf{r},\omega)$ being the medium complex dielectric function at point $\mathbf{r}$ and $\delta_{ij}^T(\mathbf{r}-\mathbf{r}')$ being the transverse part of the delta-function (see, e.g. [46], p. 502). Note, application of the fluctuation–dissipation theorem and Eq. (14.10) for the field arising in the result of spontaneous transitions means treating spontaneous transition as a kind of stimulated transition due to perturbation from electromagnetic vacuum fluctuations. Using Eq. (14.10), the spontaneous decay rate Eq. (14.9) reduces to the expression,

$$W(\mathbf{r},\omega)=\frac{2\omega^2}{\hbar}\mu_i\mu_j\mathrm{Im}G_{ij}^T(\mathbf{r},\mathbf{r},\omega). \tag{14.12}$$

Now the properties of the decay rate modified by the space inhomogeneities are defined by the properties of the Green's function. With the assumption that the dielectric function for any material object must tend to the free-space value of unity as $\omega\to\infty$, the Green's function obeys relations similar to Kramers–Kronig type. Then the statement Eq. (14.8) directly follows from Eq. (14.12) and the free space decay rate (Eq. (13.22)).

By definition, Greens's function G is introduced for every linear differential operator $\hat{L}$ in the way that $\hat{L}G(x,x')=\delta(x-x')$ holds. Then the equation,

$$\hat{L}G(x,x')=f(x), \tag{14.13}$$

has the solution,

$$u(x) = \int G(x, x')f(x')\mathrm{d}x'. \tag{14.14}$$

Green's functions play an important role in electrodynamics [47, 48]. In the general case, $\mathbf{G}(\mathbf{r}, \mathbf{r}')$ is a tensor defining the field $\mathbf{E}(\mathbf{r})$ at a point $\mathbf{r}$ from a radiating dipole $\boldsymbol{\mu}$ located at the point $\mathbf{r}'$,

$$\mathbf{E}(\mathbf{r}) = \omega^2 \mu_0 \mu \mathbf{G}(\mathbf{r}, \mathbf{r}')\boldsymbol{\mu}. \tag{14.15}$$

For free space it takes the scalar form,

$$G_0(\mathbf{r}, \mathbf{r}') = \frac{\exp(\pm ik|\mathbf{r} - \mathbf{r}'|)}{4\pi|\mathbf{r} - \mathbf{r}'|}. \tag{14.16}$$

Derivation of the generic relation Eq. (14.12) can be found in the book by Novotny and Hecht [47]. Note, in Eq. (14.15) μ_0 and μ are vacuum and medium permeabilities, respectively.

14.7 Local density of states: operational definition and conservation law

Quite surprisingly, the concept of the density of states for a finite size structure still lacks a simple, concise definition.

G. D'Aguanno et al. 2004 [49]

In the previous sections of this chapter we have seen that spontaneous decay rate in complex structures becomes a function of the position of a probe emitter. It can be foreseen that the properties of space, at least within the distance of approximately several wavelengths of emitted radiation, will define the spontaneous decay rate. This leads to the notion of the local density of states $D(\mathbf{r}, \omega)$ which could give the relation for spontaneous decay rate $W(\mathbf{r}, \omega)$ at a given point $\mathbf{r}$ modified with respect to vacuum rate $W_0(\omega)$ by means of space inhomogeneity, i.e.

$$W(\mathbf{r}, \omega) = \frac{2\pi}{\hbar} D(\mathbf{r}, \omega)|\langle E_1, 0|H_{\text{int}}|E_0, 1\rangle|^2. \tag{14.17}$$

D'Aguanno with co-workers [49] proposed a reasonable operational approach to the local density of photon states (DOS) definition through the power emitted by a classical dipole at a point $\mathbf{r}$ with respect to the power emitted by the same dipole in a vacuum. For a dipole oriented along the x-axis, this power is proportional to the imaginary part of Green's function $\mathrm{Im}[G_{\hat{\mathbf{x}}\hat{\mathbf{x}}}(\mathbf{r}, \mathbf{r}, \omega)]$. Here $\hat{\mathbf{x}}$ is the unit vector parallel to the dipole. Then based on Eq. (14.12) the recipe for calculation of the local density of photon states follows,

$$D(\mathbf{r}, \omega) \equiv D_{\text{vacuum}}(\omega)\frac{W(\mathbf{r}, \omega)}{W_0(\omega)} = D_{\text{vacuum}}(\omega)\frac{6\pi c}{\omega}\mathrm{Im}G(\mathbf{r}, \mathbf{r}, \omega), \tag{14.18}$$

whence recalling

$$D_{\text{vacuum}}(\omega) = \frac{\omega^2}{\pi^2 c^3},$$

we arrive at the relation for local DOS in terms of the Green's function,

$$D(\mathbf{r},\omega) = \frac{6\omega}{\pi c^2}\text{Im}G(\mathbf{r},\mathbf{r},\omega). \tag{14.19}$$

Thus calculation of the local density of states can be performed based on finding the proper Green's function which in turn depends on the local value of the dielectric function $\varepsilon(\mathbf{r},\omega)$ as well as the electromagnetic field propagation for a point-like source located at the point under consideration. The latter problem is hard to solve for many practically important microstructures, e.g. metal tips or faceted nanoparticles. A few examples of such singularities will be discussed in Chapter 16.

The above definition of local DOS when combined with the Barnett–Loudon sum rule Eq. (14.8) allows a statement on the sum rule for local density of states to be made. This is possible because the Barnett–Loudon sum rule was derived for a hypothetical probe quantum system without any pre-condition for frequency-dependent matrix element, but was exclusively based on the properties of Green's function and the causality principle. The local DOS sum rule then reads,

$$\int_0^\infty \frac{D(\mathbf{r},\omega) - D_{\text{vacuum}}(\omega)}{D_{\text{vacuum}}(\omega)} d\omega = 0, \quad D_{\text{vacuum}}(\omega) = \frac{\omega^2}{\pi^2 c^3}. \tag{14.20}$$

This means that *for every given point in space a deviation from the vacuum density of states will be compensated by the opposite deviation at other frequencies.* Figure 14.4 confirms this statement as well as the sum rule for decay rates. The reader is encouraged to check conformity with this rule of the Bunkin–Oraevsky formula for decay rate in a microcavity (Problem 6).

The sum rule for radiative rate can be directly applied to spontaneous emission of photons. Its extension towards a local DOS sum rule gives rise to further extension to the processes of *photon scattering* since the latter is proportional to the local density of states as well. Therefore, one can see *the modification of resonant and non-resonant (Raman) scattering can be locally modified only within a finite frequency interval and will be necessarily compensated by the opposite modification in the other frequency intervals.*

14.8 A few hints towards understanding local density of states

The Green's function entering into Eqs. (14.12) and (14.19) defines the local density of photon states and the local probability of emission and scattering of photons. It provides the route to elucidation of the local DOS in complex and inhomogeneous structures. However, it is not instructive enough for visual representation of the local DOS. Therefore, in this section a few qualitative examples will be given which are helpful for certain intuitive understanding of the local density of states notion. In general, local density of photon states implies characterization of spatial distribution of probe electromagnetic radiation in the portion of space under consideration. Higher local density of states can be characterized in terms of local concentration of probe light, as well as promoted tunneling of light through the certain portion of space bounded by barriers where only evanescent waves are possible.

The first hint is *concentration of electromagnetic energy* in a given portion of space when light flows throughout in a steady-state regime. A microcavity resembles a model system with the Purcell factor in Eq. (14.1) as a measure of local density of states enhancement with respect to the vacuum DOS value given by Eq. (13.4). Noteworthy, it is the product of the cavity Q-factor and the spatial "squeezing" of the electromagnetic mode under consideration, V/λ^3. Many examples have been given in Chapter 8 of spatial field profiles in model inhomogeneous media. Each of those examples means modified local DOS at a frequency for which concentration or depletion of the electric field occurs with respect to free space. The same reasoning is applicable to complex field profiles considered for metal dielectric nanostructures in Chapter 11. Thus one can see that an intuitive hint at the physical content of the local DOS has been given in the pioneering Purcell paper (Fig. 14.1).

The second hint is promotion of light tunneling through a portion of space with higher local DOS. Again, a cavity is a very clear primary example. In Chapter 3 (Figs. 3.18 and 3.19), resonant tunneling has been considered in wave mechanics and in wave optics and described in terms of Q-factors. High local density of states are inherent for frequencies where light successfully traverses a portion of space by means of evanescent waves under barriers and standing waves inside the cavity. Again, in conformity with the previous paragraph, light energy concentration occurs and high Q-factor is inherent. Further examples of tunneling are all cases of coupling of a linear waveguide to cavities, including high-Q whispering gallery modes in a microdisk or in a microsphere. Such schemes were discussed in Chapter 9. For example, for the coupling scheme in Figure 9.9(c) efficient tunneling of light into the cavity for the frequency resonant with a whispering gallery mode is indicative of the high local DOS. Alternatively, such tunneling can be treated as light scattering initiated by a defect in a linear waveguide into the whispering gallery mode. Since scattering is proportional to the local DOS, it is clear that it becomes rather efficient when the final state (a whispering gallery mode) features high local DOS (i.e. high Q-factor and high spatial "squeezing" factor, as was emphasized by E. M. Purcell).

The reader is proposed to examine for themselves the cases of higher local DOS in sample structures discussed in Chapter 8, which offer higher transmission of light in spite of irregular displacement of scatterers. We now turn to the important precursor of the Barnett–Loudon sum rule, more explicitly, of the local density of states conservation rule as defined by Eq. (14.20). When considering the general constraints on wave propagation in irregular structures (Section 8.7) we observed that the conservation law (Eq. 8.85) occurs and the $\rho(\omega)$ function demonstrates the remarkable property of coexisting positive and negative deviations from its average value (see Fig. 8.30). That conservation law did not imply an emission of photons but was based exclusively on a classical electrodynamical consideration of electromagnetic radiation in complex media. Note that in Figure 8.30 high

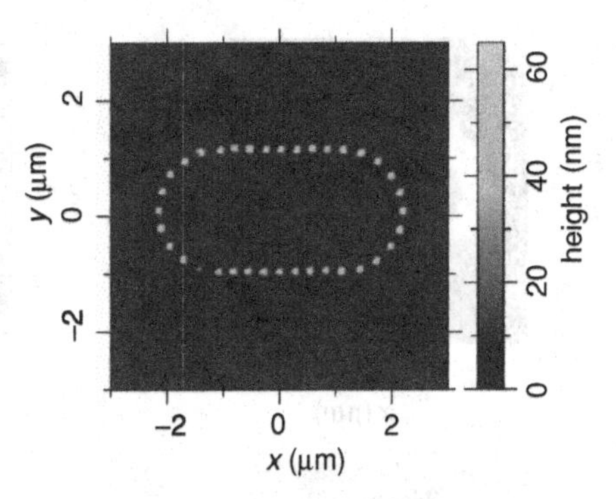

Fig. 14.9 **Topographical image obtained by atomic force microscopy of a corral formed by nanometer sized gold pillars [50].**

transmission remarkably correlates with high $\rho(\omega)$ values. This function can be treated either as one-dimensional density of electromagnetic modes (=photon DOS) or as *partial* three-dimensional density of modes (=partial photon DOS) for the particular propagation direction.

Enhanced tunneling of light into the areas with higher local DOS suggests a possible technique of local DOS mapping using a source of evanescent electromagnetic radiation scanned over the surface under consideration. This is exactly the idea of the SNOM, scanning near-field optical microscope. The SNOM principle was discussed in detail in Chapter 10. In the same manner as a tunneling *electron* microscope probes the local *electron* DOS over the surface under consideration, a scanning near-field *optical* microscope probes the local photon DOS. Therefore "SNOMography" can be efficiently used to examine the local DOS in complex structures. A representative example is given in Figures 14.9 and 14.10.

To perform SNOMographical imaging of a local photon DOS, Chicanne *et al.* [50] fabricated a well-defined topological system consisting of identical gold nanometer-sized pillars developed on a substrate by means of electron beam lithography in the form of a corral (Fig. 14.9). This test object was explored by means of SNOM imaging using leakage of an evanescent field from a SNOM tip. The photon local DOS defines how the fact of scanning close to the surface drives the field radiated by the dipole at the end of the tip through the substrate; depending on the tip position, scattering channels may turn out to be open or closed. Higher local DOS offers a higher probability of scattering towards the substrate. Figure 14.10 represents the result of SNOM imaging for the two polarizations of light along with the calculated local DOS. The good agreement is remarkably evident. Noteworthy are correlation of local DOS with geometrical structure as well as its strong dependence on light polarization.

14.9 Thermal radiation in mesoscopic structures

As we have seen in Chapter 2 (Section 2.2) the very notion of the density of states in physics was derived from the problem of black body radiation, i.e. equilibrium electromagnetic

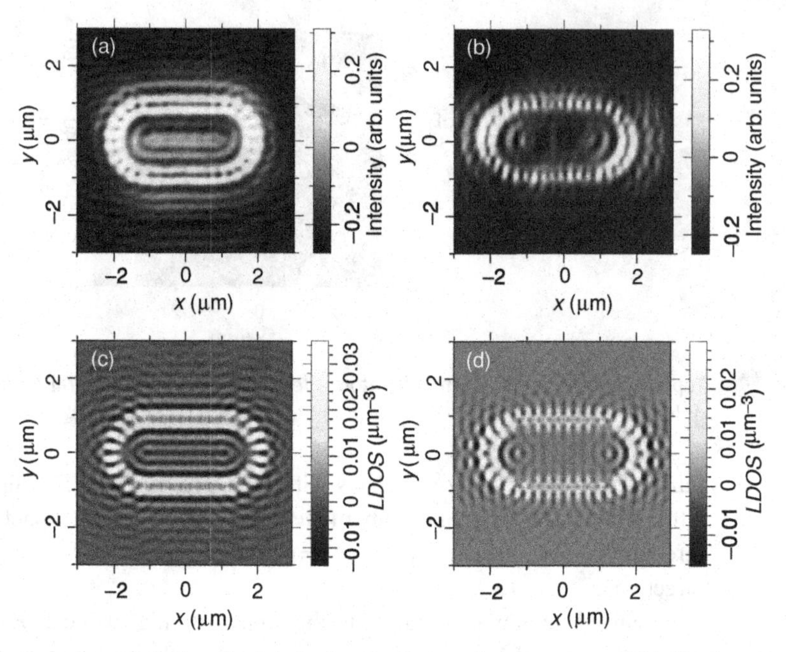

Fig. 14.10 SNOM images of a corral taken for (a) dipole orientation along x-axis, and (b) dipole orientation along y-axis. polarizations: (c), (d) the relevant calculated local DOS distribution at the height of $z = 120$ nm over the surface [50].

radiation in a cavity. First Rayleigh proposed to count modes in a cavity to get mode density, $D(k) = k^2/\pi^2$ and Planck introduced energy quanta $E = \hbar\omega$. Finally, Bose assigned the basic content to mode counting and transformed Rayleigh's hint into a solid physical notion. Recalling the discussion in Section 2.2 (see Chapter 2 for references to the original works), the energy density of thermal radiation. i.e. electromagnetic equilibrium radiation,

$$u(\omega) = \frac{\hbar\omega^3}{\pi^2 c^3}\left[\exp\left(\frac{\hbar\omega}{k_B T}\right) - 1\right]^{-1}, \tag{14.21}$$

resembles the product of the energy quantum $E = \hbar\omega$, density of states, $D(\omega) = \omega^2/(\pi^2 c^3)$ and distribution function,

$$F(E) = \left[\exp\left(\frac{E}{k_B T}\right) - 1\right]^{-1}. \tag{14.22}$$

Therefore, in complex structures with modified DOS, thermal radiation will modify accordingly. This phenomenon is well established and has even been proposed as a reasonable approach to the very definition of local DOS by Joulain *et al.* [51]. These authors also verified the feasibility of detecting such local DOS with an apertureless SNOM technique to show that a thermal near-field emission spectrum above a sample should be detectable and that this measurement could give access to the electromagnetic local DOS. In the context of calculational techniques, such a definition again reduces to finding outthe proper Green's function. Dedkov and Kyasov [52] derived density of states for a space one half of which consists of a medium with complex dielectric permittivity and magnetic permeability while the other half is vacuum,

$$D(z,\omega) = D_{\text{vacuum}}(\omega)\frac{1}{z^3}\left[\operatorname{Im}\left(\frac{\varepsilon(\omega)-1}{\varepsilon(\omega)+1}\right) + \operatorname{Im}\left(\frac{\mu(\omega)-1}{\mu(\omega)+1}\right)\right]. \tag{14.23}$$

Here z is the distance from a point in the vacuum half-space to the interface with the medium.

Consider a few experimental observations of modified thermal emission in mesoscopic structures. Lin *et al.* [53] observed both suppression and enhancement of thermal radiation in three-dimensional silicon photonic crystals for wavelengths of the order of 10 μm. Suppression occurs for band gaps whereas enhancement was observed in the pass bands in qualitative agreement with the density of states effect (see, e.g. Fig. 14.4). Maryama *et al.* reported on thermal radiation from two-dimensionally confined modes in microcavities [54]. A number of examples of modified thermal radiation from spherical microparticles have also been reported [55]. Nemilentsau *et al.* [56] predicted strong modification of thermal radiation from metallic carbon nanotubes in the terahertz range, based on local density of states evaluation by means of Green's function analysis (see Eq. (14.18)).

Along with mesoscopic structures exhibiting modified density of states, plasma should be involved in considerations as a peculiar continuous medium with an unusual density of electromagnetic modes described by Eq. (14.5). Inserting this into Eq. (14.21) gives,

$$u_{\text{p}}(\omega) = u(\omega)\sqrt{1-\frac{\omega_{\text{p}}^2}{\omega}} = \frac{\omega\sqrt{\omega^2-\omega_{\text{p}}^2}}{\pi^2 c^3}\hbar\omega\left[\exp\left(\frac{\hbar\omega}{k_{\text{B}}T}\right)-1\right]^{-1}, \tag{14.24}$$

where the plasma frequency ω_{p} is defined by Eq. (3.54). This means there is no thermal radiation in plasma for $\omega < \omega_{\text{p}}$ since there is no electromagnetic mode available. For $\omega > \omega_{\text{p}}$ thermal radiation is inhibited as compared to a vacuum. For $\omega \gg \omega_{\text{p}}$ thermal radiation density tends to that in a vacuum. Therefore, plasma is expected to exhibit sub-Planckian thermal properties everywhere through the spectrum. Interestingly, it seems the density of states effect on plasma properties has not been considered to date in spite of the fact that solar radiation which allows us to survive comes essentially from plasma. Noteworthy, plasma does not meet the Barnett–Loudon sum rule which is probably because of the negative dielectric function for $\omega < \omega_{\text{p}}$ which has not been implied when deriving this rule in Section 14.6.

14.10 Density of states effects on the Raman scattering of light

Raman scattering is viewed in quantum electrodynamics as virtual excitation of a molecule or a solid by light with frequency ω with emission of photons whose frequency satisfies a condition $\omega' = \omega \pm \Omega$, where Ω is the frequency of the intrinsic vibrations in a molecule or in a crystal lattice. In Chapter 13 (Section 13.7) scattering rate was shown to be proportional to the density of final states (Eq. 13.26). Therefore it will be equally modified in every mesoscopic structure among those considered previously with respect to spontaneous emission modification. It actually does. For example, for microcavities, modified Raman scattering have been known about since 1984 [57]. Multiple examples were reported on the whispering gallery mode effect on Raman spectra from solid and liquid microdroplets [58].

Rakovich *et al.* [59] reported on enhancement of Raman scattering from semiconductor quantum dots attached to a dielectric microsphere. A microcavity–quantum dot system was studied consisting of a melamine formaldehyde latex microsphere coated by CdTe colloidal quantum dots. It was found that the cavity-induced enhancement of the Raman scattering allows the observation of Raman spectra from only a single layer of CdTe quantum dots. At the same time, periodic structure with very narrow peaks in the luminescence spectra of a single microsphere was detected, arising from the coupling between the emission from the quantum dots and spherical cavity modes.

In spite of the apparent similarity in density of states effects on emission and scattering of photons, only recently has it been introduced into systematic consideration for photon scattering [60]. We shall see in Chapter 16 this is of principal importance for understanding the ultimate plasmonic enhancement of Raman scattering near metal nanobodies.

14.11 Directional emission and scattering of light defined by partial density of states

In mesoscopic structures with an angular-dependent density of states, emission and scattering of light will happen in accordance with the partial density of states for the particular direction. Photonic crystal slabs represent a typical example of such structures. In the case of an incomplete (non-omnidirectional) photonic band gap in three-dimensional photonic crystal samples of finite size, spontaneous emission features complex angular distribution. This is pronounced both for three- and two-dimensional photonic crystal slabs. For three-dimensional periodic structures several directions are typically pronounced where light is emitted more efficiently by the emitters embedded therein. For two-dimensional periodic slabs like, e.g. honeycomb structures, pillar structures or nanoporous slabs with regular cylindrical pores, band gaps develop in the plane of periodicity whereas in the direction nomal to that plane the medium is continuous. In that direction spontaneous emission intensity is enhanced, whereas in the plane of periodicity and for grazing angles it is inhibited. A typical example of directional emission of light from a two-dimensional photonic crystal slab is shown in Figure 14.11 for a nanoporous alumina slab containing Eu^{3+} ions. Fabrication of such template-free periodic structures was discussed in Section 7.9 (Fig. 7.18).

One can see, for a nanoporous alumina Al_2O_3 slab the emission indicatrix essentially differs from that for a reference thin film sample by a pronounced maximum in the emission along the pore axes. Similar results have been reported by many groups [62–64]. The phenomenon has even been proposed for enhancement of light emission efficiency in semiconductor light emitting diodes. InGaN/GaN light emitting diodes1.5 times brighter

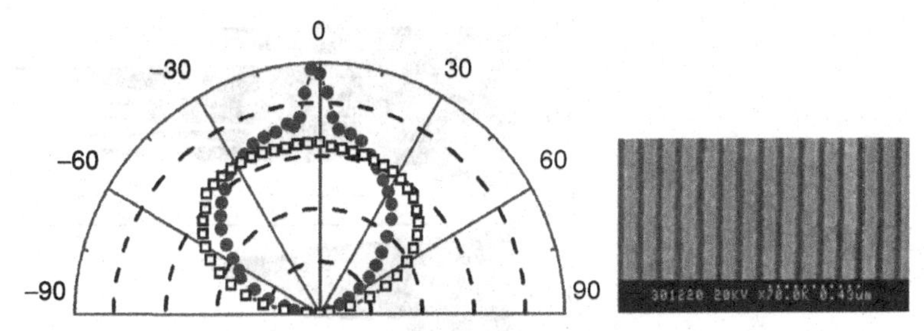

Fig. 14.11 Luminescence indicatrices of Eu^{3+} ions in titania xerogel embedded in a nanoporous alumina slab (black circles) and on a silicon surface (squares). Right-hand panel shows a portion of cross-section of the nanoporous alumina slab. Adapted from [61].

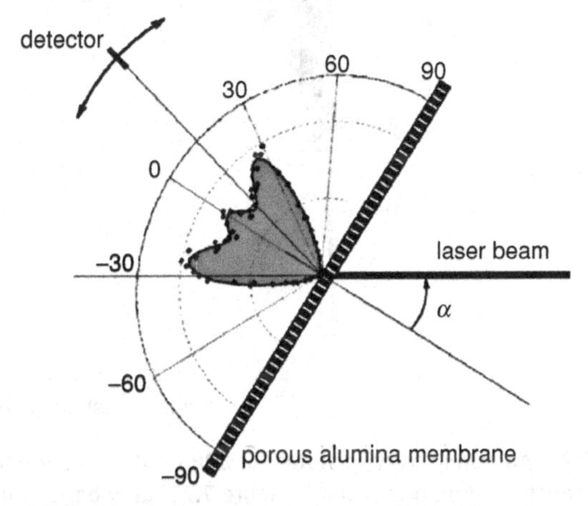

Fig. 14.12 Anisotropic scattering of light in a nanoporous alumina slab. A slab and pores geometry are presented along with transmitted light indicatrix for laser light incident at an angle 30° with respect to the pore axes [68]. See text for detail.

have been fabricated using a two-dimensional periodic upper GaN layer [65]. Anisotropic light emission finds reasonable explanation in terms of angular-dependent partial density of photon states. Note, spontaneous emission decay rate will always be independent of the observation angle as well as of the light harvesting condition, provided that the same set of emitters is examined. Interestingly, the radiophysical counterpart of directional emission relevant to a classical dipole adjacent to a photonic crystal slab has been reported earlier [66]. Periodic structures are not the exclusive design for directional spontaneous emission. This can also be obtained by placing emitters in front of a diffraction grating [67].

In photonic crystals featuring redistribution and modification of photon density of states, resonant (Rayleigh) light scattering modifies accordingly. In a two-dimensional photonic crystal with vertical cylindrical pores or pillars scattering occurs predominantly along the pores or pillars, since there is no density of states inhibition in this direction. A representative example is given in Figure 14.12 where again, template-free electrochemically developed nanoporous alumina is used.

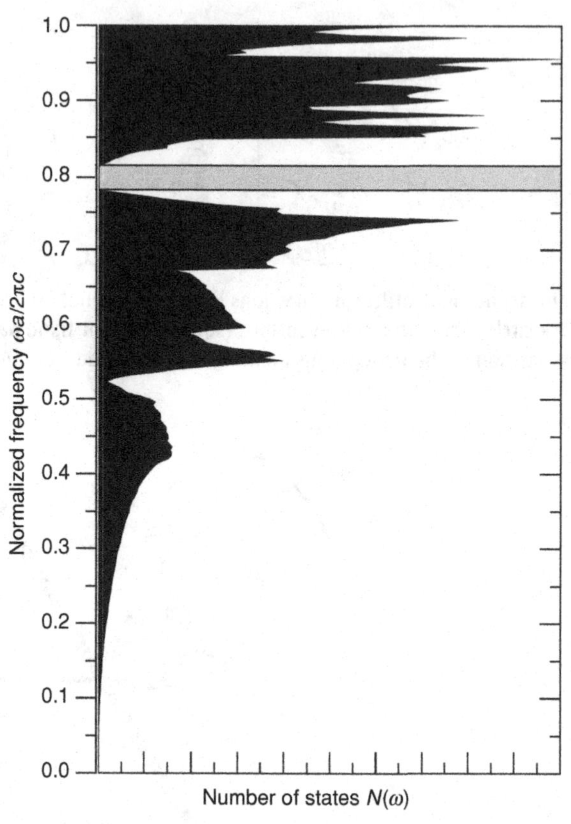

Fig. 14.13 Calculated photon density of states for a three-dimensional photonic crystals whose design and band structure were presented in Figure 7.23. Gray band indicates the band gap. Reprinted with permission from [70]. Copyright 2007, Elsevier B.V.

There are three peaks in the scattering indicatrices: the first one in the initial direction of light propagation, the second in the "mirror reflection direction" and the third in the direction coinciding with the pore axes, i.e. normal to the sample plane. Mechanisms of peak formation are interpreted as follows. The first peak (with the maximum near $-30°$) is light passed through the sample in the initial direction. The second (with the maximum near $+30°$) is light reflected from a set of regular parallel pores. The third peak appears because porous alumina membranes are not perfect (rough pore surfaces, reflection index heterogeneities). Light diffusely scatters on these defects under *non-isotropic density of photon states* conditions. The probability of light scattering in a certain direction is proportional to the density of states in this direction. Density of states has a maximum value in the direction along the pores and a minimum in the perpendicular directions for the particular spectral range under consideration. Therefore light mainly scatters in the direction along the pore's axes and the middle peak develops in this direction.

For three-dimensionally periodic structures, recently D. N. Chigrin developed an approach treating the emission indicatrix in terms of interference of Bloch waves [69]. In the context of our discussion, this approach is equally applicable to angular distribution of scattered light under condition of multiple scattering in periodic structures.

Problems

1. Looking at Purcell's paper in Figure 14.1 find out the density of photon states in the formula and explain the difference with respect to Eq. (13.4).

2. Implying that modification of decay rate calculated in Figure 14.4 for a quantum emitter inside a photonic crystal results from photon density of states modification, evaluate the spectral dependence of the density of states inside, at the edge and outside the band gap region. Starting from the vacuum density of states, plot the approximate dependence of photon density of states on light frequency in a photonic crystal.

3. Figure 14.13 represents calculated photon density of states for the three-dimensional photonic crystal whose spatial structure and band structure have been shown in Figure 7.23. Superimpose the density of states graph with the band structure diagram and evaluate correlation between (a) band gap and density of states gap, (b) linear dispersion law in the band structure for longer wavelengths and vacuum-like density of states, (c) density of states spectral features and Barnett–Loudon sum rule. Predict radiative decay rates for different frequencies.

4. Compare contributions to linewidth modification of a probe atom in plasma coming from the electron and ion components of the plasma.

5. Estimate the atomic line shrinkage for an electron density in plasma equal to 10^{18} cm^{-3}.

6. Check conformity of the Bunkin–Oraevsky formula (Eq. 14.2) for the decay rate of an atom in a cavity with the sum rules for radiative rates (Eq. 14.8) and density of states (14.20).

7. Recall all optical processes defined by the photon density of states.

8. Recall all examples known to you of enhancement and inhibition of spontaneous emission of radiation.

9. Elaborate a construction of an incandescent lamp with higher output in the visible.

10. Evaluate modification of thermal radiation from a plasma containing electrons and ions ($m = 50$, $Z = 1$), with concentration 10^{19} cm^{-3} at temperature 10^6 K.

11. Explain why the decay law of spontaneous emission for complex structures with angular-dependent partial density of states turns out to be independent of observation angle and solid angle where light is harvested by a detector.

References

[1] W. E. Lamb. Anti-photon. *Appl. Phys. B*, **60** (1995), 77–80.

[2] E. M. Purcell. Spontaneous emission probabilities at radiofrequencies. *Phys. Rev.*, **69** (1946), 681.

[3] A. N. Oraevskii. Spontaneous emission in a cavity. *Physics – Uspekhi*, **37** (1994), 393–405.

[4] F. V. Bunkin and A. N. Oraevskii. Spontanoeus emission in a cavity. *Izvestia Vuzov*, **2** (1959), 181–188. – in Russian.

[5] P. Goy, J. M. Raimond, M. Gross and S. Haroche. Observation of Cavity-Enhanced Single-Atom Spontaneous Emission. *Phys. Rev. Lett.*, **50** (1983), 1903–1906.

[6] F. de Martini, G. Innocenti, G. R. Jacobowitz and P. Mataloni. Anomalous Spontaneous Emission Time in a Microscopic Optical Cavity. *Phys. Rev. Lett.*, **59** (1987), 2955–2958.

[7] M. V. Artemyev, U. Woggon , R. Wannemacher and H. Jaschinskii. Light trapped in a photonic dot: microspheres act as a cavity for quantum dot emission. *Nanoletters*, **1** (2001), 309–313.

[8] W. Jhe, A. Anderson, E. A. Hinds, D. Meschede, L. Moi and S. Haroche. Suppression of spontaneous decay at optical frequencies: Test of vacuum-field anisotropy in confined space. *Phys. Rev. Lett.*, **58** (1987), 666–669.

[9] M. Bayer. Inhibition and enhancement of the spontaneous emission of quantum dots in structured microcavities. *Phys. Rev. Lett.*, **86** (2001), 3168–3171.

[10] R. P. Stanley, R. Houdré, C. Weisbuch, U. Oesterle and M. Ilegems. Cavity-polariton photoluminescence in semiconductor microcavities: Experimental evidence. *Phys. Rev. B*, **53** (1996), 10995–11007.

[11] V. Pellegrini, A.Tredicucci, C. Mazzoleni and L. Pavesi. Enhanced optical properties in porous silicon microcavities. *Phys. Rev. B*, **52** (1995), R14328–R14331.

[12] Y. P. Rakovich, J. F. Donegan, M. Gerlach, A. L. Bradley, T. M. Connolly, J. J. Boland, N. Gaponik and A. Rogach. Fine structure of coupled optical modes in photonic molecules. *Phys. Rev. A*, **70** (2004), 051801(R).

[13] P. R. Berman (Ed.) *Cavity Quantum Electrodynamics* (New York: Academic, 1994).

[14] Y. Yamamoto, F. Tassone and H. Cao. *Semiconductor Cavity Quantum Electrodynamics* (Berlin: Springer-Verlag, 2000).

[15] K. H. Drexhage. Influence of a dielectric interface on fluorescence decay time. *J. Luminescence*, **1–2** (1970), 693–701.

[16] W. L. Barnes. Fluorescence near interfaces: the role of photonic mode density. *J. Mod. Optics*, **45** (1998), 661–699.

[17] V. P. Bykov. Spontaneous emission in a periodic structure. *Sov. Phys. JETP.*, **35** (1972), 269–273.

[18] E. Yablonovitch. Inhibited spontaneous emission in solid state physics and electronics. *Phys. Rev. Lett.*, **58** (1987), 2059–2062.

[19] V. P. Bykov. *Light emission by atoms near material bodies* (Moscow: Nauka, 1986) – in Russian.

[20] V. P. Bykov. *Radiation of Atoms in a Resonant Environment* (Singapore: World Scientific, 1993)

[21] P. Lambropoulos, G. M. Nikolopoulos, T. R. Nielsen and S. Bay. Fundamental quantum optics in structured reservoirs. *Rep. Prog. Phys.*, **63** (2000), 455–503.

[22] D. S. Mogilevtsev and S. Ya. Kilin. *Quantum Optics Methods of Structured Reservoirs* (Minsk: Belorusskaya Nauka, 2007) – in Russian.

[23] V. N. Bogomolov, S. V. Gaponenko, I. N. Germanenko, A. M. Kapitonov, E. P. Petrov, N. V. Gaponenko, A. V. Prokofiev, A. N. Ponyavina, N. I. Silvanovich and S. M. Samoilovich. Photonic band gap phenomenon and optical properties of artificial opals. *Phys. Rev. E*, **55** (1997), 7619–7626.

[24] S. V. Gaponenko, V. N. Bogomolov, E. P. Petrov, D. A. Yarotsky, I. I. Kalosha, A. M. Kapitonov, A. A. Eychmueller, A. L. Rogach, J. McGilp, U. Woggon and F. Gindele. Spontaneous emission of dye molecules, semiconductor nanocrystals, and rare-earth ions in opal-based photonic crystals. *J. Lightwave Technol.*, **17** (1999), 2128–2137.

[25] E. P. Petrov, V. N. Bogomolov, I. I. Kalosha and S. V. Gaponenko. Spontaneous emission of organic molecules embedded in a photonic crystal. *Phys. Rev. Lett.*, **81** (1998), 77–80.

[26] E. P. Petrov, D. A. Ksenzov, T. A. Pavich, M. I. Samoilovich and A. V. Gur'yanov. Time-resolved luminescence of Eu complexes in bulk and nanostructured dielectric media. *Physics, Chemistry and Application of Nanostructures* (Eds. V. E. Borisenko, S. V. Gaponenko and V. S. Gurin), (Singapore: World Scientific, 2003), pp. 43–46.

[27] H. Yang and S.-Y. Zhu. Spontaneous emission from a two-level atom in a three-dimensional photonic crystal. *Phys. Rev. A*, **62** (2000), 013805.

[28] Zh.-Y. Li, L.-L.Lin and Zh.-Q. Zhang. Spontaneous emission from photonic crystals: Full vectorial calculations. *Phys. Rev. Lett.*, **84** (2000), 4341–4344; Weak photonic band gap effect on the fluorescence lifetime in three-dimensional colloidal photonic crystals. *Phys. Rev. B*, **63** (2001), 125106.

[29] S.-Y. Zhu, G.-X. Li, Y.-P. Yang and F.-L. Li. Spontaneous emission in three-dimensional photonic crystals with an incomplete band gap. *Europhys. Lett.*, **62** (2003), 210–214.

[30] X.-H. Wang, R. Wang, B.-Y. Gu and G.-Zh. Yang. Decay Distribution of Spontaneous Emission from an Assembly of Atoms in Photonic Crystals with Pseudogaps. *Phys. Rev. Lett.*, **88** (2002), 093902.

[31] S. V. Gaponenko, A. M. Kapitonov, V. N. Bogomolov, A. V. Prokofiev, A. Eychmuller and A. L. Rogach. Electrons and photons in mesoscopic structures: Quantum dots in a photonic crystal. *JETP Lett.*, **68** (1998), 142–147.

[32] M. J. A. de Dood, B. Gralak, A. Polman and J. G. Fleming. Superstructure and finite-size effects in a Si photonic woodpile crystal. *Phys. Rev. B*, **67** (2003), 035322.

[33] A. F. Koenderink, M. Kafesaki, C. M. Soukolis and V. Sandoghdar. Spontaneous emission in the near field of two-dimensional photonic crystals. *Opt. Lett.*, **30** (2005), 3210–3212.

[34] M. Fujita, S. Takahashi, Y. Tanaka, T. Asano and S. Noda. Simultaneous inhibition and redistribution of spontaneous light emission in photonic crystals. *Science*, **308** (2005), 1296–1298.

[35] H. P. Urbach and G. L. J. A. Rikken. Spontaneous emission from a dielectric slab. *Phys. Rev. A*, **57** (1998), 3913–3930.

[36] D. Toptygin and L. Brand. Fluorescence decay of DPH in lipid membranes: Influence of the external refractive index. *Biophys. Chem.*, **48** (1993), 205–220.

[37] E. P. Petrov, J. V. Kruchenok and A. N. Ruvimov. Effect of the external refractive index on fluorescence kinetics of perylene in human erythrocyte ghosts. *J. Fluoresc.*, **9** (1999), 111–121.

[38] G. Lamouche, P. Lavallard and T. Gacoin. Optical properties of dye molecules as a function of the surrounding dielectric medium. *Phys. Rev. A*, **59** (1999), 4668–4674.

[39] F. J. P. Schuurmans, P. de Vries and A. Lagendijk. Local-field effects on spontaneous emission of impurity atoms in homogeneous dielectrics. *Phys. Lett. A*, **264** (2000), 472–477.

[40] K. Cho. *Optical Response of Nanostructures. Nonlocal Microscopic Theory* (Berlin: Springer, 2003).

[41] G. Allan and C. Delerue. Confinement effects in PbSe quantum wells and nanocrystals. *Phys. Rev. B*, **70** (2004), 245321.

[42] T. C. Killian, S. Kulin, S. D. Bergeson, L. A. Orozco, C. Orzel and S. L. Rolston. Creation of an ultracold neutral plasma. *Phys. Rev. Lett.*, **83** (1999), 4776–4779.

[43] M. P. Robinson, B. Tolra, M. W. Noel, T. F. Gallagher and P. Pillet. Spontaneous evolution of Rydberg atoms into an ultracold plasma. *Phys. Rev. Lett.*, **85** (2000), 4466–4468.

[44] S. M. Barnett and R. Loudon. Sum rule for modified spontaneous emission rates. *Phys. Rev. Lett.*, **77** (1996), 2444–2448.

[45] L. D. Landau and E. M. Lifshitz. *Statistical Physics* (Oxford: Pergamon, 1980), Pt. 2, Chap. 8.

[46] P. W. Milonni. *The Quantum Vacuum* (Boston: Academic Press, 1994).

[47] L. Novotny and B. Hecht. *Principles of Nano-Optics* (Cambridge: Cambridge University Press, 2006).

[48] C. T. Tai. *Diadic Green's Functions in Electromagnetic Theory* (New York: IEEE Press, 1993).

[49] G. D'Aguanno, N. Mattiucci, M. Centini, M. Scalora and M. J. Bloemer. Electromagnetic density of modes for a finite-size three-dimensional structure. *Phys. Rev. E*, **69** (2004), 057601.

[50] C. Chicanne, T. David, R. Quidant, J. C. Weeber, Y. Lacroute, E. Bourillot, A. Dereux, G. Colas des Francs and C. Girard. Imaging the Local Density of States of Optical Corrals. *Phys. Rev. Lett.*, **88** (2002), 097402.

[51] K. Joulain, R. Carminati, J.-P. Mulet and J.-J. Greffet. Definition and measurement of the local density of electromagnetic states close to an interface. *Phys. Rev. B*, **68** (2003), 245405.

[52] G. V. Dedkov and A. A. Kyasov. On the structure of far-field and near-field equilibrium electromagnetic radiation near a plane border of semi-space filled with a homogeneous dielectric (magnetic) medium. *Technical Physics Lett.*, **32** (2006), 78–83.

[53] S.-Y. Lin, J. G. Fleming, E. Chow, J. Bur, K. K. Choi and A. Goldberg. Enhancement and suppression of thermal emission by a three-dimensional photonic crystal. *Phys. Rev. B*, **62** (2000), R2243–R2246.

[54] S. Maruyama, T. Kashiwa, H. Yugami and M. Esashi. Thermal radiation from two-dimensionally confined modes in microcavities. *Appl. Phys. Lett.*, **79** (2001), 1393–1395.

[55] V. V. Datsyuk and I. A. Izmailov. Optics of microdroplets. *Physics – Uspekhi*, **44** (2001), 1061–1073.

[56] A. M. Nemilentsau, G. Ya. Slepyan and S. A. Maksimenko. Near-field and far-field effects in thermal radiation from metallic carbon nanotubes. *Phys. Rev. Lett.*, **99** (2007), 147403.

[57] R. Turn and W. Kiefer. Observations of structural resonances in the Raman spectra of optically levitated dielectric microspheres. *J. Raman Spectrosc.*, **15** (1984), 411–413.

[58] V. V. Datsyuk and I. A. Izmailov. Optics of microdroplets. *Physics – Uspekhi*, **44** (2001), 1061–1073.

[59] Yu. P. Rakovich, J. F. Donegan, N. Gaponik and A. L. Rogach. Raman scattering and anti-Stokes emission from a single spherical microcavity with a CdTe quantum dot monolayer. *Appl. Phys. Lett.*, **83** (2003), 2539–2541.

[60] S. V. Gaponenko. Photon density of states effects on Raman scattering in mesoscopic structures. *Phys. Rev. B*, **65** (2002), 140303 (R).

[61] N. V. Gaponenko, I. S. Molchan, S. V. Gaponenko, A. V. Mudryi, A. A. Lyutich, J. Misiewicz and R. Kudrawiec. Luminescence of the Eu and Tb ions in the structure "microporous xerogel/mesoporous anodic alumina". *J. Appl. Spectr.*, **70** (2003), 57–61.

[62] M. Boroditsky, R. Vrijen, T. F. Krauss, R. Coccioli, R. Bhat and E. Yablonovitch. Spontaneous emission extraction and Purcell enhancement from thin-film 2-D photonic crystals. *J. Lightwave Technol.*, **17** (1999), 2096–2112.

[63] R. K. Lee, Y. Xu and A. Yariv. Modified spontaneous emission from a two-dimensional photonic bandgap crystal slab. *J. Opt. Soc. Amer. B*, **17** (2000), 1438–1442.

[64] A. L. Fehrembach, S. Enoch and A. Sentenac. Highly directive source devices using slab photonic crystal. *Appl. Phys. Lett.*, **79** (2001), 4280–4282.

[65] J. J. Wierer, M. R. Krames, J. E. Epler, N. F. Gardner, M. G. Craford, J. R. Wendt, J. A. Simmons and M. M. Sigalas. InGaN/GaN quantum-well heterostructure light-emitting diodes employing photonic crystal structures. *Appl. Phys. Lett.*, **84** (2004), 3885–3887.

[66] E. R. Brown, C. D. Parker and E. Yablonovitch. Radiation properties of a planar antenna on a photonic-crystal substrate. *J. Opt. Soc. Amer. B*, **10** (1993), 404–407.

[67] P. Andrew and W. L. Barnes. Molecular fluorescence above metallic gratings. *Phys. Rev. B*, **64** (2001), 125405.

[68] A. A. Lutich, S. V. Gaponenko, N. V. Gaponenko, I. S. Molchan, V. A. Sokol and V. Parkhutik. Anisotropic light scattering in porous materials: A photon density of states effect. *Nano Letters*, **4** (2004), 1755–1759.

[69] D. N. Chigrin. Spatial distribution of the emission intensity in a photonic crystal: Self-interference of Bloch eigenwaves. *Phys. Rev. A*, **79** (2009), 033829.

[70] K. Busch, G. von Freymann, S. Linden, S. F. Mingaleev, L. Tkeshelashvili and M. Wegener. Periodic nanostructures for photonics. *Phys. Rep.*, **444** (2007), 101–202.

15 Light–matter states beyond perturbational approach

In nanostructures where confinement of a light wave is manifested, the conditions of strong light–matter interaction become feasible. In this case, the approach based on the environment-sensitive probability of quantum transitions is no longer applicable. Instead, joint states of light and matter should be considered and their time evolution explored. The content of the previous Chapters 13 and 14 can then be treated as a perturbative description of light–matter interactions with a limited range of applicability. In this chapter a few representative examples are given where joint light–matter states bring an unprecedented flavor of nanophotonic engineering to frozen excited states for quantum memory devices and single photon sources for quantum computing.

15.1 Cavity quantum electrodynamics in the strong coupling regime

The regime of strong coupling of a quantum system with an electromagnetic field can be performed in a cavity with very high Q-factor, where light can survive for a reasonable time until it either leaves the cavity or dies through being absorbed by cavity imperfections. In a simplified picture, a photon once emitted stays within the cavity so long that it is absorbed again by the emitter (an imaginary two-level system, an atom or a quantum dot). Thus spontaneous emission appears to become reversible. In a more accurate presentation, an "atom+field" state develops which evolves via oscillations between the state $|E, 0\rangle$ ("excited atom, depleted cavity") to the state $|G, 1\rangle$ ("de-excited atom, a photon in the cavity"). Here "depleted" implies there is no photon in the cavity and "de-excited" means "being in the ground state". This phenomenon is referred to as vacuum *Rabi oscillations* since it represents in essence the phenomenon known in nonlinear atomic spectroscopy for atoms interacting with strong fields.[1]

Optical Rabi oscillations are temporal oscillations of population inversion in a two-level system driven by a *strong* resonant optical field on a timescale shorter than the dephasing time, with an oscillation frequency g proportional to the transition dipole moment μ and the electric-field amplitude $|\mathbf{E}|$, $g = \mu|\mathbf{E}|/\hbar$ [1, 2]. When a quantum emitter, e.g. an atom, is coupled to high-Q cavity mode radiation, Rabi oscillations occur between the two states

[1] Isidor Rabi (1898–1988) was an American physicist. He formulated basic ideas in radiospectroscopy and magnetic resonance spectroscopy for atoms, nuclei and molecules and received the Nobel prize in 1944 for these achievements.

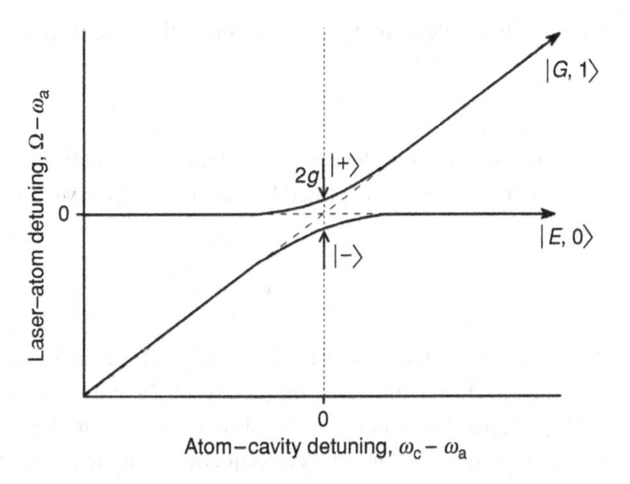

Fig. 15.1 **"Atom+field" diagram for the two lowest excited states of the atom–cavity system. $|G, 1\rangle$ is the superposition state with one photon in the cavity and the atom in the ground state. $|E, 0\rangle$ is the superposition state with no photon in the cavity and the atom in the excited state. The dashed lines show the uncoupled states.**

of the "atom+cavity field" system, $|E, 0\rangle$ and $|G, 1\rangle$. No external driving field is necessary. Vacuum radiation contained in the strong cavity mode and a photon emitted by an atom into that mode form the field involved in the interaction. In a sense, Rabi oscillations for a quantum emitter coupled to a high-Q cavity mode manifest nonlinear spectroscopy with a single atom and a single photon.

A simplified consideration of an atom and a cavity in the strong coupling regime is as follows [3]. Let an atom be characterized by the transition frequency $\omega_a = (E_E - E_G)/\hbar$ with the excited state energy E_E and the ground state energy E_G and the spontaneous emission rate in a vacuum W defining the emission line full width on the frequency scale. A cavity is characterized by the resonant frequency ω_c and the Q-factor which defines the cavity resonance width $\Delta\omega$ on the frequency scale as,

$$Q = \frac{\omega_c}{\Delta\omega}. \tag{15.1}$$

Then the "atom+field" state in the cavity is characterized by the diagram shown in Figure 15.1. The resonant frequency splitting develops for the coupled oscillators (atom and field) to give the two modified frequencies Ω_+, Ω_- of the coupled "atom+cavity" system in the form,

$$\Omega_\pm = \frac{1}{2}(\omega_a + \omega_c) \pm \sqrt{g^2 + \left(\frac{\omega_c - \omega_a}{2}\right)^2}. \tag{15.2}$$

For the initial resonance condition, $\omega_c = \omega_a$,

$$\Omega_\pm = \omega_a \pm g \tag{15.3}$$

holds and the splitting value equals $2g$. This value is referred to as the *vacuum Rabi frequency*. It defines the atom – field coupling strength. It is determined by interaction of an atom with an electromagnetic vacuum field contained in the resonant cavity mode as,

$$g = \frac{1}{\hbar}\mu|\mathbf{E}_{\rm vac}|. \tag{15.4}$$

Here μ is the transition dipole moment and $|\mathbf{E}_{\rm vac}|$ is the vacuum field (not to be confused with E denoting energy!) defined by the cavity mode vacuum energy $\hbar\omega_{\rm c}/2$ and the cavity volume V as,

$$|\mathbf{E}_{\rm vac}|^2 = \frac{1}{2}\hbar\omega_{\rm c}\frac{1}{\varepsilon\varepsilon_0 V}, \tag{15.5}$$

where ε is the relative dielectric permittivity of the cavity material, and ε_0 is the dielectric constant. Physically the Rabi frequency of $2g$ corresponds to the exchange rate of the single quantum energy between the atom and the cavity field. Note, even for large detuning, $|\omega_{\rm c} - \omega_{\rm a}| \gg g$, atom and cavity frequencies slightly deviate from their intrinsic values $\omega_{\rm a}, \omega_{\rm c}$.

In the absence of losses the model of the atom–single-mode field interaction is known as the *Jaynes–Cummings model* [4]. It is analytically solvable and very well known in quantum optics. It describes essentially the case of strong atom–field coupling and allows us to demonstrate such specifically quantum effects as vacuum Rabi oscillations and "revivals" of the atomic level population. One has to add that, in the limit of very strong coupling, the simplest version of the Jaynes–Cummings model does not hold any more. One should consider the complete version of it (non-conserving the number of photons, i.e. one in which the "rotating wave approximation" is not used) [5].

The atom–cavity *strong coupling condition* simply reads,

$$2g \gg W, \Delta\omega, \tag{15.6}$$

and *means the energy exchange rate between the atom and the cavity field dominates over both the atomic decay rate and the cavity decay rate.* From the above consideration it is clear that the strong coupling regime can be performed by means of a high-Q cavity of small volume. For atom–cavity detuning $\omega_{\rm c} - \omega_{\rm a} \ll -2g$ the joint atom–field state $|+\rangle$ is atomic-like, i.e. the probability that the atom is excited is much greater than the probability of finding a photon in a cavity, whereas the $|-\rangle$ state is the cavity-like one. For $\omega_{\rm c} - \omega_{\rm a} \gg 2g$, the reverse occurs. At resonance, for $\omega_{\rm c} = \omega_{\rm a}$, these states are the linear superposition of an excited atom, empty cavity mode product state $|E, 0\rangle$ and a de-excited atom, one-photon-cavity mode product state $|G, 1\rangle$.

The first experimental observation of vacuum Rabi splitting for atoms in a cavity dates back to the 1980s. Nowadays it can be detected at a single atom level [6]. For an atom–cavity system typically an atom flow occurs through the cavity volume, every atom spending some time within the cavity. More accurate measurements could be performed with a localized quantum emitter which becomes affordable with ultracold atoms in a trap [7].

A semiconductor photonic crystal cavity, discussed in Chapter 9 (see Fig. 9.1 and 9.2), offers Q-factors as large as 10^4–10^5 while keeping the internal cavity volume less than 1 μm^3. Emitters in the form of quantum dots with a discrete luminescence spectrum (see

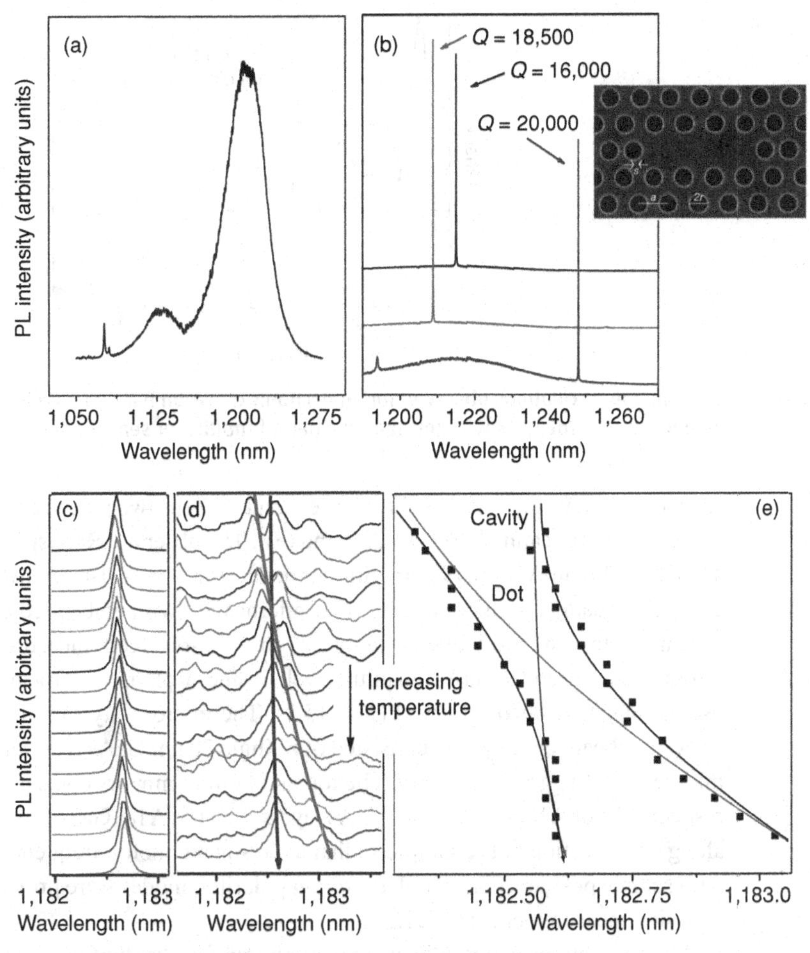

Fig. 15.2 **Observation of vacuum Rabi splitting for InAs quantum dots in an array of semiconductor photonic crystal microcavities. (a) quantum dot ensemble spontaneous emission spectrum in free space, showing both the lowest (1 175–1 250 nm) and first excited (1 100–1 150 nm) transitions; (b) representative photoluminescence spectra of quantum dots from three microcavities; (c)–(e) show dot-cavity anticrossing observed while scanning temperature in 1 K steps from 13 K in the top to 29 K in the bottom. (c) high-power spectra; (d) middle-power spectra; (e) the two coupled-system peaks (black squares, black lines are guides) are plotted as a function of temperature, and compared with the scan rates of an uncoupled dot and an empty cavity (gray lines labeled "Dot" and "Cavity", respectively). The top-right insert shows a microcavity design. Reprinted with permission from [8]. Copyright 2004 *Nature* Publ. Group.**

Chapter 5) can be embedded in the course of the single fabrication route. In the experiments by Yoshie and co-workers [8] the mode volume $V < 0.1\,\mu\text{m}^3$ was performed in a GaAs two-dimensional periodic photonic crystal slab with InAs hut-like quantum dots buried therein. Temperature dependence of the quantum dot emission spectrum offered a way to scan dot spectrum versus cavity resonance. At higher pump energies (a 770 nm cw-laser, $0.7\,\text{mW}/\mu\text{m}^2$) emission spectrum does not show splitting (Fig. 15.2c). At middle pump

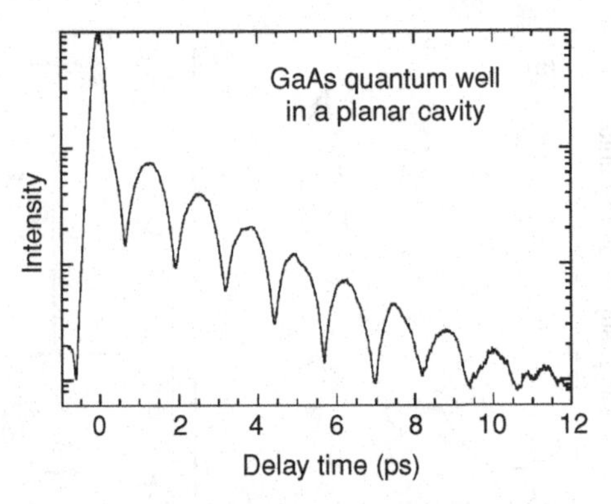

Fig. 15.3 **Vacuum Rabi oscillations observed for an excitonic emission band in a GaAs quantum well inside a planar Fabry–Perot $\lambda/2$ cavity formed by a pair of multilayer semiconductor Bragg reflectors [10].**

energies (0.8 $\mu W/\mu m^2$) Rabi splitting is evident with the two peaks repelling each other with the temperature scanned (Fig. 15.2(d) and (e)). The observed Rabi splitting value was $2g = 41\,\text{GHz} = 170\,\text{meV} = 0.192\,\text{nm}$. Similar results have been also reported by A. Forchel wih co-workers using a semiconductor pillar microcavity with embedded quantum dots [9].

Yamamoto with co-workers reported on time-resolved measurements of excitonic luminescence from a GaAs quantum well inside a planar microcavity. Dedicated Rabi oscillations have been observed (Fig. 15.3). The microcavity sample used was grown by molecular beam epitaxy, and consisted of a single 20 nm GaAs quantum well located in the middle $\lambda/2$ Bragg mirror cavity. The top and bottom mirrors consist of 15.5 and 30 pairs, respectively, of $Al_{0.5}Ga_{0.5}As$ and AlAs layers. The top $Al_{0.3}Ga_{0.7}As$ spacer layer is tapered along one direction of the sample so that the resonant photon frequency of the cavity varies with sample position. The coupled exciton polariton modes were probed by absorption and emission measurements at 4.2 K.

Rabi oscillations in a nonlinear-optical regime for quantum dots in strong fields with no cavity have also been observed [11]. Theoretical aspects of this regime are discussed, e.g. in [12, 13].

15.2 Single-atom maser and laser

The first maser was created by Charles Townes and co-workers in 1954 [14]. It consisted of a microwave cavity through which a beam of excited ammonia molecules was sent. Its optical analog, a laser was created for the first time by T. Maiman using a ruby crystal with silver mirrors deposited on its facets in 1960 [15]. Both devices are based on excitation by a statistically large ensemble of molecules, atoms or elementary excitations in solids. Even microdevices like semiconductor lasers are basically macroscopic devices.

Strong light–matter coupling in a microcavity has led to the idea of micromasers and microlasers operating at a single-atom level. Using a high quality cavity and slow atom flow through it, making use of a single atom interacting with the single cavity mode becomes feasible. It is necessary that the time interval between successive atoms entering the cavity should be greater than both the time the atom spends in a cavity and the time of radiation decay outside the cavity. Every atom when in the cavity is to perform multiple Rabi oscillations. Actually, a micromaser has been performed basically owing to two advances: the ability to prepare Rydberg states with large electric dipole moments and small linewidths, and the development of high-Q superconducting microwave cavities [16].

The idea of a single-atom laser was suggested for the first time by Mu and Savage in the seminal paper in 1992 [17] and immediately gained a lot of interest since such a laser promises creation of truly quantum states of light (see paper [18] as well as books [3, 19] and references therein). Its experimental realization based on ultimate technical performance was reported for the first time by H. J. Kimble with coworkers in 2003 [20]. The key issue was to trap a single atom with strong optical resonance inside a high quality cavity. A single cesium atom was trapped for a period of 0.05 s inside a 42 μm long cavity formed by two spherical mirrors with finesse higher than 10^5 at a transition wavelength of 935 nm. The Rabi frequency was about 100 MHz which was several times larger than both the excited state intrinsic decay rate and the cavity decay rate, in accordance with the strong coupling condition Eq. (15.6). The single atom laser exhibited thresholdless operation and sub-Poissonian photon statistics. The latter means it really offers quantum light which is "quieter" than ordinary laser radiation; this is important in quantum information devices.

15.3 Light–matter states in a photonic band gap medium

In the 1970s, V. P. Bykov suggested that there would be no spontaneous decay of an atom or a molecule in a periodic structure since such a structure can potentially feature the gap in the electromagnetic mode spectrum [21]. This seminal prediction did stimulate vast activity in the fields of classical and quantum electrodynamics of periodic dielectrics, resulting in elaboration of the solid concept of photonic crystals as well as in formulation of new ideas for optical circuitry implementation. These trends and major results of classical electrodynamics have been the subject of Chapter 7 and Chapter 9. Inhibition of the spontaneous decay of a quantum system embedded in a photonic crystal with an omnidirectional band gap has been predicted within the framework of the perturbational approach to light–matter interaction in which the probability of spontaneous decay can be introduced following a Weisskopf–Wigner approximation. The complex environment modifies the decay rate has been discussed through Chapter 13. In this consideration, "no final photon state available" means no decay will ever happen. In Chapter 13 we have seen that even an incomplete gap brings new light on atomic decay resulting in non-Markovian decay when the lifetime can not be properly assigned to an excited state. The complete band gap with a finite spectral range of zero density of photon states brings even more complexity and needs a thorough non-perturbative consideration of light–matter states.

In quantum-optical language, the problem of a quantum system interacting with a reservoir featuring a discontinuous density of photon states should be analyzed. This problem is rather specific and very complicated. Note, even a high-Q microcavity showing sharp

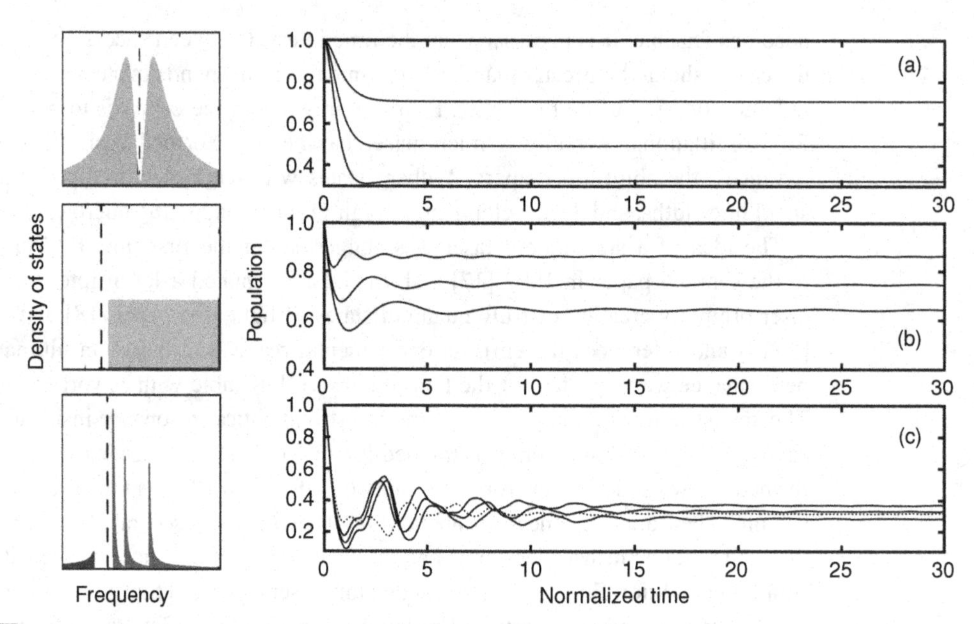

Fig. 15.4 **Population dynamics of a two-level atom in a photonic crystal with various photon density of states functions. (a), (b) and (c) panels correspond to the density of states shown in the proper graph in the left-hand column. Position of the atom transition frequency is shown by vertical dashed lines. Different curves in (a)–(c) right-hand graphs correspond to (a) different atom–field interaction strength and (b) and (c) to different positions of the atom transition frequency with respect to the band gap edge. Adapted from [25].**

resonant enhancement in density of states does not offer discontinuity in the density of states spectrum. In this section, we shall trace the properties of atom–field bound states in the photonic band gap environment based on a treatise performed by D. S. Mogilevtsev and S. Ya. Kilin [22–25]. These authors developed the collective operators method suitable for analyzing the problem of a quantum two-level system in a reservoir with arbitrary spectral behavior of photon density of states, including the cases of discontinuities inherent in perfect photonic crystals.

Consider in detail spontaneous emission and the process of "freezing" of atom excited states for a number of model density of states featuring a gap. In Figure 15.4 examples of the atomic upper-state population are shown for the three different types of photon density of states shown in the left-hand part. The uppermost case corresponds to a narrow dip in the density of states spectrum. Looking at Figure 15.4(a) one can distinguish three different stages of population dynamics. The initial stage resembles the usual decay in free space. Remarkably, duration of this stage is defined by the atomic transition frequency detuning from the band edge, whereas the decay rate does not. A response of the reservoir has not yet been felt by the atom. Then the transitional period comes, and, finally, the population is "frozen". Figure 15.4(b) also shows an additional stage, namely, slowly decaying oscillations on the way to the "frozen" state. These oscillations can be attributed to interaction of the atom–field bound state with long-lived collective excitation of the reservoir. The dynamics of the population in the process of atomic interaction with a more exquisitely structured reservoir with the density-of-states (the lowermost panel in the left-hand part in Fig. 15.4 and Fig. 15.4(c)) show the same stages of evolution. However, the behavior during the second and third stages is more complicated, and the oscillations during the third stage do not look monochromatic any more. The lowermost case resembles a finite band gap along with the sharp increase in the density of states near the band edges. This type of density of states redistribution is inherent in realistic photonic crystals.

The frozen excited atom–field bound state can be treated as localization of the electromagnetic field in the near vicinity of the excited atom. In a sense it is similar to localization of radiation in a high-Q microcavity. The field is localized within a few unit cells of the photonic crystal around the emitter. Such a cavity exists due to strong coupling between the emitter and the field and has a very high Q-factor which is limited by nonideality of a real photonic crystal. This natural radiation concentration effect has been suggested as the basis for a single emitter reversible quantum-optical memory cell and for all-optical switches in photonic circuitry.

The above consideration implies zero temperature of the reservoir to which an excited atom is coupled. In the more probable case of finite temperature, excitation of thermal photons in reservoir modes should be accounted for. Thermal radiation does affect the stability of the frozen bound atom–field state and results in a finite time for its survival [26].

15.4 Single photon sources

There is a challenge for modern quantum optics to design the light sources which could generate the requested number of photons on demand. Such a source will generate the radiation state with the defined photon number in the mode under consideration. Such states are inherent in a quantum oscillator assigned to the field mode as was considered in Section 13.1. In quantum optics these stage are referred to as *Fock states* or *number-states*.[2] A 'number-state' is a quantum state in which the number of photons, n, is a precisely fixed integer. Note that number-states of electrons, atoms and molecules are quite feasible, whereas number-states of light are more exotic. Except for the trivial vacuum-state ($n = 0$), number-states of light are very hard to obtain in practice. In what follows we briefly consider the problem of a single photon source [27].

A laser delivers light pulses that seem extremely regular in amplitude, but this is only because they contain so many photons that their number fluctuations become negligible on a macroscopic scale. Indeed, such pulses are just superpositions of many number-states with different ns. Restricting fluctuations in the number of photons seems easiest for states with $n = 1$, single-photon states. Recent advances in nanophotonics including the fabrication, manipulation and characterization of individual nano-objects, molecules and nanocrystals have opened new routes for the production of number-states containing a single photon.

A microscopic emitter of light, e.g. an atom, can realize its potential as a single-photon source if it is coupled to a resonant high-Q cavity, which can fulfill several functions. First,

[2] V. A. Fock (1898–1974) was a Russian physicist known for his distinguished contribution to quantum theory.

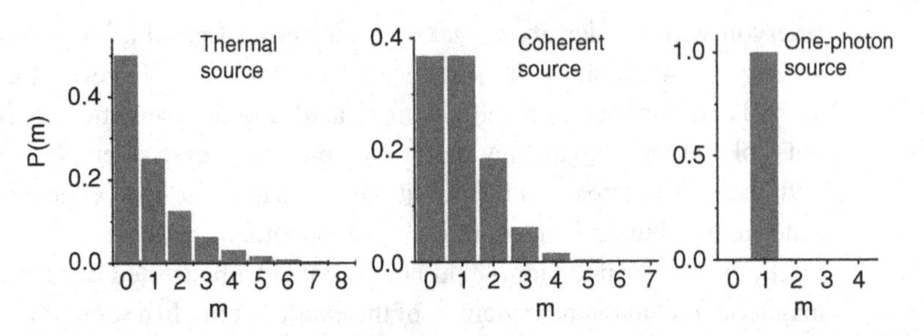

Fig. 15.5 **Probability distribution of the number of photons for three sources with an average photon number $\langle n \rangle = 1$. The thermal source presents large number fluctuations due to the Bose–Einstein statistics of black-body radiation. The coherent light source presents a Poisson distribution, narrower than that of thermal light, but still with strong number fluctuations, called photon noise. An ideal squeezed-light source delivers a number-state with $m = 1$. A single-photon source can match this distribution by delivering single photons at regular time intervals. Adapted from [27].**

it can enhance the spontaneous emission rate and thereby the rate of photon generation. Second, it can deliver the emitted photons into a well-defined spatial mode to improve light harvesting efficiency and to facilitate further manipulations. Third, a cavity can squeeze the spectral range of the emission.

It is instructive to compare the fluctuations of light states with the average photon number 1 provided by different sources (Fig. 15.5). For thermal radiation, the photon number distribution takes the Bose form,

$$P(m) = \frac{\langle n \rangle^m}{(1 + \langle n \rangle)^{m+1}}, \tag{15.7}$$

where $\langle n \rangle$ is the average number of photons in the mode. For thermal radiation the state with zero photons ($m = 0$) always has the largest probability of occupation. This distribution is far from that of a desired source of single photons, which should feature a sharp maximum at $m = 1$.

The number of photons in a coherent state is a variable that fluctuates according to a *Poisson distribution*. The probability of finding *m* photons in the mode is given by,

$$P(m) = \frac{\langle n \rangle^m}{m!} \mathrm{e}^{-\langle n \rangle}, \tag{15.8}$$

and the variance is $\Delta n^2 = \langle n^2 \rangle - \langle n \rangle^2 = \langle n \rangle$. These statistics are very different from those of thermal light. The maximum probability is to find $\langle n \rangle$ photons in the mode. The resulting noise is called *shot noise* (or photon noise) and is an absolute minimum for the noise of a macroscopic laser. The probability distribution is still rather different from the desired one.

Narrower distribution can be achieved by means of the so-called *squeezed states* of light. In accordance with the complementarity principle, fluctuations of conjugate variables meet Heisenberg relations. Principally, the states become possible where fluctuations in one of the variables can be reduced (as compared with those of a coherent state) at the expense of increased fluctuations of the conjugate variable. This is referred to as a squeezed state. A squeezed state with reduced amplitude fluctuations (i.e. photon number fluctuations) will therefore exhibit enhanced phase noise. An ideal amplitude-squeezed light source would deliver a regular stream of photons at regular time periods. The fluctuations in the numbers of photons emitted by a squeezed source are weaker than those of a coherent state. They are sub-Poissonian, and their deviation from Poisson statistics can be characterized by the time-dependent Mandel parameter,

$$M(\Delta t) = \frac{\langle n^2 \rangle_{\Delta t} - \langle n \rangle_{\Delta t}^2}{\langle n \rangle_{\Delta t}} - 1, \tag{15.9}$$

where the notation $\langle \ldots \rangle_{\Delta t}$ means 'averaged over a time interval Δt'. The Mandel parameter of a Poisson source is zero and for an ideal single-photon source it equals -1.

Semiconductor nanocrystals, discussed in detail in Chapter 5, are promising candidates for single-photon light sources. They are easy to manipulate and to couple to efficient collecting optics in a room-temperature microscope and have better stability than single organic chromophores. However, blinking limits their practical applications as sources of single photons [28]. Promising results in terms of high-quality second-order correlation function have already been reported for a single semiconductor quantum dot in a pillar microcavity [29]. Single photon sources are important for applications in quantum computing, quantum cryptography, ultimate experiments in quantum science testing the basic and the questionable notions and challenging experiments like quantum teleportation.

Problems

1. Try to explain why oscillations for an "emitter+cavity" system inherent in the strong coupling regime acquired the notation *vacuum* Rabi oscillations.

2. Calculate the vacuum field in a cavity in Figure 15.2 using Eq. (15.5) and the cavity parameters, $\varepsilon = 13$, $V = 0.04\ \mu\text{m}^3$.

References

[1] L. Allen and J. H. Eberly. *Optical Resonance and Two-Level Atoms* (New York: Wiley, 1975).

[2] P. A. Apanasevich. *Basics of the Light–Matter Interaction Theory* (Minsk: Nauka i Tekhnika, 1977) – in Russian.

[3] P. R. Berman (Ed.) *Cavity Quantum Electrodynamics* (New York: Academic, 1994).

[4] E. T. Jaynes and F. W. Cummings. Comparison of quantum and semiclassical radiation theories with application to the beam maser. *Proc. IEEE*, **51** (1963), 89–96.

[5] I. D. Feranchuk, L. I. Komarov and A. P. Ulyanenkov. Two-level system in a one-mode quantum field: numerical solution on the basis of the operator method. *J. Phys. A*, **29** (1996), 4035–4040.

[6] R. J. Thompson, G. Rempe and H. J. Kimble. Observation of normal-mode splitting for an atom in an optical cavity. *Phys. Rev. Lett.*, **68** (1992), 1132–1135.

[7] H. J. Metcalf and P. van der Straten. *Laser Cooling and Trapping* (Berlin: Springer-Verlag, 1999).

[8] T. Yoshie, A. Sherer, J. Hendrikson, G. Khitrova, H. M. Gibbs, G. Rupper, C. Ell, O. B. Schekin and D. G. Deppe. Vacuum Rabi splitting with a single quantum dot in a photonic crystal nanocavity. *Nature*, **432** (2004), 200–203.

[9] J. P. Reithmaier, G. Sęk, A. Löffler, C. Hofmann, S. Kuhn, S. Reitzenstein, L. V. Keldysh, V. D. Kulakovskii, T. L. Reinecke and A. Forchel. Strong coupling in a single quantum dot–semiconductor microcavity system. *Nature,* **432** (2004), 197–200.

[10] Y. Yamamoto, F. Tassone and H. Cao. *Semiconductor Cavity Quantum Electrodynamics* (Berlin: Springer, 2000).

[11] P. Borri, W. Langbein, S. Schneider, U. Woggon, R. L. Sellin, D. Ouyang and D. Bimberg. Rabi oscillations in the excitonic ground-state transition of InGaAs quantum dots. *Phys. Rev. B*, **66** (2002), 081306(R).

[12] G. Ya. Slepyan, A. Magyarov, S. A. Maksimenko, A. Hoffmann and D. Bimberg. Rabi oscillations in a semiconductor quantum dot: Influence of local fields. *Phys. Rev. B*, **70** (2004), 045320.

[13] D. S. Mogilevtsev, A. P. Nisovtsev, S. Ya. Kilin, S. B. Cavalcanti, H. S. Brandi and L. E. Oliveira. Driving-dependent damping of Rabi oscillations in two-level semiconductor systems. *Phys. Rev. Lett.*, **100** (2008), 017401.

[14] J. P. Gordon, H. J. Zeiger and C. H. Townes. Molecular microwave oscillator and new hyperfine structure in the microwave spectrum of NH_3. *Phys. Rev.*, **95** (1954), 282–284.

[15] T. Maiman. Stimulated optical radiation in ruby. *Nature*, **187** (1960), 493–496.

[16] D. Meschede, H. Walther and G. Müller. One-atom maser. *Phys. Rev.*, **54** (1985), 551–554.

[17] Y. Mu and C. M. Savage. One-atom lasers. *Phys. Rev. A*, **46** (1992), 5944–5954.

[18] S. Ya. Kilin and T. B. Karlovich. Single-atom laser: Coherent and nonclassical effects in the regime of a strong atom-field correlation. *JETP*, **95** (2002), 805–819.

[19] S. M. Dutra. *Cavity Quantum Electrodynamics: The strange theory of light in a box* (New York: Wiley & Sons, 2005).

[20] J. McKeever, A. Boca, A. D. Boozer, J. R. Buck and H. J. Kimble. Experimental realization of a one-atom laser in the regime of strong coupling. *Nature*, **425** (2003), 268–271.

[21] V. P. Bykov. Spontaneous emission in a periodic structure. *Sov. Phys. JETP.*, **35** (1972), 269–273.

[22] S. Ya. Kilin and D. S. Mogilevtsev. Freezing of decay of quantum system with a dip in the spectrum of the heat bath–coupling constants. *Laser Physics*, **2** (1992), 153–161.

[23] D. S. Mogilevtsev and S. Ya. Kilin. The method of atomic–field collective operators in problems of interaction of atoms with a complex–structure field reservoir. *Opt. Spectr.*, **93** (2002), 405–412.

[24] D. S. Mogilevtsev and S. Ya. Kilin. *Quantum Optics Methods of Structured Reservoirs* (Minsk: Belorusskaya Nauka, 2007) – in Russian.

[25] D. Mogilevtsev, F. Moreira, S. B. Cavalcanti and S. Kilin. The collective operator method for realistic photonic crystals. *Laser Phys. Lett.*, **3** (2006), 327–344.

[26] D. Mogilevtsev, F. Moreira, S. B. Cavalcanti and S. Kilin. Field–emitter bound states in structured thermal reservoirs. *Phys. Rev. A*, **75** (2007), 043802.

[27] B. Lounis and M. Orrit. Single-photon sources. *Rep. Progr. Physics*, **68** (2005), 1129–1179.

[28] I. S. Osad'ko. Blinking fluorescence of single molecules and semiconductor nanocrystals. *Physics – Uspekhi*, **49** (2006), 27–43.

[29] C. Santori, D. Fattal, J. Vučković, G. Solomon and Y. Yamamoto. Indistinguishable photons from a single-photon device. *Nature*, **419** (2002), 594–597.

16 Plasmonic enhancement of secondary radiation

Nanostructures with characteristic surface relief of the order of 10–100 nm are known to modify the spatial distribution of an incident electromagnetic field. Local field enhancement results in enhanced absorption of photons by molecules or nanocrystals adsorbed at the surface. The effect is extremely pronounced in metal–dielectric structures because of surface plasmon resonance. A systematic application of field enhancement in Raman scattering enhancement and in photoluminescence enhancement with respect to molecular probes is followed nowadays by application of the effect with respect to nanocrystals (quantum dots) adsorbed at metal–dielectric nanotextured surfaces. It is the purpose of the present chapter to review mechanisms of photoluminescence enhancement and Raman scattering enhancement and factors in the context of their application to enhanced luminescence of molecules and quantum dots and Raman scattering. We consider not only local field enhancement in terms of the excitation process but also the photon density of states enhancement effect on photon emission processes with Raman scattering as a specific photon emission process. In this consideration, scattering of light experiences enhancement as does spontaneous emission. Therefore field enhancement and density of states effects should manifest themselves in the same manner in photoluminescence and scattering processes. Differences in scattering and luminescence enhancement are due to quenching processes which are crucial for luminescence and less pronounced for scattering. We consider ultimate experiments on single molecule detection by means of enhanced Raman scattering and photoluminescence enhancement of atoms, molecules and quantum dots, and the approaches to efficient substrates fabrication for the purposes of ultrasensitive spectroscopy. It is advisable to read Chapter 11 on metal–dielectric nanostructures and Chapters 13 and 14 for an introduction to quantum electrodynamics of light–matter interaction and photon density of states effects.

16.1 Classification of secondary radiation

When light flux is traversing a portion of space where at least a single quantum system exists that can potentially interact with it, the three distinctive types of secondary radiation produced by that interaction can be identified (Fig. 16.1). These are elastic scattering, inelastic scattering and spontaneous emission. In terms of photons, these events can be treated as follows. Photon scattering proceeds via virtual excitation of a quantum system under consideration with immediate emission of a new photon, either with the same (elastic scattering) or with a different (inelastic scattering) frequency. Photon spontaneous emission occurs via real excitation of a quantum system and subsequent emission of a photon in the

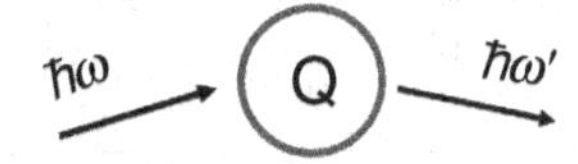

- <u>Elastic scattering (Rayleigh)</u>
  Excitation (virtual) $\hbar\omega \rightarrow$ emission $\hbar\omega$
- <u>Inelastic scattering (Raman)</u>
  Excitation (virtual) $\hbar\omega \rightarrow$ emission $\hbar\omega'$
- <u>Spontaneous emission</u>
  Excitation (real) $\hbar\omega \rightarrow$ emission $\hbar\omega'$

Fig. 16.1 **Three types of secondary radiation produced by light interaction with a quantum system.**

course of a spontaneous decay of the excited state. The frequency of the emitted photon, ω', can either be the same (resonant luminescence) or different (Stokes luminescence when $\omega' < \omega$ which is the most typical case, and anti-Stokes luminescence when $\omega' > \omega$). When molecular luminescence is considered, secondary emission is traditionally treated as *fluorescence* for allowed singlet–singlet transitions (typical lifetimes in the nanosecond range) and *phosphorescence* for forbidden triplet–singlet transitions (typical lifetimes are in the millisecond range). The remaining portion of light flux which does not experience interaction with the quantum system is called transmitted radiation.

Transmitted radiation, scattered radiation and spontaneously emitted radiation are the three types of secondary radiation resulting in light traversal of a space where possible interaction with matter exists. This simple picture holds until the weak light–matter coupling is applicable. Otherwise, light–matter states develop as is the case for microcavities and photonic crystals. In addition to these light–matter states, yet another example of a light–matter strong coupling regime is well known in the optics of continuous media. This is *polariton* formation in the case of strong exciton–photon coupling in semiconductor and dielectric crystals. In all cases of strong light–matter radiation distinctive classification of secondary radiation becomes questionable. For example, for polaritons in a crystal, for coupled light–matter states in a microcavity, or in a photonic crystal a noticeable slowing down of light propagation can be misinterpreted as resonant luminescence. Light–matter states in microcavities and photonic crystals in the strong coupling regime have been the subject of Chapter 15. Polaritons in crystals are considered in detail in many textbooks on crystal optics [1]. Note, sometimes light–matter states in microcavities and photonic crystals are referred to as localized polaritons.

16.2 How emission and scattering of light can be enhanced

All three types of secondary radiation shown in Figure 16.1 can be described by a single formula coupling outcoming radiation intensity $I(\omega')$ with incoming radiation intensity $I_0(\omega)$ and density of states at the final frequency $D(\omega')$ via the proper interaction term. The outcoming intensity is a simple product of the two above mentioned quantities and the term

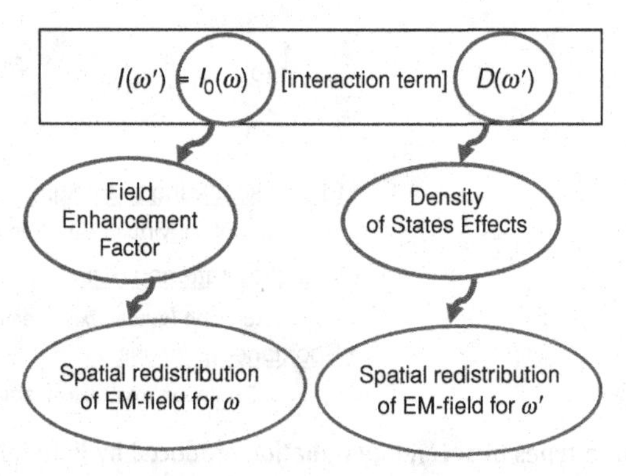

Fig. 16.2 **A general formula for secondary radiation in the weak light–matter coupling regime with highlighted ways towards its enhancement.**

describing the specific light–matter interaction event under consideration (Fig. 16.2). The interaction term includes both absorptive and emitting events in the case of spontaneous emission of light, whereas in the case of spontaneous scattering of light it describes only the event of photon scattering.

The simple and general formula shown in Figure 16.2 has been discussed in detail in Chapter 14 for spontaneous emission and spontaneous scattering of photons. It basically dates back to the very dawn of quantum electrodynamics manifested by the pioneering paper by Dirac [2]. There are two ways to enhance the emission of secondary radiation by a quantum system. *The first way* is to find a way to enhance the incident photon density at the position of the emitter in question. A straightforward pump-up by means of an incident light source power supply is not meant here. Instead, incident light concentration and accumulation at certain hot points within the metal–dielectric nanostructures, inside a microcavity or within a photonic crystal defect should be exploited purposefully. The proper enhancement factor is referred to as the *local field enhancement factor*. It has been described in detail in Section 11.1 where seven to eight orders of intensity enhancement have been demonstrated in model metal–dielectric nanostructures for frequencies near surface plasmon resonance, with an instructive explanation in terms of local high-Q areas formation. Therefore in Figure 16.2 the incident field enhancement is indicated as spatial redistribution of the electromagnetic field for the frequency ω.

The second enhancement factor is to make use of the possible density of photon states concentration for the frequency ω'. This can be performed by development of local areas in space with high Q-values for ω'. Local density of states can only be redistributed over a frequency range with concentration in certain narrow intervals at the expense of depletion otherwise. This universal sum rule was discussed in Sections 14.6 and 14.7.

To summarize, one can see that there are two complementary ways to enhance responsivity of a quantum system to incident light. Both ways are related to redistribution of electromagnetic radiation at ω and ω' frequencies. The difference is in *actual* incident field concentration in the case of the local-field enhancement factor and in the *vacuum* field concentration (or imaginary *probe* field concentration) as a measure of local density of state enhancement factor. Notably, both ways are the subject of basic restrictions. "One

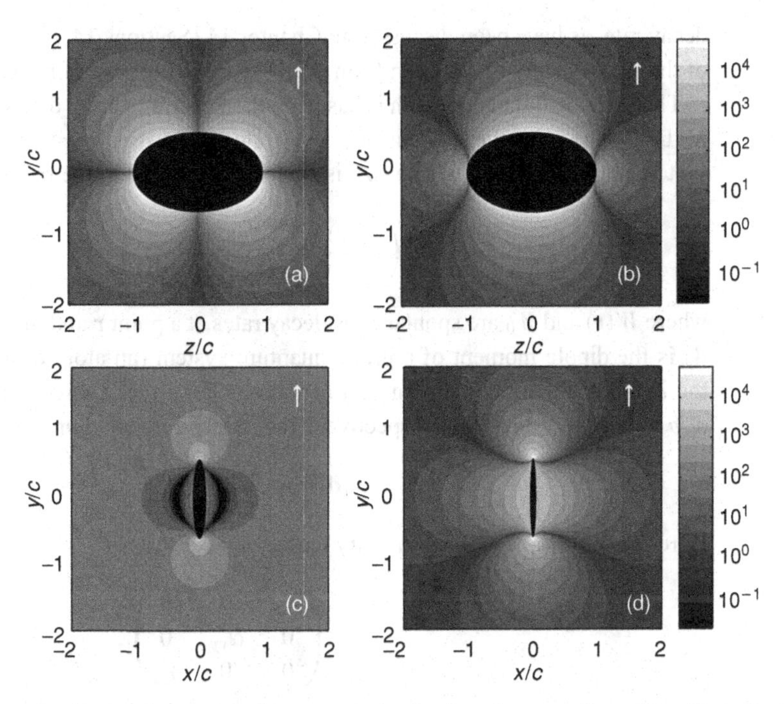

Fig. 16.3 Mappings of radiative decay rate of an atomic dipole placed near a silver ellipsoid. Radiative decay rate is normalized with respect to radiative rate in a vacuum. An ellipsoid has dimensions a, b, c in x, y, z directions, respectively. Dipole momentum orientation is along the y-axis (shown by the arrow). Silver dielectric function is $\varepsilon = -15.37 + i0.231$, emission wavelength is 632.8 nm. Panels (a) and (b) represent data for dipole position in the $x = 0$ plane for the two different ellipsoid dimensions: (a) $b/c = 0.6$, $a/c = 0.105$; (b) $b/c = 0.6$, $a/c = 0.046$. Panels (c) and (d) represent data for dipole position in the $z = 0$ plane for the two different ellipsoid dimensions: (c) $b/c = 0.6$, $a/c = 0.105$; (d) $b/c = 0.6$, $a/c = 0.046$. Reprinted with permission from [3]. Copyright 2007, Elsevier B.V.

photon in – one photon out" is the general rule of thumb based on energy conservation. Therefore the notion of "enhancement" used in this chapter must never be misinterpreted as amplification or optical gain.

16.3 Local density of states in plasmonic nanostructures

Nanometer sized metal particles capable of supporting surface plasmon resonance represent strong singularities with respect to propagation of electromagnetic waves. In Chapter 11 (Section 11.1) we have seen that the electric field amplitude can experience local enhancement by more than three orders of magnitude (see Fig. 11.3), the same is valid for the near-field scattering cross-section (Fig. 11.4). Local density of photon states does experience strong modification accordingly. It happens in the spectral range around the surface plasmon resonance. Figure 16.3 represents the calculation of radiative decay rate reported by D. V. Guzatov and V. V. Klimov [3] for a probe dipole near a silver ellipsoid. Radiative decay rate, as have been discussed in Chapter 14 (Sections 14.7, 14.8), is a direct measure of the local density of states and can even serve the purpose of its definition.

The computational approach is based on the generic relation derived by V. V. Klimov and M. Ducloy between the spontaneous decay rate and the dipole moments of a probe atom-like system and a nanobody whose size is small as compared to the radiation wavelength [4],

$$\left(\frac{W(\mathbf{r})}{W_0}\right)^{\text{radiative}} = \frac{|\mathbf{d}_0 + \delta\mathbf{d}|^2}{|\mathbf{d}_0|^2}, \tag{16.1}$$

where $W(\mathbf{r})$ and W_0 are spontaneous decay rates at a point $\mathbf{r}$ and in a vacuum, respectively, $\mathbf{d}_0$ is the dipole moment of a probe quantum system (an atom or a molecule), and $\delta\mathbf{d}$ is the induced dipole momentum of a nanobody. For a metal three-axial ellipsoid with axes a, b, c in x, y, z directions respectively, the induced dipole moment reads [3],

$$\delta\mathbf{d} = \hat{\alpha} \cdot \nabla(\mathbf{d}_0 \cdot \nabla)J. \tag{16.2}$$

Here $\hat{\alpha}$ is the ellipsoid polarizability tensor,

$$\hat{\alpha} = \begin{pmatrix} \alpha_{xx} & 0 & 0 \\ 0 & \alpha_{yy} & 0 \\ 0 & 0 & \alpha_{zz} \end{pmatrix}, \tag{16.3}$$

with the component defined by the relations,

$$\begin{aligned} \alpha_{xx} &= \frac{1}{4\pi}(1 - \varepsilon_{xx})\left(\frac{\varepsilon - 1}{\varepsilon - \varepsilon_{xx}}\right)V, \quad \varepsilon_{xx} = 1 - 2(abcI_a), \\ I_a &= \int_0^\infty \frac{du}{(a^2 + u)R(u)}, \quad R(u) = [(a^2 + u)(b^2 + u)(c^2 + u)]^{1/2}. \end{aligned} \tag{16.4}$$

Components α_{yy} and α_{zz} can be obtained by cyclic replacement of all parameters and indices (a, b, c and x, y, z). J is defined as,

$$\begin{aligned} J &= \frac{3}{4}[I(\xi) - x^2 I_a(\xi) - y^2 I_b(\xi) - z^2 I_c(\xi)], \\ I(\xi) &= \int_\xi^\infty \frac{du}{R(u)}, \quad I_a(\xi) = \int_\xi^\infty \frac{du}{(a^2 + u)R(u)}. \end{aligned} \tag{16.5}$$

One can see from Figure 16.3, there are local areas in the near vicinity to an ellipsoid where decay rate can by enhanced by the factor of more than 10^4, the effect being dependent on an ellipsoid aspect ratio. For a higher aspect ratio (elongated ellipsoids) the enhancement rises further.

For spherical particles, enhancement is much less pronounced. Its spectral dependence for the two different orientations of dipole momentum is presented in Figure 16.4 for gold and silver nanoparticles as calculated by D. V. Guzatov. Note that silver appears to be considerably more efficient in radiative rate enhancement as compared to gold. A gold particle for tangential dipole orientation does not offer enhancement at all. The correlation is remarkable between the spectral dependence of radiative decay rate

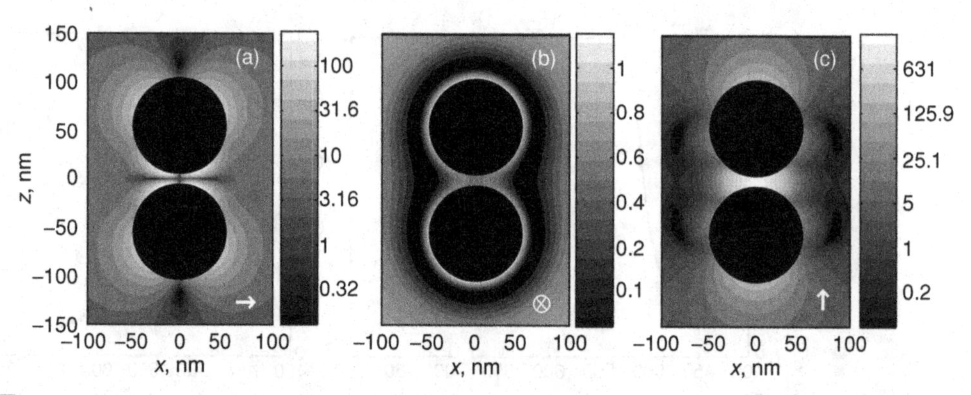

Fig. 16.5 **Mapping of spontaneous decay rates of an atomic dipole placed near a pair of silver spherical nanoparticles [6]. Radiative decay rate is normalized with respect to radiative rate in a vacuum. Nanoparticle diameter is 100 nm. Dipole momentum orientation is shown by the arrows and corresponds to (a) x, (b) y, and (c) z axes.**

field distribution in metal–dielectric nanostructures (Section 11.1) and local density of states modifications shown in Figures 16.3–16.5, we arrive at the notion of so-called "hot spots" in plasmonic nanostructures. These are such places on a nanotextured metal surface, or near metal nanobodies, where simultaneous spatial redistribution of an electromagnetic field occurs both at the frequency of the incident radiation ω and at the frequency of scattered radiation ω'. The first effect is the so-called field enhancement factor, whereas the second is local density of state enhancement. Enhancement of photon local density of states (LDOS) starting from the pioneering paper by E. M. Purcell [7] can be interpreted as development of a certain Q-factor in the space region where a test emitter (atom or other quantum system) is placed. Since the Q-factor implies the possibility of a system to accumulate energy (Q value equals the ratio of energy accumulated in the system to the portion of energy the system loses in a single oscillation period), formation of high local density of state areas in many instances can be treated as development of multiple microcavities at the frequency ω' over a nanotextured metal surface. Experimentally, mapping of the surface distribution of high LDOS areas can be performed by means of scanning near-field microscopy, as has been discussed in Section 14.8. However, to the best of the author's knowledge it has never been applied to plasmonic structures used in the surface enhancement of photoluminescence and Raman scattering.

Simple nanostructures exhibiting plasmonic enhancement of spontaneous emission and Raman scattering of light were shown in Figure 6.8. These are isolated and aggregated nanoparticles of gold and silver. Further examples of plasmonic nanostructures are shown in Figure 16.6. The left panel shows a typical nanotextured thin film of silver on a dielectric substrate fabricated by means of vacuum deposition and annealing. The right panel shows an example of a regular nanostructure developed by means of gold deposition on top of close-packed silica balls. Noble metals like Ag, Au, Pt and Ni are used. Although other metals can also be exploited, their strong oxidation in air prevents desirable proximity of a probe to the metal surface.

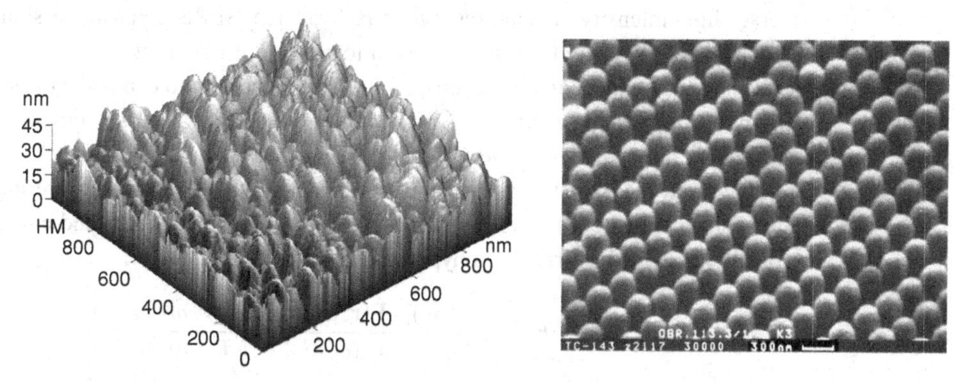

Fig. 16.6 **Representative metal–dielectric structures exhibiting enhancement of Raman scattering and photoluminescence. Left: Silver nanotextured film on a dielectric substrate. Note different vertical scale as compared to in-plane scale. Imaging by means of atomic force microscopy. Courtesy of N. D. Strekal. Right: Regular nanotextured gold surface formed by means of gold deposition over close-packed silica dielectric globules, globule diameter is about 250 nm. Imaging by means of scanning electron microscopy.**

"Hot spots" promise enormous enhancement of the secondary radiation emitted by molecules or other species located in the proper positions within plasmonic nanostructures. When multiplying incident field enhancement (see Section 11.1 for values) and LDOS enhancement discussed in the previous section, one can achieve a product of the order of 10^{10}. Such values can never be obtained for photoluminescence enhancement but these can even be exceeded in surface-enhanced Raman experiments. Photoluminescence and Raman signal enhancements will be discussed in more detail in Sections 16.5 and 16.6, respectively.

Notably, enhancement of secondary radiation occurs within the framework of the linear light–matter interaction which is unambiguously implied by the linear input–output relation, as shown in Figure 16.2. Therefore, local-field enhancement for incident light can not be interpreted as surface redistribution of incident light, i.e. as a kind of local light "microfocusing", as commonly anticipated by many authors. Since surface-enhanced Raman scattering and surface-enhanced luminescence are considered within the framework of linear light–matter interaction (contrary to e.g. surface enhanced second harmonic generation) the total signal harvesting from a piece of area containing statistically large numbers of molecules will be same independently of surface redistribution of light intensity, because total incident light intensity integrated over the piece of area remains the same. Within the framework of linear light–matter interaction, strong signal enhancement for secondary radiation by means of incident field enhancement can only be understood in terms of high local Q-factors for incident light, i.e. in terms of light *accumulation near the surface* rather than light *redistribution over the surface*. A Q-fold rise in light intensity then occurs near hot points as it happens in microcavities and Fabry–Perot interferometers. However, accumulation of light energy needs a certain time. Therefore huge signals can develop only after a certain time which is necessary for transient processes to finish, resulting in a steady increase of incident light intensity near hot points as compared to the average light intensity in incoming light flux. Transient SERS experiments should therefore be performed to clarify Q-factor effects in formation of hot points.

We now arrive at a consistent consideration of hot spots and can make the statement that *hot spots are local areas in plasmonic nanostructures where high Q-factors develop both for incident light frequency and for emitted (or scattered) light frequency.*

Many authors ignore the role of density of states effects on Raman scattering. To explain huge enhancement exceeding 10^{10} times, enhancements of incident and emitted fields are typically discussed to arrive at a hypothetical enhancement factor,

$$F = \frac{|\mathbf{E}(\mathbf{r},\omega)|^2 |\mathbf{E}(\mathbf{r},\omega')|^2}{|\mathbf{E}_0(\omega)|^2 |\mathbf{E}_0(\omega')|^2} \approx \left| \frac{\mathbf{E}(\mathbf{r},\omega)}{\mathbf{E}_0(\omega)} \right|^4, \tag{16.6}$$

instead of the correct one,

$$F = \frac{|\mathbf{E}(\mathbf{r},\omega)|^2}{|\mathbf{E}_0(\omega)|^2} \frac{D(\mathbf{r},\omega')}{D_0(\omega')}. \tag{16.7}$$

In this connection it should be outlined that local DOS enhancement in a sense does account for concentration of the electromagnetic field at ω'. Therefore the DOS enhancement factor to a large extent meets the temptation to account for field modification for emitted light. However, local DOS enhancement unambiguously implies either a vacuum electromagnetic field or a non-existing field which were emitted by a classical dipole placed in that location. Concentration of a real emitted field, appreciated by many authors, does actually occur only in the close subwavelength-scale vicinity of a nanobody and can not contribute to light harvesting in typical far-field experiments on Raman scattering and luminescence enhancement. LDOS enhancement means the unmeasurable concentration of an electromagnetic vacuum field rather than emitted light concentration. The latter can actually contribute to Raman scattering enhancement but only in the form of *stimulated* Raman scattering *by another molecule located nearby.*

A huge incident field may in certain experimental situations result in a *non-linear* response of an atom or a molecule. A few milliwatts per mm^2 results in many megawatts per cm^2 in a hot point which is enough to observe nonlinearities. We have seen that a huge concentration of emitted field can result in stimulated emission and scattering. Neither stimulated secondary emission nor nonlinear responsivity of an atom or a molecule in plasmonic nanostructures has been introduced in the theoretical considerations when speaking about luminescence and scattering enhancement in nanoplasmonics. These remain as further factors in the theory and in experimental performance to be perceived for superior optical response of plasmonic structures.

16.5 Raman scattering enhancement in metal–dielectric nanostructures

In 1974 M. Fleischmann with co-workers reported on the enormous enhancement of Raman scattering for molecules adsorbed on a roughened silver electrode [8]. The authors ascribed the effect to a higher amount of adsorbed molecules owing to larger surface area. However, it was later shown that the increase in surface area is not as significant as the increase in the Raman signal [9, 10]. In a few years the phenomenon of enormously enhanced Raman scattering had been reproduced by many authors for various types of nanotextured noble metal surface [11]. Later on a similar effect was reproduced for organic molecules mixed with colloidal metal particles in a solution [12]. The general phenomenon acquired the notation "Surface Enhanced Raman Scattering" (SERS). Immediate theoretical explanations were proposed in terms of local incident field enhancement and the relevant electromagnetic theory of SERS has been developed [13,14]. Various SERS-active substrate designs, adsorption details and experimental advances result in macroscopically averaged enhancements of up to 10^6, reproduced by many groups. In many experiments, not only absolute values of scattered intensity increase, but the shape of the Raman spectrum exhibits differences from that obtained for the same molecules in solution. In other words, different Raman lines experience different enhancement factors. Spectral modifications of a Raman signal are usually attributed to chemical mechanisms, i.e. bonding interaction between an adsorbed molecule and surface metal atoms and charge transfer processes between a molecule and metal surface. The possible role of chemical factors can severely diminish the analytical prospective of SERS as an ultimate technique in molecular spectroscopy.

In 1997 a breakthrough in SERS sensitivity was announced by two groups independently reporting on single molecule detection by means of SERS [15, 16]. These experiments provided evidence that local enhancement factors of scattering rate can reach values up to 10^{14}–10^{15} times. This is significantly greater than the previously observed ensemble-averaged enhancement factors of the order of 10^6–10^8. These findings did stimulate further experimental studies and discussions aimed at establishing the mechanisms of huge local enhancement factors in SERS [17]. Purely electromagnetic theory based on incident field enhancement can not explain single molecule Raman detection experiments. In 2002, a density of states contribution was proposed in addition to standard electromagnetic SERS theory [18]. For the particular case of a metal tip with radius R for the spherical edge, the Raman scattering enhancement factor arising from the density of states contribution was found to be approximately $(\lambda/R)^4$ for radiation wavelength λ [19].

The first consistent calculations of Raman scattering enhancement factors including both incident field enhancement and local DOS enhancement are shown in Figure 16.7 for a probe dipole located near a spherical and a spheroidal silver nanoparticle. The excitation light wavelength was chosen to obtain the highest incident field enhancement factor.

One can see, for a spheroidal particle it can exceed the value of 10^{10} when both dipole and incident field are oriented along the longer spheroid axis (Fig. 16.7(a)), whereas in the most unfavorable situation of a spherical particle and orthogonal dipole and field orientations it is no more than 30 (Fig. 16.7(b)). In Figure 16.7(c) and (d) enhancement factors are calculated for the same parameters of the problem but with local density of states involved in the calculations. One can see, LDOS enhancement offers at least 10^3-fold enhancement on average. Total enhancement at the edge of a spheroidal particle can exceed 10^{14} which is enough to obtain the single molecule detection regime in Raman spectroscopy. The relative contribution from the two enhancement factors can differ for different original wavelength used for molecule excitation. Probably at a certain combination of plasmon

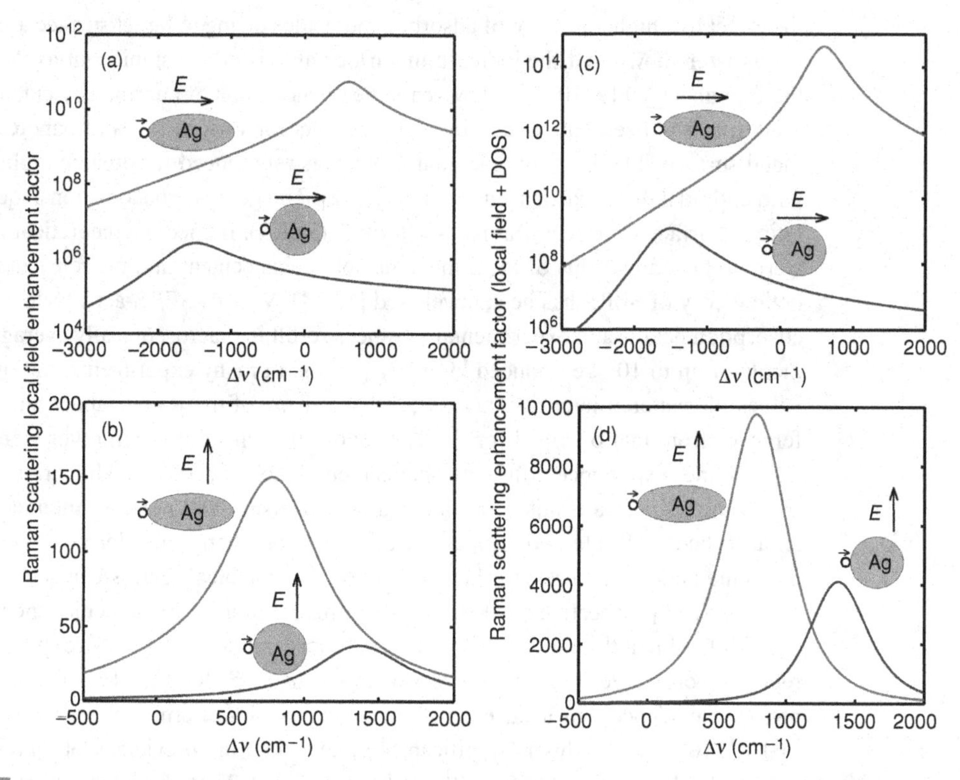

Fig. 16.7 **Calculated Raman scattering cross-section enhancement factors for a molecule with polarizability $\alpha = 10^3$Å attached to a spheroidal and spherical silver nanoparticle as a function of spectral shift $\Delta\nu = (\omega - \omega')/2\pi$ [20]. Molecule position, its dipole moment orientation and incident field orientation are indicated near every curve. Panels (a) and (b) show enhancement based on electromagnetic theory of SERS. Panels (c) and (d) show enhancement including both incident field and local DOS factors. Wavelength of incident light is (a), (c)** 375 nm **and (b), (d)** 338 nm.

resonance frequency, excitation frequency and detection frequency, contributions from incident field enhancement and DOS enhancement can become equal, or even the DOS contribution could dominate over the incident field enhancement. Recent experiments on thorough analysis of SERS excitation spectra have shown that the highest enhancement occurs when the excitation wavelength is in between the extinction maximum (defined by the surface plasmon resonance) and the Raman shifted wavelength [21]. This finding can be treated as in favor of the above model. In this case the spectral position between the extinction maximum and the Raman wavelength may correspond to the optimal trade-off between incident field concentration and local DOS redistribution. Further experiments towards understanding SERS mechanisms in detail should include well-defined geometrical nanostructures which are more feasible for modeling, precise control of molecule location, analysis of polarization features and angular distribution of the scattered radiation. Regular periodic-like structures were found to offer much higher enhancement as compared to random ones [22]. Certain evidence on inhomogeneous angular distribution of scattered light has been reported as well [23]. The observed dominating directions can be interpreted in favor of the angular-dependent partial density of states, which again makes a proposal on DOS enhancement of Raman signal plausible.

Noteworthy, spectral dependence of the SERS enhancement as a product of local incident field and local DOS factors presented in Figure 16.7(c) and (d) can in many instances explain different enhancement factors for different Raman lines, thus making chemical mechanisms not necessarily involved.

The above consideration is valid not only for Raman spectroscopy but for all versions of vibrational spectroscopies, e.g. it can be applied for single quantum dot vibrational spectroscopy. It is also valid for Mandelstam–Brillouin scattering as well as for Rayleigh scattering. The results are considered as a first step towards an extensive theory for single molecule Raman detection. Further steps should involve more complicated geometries and optimal displacement of excitation and scattering wavelength with respect to plasmon resonance. In particular, coupled metal nanoparticles are believed to promise a superior enhancement factor when a dipole is located between them and orientated along the line connecting the centers of the particles. L. Brus with co-workers observed superior Raman enhancement of single Rhodamine 6G molecules by silver nanoparticle junctions [24].

16.6 Luminescence enhancement in metal–dielectric nanostructures

Though spontaneous emission intensity, similar to Raman scattering, is proportional to the product of incident field and photon DOS, it does not mean that structures showing a large SERS signal will be at the same time efficient in photoluminescence enhancement if the probe SERS molecules are replaced by a fluorescent probe (molecules, ions, quantum dots). There is the principal difference between photon scattering and photon spontaneous emission events. Photon scattering is an instantaneous event whereas spontaneous emission of photons is characterized by a finite internal relaxation rate and a finite excited state lifetime. Proximity of a fluorescent probe to metal particles or metal surface promotes a rapid non-radiative relaxation path which in most cases predominates over radiative lifetime and manifests itself as strong luminescence quenching. Quenching overthrows enhancement in most experiments which is well known from SERS studies since the 1980s. Luminescence quenching has even been treated as a helpful phenomenon in elucidating Raman lines typically masked by much stronger luminescence bands.

Quenching near a metal nanobody occurs by means of rapid non-radiative resonance energy transfer from an excited probe system (an atom, a molecule, a quantum dot) to a metal nanoparticle or a metal roughened surface. This transfer is severely promoted by the strong optical absorption inherent in plasmonic structures. The familiar analog is resonant energy transfer between a pair of molecules or quantum dots with small distance and overlapping emission spectrum of excited (a donor) and absorption spectrum of non-excited (an acceptor) counterparts. For molecules, this phenomenon was considered for the first time by Th. Förster in 1946 [25]. This process is drastically distance

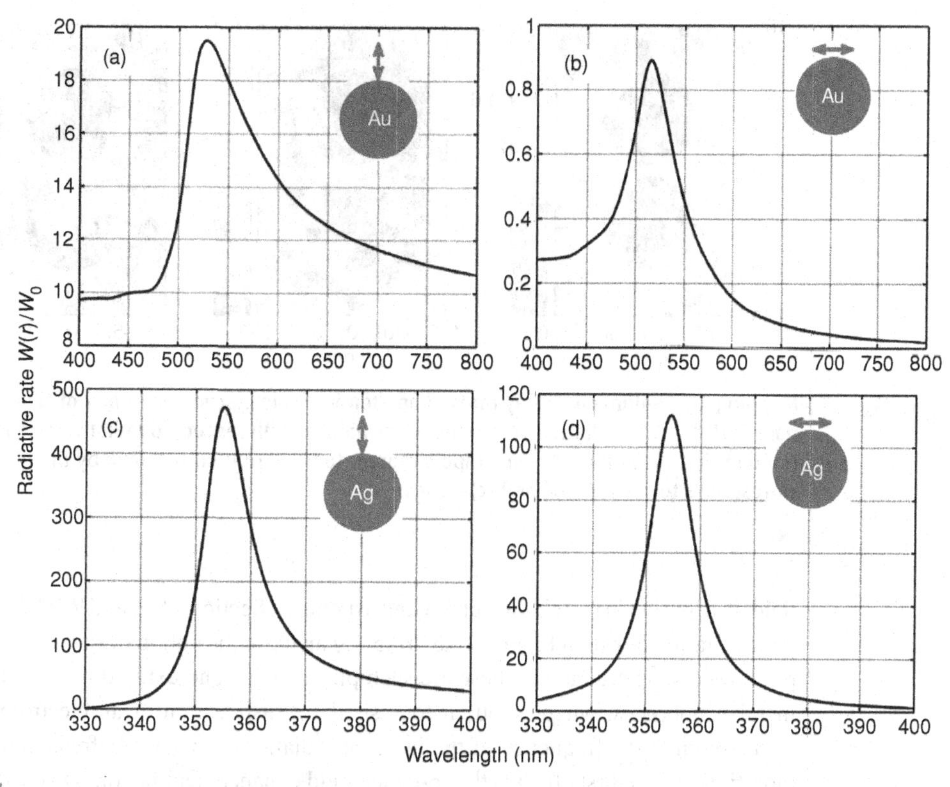

Fig. 16.4 **Radiative rate enhancement of a probe dipole near a spherical nanoparticle of (a) and (b) gold and (c) and (d) silver. Dipole orientation is shown by the arrows. Adapted from [5].**

and the optical response functions of gold and silver. This becomes evident when Figure 16.4(a) and (b) is compared with Figure 6.11 and Figure 16.4(c) and (d) is compared with Figure 6.3 and Figure 6.6(b).

A junction between two metal nanoparticles offers strong modification of radiative decay rates even for spherical nanoparticles. In this case maximal enhancement occurs between the particles and strongly depends on dipole orientation (Fig. 16.5). The most favorable is dipole orientation along the axis connecting the centers of the two particles (Fig. 16.5(c)). In this case more than 10^3-fold enhancement of radiative decay rate becomes feasible. The most unfavorable situation is a dipole orientation normal to the plane in which particles are located (Fig. 16.5(b)). This orientation gives rise to inhibition of spontaneous decay rather than its enhancement.

16.4 "Hot spots" in plasmonic nanostructures

Based on the above considerations of the secondary emission of radiation in terms of the schemes shown in Figures 16.1 and 16.2, and using the modeling results for electromagnetic

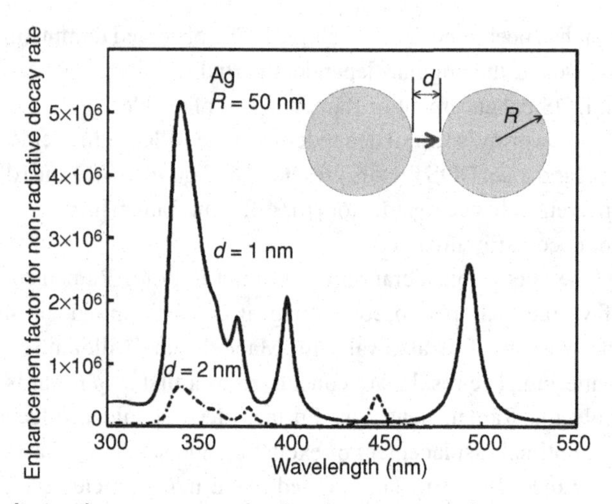

Fig. 16.8 **Calculated non-radiative decay rate normalized with respect to radiative decay rate in a vacuum for a probe dipole located between two spherical silver nanoparticles. The dipole is oriented along the axis connecting the particles centers. Particle radius is 50 nm. The results are shown for interparticle distance 1 nm (solid line) and 2 nm (dash-dot line). Adapted from [29].**

dependent as,

$$\frac{W^{\mathrm{nrad}}}{W_0} = \left(\frac{R_{\mathrm{F}}}{R}\right)^6, \tag{16.8}$$

where W^{nrad} is the non-radiative energy transfer rate, W_0 is the donor radiative decay rate in a vacuum, R is the donor–acceptor distance, and R_{F} is a characteristic distance referred to as the *Förster radius*. It typically measures a few nanometers. For this spacing non-radiative energy transfer rate from donor to acceptor becomes equal to the donor intrinsic radiative decay rate. Explicit expressions for R_{F} can be found, e.g. in the books [26, 27]. Resonant energy transfer in semiconductor quantum dots is thoroughly established as well. For an emitter near metal nanoparticles, non-radiative energy transfer has been considered for a single metal nanoparticle [28] and for a two-particle cluster [29]. For analytical expressions the reader is referred to these original works. Here we restrict ourselves to presentation of calculated nonradiative decay rate for a dipole near a pair of metal nanoparticles for location and orientation corresponding to the optimal radiative rate enhancement (see Fig. 16.5(c)). As can be seen from Figure 16.8, non-radiative decay exhibits strong dependence on wavelength as well as on interparticle distance. Comparing absolute values of rate enhancement for radiative (Fig. 16.5(c)) and non-radiative (Fig. 16.8) decays one can see that the latter can easily overtake the former.

In luminescence spectroscopy, to make use of enhancement factors one has to engineer the optimal topology of a probe–metal nanostructure to get positive balance of competing enhancement/quenching effects. In other words, the luminescent probe is to be displaced at a certain distance on metal nanobodies at a point where quenching is yet negligible but field and DOS enhancement still present. This is the case if probes are dispersed in a matrix with nanosize metal colloids at low concentration. Another case involves using a dense metal

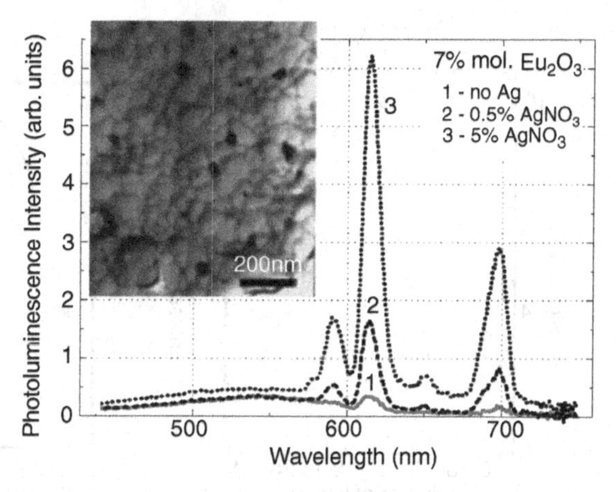

Fig. 16.9 **Photoluminescence spectra of Eu^{3+} ions in sol–gel films containing silver nanoparticles. Successive growth of Ag salt content gives rise to a considerable increase in photoluminescence intensity correlating with formation of silver nanoparticles. These are seen in the inset as dark spots. Adapted from [30].**

surface but adjusting a spacer between a probe and the surface. Such experiments have been performed by many groups for luminescent ions and molecules during past decades. Controllable spacing in this case is performed using multilayer Langmuir–Blodgett or polyelectrolyte films. A few groups have performed fine tip-enhanced spectroscopy where a sharp metal tip is scanned over a probe molecule to evaluate its effect on fluorescence. In what follows a few representative examples of these experiments will be provided.

Figure 16.9 shows photoluminescence data for sol–gel films in which Eu^{3+} ions are embedded and silver nanocolloids are developed by means of special heat treatment. The overall luminescence intensity increases by more than ten times in direct correlation with the presence of silver colloids. Positive balance of enhancement/quenching is due to average spacing between ions and colloids in the film.

Figure 16.10 presents photoluminescence enhancement of semiconductor nanocrystals and organic fluorophores. In this experiment nanotextured gold and silver surfaces were used, the spacing being provided by multiple polyelectrolyte layers. This techniques offers nearly 1 nm precision by sequential development of a layer-by-layer spacer. Optimal spacing was found to be approximately 12 nm. Semiconductor nanocrystals (so-called quantum dots) are novel luminescent species whose absorption and emission spectra as well as transition probabilities are essentially controlled by quantum confinement of electrons and holes (see Chapter 5). These factors offer tuneability of emission spectra simply by size variation. Simultaneously, superior stability of luminescent properties has been found which is more than 100 times higher than that of traditional organic luminophores like rhodamines or fluorescein. Quantum dots are therefore promising candidates for novel commercial luminophores, as well as for novel bioluminescent labels at the single molecule level.

For fluorescein-labeled protein (Fig. 16.10(b)) nine-fold enhancement has been observed. It is clearly seen that luminescence intensity is sensitive to the distance between the silver

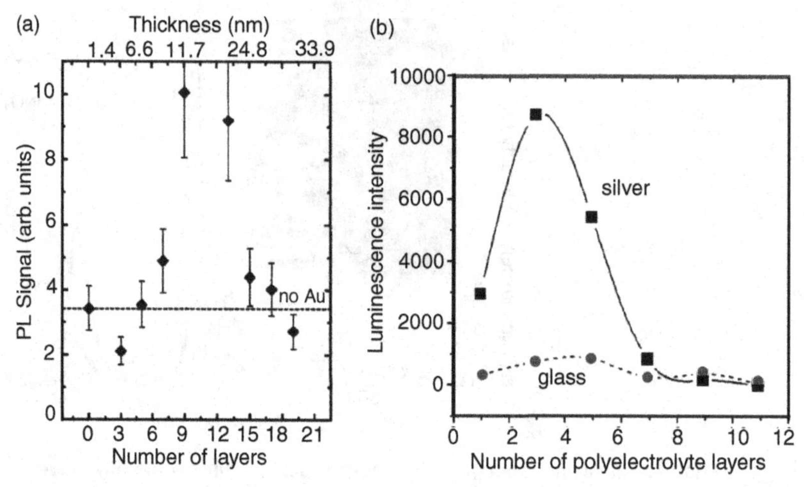

Fig. 16.10 **Photoluminescence intensity as a function of the number of polyelectrolyte layers forming a dielectric spacer over a metal surface. (a) Core–shell CdSe/ZnS semiconductor nanocrystals with mean diameter 4 nm on a nanotextured gold surface [31]; (b) Albumin-fluorescein isothiocyanate conjugate of bovine serum on a nanotextured silver surface [32].**

island film and the fluorophore and exhibits a maximum at around three polyelectrolyte layers, which corresponds to a spacer thickness of about 4.2 nm. For semiconductor core–shell quantum dots the optimal distance has been found to be about 10 nm for CdSe/ZnS dots adsorbed on a gold colloidal film. This discrepancy could be due to the different metal–fluorophore system, and in particular due to a different length scale of surface roughness and also due to the relatively large size of labeled protein molecules (which are oblate ellipsoids with dimensions of 140 nm × 4 nm) in comparison with the 4 nm size of nanocrystals. Progress in synthesis of semiconductor nanoparticles in various ambient environments, understanding of their optical properties combined with an idea of using quantum dots as efficient luminophores in light emitting devices and as fluorescent labels in high sensitivity biospectroscopy do stimulate extensive experiments on purposeful application of field enhancement and DOS effects for quantum dot based nanostructures.

Among extensive experimental research on plasmonic enhancement of photoluminescence a few selected experiments are worthy to be highlighted. Using specially annealed gold films on a dielectric substrate, N. Strekal with co-workers demonstrated selective enhancement of either Raman scattering or fluorescence (50-fold enhancement) depending on whether a spacer is absent, or present, between the adsorbate and the metal surface [33]. T. Ozel *et al.* demonstrated that in the case of two different emission bands, one of these bands can be selectively enhanced whereas another one shows partial inhibition [34]. These data were shown in Figure 5.45. Different effects of metallic environment on different emission bands makes the contribution from local DOS plausible, since enhancement through the common absorption channel remains the same for both bands. In the presence of luminescence quenching from a competing non-radiative decay path, lifetime measurements can not give evidence in favor of or contrary to the density of state enhancement factor. The question then arises whether it is possible or not to get an experimental hint

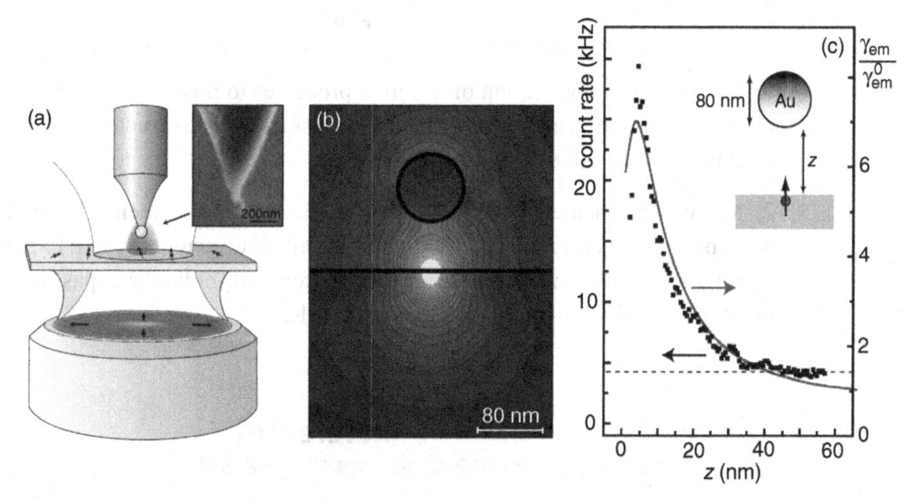

Fig. 16.11 **Distance-dependent spontaneous emission rate of a single molecule in the presence of a gold spherical nanoparticle [40]. (a) Sketch of the experimental arrangement. Inset: SEM image of a gold particle attached to the end of a pointed optical fiber. (b) Field distribution $|\mathbf{E}|^2$ (factor of 2 between successive contour lines) of an emitting dipole (wavelength is 650 nm) located 2 nm beneath the surface of a glass substrate and faced by a gold particle separated by a distance of $z = 60$ nm from the glass surface. (c) Fluorescence rate as a function of particle–surface distance z for a vertically oriented molecule (solid curve: theory, dots: experiment). The horizontal dashed line indicates the background level.**

on density of state contribution. In this context an emission indicatrix can be helpful since partial density of states in complex nanostructures can be strongly anisotropic. Several authors did actually observe pronounced directionality of spontaneous emission for luminescent probes attached to a nanotextured metal surface [35, 36] thus providing arguments in favor of density of states contribution to luminescence enhancement. Aslan *et al.* [37] proposed extension of the standard photoluminescence enhancement explanation in terms of non-radiative energy transfer to surface plasmons with subsequent radiative decay of plasmons into propagating electromagnetic modes. Should this be the case, non-radiative energy transfer from a fluorophore to metal will not inhibit the overall emission efficiency and photoluminescence enhancement will more readily occur.

Several groups have performed tip-enhanced spectroscopy using precise manipulation of a metal tip or nanoparticle by means of an atomic force microscopy cantilever [38, 39]. In this case spacing between a fluorophore and a metal nanobody is controlled by nanomanipulation with the metal nanobody, the fluorophore being immobilized in a matrix or on a surface. Using this approach Anger, Bharadwaj and Novotny compared distance-dependent emission rate from a molecule near a gold nanoparticle with that calculated based on modified excitation rate, radiative and non-radiative processes promoted by metal proximity [40]. Their results are summarized in Figure 16.11. Fine agreement of theoretical expectations and experimental observation may be clearly seen. The optimal distance for enhancement was found to be between 5 to 10 nm which also agrees with the data previously reported for ensemble-averaged enhancement (see, e.g. Fig. 16.10).

Problems

1. Consider the contribution of different processes to possible Raman scattering enhancement and luminescence enhancement in plasmonic nanostructures and explain why the Raman signal is much more readily enhanced.

2. Many experimenters have found that plasmonic enhancement of photoluminescence is more pronounced for molecules with intrinsically low quantum yield. Explain this observation. Hint: consider the relative contribution from non-radiative bypass to the overall decay rate for originally high and low quantum yields.

References

[1] C. Klingshirn. *Semiconductor Optics* (Berlin: Springer, 1995).

[2] P. A. M. Dirac. The quantum theory of the emission and absorption of radiation. *Proc. Royal Soc. London, Ser. A*, **114** (1927), 243–265.

[3] D. V. Guzatov and V. V. Klimov. Radiative decay engineering by triaxial nanoellipsoids. *Chem. Phys. Lett.*, **412** (2005), 341–346.

[4] V. V. Klimov and M. Ducloy. Spontaneous emission rate of an excited atom placed near a nanofiber. *Phys. Rev. A*, **69** (2004), 013812.

[5] D. V. Guzatov. *Calculation of Light Scattering by a System "Atom (molecule) + nanosphere" with Field Localization and Field Enhancement Effects.* Technical report. Minsk: Stepanov Institute of Physics, 2008 (unpublished).

[6] D. V. Guzatov. *Spontaneous Radiation of Atoms and Molecules Near Nano-objects With Complex Configurations*. Ph.D. thesis, P. I. Lebedev Physics Institute Moscow (2007). (in Russian).

[7] E. M. Purcell. Spontaneous emission probabilities at radiofrequencies. *Phys. Rev.*, **69** (1946), 681.

[8] M. Fleischmann, P. J. Hendra and A. J. McQuilaan. Raman spectra of pyridine adsorbed at a silver electrode. *Chem. Phys. Lett.*, **26** (1974), 163–166.

[9] D. L. Jeanmaire and R. P. van Duyne. Surface Raman spectroelectrochemistry. *J. Electroanal. Chem.*, **84** (1977), 1–20.

[10] M. G. Albrecht and J. A. Creighton. Anomalously intense Raman spectra of pyridine at a silver electrode. *J. Amer. Chem. Soc.*, **99** (1977), 5215–5217.

[11] R. K. Chang and T. E. Furtak (Eds.). *Surface Enhanced Raman Scattering* (New York: Plenum Press, 1982).

[12] A. V. Baranov and Ya. S. Bobovich. Giant combinational scattering as structural-analytical method in material research. *Optics and Spectroscopy*, **52** (1982), 385–387.

[13] T. Gersten and A. Nitzan. Spectroscopic properties of molecules interacting with small dielectric particles. *J. Chem. Phys.*, **75** (1981), 1139–1152.

[14] J. Gersten and A. Nitzan, in: Surface Enhanced Raman Scattering, edited by R. K. Chang and T. E. Furtak (New York: Plenum, 1982), 89–110.

[15] K. Kneipp, Y. Wang, H. Kneipp, L. T. Perelman, I. Itzkan, R. R. Dasari and M. S. Feld. Single molecule detection using Surface-Enhanced Raman Scattering (SERS). *Phys. Rev. Lett.*, **78** (1997), 1667–1670.

[16] S. Nie and S. R. Emory. Probing single molecules and single nanoparticles by surface-enhanced Raman scattering. *Science*, **275** (1997), 1102–1105.

[17] K. Kneipp, M. Moskovits and H. Kneipp (Eds.). *Surface-Enhanced Raman Scattering* (Berlin: Springer-Verlag, 2006).

[18] S. V. Gaponenko. Effects of photon density of states on Raman scattering in mesoscopic structures. *Phys. Rev. B*, **65** (2002), 140303(R).

[19] V. S. Zuev and A. V. Frantsesson. Possible interpretation of the Raman scattering enhancement near a nano-needle. *Optics and Spectroscopy*, **93** (2002), 117–127.

[20] S. V. Gaponenko and D. V. Guzatov. Possible rationale for giant enhancement factors in surface enhanced Raman scattering. *Chemical Physics Lett.*, **477** (2009), 411–414.

[21] A. D. McFarland, M. A. Young, J. A. Dieringer and R. P. Van Duyne. Wavelength-scanned surface-enhanced Raman excitation spectroscopy. *J. Phys. Chem. B*, **109** (2005), 11279–11285.

[22] S. V. Gaponenko, A. A. Gaiduk, O. S. Kulakovich, S. A. Maskevich, N. D. Strekal, O. A. Prokhorov and V. M. Shelekhina. Raman scattering enhancement using crystallographic surface of a colloidal crystal. *JETP Lett.*, **74** (2001), 309–313.

[23] J. J. Baumberg, T. A. Kelf, Y. Sugawara, S. Cintra, M. E. Abdelsalam, P. N. Bartlett and A. E. Russell. Angle-resolved surface-enhanced Raman scattering on metallic nanostructured plasmonic crystals. *Nano Lett.*, **5** (2005), 2262–2267.

[24] A. M. Michaels, J. Jiang and L. Brus. Ag nanocrystal junctions as the site for Surface-Enhanced Raman Scattering of single Rhodamine 6G molecules. *J. Phys. Chem. B*, **104** (2000), 11905–11911.

[25] Th. Förster. Energiewanderung und Fluoreszenz. *Naturwissenshaften*, **33** (1946), 166–175.

[26] J. R. Lakowicz. *Principles of Fluorescence Spectroscopy* (New York: Kluwer Academic, 1999).

[27] L. Novotny and B. Hecht. *Principles of Nano-Optics* (Cambridge: Cambridge University Press, 2006).

[28] R. Ruppin. Decay of an excited molecule near a metal particle. *J. Chem. Phys.*, **76** (1982), 1681–1687.

[29] V. V. Klimov and D. V. Guzatov. Optical properties of an atom in the presence of a cluster consisting of two nanospheres (invited paper). *Quantum Electronics*, **37** (2007), 209–231.

[30] S. V. Serezhkina. *Thermostimulated Processes in Nanocomposite Films Based on Silicon and Germanium Oxides.* Ph.D. thesis, Belarussian State University, Minsk (2006).

[31] O. S. Kulakovich, N. D. Strekal, A. Yaroshevich, S. Maskevich, S. Gaponenko, I. Nabiev, U. Woggon and M. Artemyev. Enhanced luminescence of CdSe quantum dots on gold colloids. *Nano Letters*, **2** (2002), 1449–1452.

[32] O. S. Kulakovich, N. D. Strekal, M. V. Artemyev, A. S. Stupak, S. A. Maskevich and S. V. Gaponenko. Improved method for fluorophore deposition atop a polyelectrolyte

spacer for quantitative study of distance-dependent plasmon-assisted luminescence. *Nanotechnology*, **17** (2006), 5201–5206.

[33] N. Strekal, A. Maskevich, S. Maskevich, J. -C. Jardillier and I. Nabiev. Selective enhancement of Raman or fluorescence spectra of biomolecules using specially annealed thick gold films. *Biopolymers (Biospectroscopy)*, **57** (2000), 325–328.

[34] T. Ozel, I. M. Soganci, S. Nizamoglu, I. O. Huyal, E. Mutlugun, S. Sapra, N. Gaponik, A. Eychmuller and H. V. Demir. Giant enhancement of surface-state emission in white luminophor CdS nanocrystals using localized plasmon coupling. *New J. Phys.*, **10** (2008), 083035.

[35] K. Aslan, S. N. Malyn and C. D. Geddes. Metal-enhanced fluorescence from gold surfaces: angular dependent emission. *J. Fluoresc.*, **17** (2007), 7–13.

[36] K. Ray, M. H. Chowdhury and J. R. Lakowicz. Aluminum nanostructured films as substrates for enhanced fluorescence in the ultraviolet-blue spectral region. *Analytical Chemistry*, **79** (2007), 6480–6487.

[37] K. Aslan, S. N. Malyn and Ch. D. Geddes. Plasmon radiation in SEF as the reason of angular dependence: Angular-dependent metal-enhanced fluorescence from silver island films. *Chem. Phys. Lett.*, **453** (2008), 222–228.

[38] S. Kühn, U. Hakanson, L. Rogobete and V. Sandoghdar. Enhancement of single-molecule fluorescence using a gold nanoparticle as an optical nanoantenna. *Phys. Rev. Lett.*, **97** (2006), 017402.

[39] A. Bek, R. Jansen, M. Ringler, S. Mayilo, Th. A. Klar and J. Feldmann. Fluorescence enhancement in hot spots of AFM-designed gold nanoparticle sandwiches. *Nano Letters*, **8** (2008), 485–490.

[40] P. Anger, P. Bharadwaj and L. Novotny. Enhancement and quenching of single molecule fluorescence. *Phys. Rev. Lett.*, **96** (2006), 113002.

Author index

Subject index

Printed in the United States
By Bookmasters